The Stanford GraphBase

A Platform for Combinatorial Computing

DONALD E. KNUTH
Stanford University

ACM Press
New York, New York

ADDISON-WESLEY PUBLISHING COMPANY

Reading, Massachusetts • Menlo Park, California • New York
Don Mills, Ontario • Wokingham, England • Amsterdam • Bonn
Sydney • Singapore • Tokyo • Madrid • San Juan • Milan • Paris

TEX is a trademark of the American Mathematical Society.
METAFONT is a trademark of Addison-Wesley Publishing Company.
PostScript and Adobe PhotoShop are trademarks of Adobe Systems Incorporated.

Library of Congress Cataloging-in-Publication Data

Knuth, Donald Ervin, 1938-
   The Stanford GraphBase : a platform for combinatorial computing /
Donald E. Knuth.

    viii,576 p. 24 cm.
    Includes bibliographical references and indexes.
    ISBN 0-201-54275-7
    1. Stanford GraphBase. 2. Combinatorial analysis--Data
processing. 3. CWEB (Computer system). I. Title.
QA164.K6 1993
511'.5'02853--dc20
                        93-4233
                        CIP

*QA
164
K6
1993*

The following copyright notice is included in all files of the Stanford GraphBase:

© 1993 Stanford University

This file may be freely copied and distributed, provided that no changes whatsoever are made. All users are asked to help keep the Stanford GraphBase files consistent and "uncorrupted," identical everywhere in the world. Changes are permissible only if the modified file is given a new name, different from the names of existing files in the Stanford GraphBase, and only if the modified file is clearly identified as not being part of that GraphBase. (The CWEB system has a "change file" facility by which users can easily make minor alterations without modifying the master source files in any way. Everybody is supposed to use change files instead of changing the files.) The author has tried his best to produce correct and useful programs, in order to help promote computer science research, but no warranty of any kind should be assumed.

Usage of those files in derived works is otherwise unrestricted.

All portions of the present book that are not distributed as part of the Stanford GraphBase files are copyright © 1994 by the ACM Press, a division of the Association for Computing Machinery, Inc. (ACM). All rights for those portions (including the special indexes) are reserved. No part of this additional information may be reproduced, stored in a retrieval system or transmitted, in any form or by any means, electronic, mechanical, photocopying, recording, or otherwise, without the prior written permission of the publisher. Printed in the United States of America.

ISBN 0-201-54275-7
1 2 3 4 5 6 7 8 9 10--MA--9796959493

3 3001 00786 3045

PREFACE

This book has two main goals. First, it provides dozens of examples of what I like to call "literate programs" — computer programs that aspire to be works of literature, to be read and enjoyed by human beings although they can also be read and interpreted by machines. Most of these programmatic essays are short stories, about 10 pages long; some are longer, but consist of small parts that can be understood independently. I have tried to make new contributions to the exposition of several important algorithms and data structures, so that students of computer science might find this book a useful supplement to their other texts.

Second, the programs in this book should be useful to anyone who develops methods of combinatorial computing. Researchers who work on combinatorial algorithms need standard sets of data for use in benchmark tests and for comparison of competing approaches. I want to make it easy for people to say, "Algorithm x on graph y costs z," for many different combinations of x and y, where x, y, and z are well defined. I have tried to gather sets of data that are both interesting and applicable to a wide variety of problem domains. By making these programs and datasets widely available, I am hoping to counteract the crisis of confidence that has unfortunately been growing between theoretical computer scientists and the programmers-on-the-street who have real problems to solve.

A third purpose of the programs in this book is to have fun. (Indeed, pleasure has probably been the main goal all along. But I hesitate to admit it, because computer scientists want to maintain their image as hard-working individuals who deserve high salaries. Sooner or later society will realize that certain kinds of hard work are in fact admirable even though they are more fun than just about anything else — see [15].) A dozen or so of the GraphBase routines are demonstration programs that can lead to many hours of amusement.

This book is by no means a systematic text about combinatorial computing, but it does discuss many of the most significant and most beautiful combinatorial algorithms that are presently known. Included, for example, are:

- Dijkstra's algorithm for shortest paths;
- Hopcroft and Tarjan's algorithms for strong components, biconnected components, and topological sorting;
- the Hungarian algorithm for the assignment problem;
- heuristic algorithms, like the method of "stratified greed," for approximate solutions to intractable optimization problems;
- algorithms for generating all permutations, combinations, partitions, and bifurcations;
- a robust and efficient algorithm for Delaunay triangulations and Voronoi diagrams;
- Walker's alias method for nonuniform random number generation;
- the construction of expander graphs;
- logic circuits for parallel multiplication;
- a variety of algorithms to find minimum spanning trees.

The programs also implement important data structures, such as binary heaps, Fibonacci heaps, plane triangulations, and binomial queues.

I began the work that led to this book almost 20 years ago when I decided to use five-letter words of English as the basis for examples in Volume 4 of *The Art of Computer*

Programming. I was inspired to acquire additional datasets when I learned about a standard test suite of linear programming problems collected during the 1970s at Stanford's Computer Science and Operations Research departments by Michael Saunders and other coworkers of George Dantzig. Their collection was subsequently enlarged by David Gay at AT&T Bell Labs; now it contains more than 80 sets of data, accessible worldwide via email and ftp as part of the netlib distribution service (see [5]). Researchers have found — not surprisingly — that real-world datasets for linear programming behave quite differently from randomly generated test problems. Hence the netlib collection has led to important algorithmic advances and to a far richer and more relevant theory than a purely academic approach would have produced. I am convinced that an analogous collection of test data for combinatorial applications will have a similarly beneficial impact. Already netlib includes a library of traveling salesrep problems [28].

Many specific combinatorial questions can be asked about the graphs and networks defined in the Stanford GraphBase, leading to an almost inexhaustible supply of challenging computational tasks. I hope that many of these problems will appeal to researchers in the field and will stimulate a friendly rivalry in which new and improved algorithms are discovered. Examples of my own initial entries into such competitions appear in the programs called ASSIGN_LISA and MILES_SPAN. Can anybody solve the assignment problems or minimum-spanning-tree problems described there, using substantially fewer memory references than needed by those programs? Additional challenges appear in programs called ECON_ORDER and FOOTBALL, which tackle optimization problems whose best possible solutions are not completely known. How much computer time is needed to find the optimum linear arrangements of economic sectors? Is there a sequence of football scores that ranks Stanford over Harvard by more than 2279 points, or Harvard over Stanford by more than 2173? What bounds can be given for the optimum solutions to such problems? The answers to specific questions like these should advance our understanding of significant combinatorial methods. The task of drawing the GraphBase graphs æsthetically is also quite important.

Many people have helped me develop the Stanford GraphBase during the past several years. In particular, my secretary, Phyllis Winkler, made key contributions. Significant help came from my former graduate students Ramsey Haddad and Pang-Chieh Chen, as well as a number of former Stanford undergrads: Dave Alexander, Jim Austgen, Laie Caindec, Christine Chang, Stephanie Gore, Dylan Kohler, Sol Lederman, Larry Murk, Jinny Shinsato, Brian Shriver, Ali Tabibian, and Lei Zhu; these students helped with data entry and wrote initial prototypes of many of the programs in this book as we began to explore the characteristics of interesting nonrandom graphs. I also wish to thank the National Science Foundation for supporting much of our activity, and Sun Microsystems for providing a computer. Tomas Rokicki helped me design the Makefile and test the routines on a variety of systems. Special thanks go to John Hobby for his METAPOST system [10], which made the preparation of illustrations for this book a pleasure instead of a chore.

I've tried to make this book error-free, but I must have blundered at least once. Therefore I will gratefully pay $2.56 to the first person who finds and reports anything herein that is mathematically, historically, typographically, or politically incorrect.

—Donald E. Knuth
Stanford, California

CONTENTS

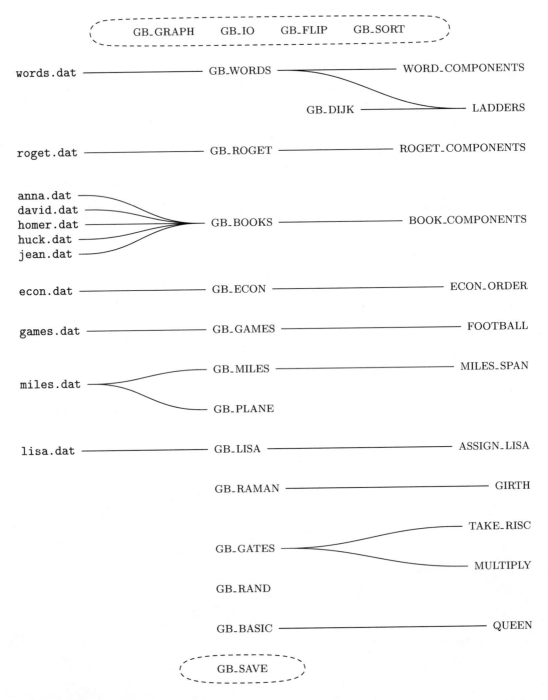

Figure 1. Major constituents of the Stanford GraphBase.

CHAPTER 1: *OVERVIEW*

The Stanford GraphBase is a collection of datasets and computer programs that generate and examine a wide variety of graphs and networks. Figure 1 shows all of its major constituents and their interrelations. Files of data, with names like words.dat and lisa.dat, appear at the left of the illustration; generator routines, with names like GB_WORDS and GB_LISA, appear in the center; demonstration programs, with names like WORD_COMPONENTS and ASSIGN_LISA, appear at the right.

The computer programs are written in CWEB, a combination of the C programming language and the TEX typesetting system. CWEB is a prototypical system for "literate programming" [20]; its conventions facilitate the writing of correct programs that are relatively easy to read and to understand. Chapter 4 explains more about CWEB.

All 31 of the GraphBase programs appear later in this book. In most cases, they can be read quickly, since their median length is only 12 pages. Three of them — MILES_SPAN, GB_GATES, and especially GB_BASIC — are somewhat longer, but consist of independent small pieces. Thus the GraphBase programs are not at all monolithic in nature; they are composed of small, relatively simple building blocks.

The twelve data files *.dat occupy 355,545 bytes of memory altogether. The program files, which have names of the form *.w, occupy approximately 780 kilobytes. (These are "metric" kilobytes, units of 1000 bytes, not 1024.) When we add the space occupied by a few auxiliary files needed for installation and special effects, we find that the total storage requirement for the entire Stanford GraphBase is less than 1.2 megabytes; therefore everything fits comfortably on a single floppy disk. Chapter 3 explains how to acquire and install the GraphBase files.

Four of the GraphBase programs — GB_GRAPH, GB_IO, GB_FLIP, and GB_SORT — are shown in an oval near the top of Fig. 1. They form the *kernel* of the system. Every program that uses the conventions of this book will include the subroutines and data structures of GB_GRAPH; these conventions are explained at the beginning of Chapter 2. Programs that input *.dat files will also include GB_IO. Programs might also incorporate the machine-independent random number generator of GB_FLIP and/or the efficient sorting subroutine of GB_SORT.

Thirteen of the programs — GB_WORDS, GB_ROGET, GB_BOOKS, GB_ECON, GB_GAMES, GB_MILES, GB_PLANE, GB_LISA, GB_RAMAN, GB_GATES, GB_RAND, GB_BASIC, and GB_SAVE — are listed in the central column of Fig. 1. These are the *generator* modules. Every program that uses a particular kind of graph will obtain the graph by including the appropriate generator module, which appears in a runtime library together with the kernel routines. Many of the generator modules need special data in order to do their work, and Fig. 1 lists such datasets in the left column. For example, the GB_ECON generator relies on econ.dat, a data file that contains detailed information about the United States economy.

Most of the generator modules are accompanied by demo programs so that users can see exactly how the graphs can be created initially and used creatively. For example, the demo program for GB_ECON is ECON_ORDER. The author has spent many a pleasant hour playing with these demos, and readers of this book might find them equally enjoyable. But of course we should also note (for the sake of research sponsors and auditors) that the demo programs have a serious expository purpose. And it should also be clear that the GraphBase generators can be used in thousands and thousands of interesting ways besides the simple applications demonstrated here.

One GraphBase program, called TEST_SAMPLE, is not mentioned in Fig. 1 because it is used only when the Stanford GraphBase is first being installed on a new computer. Ordinary demo programs use only one GraphBase generator, but TEST_SAMPLE is different: It includes all thirteen generators, because it tests the characteristics of graphs that are likely to be defective when the system isn't working properly. The existence of TEST_SAMPLE proves that it is possible to load all the generators simultaneously without interference; but that program does occupy about 254 kilobytes of program space on a typical (SPARC) computer. By contrast, most of the ordinary demo programs occupy only about 60 kilobytes, not counting the space allocated dynamically for graph data, because they need only the kernel routines and one generator.

1.1 WORDS

Let's look now at LADDERS, a word game that is the second demo program in Fig. 1. The examples of this chapter are presented as interactive dialogs that might occur on a typical computer terminal, using `typewriter type` to represent things typed by the machine and `slanted typewriter type` to represent things typed by the user. UNIX conventions will be assumed for convenience. If you have installed the GraphBase and said `make ladders`, as described in Chapter 3, you are ready to play the game.

```
tutorial> ladders

Starting word: Eh?
(Please type five lowercase letters and RETURN.)
Starting word: words
    Goal word: graph
        0 words
        1 wolds
        2 golds
        3 goads
        4 grads
        5 grade
        6 grape
        7 graph

Starting word:
```

The task of LADDERS is to form what Lewis Carroll called "word links" or "doublets" and what puzzlists nowadays call "word ladders" [6, 16]: sequences of words in which each one after the first differs from the previous in only a single position. The program uses a dictionary of 5757 words, representing every five-letter word of English. More precisely, `words.dat` contains every English five-letter word that the author encountered during a 20-year period when he was compiling a personal dictionary of such words. (Chapter 2 explains more about `words.dat`.) The GB_WORDS generator takes those 5757 words and makes them into an undirected graph with 5757 vertices; two vertices are adjacent in the graph when the corresponding words differ in exactly one position. The graph has 14,135 edges, that is, 14,135 pairs of adjacent vertices.

Since $26^5 = 11,881,376$ five-letter words are theoretically possible, the chance that a randomly chosen sequence of five letters will lie in `words.dat` is $5757/26^5$, about 1 in 2000. A random selection of 5757 strings of five letters, considering all $\binom{11881376}{5757}$ choices to be equally likely, would lead to only $5 \times 25 \times 5757 \times 5756/(2 \times 11881375) \approx 174.3$ edges on the average. Thus the words in `words.dat` are definitely nonrandom; they join up with each other in unexpected ways.

One of the consequences of the large number of edges is that paths exist between most pairs of five-letter words. LADDERS will find a shortest such path. The GB_DIJK module shown in Fig. 1 is a general-purpose set of routines that will find shortest paths efficiently in any GraphBase graph; thus LADDERS serves as a demonstration program for GB_DIJK as well as for GB_WORDS.

Here are some additional examples:

```
        Starting word: flour
           Goal word: bread
                  0 flour
                  1 floor
                  2 flood
                  3 blood
                  4 brood
                  5 broad
                  6 bread

        Starting word: chaos
           Goal word: order
                  0 chaos
                  1 choos
                  2 chops
                  3 coops
                  4 comps
                  5 comes
                  6 comer
                  7 coder
                  8 cider
                  9 aider
                 10 adder
                 11 odder
                 12 order

        Starting word: pound
           Goal word: marks
        Sorry, there's no ladder from pound to marks.
```

Some five-letter words are `aloof`; they have no neighbors in the graph.* In fact, quite a few words fall in this category, 671 altogether. Some of them are common words that have been in the language for centuries: `their`, `again`, `ahead`, `below`, `earth`, `human`, `young`, `music`, `ocean`, `laugh`, `sugar`, `extra`, `doubt`; `truly`, `false`, `maybe`; `first`, `third`, `ninth`. Most, however, are loan words from other languages, like `delta`, `theta`, `kappa`, `sigma`, `omega`, `psalm`, `rabbi`, `aleph`, `matzo`, `pizza`, `opera`, `piano`, `polka`, `waltz`, `rumba`, `adios`, `rodeo`, `junta`, `llama`, `koala`, `okapi`, `bwana`, `sahib`, `sheik`, `jihad`, `mecca`, `khans`, `imams`, `gurus`, `khaki`, `batik`, `sushi`, `futon`, `haiku`, `ninja`, `lanai`, `hogan`, `igloo`, `tepee`, `dacha`, `gulag`, `vodka`, `kefir`, `ruble`, `kopek`, `zloty`.

Some of the isolated words, like `gauss`, come from proper names. Proper names such as Knuth and other words that must be capitalized are not allowed in `words.dat`; but `gauss` is legitimate because it is a unit of magnetic flux. The names of other scientists (`fermi`, `henry`, `hertz`, `petri`, `tesla`, `weber`) have snuck into `words.dat` via a similar route, but all of those just listed are neighborless in the graph. So are a variety of words that imitate sounds: `aargh`, `achoo`, `miaow`, `oomph`, `pffft`, `pssst`, `woosh`, `yecch`. Still more examples come from acronyms (`radar`, `scuba`, `snafu`) and/or technology (`defun`, `deque`, `email`, `fanin`, `ftped`, `gluon`, `ioctl`, `pixel`). These words are certainly interesting to crossword puzzle fans, but they're useless in word ladders.

The average degree of a vertex in this graph is $2 \times 14135/5757 \approx 4.91$, and the actual distribution of degrees is

degree	number	example	degree	number	example	degree	number	example
1	774	which	9	213	trees	17	32	makes
2	727	could	10	188	words	18	32	miles
3	638	these	11	162	means	19	20	later
4	523	there	12	120	years	20	8	bears
5	428	while	13	116	paper	21	6	lines
6	329	vowel	14	102	looks	22	4	wares
7	280	sound	15	75	water	23	2	mates
8	249	right	16	53	parts	24	3	mines

There also are two words of degree 25, namely `bares` and `cores`; no words have degree 26 or more. The 25 neighbors of `bares` all have degree ≥ 6, and their average degree is 16.24; the 25 neighbors of `cores` all have degree ≥ 9, and their average degree is 14.4. So they belong to a dense part of the graph for which the word `cores` is quite appropriate.

Some words of degree 1 are neighbors of each other. There are 103 such pairs, of which the most noteworthy are perhaps those that mix cultures: `alpha` — `aloha`, `baize` — `maize`, `odium` — `opium`, `kudos` — `judos`, `droid` — `druid`, `sutra` — `supra`, `deuce` — `deice`, `gonad` — `monad`. Similarly, the graph of five-letter words has 42 components of size three (`chain` — `chair` — `choir`, `alike` — `alive` — `olive`, etc.), 13 of size four (`anion` — `onion` — `union` — `unwon`), 6 of size five, 4 of size six, 6 of size seven, 1 of size eight, 3 of size fifteen, and 3 more of respective sizes 17, 19, 24. The word `pound`, which we were unable to change into `marks` in the example on the previous page, belongs to the

---

* Note to South African readers: Sorry, I have never seen a kloof.

17-word component, which has diameter 6:

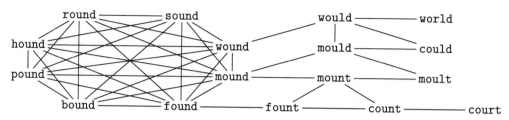

Figure 2. The words reachable from `pound`.

All components of the graph that have sizes 2 to 24 are displayed by the demonstration program WORD_COMPONENTS. Such words participate in relatively few ladders.

On the other hand, the vast majority of all five-letter words belong to a giant component, of size 4493. If you can move at least two steps from a given word, changing two different letters, chances are excellent that your word belongs to the giant component, so there will be a word ladder connecting it to any other word of the giant component.

A variant of WORD_COMPONENTS called WORD_GIANT creates the graph that contains only the words of the giant component. This graph on 4493 vertices contains 13,619 edges.

The giant component, in turn, contains a giant biconnected component of 3708 vertices. The diameter of the giant component is 29, which is the shortest distance between `amigo` and `signs`. All paths from `amigo` to `signs` must enter the giant biconnected component at `arise` and leave it at `begot`; the distance between articulation points `arise` and `begot` is 14. The maximum distance, 29, also occurs between `amigo` and `repro`, in which case the articulation points `arise` and `retro` are 18 steps apart in the giant bicomponent. (See program BOOK_COMPONENTS for a discussion of bicomponents and articulation points.)

The shortest paths found by LADDERS sometimes involve words that are rather rare. For example, the 7-step path from `words` to `graph` in our first example involved `wolds` and `goads`. We can restrict consideration to the 3000 most common words if we start over with a nonstandard option on the command line, as follows:

```
tutorial> ladders -n3000
Starting word: words
   Goal word: graph
          0 words
          1 cords
          2 corps
          3 coops
          4 crops
          5 drops
          6 drips
          7 grips
          8 gripe
          9 grape
         10 graph
```

The graph now has 3000 vertices and 5209 edges, with a giant component of size 2056. The shortest path from `words` to `graph` has become three steps longer than before, but each of its steps now involves a fairly common word. (The shortest path from `chaos` to `order` has similarly grown to length 13: `chaos, chats, coats, costs, hosts, hoses, roses, rises, rides, rider, eider, elder, older, order`.) If we limit consideration to only 2000 words, it turns out that 20 steps are needed to make `words` into `graph`, while `order` has become unreachable from `chaos`. The giant component of the 2000-word graph contains 1186 vertices, and its diameter is 36 — the distance from `harsh` to `mercy`. The diameter of the 3000-word graph turns out to be even larger: The distance from `carve` to `posse` is 37.

Appendix D, on pages 564–567, lists the 3000 most common five-letter words of English in alphabetical order, together with their rank according to GB_GRAPH's default conventions.

GB_GRAPH can select the n most common words because `words.dat` contains frequency information as well as the words themselves. Chapter 2 explains how frequency counts have been compiled from a variety of sources so that different kinds of vocabulary can be given greater weight. It's possible, for example, to reflect British or American or Australian usage or to emphasize words found in computer texts. One of the available variations restricts the graph to the 1466 five-letter words found in the Revised Standard Version of the Bible. In this case the graph contains only 1596 edges, so it is quite sparse; but it still contains interesting word ladders. Here, for example, is a minimal path from `sword` to `peace` that uses only Biblical words:

`sword, swore, store, stare, stars, stays,`

`slays, plays, plans, plane, place, peace.`

A denser graph can be obtained from `words.dat` in various ways. For example, we can say that two words are adjacent whenever they have at least four letters in common, without regard to the order of the letters. Then we can get from `words` to `graph` in just four steps:

`words, sward, pards, harps, graph.`

Or if we restrict consideration to the 1000 most common words, five steps suffice:

`words, worst, ports, parts, grasp, graph.`

The full 5757-word graph under this interpretation has 128,223 edges, not 14,135, so the average degree is $2 \times 128223/5757 \approx 44.5$. Only 13 words are isolated: `bibbs`, `flyby`, `jazzy`, `jujus`, `klutz`, `kudzu`, `macaw`, `mezzo`, `oxbow`, `pffft`, `scuzz`, `whizz`, `zowie`. There also are 2 components of size two: {`pizza, spazz`}, {`ahhhh, ohhhh`}. But everything else belongs to a giant component of size 5740. Even when we consider small subsets such as the 200 most common five-letter words, the graph defined in this way has average degree 2.71 and a relatively large component of size 131.

Lewis Carroll himself suggested that word ladders be allowed to include steps in which consecutive words either differ from each other in one position or are anagrams of each other. This convention made it possible for him to turn `iron` into `lead` [4].

With a similar procedure, we can use `words.dat` to define an interesting *directed* graph: One word leads to another if its final four letters appear in the other. This rule defines

94,084 arcs on 5757 vertices. The shortest directed path from `words` to `graph` is

$$\texttt{words} \rightarrow \texttt{dross} \rightarrow \texttt{soars} \rightarrow \texttt{orcas} \rightarrow \texttt{chars} \rightarrow \texttt{sharp} \rightarrow \texttt{graph};$$

and we can go back to `words` in five steps,

$$\texttt{graph} \rightarrow \texttt{harps} \rightarrow \texttt{prats} \rightarrow \texttt{astro} \rightarrow \texttt{trows} \rightarrow \texttt{words}.$$

There is a giant strong component containing 4585 vertices. There are no directed paths between `words` and `graph` if we confine ourselves to the most common 1000 words, but with 2000 words we can go one way, as follows:

$$\texttt{words} \rightarrow \texttt{drops} \rightarrow \texttt{ports} \rightarrow \texttt{roast} \rightarrow \texttt{oaths} \rightarrow \texttt{trash} \rightarrow \texttt{sharp} \rightarrow \texttt{graph}.$$

The graph of five-letter words can, of course, be used to illustrate many combinatorial problems in addition to the problem of shortest paths. For example, we can consider minimum Steiner trees. It turns out that the shortest way to connect up the three words `blood`, `sweat`, and `tears` requires 15 edges; here is one of the optimum solutions:

```
blood
    brood                        sweat
      broad                    sweet
        bread            tweet
      tread  tweed
        treed
        trees
        tries
        trims
        trams
        teams
        tears
```

We can try to color word graphs with fewest colors; we can seek maximum independent sets of words; and so on.

The words can also be used in covering problems. Question: Is there a set of five five-letter words that covers 25 letters of the alphabet? Answer: No, at least not in `words.dat`. But we *can* cover 24 letters with five words, for example with {`becks`, `fjord`, `glitz`, `nymph`, `squaw`}. And there are many, many ways to cover 20 letters with four words: {`rough`, `black`, `empty`, `words`}, {`lucky`, `mixed`, `graph`, `towns`}, {`angry`, `depth`, `flows`, `quick`}, {`rocky`, `plans`, `fixed`, `thumb`}, etc. — 1,781,773 solutions altogether. Moreover, 1684 of these, including {`third`, `flock`, `began`, `jumps`}, hit 20 of the first 21 letters of the alphabet.

If we say two words are equivalent if they contain the same letters, the 5757 members of `words.dat` define 4522 equivalence classes. For example, `motet` ≡ `totem`; `ruble` ≡ `bluer`; `petri` ≡ `tripe`; `gnats` ≡ `tangs` ≡ `angst`; `cruel` ≡ `lucre` ≡ `ulcer`; `scion` ≡ `icons` ≡ `coins` ≡ `sonic`; `artsy` ≡ `trays` ≡ `stray` ≡ `satyr`. Most classes, 3649 altogether, contain only a single element; these are words that have no anagrams.

Are there four five-letter words that collectively make an anagram of 'the Stanford Graph-Base', ignoring case distinctions? Sure, lots. Even if we limit ourselves to the 1000 most common words, we get 94 solutions, such as 'poets began harsh draft' or 'hated beans grasp forth'. There are 24,981 solutions when the entire vocabulary of words.dat is allowed.

Symmetrical word squares such as

water	parts	villa	darts	panes	strip
alive	alarm	ivies	adore	adult	twine
tides	radio	liras	roman	nudie	rides
event	trick	lease	traps	elite	inert
rests	smoke	asset	sense	steed	pesto

aren't as hard to find as they might seem: There are 541,968 of them. In fact, all but 23 of the 5757 words are present in at least one such square. M. D. McIlroy has made a similar study of 7×7 word squares, which are much rarer, and found 51 solutions [25]. The problem of finding word squares provides an excellent example of the backtrack technique called preclusion (see [7, 21]).

We can even make symmetrical $5 \times 5 \times 5$ word cubes. Two of the nicest are

aster	scale	tacos	elope	reset
scale	codex	adopt	leper	extra
tacos	adopt	cover	opera	strap
elope	leper	opera	perks	erase
reset	extra	strap	erase	taper

and

types	yeast	pasta	ester	start
yeast	earth	armor	stole	three
pasta	armor	smoke	token	arena
ester	stole	token	elect	rents
start	three	arena	rents	tease

If a_{ijk} denotes the letter in row i, column j, and layer k, the symmetry condition is

$$a_{ijk} = a_{ikj} = a_{jik} = a_{jki} = a_{kij} = a_{kji}.$$

Thus every "slice" of the word cube is a symmetrical word square. The words of words.dat lead to a total of 83,576 such cubes, although almost all solutions involve at least one rarish word.

Suppose we regard a word as a vector of numbers, with a = 1, b = 2, ..., z = 26. Then the only word containing no "odd" letters is pffft. We can add or subtract words by doing arithmetic modulo 26 in each component (so that z is equivalent to zero). For example,

$$\text{words} + \text{spins} = \text{pearl}; \qquad \text{dance} + \text{oomph} = \text{spasm};$$
$$\text{allow} + \text{mirth} = \text{nudie}; \qquad \text{putty} + \text{klutz} = \text{agony};$$
$$\text{email} = \text{bales} + \text{clods} = \text{chute} + \text{befog} = \text{bliss} + \text{carps}$$
$$= \text{inane} - \text{dazes} = \text{shuck} - \text{nutty};$$
$$(\text{power} + \text{visit})/2 = \text{slugs}; \qquad (\text{vague} + \text{risky})/2 = \text{tempo}.$$

If we decide to define a graph on 5757 vertices by saying that two words are adjacent if and only if their sum is also a word, we get 5651 edges and a giant component of size 4061.

Alternatively, we might say two words are adjacent if and only if their vectors are orthogonal (the dot product is zero), modulo a given parameter m; then there are respectively (8704749, 5510006, 4344971, 3320182, 2889283, 2367939, 2171241, 1838641, 1741727, 173821, 15310) edges when $m = (2, 3, 4, 5, 6, 7, 8, 9, 10, 100, 1000)$. It should be interesting to find maximum cliques in such graphs.

The word pairs with smallest and largest vector dot product are

$$\texttt{cabal} \cdot \texttt{abaca} = 22 \qquad \text{and} \qquad \texttt{fuzzy} \cdot \texttt{buzzy} = 2090\,.$$

The greatest Euclidean distance between word vectors is

$$|\texttt{yucca} - \texttt{abuzz}| = \sqrt{2415}\,.$$

Another interesting alternative to the word-ladder graph arises when we consider words adjacent if and only if the Euclidean distance between them is at most $\sqrt{d}$ for some parameter d. This definition lets us control the number of edges in the graph smoothly if we want a family of graphs with varying sparseness. The number of edges is respectively $(9297, 29168, 60814, 102641)$ for $d = (10, 20, 30, 40)$.

Suppose we are looking for a graph with 1000 vertices and approximately 10,000 edges, in order to test some combinatorial algorithm. We can let the vertices correspond to the 1000 most common five-letter words, as computed by GB_WORDS, and we can say that two words are adjacent if and only if their Euclidean distance is $\leq \sqrt{68}$; then we get 9982 edges. Alternatively, it's a bit simpler to say that two words are adjacent if and only if their Manhattan distance is ≤ 13; the Manhattan distance, also called the l_1 norm, is the sum of the distances between letters in corresponding positions. Then we get 9820 edges. (The Manhattan distance between $\texttt{yucca}$ and $\texttt{abuzz}$ is

$$24 + 19 + 18 + 23 + 25 = 109\,;$$

this is the maximum Manhattan distance between any two words of $\texttt{words.dat}$.) Even simpler is to let two words be adjacent if and only if the letters in corresponding positions differ by at most 6; this is the so-called max norm or l_∞ norm. For example, the max-norm criterion makes $\texttt{tombs}$ just 1 step away from $\texttt{splat}$, although the Euclidean distance is $\sqrt{5}$ and the Manhattan distance is 5. With max norm at most 6, we get 13,057 edges, and with max norm at most 5 we get 5929. To get exactly 10,000 edges, with any of these norms, we could use *gb_linksort* (a subroutine in GB_SORT) to choose the 10,000 shortest edges by breaking ties between word pairs at equal distance in a pseudo-random but machine-independent manner.

What five-letter nonword most wants to be a word, in the sense that it is adjacent to the most words of $\texttt{words.dat}$? The unique answer is $\texttt{lates}$, which has 24 legitimate neighbors (and which the Oxford English Dictionary calls a Nile perch). Close behind, with 23 and 22, are $\texttt{lares}$ and $\texttt{cales}$.

We have discussed GB_WORDS and LADDERS at great length because they typify the generation and demonstration routines of the Stanford GraphBase. In the remainder of this chapter, we'll see that the other programs have a similar flavor and that different generators tend to produce rather different kinds of graphs.

1.2 ROGET

The graphs generated by GB_WORDS are undirected.* Now let's look at GB_ROGET, which generates graphs that have *directed* arcs between the vertices.

Peter Mark Roget published his celebrated *Thesaurus* in 1852. His goal was to produce the inverse of a dictionary: While a dictionary maps words into sets of concepts, his *Thesaurus* mapped concepts into sets of words. He divided all concepts into 1000 categories, beginning with category 1, "existence," and ending with category 1000, "temple." His book was an instant success, and he made extensive notes in preparation for an expanded edition. He died (at age 91) before finishing that task, but his son John Louis Roget carried on the work and published the second edition in 1879.

Roget's second edition improved on the first in several ways, one of which was the increased number of cross-references between categories. For example, category 1, "existence," pointed to category 494, "truth" (actual existence), and to 186, "presence" (existence in space), as well as to 66, "beginning" (coming into existence), etc. There were no references from "temple" to other categories, but there were references from "abode" and "seclusion" to "temple." The cross-references in the 1879 edition define a directed graph, whose vertices are the categories; `roget.dat` records all of the cross-reference data.

John Lewis Roget decided to subdivide a few of his father's original categories where he found "distinct ideas obviously united under one head." Thus the new edition turned out to have 1022 categories instead of 1000. He assigned numbers like '16a' to the new categories, in order to keep the original numbering scheme largely intact; but such combinations of numbers and letters are cumbersome. Therefore `roget.dat` uses a new numbering scheme: All categories have been reassigned to consecutive numbers, so that "truth" is now number 506 and "temple" is 1022.

Many of the categories come in pairs or triples, representing opposite or closely related concepts. Roget assigned adjacent numbers to such categories and arranged to have them printed in parallel columns. For example, category 1, "existence," was paired with category 2, "inexistence"; category 5, "intrinsicality," was paired with category 6, "extrinsicality." There was no explicit reference from 5 to 6 or from 6 to 5, but `roget.dat` treats their appearance in facing columns as an implicit cross-reference.

The full graph specified in `roget.dat` has 1022 categories (vertices) and 5075 cross-references (arcs). Subgraphs can be obtained in various ways. For example, we can avoid all implicit references between pairs or triples of consecutive categories by stipulating that arcs are allowed only between categories that differ numerically by 3 or more. If we do that, the number of arcs drops from 5075 to 3823 — which is still enough for plenty of interesting structure. Another way to obtain a subgraph is to discard arcs at random with a given probability and/or to choose a subset of the vertices at random.

The connectivity structure of these directed graphs can be analyzed by a demonstration routine called ROGET_COMPONENTS. For example, the UNIX command

```
tutorial> roget_components
```

---

* We did consider a directed variant briefly, but that variation is not a standard part of the Stanford GraphBase.

will print out

```
Reachability analysis of roget(1022,0,0,0)
```

followed by a detailed listing of 77 strongly connected components and the links between them. Some of the strong components contain only a single category ("humorist"); but there is a giant strong component containing 904 categories, each reachable from the others. If the 77 strong components are shrunk to points, we obtain a partial ordering with 61 arcs on 77 vertices.

Similarly

```
tutorial> roget_components -d3
Reachability analysis of roget(1022,3,0,0)
```

will exhibit 248 strong components, the largest of which has 762 vertices; 176 distinct arcs run between strong components. In this case, '-d3' has caused arcs to be suppressed unless they run between categories whose numbers are at least 3 apart. And

```
tutorial> roget_components -n500
Reachability analysis of roget(500,0,0,0)
```

finds 158 strong components in a directed graph that has 500 vertices and 1247 arcs; here 500 categories have been chosen at random from Roget's 1022. We can throw away approximately half of the arcs by saying

```
tutorial> roget_components -n500 -p32768
Reachability analysis of roget(500,0,32768,0)
```

(Each arc is discarded with probability 32768/65536.) Now only 624 arcs remain, and the number of strong components has risen to 334. Another random selection is made by changing the random "seed" as follows:

```
tutorial> roget_components -n500 -p32768 -s1
Reachability analysis of roget(500,0,32768,1)
```

This graph on 500 vertices has 646 arcs and 342 strong components.

Notice that the graph is named '*roget*(500, 0, 32768, 1)'. This example illustrates an important general feature of all graphs generated by the Stanford GraphBase: *Each graph has a characteristic name that identifies it uniquely.* Although the 500 vertices of *roget*(500, 0, 32768, 1) have been chosen "at random," and although the original arcs between those vertices have been removed "with probability 32768/65536," great care has been taken to ensure that all randomization depends, in a machine-independent fashion, on the seed value 1 that was assigned to the random number generator. Great care is also taken to ensure that the data in `roget.dat` has not been corrupted. Therefore researchers in different parts of the world who specify exactly the same GraphBase parameters will obtain exactly the same graph.

Different parameter settings will lead to a great many different graphs. In the case of *roget*, the parameters are n (the number of categories), d (the minimum distance between categories that are linked by arcs), p (the probability, times 65,536, that an arc should be discarded), and s (the seed for random numbers). The *words* generator described implicitly in Section 1.1 has its own set of parameters, and we will see many more examples momentarily. The parameters of each generator are summarized in Appendix B on pages 550–556. In each case, graphs are well defined by their parameters; there is no dependency on particular machines or operating systems. Moreover, the data files `words.dat` and `roget.dat` will never change, so there is no dependency on time or "version number" either.

1.3 BOOKS

The next generator in the GraphBase is related to words and ideas in quite another way — through works of literature. Now the vertices of the graphs represent the characters that appear in well-known novels, and the edges represent encounters between those characters. The plots of the novels are therefore represented in the structure of the graph.

Five classic works from five different cultures have been "graphed" in this fashion: Tolstoy's *Anna Karenina* (digested in `anna.dat`), Dickens's *David Copperfield* (in `david.dat`), Homer's *Iliad* (in `homer.dat`), Twain's *Huckleberry Finn* (in `huck.dat`), and Hugo's *Les Misérables* (in `jean.dat`).

The GB_BOOKS generator creates a variety of graphs from these datasets, depending on different ways to choose subsets of the characters and subsets of the book chapters. Its main subroutine has in fact eight parameters:

$$book(\langle title\rangle, n, x, \textit{first\_chapter}, \textit{last\_chapter}, \textit{in\_weight}, \textit{out\_weight}, \textit{seed}).$$

The ⟨title⟩ is either `"anna"` or `"david"` or `"homer"` or `"huck"` or `"jean"`. The *first_chapter* and *last_chapter* specify a range of chapters from which the edges will be constructed. The *in_weight* and *out_weight* associate a "weight" w with each character, using the formula

$$w = c_{\text{in}} \cdot \textit{in\_weight} + c_{\text{out}} \cdot \textit{out\_weight}$$

when the character appears in c_{in} of the selected chapters and in c_{out} of the unselected chapters. The n characters of highest weight are then selected, and the x characters of very highest weight are then rejected. This rule leaves $n - x$ characters as vertices of the graph. (The value of x is usually 0 or 1; we can set $x = 1$ in David Copperfield and Huckleberry Finn in order to exclude the main character, who is joined to just about everybody because those stories are narrated in the first person.) The *seed* parameter is used to break ties randomly in cases of equal weight.

For example, *book*(`"anna"`, 10, 0, 15, 20, 1, 1, 0) is a graph on 10 vertices, representing the 10 most prominent people of *Anna Karenina* in the sense that they appear in the most chapters of that book. There will be an edge between characters if and only if those characters meet each other between chapters 15 and 20, inclusive.

The demonstration program BOOK_COMPONENTS provides insight into the nature of these graphs by displaying their biconnected components. Two vertices belong to the same bicomponent if they are adjacent or if there is a cycle containing them both. For example, the graph just described can be analyzed as follows:

```
tutorial> book_components -tanna -n10 -f15 -l20 -V
Biconnectivity analysis of book("anna",10,0,15,20,1,1,0)
```

```
LE=Konstantin Dmitrievitch Levin, proprietor of Pokrovskoe
      [weight 103]
AN=Anna Arkadyevna Karenina, wife of AL [weight 73]
VR=Count Alexey Kirillovitch Vronsky, young officer
      [weight 67]
ST=Prince Stepan Arkadyevitch Oblonsky (Stiva), brother of AN
      [weight 64]
KI=Princess Ekaterina Alexandrovna Shtcherbatskaya (Kitty),
      wife of LE [weight 59]
DO=Princess Darya Alexandrovna Oblonskaya (Dolly), wife of ST
      [weight 46]
AL=Alexey Alexandrovitch Karenin, minister of state
      [weight 39]
KO=Sergei Ivanovitch Koznishev, half-brother of LE [weight 38]
PS=Princess Shtcherbatskaya, mother of DO and KI [weight 27]
PR=Prince Alexander Shtcherbatsky, father of DO and KI
      [weight 25]

Isolated vertex LE
Bicomponent PR also includes:
 PS (from PR; ..to KI)
 and articulation point KI
Bicomponent KI also includes:
 VR (from ST; ..to AN)
 DO (from ST; ..to AN)
 ST (from KI; ..to AN)
 and AN (this ends a connected component of the graph)
Isolated vertex AL
Isolated vertex KO
```

Each character is identified by a two-letter code for convenience. (The verbose '-V' option to BOOK_COMPONENTS has caused full names, identifications, and weights to be displayed; if '-v' had been used instead of '-V', only the names would appear, as LE=Konstantin Dimitrievitch Levin.) In the graph *book*("anna", 10, 0, 15, 20, 1, 1, 0), LE is an isolated vertex; Levin actually appears in 103 of the book's 239 chapters, but not in chapters 15–20. The graph defined by chapters 15–20 is shown in Fig. 3.

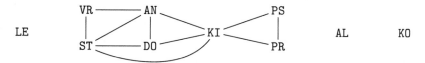

Figure 3. Encounters in *Anna Karenina*, Chapters 15–20.

BOOK_COMPONENTS displays the bicomponents in a way that allows verification of the component structure. Vertex KI is an articulation point where two bicomponents meet.

The full *Anna Karenina* graph is obtained by default if BOOK_COMPONENTS is called with no special command-line options. It has 138 vertices and 493 edges, which define 30 bicomponents; one of these is a "giant" bicomponent containing 105 vertices.

If the *book* routine of GB_BOOKS is called with parameter $n = 0$, or if n is greater than the total number of characters, N, in the specified book, the given value of n is changed to N. If *first_chapter* is 0, *first_chapter* it is reset to 1; if *last_chapter* is 0 or too large, *last_chapter* is reset to the number of the book's final chapter. Thus the subroutine call *book*("anna", 0, 0, 0, 0, 0, 0, 0) will produce a graph whose name is

$$book(\text{"anna"}, 138, 0, 1, 239, 0, 0, 0).$$

This use of 0 for default values is typical of all GraphBase generation procedures. For example, you needn't know that words.dat contains exactly 5757 words; you can ask for n words with $n = 0$ or with $n = 100,000$. The *words* routine will then change n to the maximum permissible number, 5757.

The full graph for *David Copperfield* has 87 vertices and 406 edges. With character DC omitted, it has 86 vertices and 324 edges, because David Copperfield encounters 82 of the other 86 characters (missing only four members of the Crewler family). The full graph for *Huckleberry Finn* has 74 vertices and 301 edges; 73 vertices and 248 edges remain when HF is omitted. The full graph for *Les Misérables* has 80 vertices and 254 edges.

The largest book graph comes from the *Iliad of Homer*, which defines 561 vertices and 1629 edges. However, most of the characters in Homer's story appear only as they are dying in battle.

The full Homeric graph has five isolated vertices and seven other connected components; similarly, the full Huckleberry graph has three connected components, and Victor Hugo's graph has three isolated vertices. Great novelists are evidently under no obligation to connect everything up.

GB_BOOKS also contains another routine, *bi_book*, which generates *bipartite* graphs. The vertices of one part are characters in the specified novel, as before; the vertices of the other part are chapters. In this case, edges simply run between the characters and the chapters in which they appear. The eight parameters of *book* apply also to *bi_book*. For example, the graph *bi_book*("anna", 10, 0, 15, 20, 1, 1, 0) is

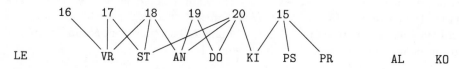

Bipartite graphs lead to interesting matching and covering problems.

1.4 ECON

Our scene now shifts abruptly from the world of literature to the world of finance. File econ.dat is filled with numerical data representing the input/output structure of the entire United States economy in 1985. The economy has been subdivided into 79 sectors, using traditional classifications made by the Bureau of Economic Analysis of the U.S. Department of Commerce. Monetary values, rounded to millions of dollars, are given for the flow of goods and services from sector to sector.

It's easiest to understand this data if we look first at a breakdown into only 8 sectors instead of 79; furthermore, we'll round off to billions of dollars instead of millions:

	Natural resources	Infrastructure	Organic products	Inorganic products	Trade	Financial services	Commercial services	Personal services	Adjustments	Users	Total
Natural resources	42	53	196	24	0	0	1	5	0	15	336
Infrastructure	18	154	60	72	53	56	24	47	0	710	1194
Organic products	23	91	249	75	22	8	22	75	0	419	984
Inorganic products	19	124	70	410	5	2	15	50	1	542	1238
Trade	9	45	41	54	10	3	6	28	0	531	727
Financial services	27	22	15	25	47	138	21	64	0	591	950
Commercial services	14	74	36	54	84	47	40	49	0	85	483
Personal services	3	14	10	24	29	9	20	33	0	685	827
Adjustments	1	11	8	10	2	3	1	0	0	420	456
Users	180	606	299	490	475	684	333	476	455	0	3998
Total	336	1194	984	1238	727	950	483	827	456	3998	

Agriculture, forestry, petroleum, and mining are combined here under the title "Natural resources"; transportation, communication, utilities, construction, and repair are collectively called "Infrastructure"; and so on. The entry in row x and column y stands for the flow from x to y. For example, wholesalers and retailers ("Trade") sold 54 billion dollars' worth of merchandise to the manufacturers of inorganic products.

In addition to the eight economic sectors, two pseudo-sectors called "Adjustments" and "Users" have been added to the data in the table so that the numbers form a proper circulation: The total amount flowing into each sector (the column sum) is equal to the total amount flowing out (the row sum). "Users" are ordinary people who consume products and work for a living. The flow of money from sectors to users is called *total final demand*, representing the dollars not paid within the economy by other sectors; the flow from users to sectors is called *value added*, representing wages and salaries. "Adjustments" cover things like inventory revaluation, imports of special materials, and government work. The sum of total final demand, which is also the sum of value added, is traditionally called Gross National Product (GNP). In 1985 the GNP was almost exactly 4 trillion dollars.

The sum of all entries in the matrix is the total flow of money, but it is not a valid measure of economic activity because it overcounts a lot of things. For example, the value of an expensive piece of metal is included in each transaction from the miner to the sheet metal plant and from there to the fabricator, before it reaches the user. Economists try to define appropriate "Adjustments" so that GNP is a reasonably informative measure that counts everything just once.

A demo program called ECON_ORDER illustrates one combinatorial application of such data. Ignoring "Adjustments" and "Users," it attempts to permute rows and columns of the matrix so that the sum of entries below the diagonal is minimized. This ordering arranges the sectors so that suppliers tend to come first and consumers last. For example, when it is applied to the 8 sectors in the previous table, it gives the following arrangement:

	Commercial services	Financial services	Trade	Natural resources	Organic products	Inorganic products	Infrastructure	Personal services
Commercial services	40	47	84	14	36	54	74	49
Financial services	21	138	47	27	15	25	22	64
Trade	6	3	10	9	41	54	45	28
Natural resources	1	0	0	42	196	24	53	5
Organic products	22	8	22	23	249	75	91	75
Inorganic products	15	2	5	19	70	410	124	50
Infrastructure	24	56	53	18	60	72	154	47
Personal services	20	9	29	3	10	24	14	33

(This arrangement of the sectors is optimal, although our intuition might suggest that "Infrastructure" is misplaced. When "Infrastructure" is subdivided into more specific sectors, some parts of it like "Repair and maintenance" and "Communications" move toward the top, while other parts like "New construction" move to the bottom. Readers of Leontief's article [23] should note that ECON_ORDER produces the *reverse* of his ordering; his tables are approximately lower triangular instead of upper triangular.)

The actual computer run looks like this:

```
tutorial> econ_order -n8
Ordering the sectors of econ(8,2,0,0), using seed 0:
 (Cautious descent method)
(The amount of feed-forward must be at least 566.)

Local minimum feed-forward is 609, found after 39 steps.
The corresponding economic order is:
 Commercial services
 Financial services
 Wholesale and retail trade
 Natural resources
 Organic products
 Inorganic products
 Infrastructure
 Personal services
```

(The computer gives amounts in millions of dollars, but they have been rounded to billions here for simplicity.) Notice the comment on line 4 that the amount below the diagonal must be at least 566. This is a simple lower bound, the sum of $\min(a_{ij}, a_{ji})$ over all $i < j$; either a_{ij} or a_{ji} must end up below the diagonal.

The problem addressed by ECON_ORDER is a generalization of the NP-complete problem FEEDBACK ARC SET [11], which is quite difficult when the number of sectors increases. Therefore ECON_ORDER uses a heuristic called "cautious descent," and it reports only an ordering that is *locally* minimum — no orderings that are "near" this one are better. The cautious descent method starts with a random permutation of the sectors and looks for local improvements, choosing if possible a nearby permutation that reduces the total feed-forward. It is called cautious because it always chooses the *least* available reduction. Another heuristic is the "greedy" method, which is used if you specify -g on the command line to ECON_ORDER; this heuristic repeatedly moves to the nearby permutation that gives the *greatest* available reduction. One goal of ECON_ORDER is to compare the method of cautious descent to the greedy algorithm. It turns out that cautious descent almost always loses badly; several runs of the greedy algorithm with different starting permutations tend to run significantly faster than one run of cautious descent and to produce at least one result that is at least as good as the one found cautiously.

The command-line options '-r100 -t5' to ECON_ORDER tell it to repeat its calculations 100 times, using the random seed 5 to control the random permutations used as starting points. This feature makes it easy to study both greedy and cautious heuristics. For example,

```
tutorial> econ_order -n8 -r100 -t5
```

always finds the optimum permutation by 100 trials of cautious descent, but takes an average of 37.52 steps per try. The alternative

```
tutorial> econ_order -n8 -r100 -t5 -g
```

uses the greedy algorithm and again finds the optimum unerringly, but with an average of only 4.48 steps per try.

The command-line option -v causes ECON_ORDER to report further details of its calculations. Alternatively, -V makes it report every step it takes. Such "verbose" options are available with most of the demo programs.

There are many ways to divide the economy into 8 sectors. The default way, illustrated in the tables shown earlier, tends to equalize the row sums; but ECON_ORDER will choose at random from all possible 8-sector partitions if you specify the command-line option -s followed immediately by a nonzero seed number. For example,

```
tutorial> econ_order -n8 -s169
```

finds the minimum feed-forward in the directed graph named $econ(8, 2, 0, 169)$. The optimum order turns out to be

```
    Inorganic resources
    Organic resources
    Mineral and chemical products
    Primary metal manufacturing
```

```
Metal equipment
Organic products
Services
Specialized manufactured goods
```

on the 8 sectors of that graph.

As n gets larger, the heuristic methods more frequently converge to locally optimum arrangements that are not in fact optimum. For example,

    tutorial> econ_order -n20 -r100

finds an ordering with feed-forward \$525,421M only 9 times out of 100; its next best solution, costing \$525,552M, is found 62 times. It takes about 665 cautious steps per trial. By contrast,

    tutorial> econ_order -n20 -r100 -g

uses only about 16 greedy steps per trial; it finds the \$525,421M solution 33 times and the \$525,552M solution 37 times. Greed clearly beats caution, at least in economics.

The full 79-sector problem can be studied by running "econ_order -r10000 -g", preferably during dinner. The best solution obtained in this way has feed-forward \$456,295M. It is first found on the 268th trial and again on the 1583rd, but only twice in 10,000 greedy trials. This solution can, in fact, be proved optimum, using sophisticated branch-and-bound techniques developed by Grötschel, Jünger, and Reinelt [8]. So theory complements greed — at least in economic ordering.

1.5   GAMES

An entirely different class of examples can be obtained from a generator called GB_GAMES, which generates graphs based on college football scores. The vertices of these graphs are U.S. colleges or universities that have prominent football teams; the edges connect teams that played each other during the 1990 season.

Such graphs tend to be "cliquey," with clusters of vertices that are almost all adjacent to each other, because most games are between teams of the same league or conference. However, a few teams (like Notre Dame) are independent of any league, and there are enough nonconference games between local rivals to make the entire graph biconnected. Some algorithms perform well on graphs like this; others perform poorly. Therefore the *games* graphs are useful components of the GraphBase library.

The games.dat file records the scores of 638 games between 120 teams. The date of each game is listed, and the home team (if any) is identified. Each team has a nickname (like Fighting Illini or Fightin' Hoosiers), which can be used to enliven computer printouts. Moreover, subjective rankings by coaches and sportswriters are included so that graphs can be restricted to the top n teams if desired. The generation routine is controlled by eight parameters,

$$games(n, ap0, upi0, ap1, upi1, \textit{first\_day}, \textit{last\_day}, seed);$$

this subroutine call produces a graph with n vertices, selected on the basis of weighted averages of Associated Press and/or United Press International ratings before and/or after the season, with ties broken if needed by appealing to random numbers defined by *seed*. All games played between those teams, starting on *first_day* and ending on *last_day*, produce edges in the graph.

The "length" of an arc from u to v is the number of points scored by u against v. Thus the length from v to u is generally different from the length from u to v, and the length data might need to be adjusted before some algorithms are applied to a *games* graph.

A demo program called FOOTBALL illustrates one whimsical use of these data. Have you ever wondered how many points Stanford would score against Harvard if the teams ever met? FOOTBALL helps answer the question, as follows:

```
tutorial> football

Starting team: Stanford
   Other team: Harvard
 Sep 22: Stanford Cardinal 37, Oregon State Beavers 3 (+34)
 Oct 13: Oregon State Beavers 35, Arizona Wildcats 21 (+48)
 Sep 15: Arizona Wildcats 25, New Mexico Lobos 10 (+63)
 Oct 06: New Mexico Lobos 48, Texas-El Paso Miners 28 (+83)
 Oct 13: Texas-El Paso Miners 12, Hawaii Rainbow Warriors 10 (+85)
 Dec 01: Hawaii Rainbow Warriors 59, Brigham Young Cougars 28 (+116)
 Nov 03: Brigham Young Cougars 54, Air Force Falcons 7 (+163)
 Oct 27: Air Force Falcons 52, Utah Utes 21 (+194)
 Sep 01: Utah Utes 19, Utah State Aggies 0 (+213)
 Nov 03: Utah State Aggies 55, New Mexico State Aggies 10 (+258)
 Nov 17: New Mexico State Aggies 43, Fullerton State Titans 9 (+292)
 Oct 27: Fullerton State Titans 35, Long Beach State Forty-Niners 37 (+290)
 Sep 22: Long Beach State Forty-Niners 28, Pacific Tigers 7 (+311)
 Sep 29: Pacific Tigers 28, Nevada-Las Vegas Rebels 37 (+302)
 Nov 03: Nevada-Las Vegas Rebels 18, Fresno State Bulldogs 45 (+275)
 Sep 01: Fresno State Bulldogs 41, Eastern Michigan Hurons 10 (+306)
 Sep 15: Eastern Michigan Hurons 45, Ohio University Bobcats 18 (+333)
 Oct 06: Ohio University Bobcats 10, Bowling Green Falcons 10 (+333)
 Sep 08: Bowling Green Falcons 21, Virginia Tech Gobblers 7 (+347)
 Nov 24: Virginia Tech Gobblers 38, Virginia Cavaliers 13 (+372)
 Sep 22: Virginia Cavaliers 59, Duke Blue Devils 0 (+431)
 Nov 03: Duke Blue Devils 57, Wake Forest Demon Deacons 20 (+468)
 Sep 29: Wake Forest Demon Deacons 52, Army Cadets 14 (+506)
 Oct 20: Army Cadets 56, Lafayette Leopards 0 (+562)
 Nov 03: Lafayette Leopards 59, Fordham Rams 14 (+607)
 Sep 29: Fordham Rams 35, Brown Bears 28 (+614)
 Nov 17: Brown Bears 17, Columbia Lions 0 (+631)
 Oct 27: Columbia Lions 17, Princeton Tigers 15 (+633)
 Nov 03: Princeton Tigers 34, Pennsylvania Red \& Blue 20 (+647)
```

```
Sep 15: Pennsylvania Red & Blue 16, Dartmouth Big Green 6 (+657)
Sep 22: Dartmouth Big Green 33, Lehigh Engineers 14 (+676)
Nov 03: Lehigh Engineers 52, Colgate Red Raiders 7 (+721)
Sep 22: Colgate Red Raiders 59, Cornell Big Red 24 (+756)
Nov 03: Cornell Big Red 41, Yale Bulldogs 31 (+766)
Nov 17: Yale Bulldogs 34, Harvard Crimson 19 (+781)
```

Thus Stanford racks up a 781-point lead, if we assume some kind of transitivity. (Of course we do not allow intermediate teams to appear more than twice in the series of games.)

An even better score is obtained if we tell the program to try harder. After the command 'football 10', a sequence of games is found that gives Stanford an 1895-point advantage. Similarly, 'football 100' shows, after a few seconds, how to accumulate 2126 points for Stanford over its friendly Eastern foe.

The problem tackled by FOOTBALL is a special case of the general unsymmetrical longest path problem, for which no good algorithms are known. An exhaustive search for the optimum sequence of games would take incredibly long, so FOOTBALL uses a greedy heuristic, analogous to the method used by ECON_ORDER. With a numeric parameter, an improved heuristic called "stratified greed" is used; this method, which is easy to implement and reasonably effective in practice, is explained in detail in the program itself. Stratified greed gives no guarantees, but it does tend to explore the most fruitful parts of a search tree. The verbose option '-v' will make FOOTBALL display some of its internal calculations when it is using this heuristic.

The numeric parameter to stratified greed, called w for width, represents the maximum number of possibilities that are explored in each "stratum" of the search tree. When w is increased, more cases are explored, but the previous cases are not necessarily a subset of the new ones; therefore the final results might actually become poorer after w has been made larger. Here, for example, are typical results achieved as a function of w when FOOTBALL is asked to rank Stanford over Harvard:

w	result	time	w	result	time	w	result	time	w	result	time
1	1573	0.3	100	2126	17	1000	2124	244	4000	2185	2049
2	1776	0.4	150	2102	26	1200	2124	312	6000	2185	4339
3	1847	0.6	200	2129	36	1400	2124	386	8000	2180	7632
4	1950	0.8	300	2081	56	1600	2147	470	10000	2180	11500
5	1951	0.9	400	2113	77	1800	2124	558	12000	2180	16500
10	1895	1.8	600	2154	124	2000	2124	659	14000	2180	22400
50	2057	8.7	800	2145	181	3000	2158	1207	16000	2180	29300

(Running times here are in cpu seconds on a SPARCstation 2, without compiler optimization.) Notice that the running time is roughly proportional to w when w is small, although it eventually becomes proportional to w^2. If more sophisticated data structures were substituted for the linear lists in the present implementation, we could bring this asymptotic behavior down to $O(w \log w)$. But the algorithm appears to reach a point of diminishing returns, after which feasible increases in w do not improve the solutions found, so it probably should never be run with extremely large w.

1.6 MILES

The GraphBase data file `miles.dat` contains highway distances between 128 North American cities, and the GB_MILES generator converts this data into a variety of interesting graphs. For example, $miles(128, 0, 0, 0, 0, 0, 0)$ produces the complete graph on 128 vertices, with the length of each edge equal to the intercity distance. These distances satisfy the triangle inequality: The distance from x to y plus the distance from y to z is never less than the distance from x to z.

Smaller and sparser graphs can be obtained by choosing parameters judiciously in the general form

$$miles(n, north\_wt, west\_wt, pop\_wt, max\_dist, max\_deg, seed);$$

here n is the number of vertices, max_dist is a cutoff distance specifying that longer edges should be suppressed, and max_deg specifies that each vertex should have at most this many neighbors. The remaining parameters specify how to choose the n cities by geographic location and/or by population or at random.

The 128 cities are shown in Fig. 4, which can be produced by applying TEX to the file `cities.texmap`. Notice that all city names begin with the letters R–Z. This is a consequence of the author's decision to consider only the last 128 cities of the 500 listed in Rand McNally's *Standard Highway Mileage Guide* (1949), so that `miles.dat` would not be too large a file.

The MILES_SPAN demonstration program illustrates one typical use of mileage graphs: It calculates a *minimum spanning tree*, a set of $n - 1$ edges that connect n points with minimum total length. For example,

```
tutorial> miles_span
The graph miles(100,0,0,0,0,10,0) has 405 edges,
  and its minimum spanning tree has length 14467.
 The Kruskal/radix-sort algorithm takes 8379 mems;
 the Jarnik/Prim/binary-heap algorithm takes 7972 mems;
 the Jarnik/Prim/Fibonacci-heap algorithm takes 11519 mems;
 the Cheriton/Tarjan/Karp algorithm takes 17090 mems.
```

What does this mean? By default, MILES_SPAN sets $n = 100$, $north\_wt = west\_wt = pop\_wt = 0$, $max\_dist = 0$ (which is equivalent to $max\_dist = \infty$, by convention), $max\_deg = 10$, and $seed = 0$; this creates the graph $miles(100, 0, 0, 0, 0, 10, 0)$, which has 100 vertices and 405 edges. (The average degree of a vertex is $2 \times 405/100 = 8.1$, not 10, because some vertices have fewer than 10 neighbors in order to keep other vertices from having more than 10.) By using command-line options -n, -N, -W, -P, -d, and/or -s followed by decimal numbers, the user can change any of the parameters except max_dist. For example,

```
tutorial> miles_span -n128
The graph miles(128,0,0,0,0,10,0) has 508 edges,
  and its minimum spanning tree has length 16598.
 The Kruskal/radix-sort algorithm takes 10714 mems;
 the Jarnik/Prim/binary-heap algorithm takes 10298 mems;
 the Jarnik/Prim/Fibonacci-heap algorithm takes 14650 mems;
 the Cheriton/Tarjan/Karp algorithm takes 23098 mems.
```

Figure 4. The 128 cities featured in `miles.dat`.

The full graph is obtained by raising *max_deg*, as follows (see Fig. 5):

```
tutorial> miles_span -n128 -d1000
The graph miles(128,0,0,0,0,127,0) has 8128 edges,
  and its minimum spanning tree has length 16598.
The Kruskal/radix-sort algorithm takes 104252 mems;
the Jarnik/Prim/binary-heap algorithm takes 80235 mems;
the Jarnik/Prim/Fibonacci-heap algorithm takes 91256 mems;
the Cheriton/Tarjan/Karp algorithm takes 274127 mems.
```

Notice that the same minimum spanning tree is obtained even though the number of edges has become much greater. Restricting the maximum degree does not affect the minimum spanning tree because the shortest edges survive the restriction. (However, if you try -d7, the spanning tree length increases to 16649; if you try -d6, the graph becomes disconnected.)

The most interesting thing about MILES_SPAN is not the results it computes but the way it computes them. Many different algorithms are known for the efficient calculation of minimum spanning trees, and four of the principal variants have been implemented in MILES_SPAN as demonstrations of how to use GraphBase graphs in typical combinatorial applications. The author has also attempted to explain those algorithms as well as possible. Thus, the MILES_SPAN program is intended to be read by people more often than it is to be executed by machines.

When the code for MILES_SPAN was being written, there was no good way to predict which of the four methods would be fastest, on graphs of moderate size such as those produced by *miles*. Nor was it clear how to make a fair comparison of the methods — a comparison that would be meaningful on more than one computer and in the future as well as today. (Consider, for example, the FOOTBALL running times listed above. They apply only to the SPARCstation 2 computer, which was wonderful in 1990 but will probably seem slow and obsolete in the year 2000.)

Experiments have shown, however, that a remarkably simple idea provides an excellent way to compare combinatorial algorithms such as this, namely the technique of *mem counting*. Each algorithm's running time is reported by MILES_SPAN in terms of mems, meaning references to memory. This measure isn't perfect, but it is far more relevant than anything else that can be measured easily. A good algorithm will consume relatively few mems, and a study of MILES_SPAN shows that mem counting does indeed provide a fair way to rate the methods against each other. From these empirical tests we can conclude that Kruskal's method and the binary heap implementation of the Jarník-Prim method are decidedly superior to the other two methods, on graphs produced by *miles*. (Kruskal's method seems to consume mems more efficiently than the others; see section 71 of the program.)

Thus MILES_SPAN is not only an exposition of four important combinatorial algorithms; it also provides a full-scale illustration of how to compare algorithms by counting mems. Sections 10 and 11 of the program contain basic propaganda about mems and explain how to implement the counts with little difficulty. Pointers to places where the philosophy of mem-counting is elaborated further, or where slightly subtle technicalities relating to mems are discussed, can be found by referring to "discussion of *mems*" in the index to programs near the end of this book.

Figure 5. A minimum spanning tree on the 128 cities of miles.dat.

1.7 PLANE

The graphs discussed so far are not guaranteed to be planar; that is, we might not be able to draw them in a plane without crossing edges. A special GraphBase generator called GB_PLANE is available to help fill the need for such graphs. It is able to create planar graphs from the data in `miles.dat` with a subroutine called *plane_miles*.

In general, *plane_miles*(n, *north_wt*, *west_wt*, *pop_wt*, *extend*, *prob*, *seed*) chooses n cities based on *north_wt*, *west_wt*, *pop_wt*, and *seed*, just as the *miles* routine does. Then it uses the latitude and longitude of those cities to determine their neighbors, defining neighbors in a technical sense called the "Delaunay triangulation." (See Section 2.8 for further information about Delaunay triangulation.) If the *extend* parameter is nonzero, *plane_miles* also appends an $(n+1)$st vertex called ∞, or `INF`, which is a neighbor to all the outermost cities — those on the convex hull. The resulting graph always has $n+1$ vertices and $3n-3$ edges joining vertices to their neighbors, and it is planar. The length of each edge is copied from `miles.dat`, except that edges involving ∞ have length 2^{28}.

For example, Fig. 6 shows the graph obtained from the 10 most populous cities in `miles.dat`, using *plane_miles*$(10, 0, 0, 1, 1, 0, 0)$. Edges that run to the boundary of the illustration go to the special vertex ∞. A thin triangle San Francisco–San Diego–San José is almost invisible in the illustration but present in the triangulation.

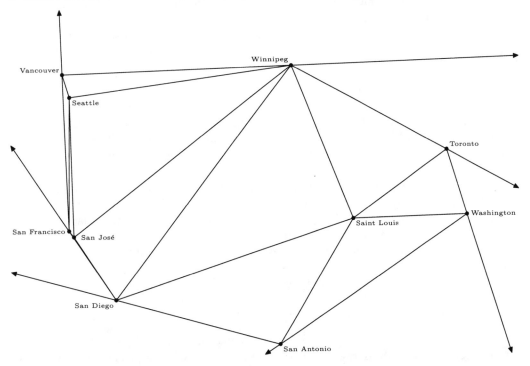

Figure 6. The Delaunay triangulation of 10 cities.

A planar graph with $n + 1$ vertices cannot have more than $3n - 3$ edges; therefore the graphs constructed by the method just described are maximal planar graphs. Nonmaximal planar graphs are obtained by making the *prob* parameter nonzero. This causes *plane_miles* to discard edges at random from the maximal planar graph that it would otherwise have produced. Each edge is discarded with probability $prob/65536$. For example, if $prob = 21845 \approx 65536/3$, about one third of the edges will be discarded and about $2n$ will remain, so the average vertex degree will be ≈ 4.

GB_PLANE contains another subroutine, *plane*(n, *x_range*, *y_range*, *extend*, *prob*, *seed*), which has similar characteristics but is not tied to `miles.dat`; therefore n can be much larger than 128 when *plane* is used. In this case, n cities are placed at random in a rectangle, with coordinates (x, y) satisfying

$$0 \le x < \textit{x\_range}, \qquad 0 \le y < \textit{y\_range}.$$

They are connected by Delaunay neighbor edges and possibly extended by a special vertex at ∞, as before. The length of an edge between (x_1, y_1) and (x_2, y_2) will be the nearest integer to

$$2^{10} \sqrt{(x_1 - x_2)^2 + (y_1 - y_2)^2}.$$

Both *x_range* and *y_range* must be at most $2^{14} = 16192$; this upper limit is also the default, which is used if *x_range* or *y_range* is zero.

GB_PLANE also contains a general-purpose *delaunay* subroutine, which finds the planar graphs defined by neighbors of any given set of points (x, y) in the square $\{0 \le x < 2^{14}, 0 \le y < 2^{14}\}$.

1.8 LISA

Now the scene changes to a dataset of an entirely different kind: The file `lisa.dat` contains a digitized version of Leonardo da Vinci's famous painting, *Mona Lisa*. The image appears in 360 rows and 250 columns of pixel values. Each pixel value represents one of 256 possible shades of gray, from 0 (black) to 255 (white).

What is `lisa.dat` doing here? The Stanford GraphBase does not claim to be a repository for high-resolution graphic images; other collections are being developed to fill that need. (They should perhaps be called GraphicBases.) Leonardo's classic portrait has, however, traditionally been used and abused in computer experiments [13, 17, 18], so there's good reason to define a standard digitization that can be used to explore combinatorial algorithms.

The generator module GB_LISA accesses this image data and converts it to graphs in several ways. One procedure, for example, is *bi_lisa*, which creates bipartite graphs. The procedure call

$$bi\_lisa(m, n, m0, n0, m1, n1, thresh, c)$$

takes a rectangular image from rows *m0* to *m1* -1 and columns *n0* to *n1* -1 of the 360×250 data in `lisa.dat` and transforms it linearly into an $m \times n$ array of pixel values between 0 and 65,535, inclusive. Then it forms a bipartite graph with $m + n$ vertices: m for the rows and n for the columns. Finally it places an edge between row k and column l if and only if the pixel value in position $[k, l]$ of the $m \times n$ image is *thresh* or more. Thus the light-colored areas of Leonardo's painting cause edges to appear in the graph. If $c \neq 0$, however, this convention is reversed, and edges come from the darker areas.

The demonstration routine ASSIGN_LISA illustrates another way to use `lisa.dat`. It treats the image data as an $m \times n$ array of input values to the so-called assignment problem, the problem of choosing at most one element in each row and at most one element in each column so that the sum of chosen elements is maximum. For example, the solution to a 36×25 assignment problem looks like this:

Here 25 pixels have been chosen — no two in the same row or column — and the sum of their brightness levels is as large as possible subject to that restriction. The illustration was prepared with the following UNIX command:

```
tutorial> assign_lisa m=36 n=25 -P
Assignment problem for lisa(36,25,255,0,360,0,250,0,22950000)
Solved in 44443 mems.
```

The `-P` option specifies that a PostScript® output file called `lisa.eps` should be produced, documenting the solution by means of a halftone image with chosen pixels framed.

Notice that the image in this example is called $lisa(36, 25, 255, 0, 360, 0, 250, 0, 22950000)$. In general, the parameters to $lisa(m, n, d, m0, m1, n0, n1, d0, d1)$ call for an $m \times n$ image to be formed from a region of size $(m1 - m0) \times (n1 - n0)$, as with bi_lisa, but the pixel values are converted to the range 0 to d instead of 0 to 65,535. Precise details are explained in the program for GB_LISA. Roughly speaking, raw pixel values are first computed in the range 0 to $255(m1 - m0)(n1 - n0)$, then they are transformed in such a way that raw values $\leq d0$ map into 0, raw values $\geq d1$ map into d, and raw values between $d0$ and $d1$ are mapped by linear interpolation.

ASSIGN_LISA uses an efficient implementation of the classic Hungarian algorithm to solve the assignment problem and reports its running time in mems. A solution to the full-scale

Figure 7. A 360×250 assignment problem based on `lisa.dat`.

problem, which uses all the data with $m = 360$ and $n = 250$, is shown in Fig. 7; this problem in much more difficult than than the 36×25 problem because it has 100 times as many entries in its matrix, so its solution takes a lot longer: 30,993,832 mems. An algorithm that solves the problem with significantly fewer mems would be of great interest.

The idea of solving an assignment problem with visual data is, of course, pretty weird. We can hardly claim that such choices of pixels have any relevance whatever to art appreciation or æsthetics. But they do have significant pedagogical value: What better way is there to comprehend a 360×250 array of numbers than to see it as a picture? The challenge of the assignment problem becomes graphically clear when we can present it graphically.

Figure 8. A different sort of assignment problem. White links tend to match up pixels of nearly equal brightness, while gray-striped links tend to match pixels of contrasting intensity.

The Mona Lisa data can be used also in assignment problems that have a more natural interpretation. Suppose we divide mn cells $[k, l]$ into two groups, according as the sum $k + l$ is even or odd. Then each cell has four neighbors of the opposite parity, except at the edges or corners where the number of neighbors is either three or two. And we can use the assignment algorithm to match the $mn/2$ even cells with neighboring odd cells, in such a way that the sum of the squares of differences in pixel values between matched cells is minimized or maximized. Minimizing this value tends to make the pairs follow contour lines of the image, while maximizing tends to make the pairs cut across those lines. For example, Fig. 8 shows both solutions, superimposed on a 30×22 matrix of pixel values. Gray-striped edges maximize the pixel differences between matched cells, while white edges minimize them. (In places where the two solutions coincide, we see only an isolated white edge, because a white edge covers a gray-striped edge in the same position.)

We can, in fact, go further and create amusing "domino portraits" of Mona Lisa, following a suggestion by Kenneth C. Knowlton [12]. The max-difference (gray-striped) pairing between even and odd cells in Fig. 8 defines a way to pack a 30×22 rectangle with 330 dominos. And 330 is exactly the number of dominos in six complete sets of "double-nines," since every such set contains 55 dominos ▗▖, ▗▖, ▗▖, ..., ▦▦, ▦▦. So we can take six complete sets and place them in such a way that the sum of squares of differences

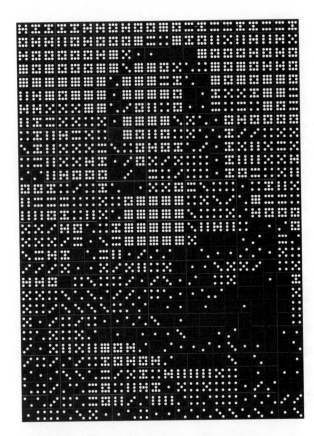

Figure 9. An optimum assignment of dominos.

between pixel values and domino pips is minimized; then the dominos will tend to reproduce the original image. This turns out to be yet another 330×330 assignment problem.

The solution is quite striking, seen from afar. (Please look at Fig. 9 from about 10 yards away.) Who knows, perhaps it even has some æsthetic value. Each of the 55 domino types appears exactly six times. The max-difference pairing of Fig. 8 was used to set up the final problem of assigning dominos to pixel pairs because, as Knowlton observed, high-contrast dominos like ▦, ▦, and ▦ tend to be the hardest to place.

Another kind of graph is created by the procedure call

$$plane\_lisa(m, n, d, m0, m1, n0, n1, d0, d1).$$

In this case, after an $m \times n$ image is constructed as before, the routine defines a set of maximal regions consisting of adjacent pixels that have equal value. Every such region then becomes the vertex of a graph; two vertices are adjacent if the corresponding regions contain adjacent pixels. It is easy to verify that the resulting graph is planar, even if some regions totally enclose others in the image. The value of d must be chosen carefully; if it is too small the graph will be trivial, but if it is too large the graph will be very much like an $m \times n$ grid. Example values of m, n, and d can be found in the table of example parameters, Appendix C, on pages 557–563.

1.9 RAMAN

Yet another kind of graph is produced by a dataless GraphBase generator called GB_RAMAN.
Unlike the irregular, organically structured graphs we have been deriving from cultural
artifacts, the results of GB_RAMAN are highly regular. Technically speaking, they are called
Ramanujan graphs based on integer quaternions; their precise mathematical properties are
explained in the program documentation. Informally, they are called "expander graphs,"
and they have important applications in communication networks and in the design of
algorithms.

A demonstration program called GIRTH illustrates several elementary characteristics of
these graphs. The user specifies two distinct prime numbers, p and q; these will define a
Ramanujan graph with approximately q^3 or $\frac{1}{2}q^3$ vertices, so q shouldn't be too large. Every
vertex will have degree $p + 1$. Ramanujan graphs are notable for having relatively large
girth and small diameter. In other words, their shortest cycle is rather large, and every
vertex is reachable from every other in a small number of steps. When all vertices have
degree $p+1$, there are $p+1$ vertices at distance 1 from a given vertex, $p(p+1)$ at distance 2,
$p^2(p+1)$ at distance 3, and so on, continuing until at least half the girth is reached. Some
of these graphs have larger girth than any other known examples having the same number
of vertices and degrees. Here is a sample run of the program:

```
tutorial> girth
This program explores the girth and diameter of Ramanujan graphs.
The bipartite graphs have q^3-q vertices, and the non-bipartite
graphs have half that number. Each vertex has degree p+1.
Both p and q should be odd prime numbers;
   or you can try p = 2 with q = 17 or 43.

Choose a branching factor, p: 5
OK, now choose the cube root of graph size, q: 37
The graph has 50616 vertices, each of degree 6, and it is bipartite.
Any such graph must have diameter >= 7 and girth <= 14;
theoretical considerations tell us that this one's diameter is <= 15,
and its girth is >= 10.
Starting at any given vertex, there are
        6 vertices at distance 1,
       30 vertices at distance 2,
      150 vertices at distance 3,
      750 vertices at distance 4,
     3570 vertices at distance 5,
    12748 vertices at distance 6,
    20276 vertices at distance 7,
    11772 vertices at distance 8,
     1306 vertices at distance 9,
        7 vertices at distance 10,
        0 vertices at distance 11.
So the diameter is 10, and the girth is 10.
```

1.10   GATES

Another generator module, GB_GATES, produces graphs that are executable — that is, they are directed acyclic graphs that define logical circuits based on fundamental Boolean operations (and, or, not, exclusive or). Some of the circuits also have feedback represented by latches. Two principal generation subroutines are provided: $risc(r)$, which creates a circuit for a small reduced-instruction-set computer having r arithmetic registers, and $prod(m, n)$, which creates a high-speed circuit for multiplication of m-bit numbers by n-bit numbers. GB_GATES also has subroutines to evaluate or partially evaluate logical circuits, when some or all of the input values are specified.

The demo programs that illustrate how to exercise these procedures are called TAKE_RISC and MULTIPLY. Let's consider TAKE_RISC first. It asks you for two numbers, then forms their product, quotient, and remainder as follows:

```
tutorial> take_risc
Welcome to the world of microRISC.

Gimme a number: 1009
OK, now gimme another: 22
The product of 1009 and 22 is 22198.
The quotient is 45, and the remainder is 19.
```

This might seem pretty trivial until you look at the program and realize what is happening: TAKE_RISC is simulating every gate inside an imaginary computer chip, clock pulse by clock pulse, as the chip executes a little program. The imaginary chip has only a primitive ability to add and subtract 16-bit numbers; all of its registers and machine operations are built up from scratch using low-level operations on single bits. Yet it computes the product, quotient, and remainder without wasting much time. (You can see a trace of the microcomputer's activity if you specify the command-line option '-v'.)

Similarly, the MULTIPLY program demonstrates the circuits generated by $prod(m, n)$:

```
tutorial> multiply 100 100
Here I am, ready to multiply 100-bit numbers by 100-bit numbers.
(I'm simulating a logic circuit with 65033 gates, depth 42.)

Number, please? 12345678987654321
Another? 1000000001
12345678987654321x1000000001=12345678999999999987654321.
```

The circuit for 100-bit by 100-bit multiplication is said to have depth 42. This means 42 levels of logic, with at most two inputs and one output per gate. (That's pretty good; nobody knows how to accomplish the same task with substantially fewer levels of logic.) Numbers to be multiplied are input and output in decimal notation, but MULTIPLY converts them to and from binary notation as it simulates each gate of the logic circuit. Therefore multiplication by a number like 1000000001 is not an especially easy task for the computer, although it's easy for us to check by hand.

Another variant of MULTIPLY creates special-purpose circuits to multiply by a constant. To try this, give an arbitrary "seed" value as a third parameter on the command line;

MULTIPLY will then choose a random constant based on the seed, and it will simplify the circuit by customizing it for that particular multiplier, as in the following example:

```
tutorial> multiply 100 100 31416
OK, I'm ready to multiply any 100-bit number by
      18991640951785083576983582758O.
(I'm simulating a logic circuit with 35268 gates, depth 41.)

Number, please? 1000000000000000000000000000000001
(Sorry, 1000000000000000000000000000000001 has more than 100 bits.)

Number, please? 100000000000000000000000000000001
100000000000000000000000000000001x18991640951785083576983582758O=
      1899164095178508357698358275801899164095178508357698358275O.
```

1.11 RAND

We have been constructing nonrandom graphs, but the Stanford GraphBase also makes it possible to create random ones. Well..., pseudo-random ones; these graphs, like all the others, are completely defined by their parameters. The graph obtained by calling *random_graph*(1000, 5000, 0, 0, 0, 0, 0, 1, 1, 1) will be the same on all computers, and it will be the same if you generate it next year. Hence you can do repeatable experiments with it.

The generator module GB_RAND provides several subroutines. First is

$$random\_graph\,(n, m, multi, self, directed, dist\_from, dist\_to, min\_len, max\_len, seed)\,,$$

which generates a random graph with n vertices and m edges or arcs — undirected edges if $directed = 0$, otherwise directed arcs. Different *seed* values generally give different results, unless n and m are very small. The *multi* and *self* parameters control whether duplicate arcs or arcs from a vertex to itself are allowed; if they are zero, such arcs are disallowed. The lengths of edges and arcs are chosen uniformly at random between *min_len* and *max_len*, inclusive.

The *dist* parameters control the distribution of arcs. If they are zero, each endpoint is chosen uniformly at random from the n vertices. But each *dist* parameter may also point to a vector of numbers that defines a probability distribution, making some vertices more popular than others.

Similarly, the procedure call

$$random\_bigraph\,(n1, n2, m, multi, dist1, dist2, min\_len, max\_len, seed)$$

creates random bipartite graphs. Its parameter conventions are essentially identical to those of *random_graph*.

Another procedure, *random_lengths*, is available to transform the lengths of the arcs or edges of graphs that have been generated by any other GraphBase routine. An arbitrary nonuniform distribution of lengths can be specified, if desired. Nonuniform random variables are generated by an efficient algorithm called Walker's alias method.

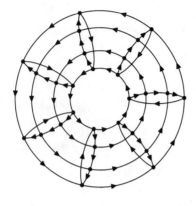

Figure 10. *board*(4, 7, 0, 0, 1, 3, 1).
(A 4 × 7 directed torus.)

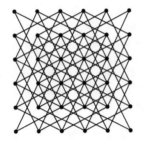

Figure 11. *board*(6, 6, , 0, 0, 5, 0, 0).
(Knight moves on a 6 × 6 chessboard.)

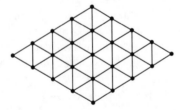

Figure 12. *simplex*(28, 4, 8, 4, 0, 0).
(A miniature "Hex" board.)

1.12 BASIC

Finally, there's a generator module called GB_BASIC that makes it easy to create most of the example graphs that appear in standard textbooks. There are graphs based on generalized chesspieces on *d*-dimensional chessboards, with coordinates optionally allowed to "wrap around"; these are generated by a routine called *board*. There are graphs based on generalized triangular configurations, generated by *simplex*. There are graphs based on the subsets of a multiset, generated by *subsets*; on the permutations of a multiset, generated by *perms*; on the partitions of an integer, generated by *parts*; or on the binary trees with *n* branches, generated by *binary*. An almost inexhaustible variety of graphs with different sorts of mathematical structure can be obtained from these routines. Figs. 10–15 illustrate a few of the possibilities.

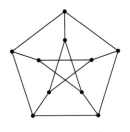

Figure 13. $subsets(2, 1, -4, 0, 0, 0, 1, 0)$.
(The Petersen graph, also called $petersen()$.)

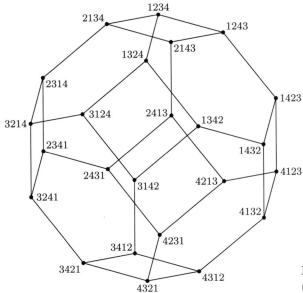

Figure 14. $perms(-3, 0, 0, 0, 0, 0, 0)$.
(Truncated octahedron, aka $all\_perms(4, 0)$.)

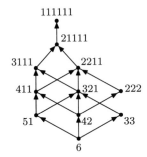

Figure 15. $parts(6, 6, 6, 1)$.
(Partitions of 6, aka $all\_parts(6, 1)$.)

GB_BASIC also makes it easy to transform one or more existing graphs into new ones. It provides subroutines for the standard operations of copying, complementing, and forming unions, intersections, line graphs, and products. It also features a general routine for induced graphs, which are obtained by omitting and/or splitting and/or collapsing vertices and/or replacing vertices by other graphs. Details appear in the program description.

Thus, for example, if you need a huge graph that is much bigger than the ones you get from *miles* (which is limited to 128 cities), you can try taking the product of a *miles* graph with a *books* graph. The vertices will then be ordered pairs consisting of cities and people, like (Saint Louis, MO, Huckleberry Finn). If you find the graph product structure too regular, you can use the *induced* procedure to make it irregular in various ways.

1.13 SAVE

The special module GB_SAVE enables you to convert any GraphBase graph into an ordinary ASCII text file. Such a file can then be saved for later use, mailed to other researchers, and/or used with other graph-manipulation systems. To convert graph *g* into ASCII form, within a C program, you simply say *save_graph*(*g*, `"filename"`). The file name conventionally ends with the suffix '`.gb`'.

Conversely, *g* = *restore_graph*(`"filename"`) takes such a file and converts it to the internal data structures used in GraphBase routines. This mechanism provides a simple interface between the Stanford GraphBase and any other system. It also facilitates the creation of additional libraries of well-defined graphs.

CHAPTER 2: *TECHNICALITIES*

Chapter 1 presented an overview of the entire Stanford GraphBase. Now we will look at the substructure.

2.1 REPRESENTATION OF GRAPHS

All graphs generated by GraphBase routines have a uniform representation that works well in many different applications. Three kinds of records — that is, three kinds of **struct** data types in the C language — are defined for the respective representations of graphs, vertices, and arcs. A C program that works with GraphBase graphs will typically declare variables such as

> **Graph** $*g$;
> **Vertex** $*u$, $*v$;
> **Arc** $*a$;

then g will point to a graph record, u and v will point to vertices, and a will point to an arc.

There is one **Graph** record for each graph. That record points to an array of **Vertex** records, and each **Vertex** in turn points to a linked list of **Arc** records. Suppose g points to a **Graph**. Then $g \rightarrow n$ is the number of vertices in the corresponding graph, and $g \rightarrow vertices$ is a pointer to the first of these vertices. (In a C or CWEB program, field $g \rightarrow n$ of the record pointed to by g is actually typed '`g->n`'; CWEB's formatter adds typographic sugar, converting the `->` notation to a short arrow $\rightarrow$ in order to indicate the tight binding between a pointer and its subfields.) Thus the vertices of the graph represented by g are pointed to by

$$g \rightarrow vertices + k, \qquad \text{for } 0 \le k < g \rightarrow n.$$

The kth vertex is $g \rightarrow vertices[k]$.

Neighboring vertices in the graph are specified by directed arcs from one vertex to another. Undirected edges between vertices are specified by two arcs, one in each direction. If v points to a **Vertex**, and if no arcs lead from v, the field $v \rightarrow arcs$ will be Λ (typed '`NULL`' in the computer file). Otherwise $v \rightarrow arcs$ will point to the first arc from v, and $v \rightarrow arcs \rightarrow next$ will either be Λ or point to the second arc from v, etc.

Suppose variable a points to an **Arc** record, representing a directed arc from some vertex v to another vertex. Then $a \rightarrow tip$ is a pointer to that other vertex, and $a \rightarrow next$ is a pointer to the next arc from v (if any).

Another important field, $v \rightarrow name$, is a string that gives symbolic significance to vertex v. More precisely, $v \rightarrow name$ points to the first character of a sequence of characters that names v; this sequence is terminated by a null character, as C strings normally are. The name of a vertex usually has no special significance to algorithms that manipulate graphs, but names help greatly when the result of a calculation is being presented for human consumption.

As a consequence of these conventions, all arcs of a graph g can be printed by the following simple fragment of a C program, assuming that the variables g, v, and a have been declared as stated earlier:

$$\textbf{for } (v = g\text{-}vertices;\ v < g\text{-}vertices + g\text{-}n;\ v\text{++})\ \{$$

$\qquad printf(\texttt{"\%s:"}, v\text{-}name);$

$\qquad \textbf{for } (a = v\text{-}arcs;\ a;\ a = a\text{-}next)\ printf(\texttt{"\textvisiblespace\%s"}, a\text{-}tip\text{-}name);$

$\qquad printf(\texttt{"\textbackslash n"});$

$\}$

If g points, for example, to the 2×3 directed grid graph

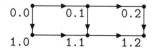

which is created by calling subroutine $board(2, 3, 0, 0, 1, 0, 1)$ of GB_BASIC, the output of that program fragment will be

```
0.0: 1.0 0.1
0.1: 1.1 0.2
0.2: 1.2
1.0: 1.1
1.1: 1.2
1.2:
```

Every **Vertex** record contains eight subfields. We have already mentioned *name* and *arcs*; the other six subfields are called *utility fields* because they can be used for many different purposes. Each utility field can either hold an integer value, or point to a string of characters, or point to a **Graph**, **Vertex**, or **Arc** record, depending on the application. The six utility fields are named u, v, w, x, y, z, and their five possible interpretations are distinguished by adding one of the respective suffixes *.I*, *.S*, *.G*, *.V*, *.A*; thus, for example, $v\text{-}w.I$ stands for the integer in utility field w of the **Vertex** record pointed to by v.

Utility field names are usually given meaningful aliases by means of macro definitions. For example, GB_GAMES defines *nickname* to mean $y.S$. If v points to the vertex representing Stanford's football team, then $v\text{-}name$ is the string `"Stanford"` while $v\text{-}nickname$ is `"Cardinal"` and $v\text{-}conference$ is `"Pacific Ten"`. Algorithms that operate on graphs often use the utility fields to define auxiliary data structures.

An **Arc** record contains an integer length field, *len*, and two utility fields, a and b, in addition to the two standard fields *tip* and *next* that were mentioned earlier. One way to use these utility fields, for example, would be to let $a.V$ point to v when a is an arc from v to u, and to let $b.A$ point to another arc that has the same destination vertex u. If we make the macro definitions

$\qquad \textbf{\#define } source \quad a.V$

$\qquad \textbf{\#define } back\_arcs \quad z.A$

$\qquad \textbf{\#define } back\_next \quad b.A$

the following program fragment will extend the GraphBase representation of g so that each vertex u has a list $u{\to}back\_arcs$ of all arcs whose tip is u:

```
for (v = g→vertices; v < g→vertices + g→n; v++) v→back_arcs = Λ;
for (v = g→vertices; v < g→vertices + g→n; v++)
   for (a = v→arcs; a; a = a→next) {
     a→source = v;
     u = a→tip;
     a→back_next = u→back_arcs;
     u→back_arcs = a;
   }
```

In an undirected graph, such a list of back arcs is never necessary, because each undirected edge between vertices u and v is represented by a pair of adjacent arcs, a and $a+1$. If $u < v$, we will have $a{\to}tip = v$ and $(a+1){\to}tip = u$. An undirected loop from v to itself is represented by two consecutive arcs, for which we have $a{\to}next = a + 1$. The GraphBase routines that generate undirected graphs will never create an **Arc** record with $a{\to}next = a + 1$ unless a and $a + 1$ are the opposite arcs of a self-loop.

So now we know about **Vertex** and **Arc** records. What about the other kind? Well, a **Graph** record contains thirteen fields. We have already discussed $g{\to}vertices$ (the address of the vertex array) and $g{\to}n$ (the number of vertices). Another field, $g{\to}m$, contains the total number of arcs; in an undirected graph this will be twice the number of edges. A fourth field, $g{\to}id$, identifies the graph by naming the parameters that were used to create it. For example, the 2×3 directed grid graph referred to earlier has $g{\to}id = $ `"board(2,3,0,0,1,0,1)"`.

The kernel program GB_GRAPH includes convenient routines to allocate storage for new graphs, new arcs, new strings, and new data structures of all kinds. Storage that has been allocated for a particular purpose is grouped by **Area** pointers, in such a way that if s is an **Area** we can deallocate all storage in group s by simply saying $gb\_free(s)$. The main storage area for a graph g contains the vertices and arcs and the strings for names; it is called $g{\to}data$. Another storage area, $g{\to}aux\_data$, holds optional information about the graph that is not strictly necessary in all applications. For example, the five-letter-word graphs created by GB_WORDS include a hash table for word lookup as part of their aux_data.

The remaining fields of a **Graph** record are six utility fields called uu, vv, ww, xx, yy, and zz, and a string called $util\_types$. The characters of $util\_types$ specify the current conventions being used in all utility fields of the graph.

Complete details about graph representation and the subroutines that facilitate the generation of new graphs appear in the kernel program GB_GRAPH. But the basic information just given should suffice for an initial reading of most of the higher-level programs.

2.2 WORDS

No two people will agree on exactly which strings of letters are English words and which are not, but `words.dat` represents the author's attempt to collect all the five-letter English words that were "real" to him in 1992. The nucleus of this collection was a dictionary for playing the game of Jotto, compiled by Michael Beeler prior to 1971 [1] and extended shortly afterward to a list of 6627 words corresponding to the contents of *Webster's 7th*

New Collegiate Dictionary. Beeler estimated that about 16,000 words would have been present if he had used an unabridged dictionary. But the 6627 five-letter words in *Webster's Collegiate* already included plenty of esoterica, so the author pared Beeler's list down by removing whatever he could not remember seeing previously. After all, just about every pronounceable combination of five letters has been used to spell something somewhere at some time in history, but a game isn't fun when the words aren't familiar.

Additional words were added to the culled file during the next 20 years whenever the author came across a bona fide five-letter word that was not present. For example, one of the missing words was `chiff`, a term well known to pipe organ enthusiasts but long overlooked by lexicographers.

Why this fascination with five-letter words? Well, the graphs of four-letter and shorter words are too small and too dense; the graphs of six-letter and longer words are too sparse. But five-letter English words define a graph that seems just right for many pedagogic purposes.

The English language is constantly changing. Therefore `words.dat` contains quite a few words that would not have belonged in such a collection in 1971: `antsy`, `condo`, `cuspy`, `decaf`, `dweeb`, `faxed`, `glitz`, `gonzo`, `neato`, `smurf`, `vapes`, etc., as well as new technical terms from computer science and mathematics such as `arity`, `bebug`, `blash`, `inode`, `muxes`, `nurbs`, `poset`, `paren`, `pixel`, `poset`, `treap`, `voxel`, `xored`.

Chapter 1 mentions some of the restrictions that were observed as the words were collected. Proper names were not allowed, although of course some names are legitimate because they are also words in their own right; consider `degas`, `hardy`, `harry`, `jenny`, `sally`, `texas`, `twain`, and names of physical units like `curie`, `gauss`, `joule`. The character set, in fact, was strictly limited to the lowercase letters `a` through `z`. Words that require accents, such as `abbés`, `blasé`, `outré`, `piñon`, `roués`, `señor`, were ineligible, as were hyphenated words. The author waited until seeing `email` and `yoyos` without hyphens before including them. Some words that once required accents (`naïve`, `rôles`, `sauté`) have become sufficiently common in unaccented form (`naive`, `roles`, `saute`) to be accepted.

Two words coined by the author in technical papers, `noads` and `triel`, were left out of `words.dat` because nobody else was known to have used them. Several lovely words of rabbit language (`hrair`, `hraka`, `tharn`) were introduced in Richard Adams's novel *Watership Down*, but they were reluctantly excluded for similar reasons; the author has never seen them outside of that novel, except on license plates.

Is `fooey` a word? It was not included because the author could find only `pfui`, in print.

Incidentally, five-letter words include many plurals of four-letter words, and it should be mentioned that no Victorian-style censorship was done. Users who ask LADDERS to find the shortest path from `shots` to `skits` might therefore be shocked, unless they curtail the vocabulary. One way to ensure that semantically unsuitable words won't defile any output from the Stanford GraphBase is to restrict consideration to vertices whose weight is 100 or more, according to the default ranking criterion explained in GB_WORDS; this can be done by using the option `-n3317` to LADDERS. (A few anatomical terms might still appear.) The output of WORD_COMPONENTS includes all 5757 words in order by default rank.

Of course the purpose of the Stanford GraphBase is to be an archival source of interesting datasets, not a reference collection of up-to-the-minute facts. Therefore `words.dat` is now complete and correct for all time, by definition. The English language will continue to evolve, but `words.dat` will forever remain `fixed`.*

The basic *words* subroutine of GB_WORDS assigns a weight to each word by using frequency information that has been included in `words.dat`. For example, the entry for `which` in the data file is

$$\text{which} * 14016,3560,4467,347,3083,756,362 \ .$$

The * means that `which` is a common word known to young people, and the seven numbers are frequency counts from seven different samples of English text. The default weight is computed as

$$100 + 4 \times 14016 + 2 \times (3560 + 4467) + (347 + 3083 + 756 + 362) = 76766 \,,$$

where the '100' corresponds to the '*'. It turns out that `which` has the highest default weight of all five-letter words; next are `there` and `their`, which weigh in at 76153 and 69545, respectively. The word `often` occurs rather often (12394), more often than `occur` does (1339). At the other extreme, 891 words have weight 0, since they aren't present in any of the seven text samples. If we plot nonzero weights versus ranks on a log-log chart, we get the following graph:

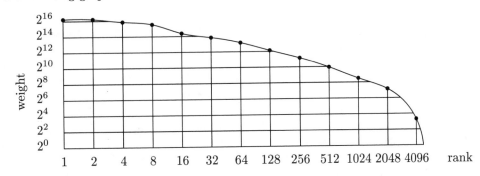

("Zipf's law" [14], which predicts that the points on such a chart will lie on a straight line, is only a rough approximation to the truth.)

Words without '+' might be rated '+', meaning that they are probably known to educated adults if not to children. The default weights of such words are computed by adding 10 instead of 100; thus the default weight in general is

$$\begin{Bmatrix} 100 \\ 10 \\ 0 \end{Bmatrix} + 4f_1 + 2f_2 + 2f_3 + f_4 + f_5 + f_6 + f_7 \,,$$

---

* I learned the word `taiga` on November 8, 1992; but for the first time in many years, I didn't file the word away in my computer. On March 19, 1993, I learned `dunch`. But please do not send me lists of words that are missing from `words.dat`; my days of dictionary building are over, and I believe we can be `happy` with the collection as it exists.

where $f_1, \ldots, f_7$ are the frequency counts in `words.dat`. All nine weight factors — which are 100, 10, 4, 2, 2, 1, 1, 1, 1 by default — can be changed by using a vector of nine numbers as the second parameter to *words*, as explained in GB_WORDS.

No model of language is known to account for the observed number of edges between the n most common English words. If the words were truly random, the expected number of edges would be only $125\binom{n}{2}/(26^5 - 1) \approx 0.0000053n^2$; the actual number, however, is two orders of magnitude larger. It ranges from about $0.00076n^2$ when $n = 1000$ to about $0.00043n^2$ when $n = 5700$. Here is a plot that shows the average degree of a vertex, which is twice the number of edges divided by n:

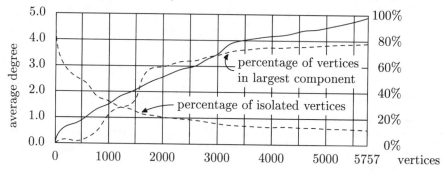

This chart also indicates the size of the "giant" component, which comprises more than half the vertices when n is greater than 1610.

The seven samples of text used for frequency data have interesting idiosyncrasies that can be used to modify the vocabulary when subsets of the full graph are being chosen. Frequency f_1 comes from *The American Heritage Word Frequency Book* by John B. Carroll, Peter Davies, and Barry Richman [3]; that book lists complete frequency data from the American Heritage Intermediate Corpus, a selection of 500-word samples from approximately one thousand texts used in American schools, grades 3 to 9. Thus f_1 reflects word usage in the elementary education system of the United States during the 1960s.

The second frequency, f_2, comes from the Brown University Corpus, which represents typical adult literature in America [22]. It is an amalgam of excerpts from newspaper stories, including letters to the editor and theater or book reviews; from novels, including transcriptions of various dialects; from handbooks; from sermons; from published papers on geology, sociology, history, mathematics, physics, chemistry, and biology; from legal briefs; from recipes; and from government proclamations and documents, including tax forms. This material now is part of ICAME, the International Computer Archive of Modern English, which resides at the Norwegian Computing Center for the Humanities, University of Bergen, Norway. (Researchers can purchase the computer files for a reasonable fee.)

Frequency count f_3 is from the Lancaster-Oslo/Bergen Corpus, also part of ICAME. Although similar to the Brown Corpus, it is based on British sources instead of American; you get to see `tyres` as well as `tires`. The level of discussion also tends to be more intellectual, something like switching TV channels from ABC to public television.

The Melbourne-Surrey Corpus, another part of ICAME, led to frequency count f_4. It comes from vintage-1980 editorials published in an Australian newspaper called *The Age*.

Frequency count f_5 comes from the complete text of the Revised Standard Version of the Bible (including the Apocrypha). The other two counts, f_6 and f_7, can be mentioned in the same paragraph with f_5 only because they, too, were compiled from complete texts instead of from samples: f_6 comes from *The TEXbook* and *The METAFONTbook* by Donald E. Knuth (Reading, MA: Addison-Wesley, 1982 and 1984); f_7 comes from Chapters 1–9 of *Concrete Mathematics* by Graham, Knuth, and Patashnik (Reading, MA: Addison-Wesley, 1988).

The job of gathering frequency data was not as simple or cut-and-dried as it might seem. For example, here's one of the sentences in the Brown Corpus: "Sun of a bich com to Sea you, the god damed theaf and lop yeard pigen tode helion." How many five-letter words do you count there? Typographic errors caused by poor keyboard entry were unfortunately not uncommon in the Brown data: 'socal' for 'social', 'sould' for 'should', 'tothe' for 'to the', 'voume' for 'volume', 'an ytime' for 'anytime', 'skiis' for 'skis', 'cocao' for 'cacao' (or maybe 'cocoa'), and 'diety' twice for 'deity'. (The latter typo, incidentally, has also crept into the UNIX spelling dictionary.) Surely `bicep` is not a word, although the Brown Corpus undoubtedly found an author who used it. In the Melbourne data, 'asses' appears where 'assess' was intended.

Some words of the American Heritage corpus were consciously omitted from `words.dat` even though that corpus is supposedly elementary. For example, one of its selections from Grade 3 material used `unkum` four times; such made-up words might be great for teachers but they don't belong in the GraphBase. Three Biblical words were dropped because they are more Hebrew than English: `algum` (2 Chronicles 9:10–11), `almug` (1 Kings 10:11–12), and `omers` (Exodus 16:22).

The hardest aspect of frequency counting was the task of distinguishing proper names from ordinary words. In f_1, capitalized words were simply not counted. For example, `theme` occurred 183 times, `Theme` 9 times, and `THEME` twice; the value of f_1 for `theme` was taken to be 183, although the word actually appeared 194 times. A similar convention was used for f_5. But in f_2, f_3, f_6, and f_7, the first word of a sentence was counted if only its first letter was capitalized. The Australian counts f_4 had to be handled specially because the relevant data was entirely uppercase in the available computer file. Everything in that corpus was first treated as an ordinary word, and then names like `April`, `CSIRO`, and `Hawke` were deleted. The five references to `jimmy` were deleted because they all referred to President Carter. Also, the fact that `Wales` occurred six times meant that the count for `south` had to be decreased from 55 to 49; the context was "New South Wales." Errors were undoubtedly made, but again we must remember that `words.dat` has now been declared correct by definition.

The default weights for combining frequency information in `words.dat` seem to correlate reasonably well with a native speaker's intuition about relative word ranks, but there are a few exceptions. For example, `magma` and `swiss` have unusually high counts in f_1; `burnt` and `didst` are unusually high in f_5; `fonts` and `macro` are abnormally high in f_6. The distinction between common ('*') and advanced ('+') and unusual ('␣') words is not clear cut; in most cases, words were designated common if and only if they appeared in a dictionary intended for fifth graders [27], while advanced words were selected in a completely subjective manner. Many words, especially derived forms like plurals or comparative adjectives like `wryer`, appear in dictionaries and in people's vocabularies more often than they appear in actual use. Conversely, dialect words like **gonna** appear more often in use than in dictionaries.

2.3 ROGET

A few details about the preparation of **roget.dat** should be stated here, for the record. The 1879 edition of Roget's *Thesaurus* was published in America in 1882, and the American edition was the source of the cross-reference data.

Several of Roget's categories actually had identical names. For example, his original category 278 was called "direction," meaning direction of motion, while 693 was "direction" meaning direction of projects; the latter was changed to "management" in a later edition, hence "management" was adopted in **roget.dat**. Roget had category 414, "discord" in music, as well as 713, "discord" among people; the latter became "dissension." Elsewhere "exclusion" was changed to "omission," "exemption" to "non-ownership," "impulse" to "spontaneity," "chance" to "non-design," "ornament" to "ornamentation," "relinquishment" to "non-retention," "sequence" to "following," "taste" to "good taste," and several category names were qualified by the word "physical." Now each category has a unique name.

Modern readers might be tempted to change "inexistence" to "nonexistence," "fetor" to "stench," etc., but the author resisted these temptations in order to preserve the charm of Roget's original vocabulary. A few category names were, however, changed from adjectives to nouns, for consistency with Roget's normal practice: "purple" became "purpleness." The synopsis of categories at the beginning of Roget's book did not always agree with the category names found in the main part of the book itself; the latter names were deemed correct.

Sometimes three categories deserve to be implicitly linked, like "beginning," "middle," and "end," or "convexity," "flatness," and "concavity." But implicit cross-references were not inserted on semantic grounds; they were considered present only when two categories were printed in adjacent columns. Thus there are arcs from "convexity" to both "flatness" and "concavity," and vice versa, but there is no arc between flatness and concavity, because Roget's book presented the material in the following way:

$$
\begin{array}{c|c|c}
\text{convexity} & \text{concavity} & \text{flatness}
\end{array}
$$

Peter Mark Roget's grandson, Samuel Romilly Roget, continued the task begun by his ancestors and published a third edition of the *Thesaurus* in 1925. To keep up with changing times, he added a few new categories like "flying." Then a fourth edition, edited by Robert A. Dutch in 1962, became perhaps the final *Thesaurus* to adhere closely to Roget's original category names. By this time the number of cross-references had increased to about 2.5 times the number in **roget.dat**. For example, Dutch's edition has references

from category 1, "existence," to 25 other categories, while the 1879 edition had just 10;
it has 10 references from category 400, "pungency," compared to 4 in 1879. Dutch said,
"The sequences of figures appearing under the heading **See:** at the foot of each numbered
section are the section numbers of all references quoted in the main head. Most readers
will have no occasion for them. They have been consolidated for the benefit of students of
language whose theoretical enquiries into the relationships of words may be assisted by such
an apparatus. The desirability of an aid of this kind has more than once been expressed."
Indeed, if such an aid had been provided in the 1879 edition, several Stanford undergrads
would not have had to work so hard to assemble the data that now appears in `roget.dat`.

Incidentally, the 1879 *Thesaurus* did not mention the word "graph"; but "graphic" was
included in the categories called "intelligibility," "painting," and "description." Also "vertex" appeared under "summit," and "arc" under under "curvature." The third edition did
include "graph" under "representation." By 1962, "graphs" appeared as part of "mathematics" under "numeration" (but not yet in the context of graph theory).

2.4 BOOKS

The files `anna.dat`, `david.dat`, `jean.dat`, `huck.dat`, and `homer.dat` begin by tabulating
the cast of characters, then they summarize each chapter by means of "encounter cliques."
For example, the twentieth chapter of *Anna Karenina* is represented by the line

$$1.20\text{:AN,DO,ST,KI;DO,KI,GR,TA}$$

in `anna.dat`. This line means that AN, DO, ST, and KI encountered each other near the
beginning of the chapter, then DO and KI were joined by GR and TA near the end.

Notice that the chapter name is given as '1.20'; this means chapter 20 of book 1. There
are 279 chapters, named 1.1 to 8.19, but they are implicitly renumbered 1–239 for para-
metric purposes in *book* and *bi_book*. The naming scheme for the 356 chapters of *Les
Misérables* needs three levels of numbers.

Sometimes a character appears alone, in sort of a soliloquy. Then a clique of size 1 is
listed. For example, the line

$$1.16\text{:VR}$$

indicates that only Vronsky makes a substantial appearance in chapter 16.

The number of cliques containing a particular pair of characters is not significant. A
single edge between AA and BB will be inserted in the *book* graph if and only if AA and BB
appear together in at least one clique. Sometimes the person who prepared the data has
chosen to follow the plot of the novel closely so that one can easily match the data file with
the book itself. But at other times the data might be "optimized" so as to be brief yet
mathematically equivalent to the book. For example, a clique AA,BB,CC,DD might be listed
even if characters AA, BB, CC, DD were never present simultaneously in any scene, provided
that each character did happen to meet the others at least once.

The notion of a "character" in a novel is usually straightforward, but in some cases we
want to generalize the idea so that a group of people can collectively act as an individual.
Thus, for example, one of the characters in *Huckleberry Finn* is TF, "Townfolk." In the
Iliad, OG stands for the Olympian gods, collectively; GS and TS are arrays of Greek soldiers
and Trojan soldiers.

Judgment calls occasionally need to be made about whether encounters are significant enough to be recorded. What should be done when AA tells BB that CC once met DD? What should be done when EE sees FF, but FF cannot see EE? In all such cases, arbitrary decisions were made and the present data files are correct by definition.

The data for *David Copperfield* was gathered by Phyllis Winkler; for *Les Misérables*, by Dylan Kohler; for *Huckleberry Finn*, by Christine Chang; for the *Iliad*, by David Alexander; for *Anna Karenina*, by the present author, who also edited the other datasets. Keeping such notes turns out to be a pleasant task. Somehow a little bit of data-gathering keeps a person's mind more alert and enhances one's appreciation of a great book.

2.5 ECON

The fiscal data in econ.dat comes from calendar year 1985, the most recent year for which annual accounts had been completed when econ.dat was prepared. Numerical results for 1985 were published by the U.S. government in January, 1990 [31].

The published statistics actually consist of two tables, not a single set of input/output flows as in econ.dat, because the Bureau of Economic Analysis is careful these days to distinguish between *industries* and *commodities*. There are 79 industries, called "Livestock and livestock products" through "Local government enterprises," and there are 79 commodities with the same names. There also are six dummy industries and dummy commodities, officially called "Noncomparable imports," "Scrap, used and secondhand goods," "Government industry," "Rest of the world industry," "Household industry," and "Inventory valuation adjustment." Thus the data in [31] is reported for 85 industries and commodities altogether, including the dummy ones. Since the dummy sectors were introduced as aids to accounting, econ.dat lumps them all together and calls them "Adjustments." There also is an 86th sector, which corresponds to econ.dat's "Users."

Table 1 of [31] contains an 86×86 matrix called "The use of commodities." Its rows are indexed by commodities and its columns by industries, and its entries $a_{i,j}$ represent the output of commodity i to industry j (which is also the input of commodity i by industry j). Special entries $a_{i,86}$ represent the total final demand for commodity i; special entries $a_{86,j}$ represent the value added by industry j; and $a_{86,86} = 0$.

Table 2 of [31] contains an 85×85 matrix called "The make of commodities." Its rows are indexed by industries and its columns by commodities, and its entries $b_{j,k}$ represent the output of industry j belonging to commodity k. Industries generally produce a primary commodity having the same name as the industry name, but they usually produce secondary byproducts as well.

Tables 1 and 2 of [31] are related to each other in the following way. Let $C_i = \sum_{j=1}^{86} a_{i,j}$ be the total output of commodity i, including its total final demand. Then we also have $C_k = \sum_{j=1}^{85} b_{j,k}$; that is, the total output of a commodity is the total of all contributions made to that commodity, summed over all industries. Let $I_j = \sum_{i=1}^{86} a_{i,j}$ be the total input of industry j, including value added. Then we also have $I_j = \sum_{k=1}^{85} b_{j,k}$, the total output of industry k, summed over all commodities.

The dollar amounts in econ.dat flow from commodity to commodity, so we want to create an 85×85 matrix $c_{i,k}$ whose rows and columns are both indexed by commodities. For this purpose, we make a weighted average of Table 1's columns, thereby distributing the dollar flow from industries back to commodities in appropriate proportions. More precisely, we define

$$c_{i,k} = \sum_{j=1}^{85} a_{i,j} b_{j,k} / I_j$$

for each pair of commodities, $1 \leq i, k \leq 85$. We also let $c_{i,86} = a_{i,86}$ be the total final demand for commodity i, and

$$c_{86,k} = \sum_{j=1}^{85} a_{86,j} b_{j,k} / I_j$$

be the total value added for commodity k. Then it is easy to check that the total flow out of commodity i is

$$\sum_{k=1}^{86} c_{i,k} = \sum_{k=1}^{85} \sum_{j=1}^{85} a_{i,j} b_{j,k} / I_j + a_{i,86} = \sum_{j=1}^{85} a_{i,j} + a_{i,86} = C_i ,$$

while the total flow in to commodity k is

$$\sum_{i=1}^{86} c_{i,k} = \sum_{i=1}^{86} \sum_{j=1}^{85} a_{i,j} b_{j,k} / I_j = \sum_{j=1}^{85} b_{j,k} = C_k .$$

Everything balances, as desired.

Two technicalities should be noted. First, two of the dummy industries, $j = 80$ ("Non-comparable imports") and $j = 81$ ("Scrap, used and secondhand goods") have $I_j = 0$, so we can't divide by I_j. But in those cases, $b_{j,k} = 0$ for all k and $a_{i,j} = 0$ for all i, so we can just leave the terms for $j = 80$ and $j = 81$ out of the sums. Second, the quantities $c_{i,k}$ are rational numbers that have been rounded to integers in `econ.dat`. The rounding was done to $c_{i,k}$ for $1 \leq i, k \leq 85$ only, then $c_{i,86}$ was fudged so that C_i was preserved. Similarly, $c_{86,k}$ was not actually computed from the formula above; it was computed as C_k minus the sum of rounded values $\{c_{1,k}, \ldots, c_{85,k}\}$.

Commodity number 11, the economic sector called "New construction," has special properties: $b_{11,11} = C_{11} = I_{11}$, and $b_{11,k} = b_{j,11} = 0$ for all $j, k \neq 11$; also $a_{11,j} = 0$ for $1 \leq j \leq 85$. Thus all the flow out of "New construction" goes to "Users," and the ECON_ORDER program will always rank this sector last when "New construction" has not been merged with another sector.

Data for a more detailed breakdown of the U.S. economy into 537 industries instead of 85 is available for the year 1977, in machine-readable form (see [30]). But the 85-industry statistics from 1985, which we have massaged and compressed into statistics for 79 commodities plus "Adjustments" and "Users," should suffice for GraphBase purposes.

2.6 GAMES

The data in `games.dat` was culled from several dozen newspapers during the final months of 1990.

Football teams affiliated with the National Collegiate Athletic Association belong to so-called divisions, depending on the facilities and scholarships available at a particular college or university. The best teams generally belong to division 1-A. There are 106 football teams in division 1-A, and they all were selected for inclusion in `games.dat`.

The author also decided to include the eight Ivy League teams, which belong to division 1-AA, because substantial computer science research is done at Columbia, Cornell, etc. But the Ivy League teams didn't play any games against division 1-A; the resulting graph was therefore disconnected. Fortunately, another prominent set of 1-AA teams, the Patriot League, was available to connect everything up. This made a grand total of 120 teams.

The best sequence of games found by FOOTBALL in attempts to rate Stanford over Harvard occurred for $w = 4000$ in the author's experiments, when the search direction was reversed as explained in Chapter 3. This sequence achieves a 2279-point spread by considering 113 teams in the following order, using the abbreviations for team names found in `games.dat`: STAN, UCLA, SDSU, UTAH, USU, LBSU, SJSU, UNLV, FULL, MSSU, MEMPH, ECAR, LTECH, AKRON, NAVY, TOL, WMICH, MIOH, BALL, WIS, MSU, MICH, MISS, GA, SMISS, LOUVL, BAMA, PSU, TEMP, BOST, WVA, PITT, NDAME, MIAMI, TEX, TCU, RICE, HOUS, BAYL, ARK, TULSA, NMSU, UTEP, USAF, OSU, PURD, WASH, CAL, OR, ASU, WSU, ORSU, AZ, USC, SYR, RUTG, KY, TENN, FLA, AUBN, FSU, SCAR, NCAR, MD, CLEM, ILL, IND, MO, COLO, KSU, ISU, OK, NEB, MINN, IOWA, NW, NIL, FRES, PAC, HI, BYU, CSU, NMEX, TTECH, SMU, VAND, LSU, TA&M, SWLA, TUL, CINCI, CMICH, KENTS, OU, BOWLG, VTECH, VA, DUKE, WAKE, ARMY, LAFAY, FORD, BROWN, PENN, DART, LHIGH, COLG, YALE, PRIN, BUCK, CORN, COLUM, HARV. Similarly, a 2173-point spread for Harvard over Stanford can be achieved with the sequence HARV, HOLY, BUCK, CORN, COLUM, LAFAY, FORD, BROWN, YALE, PRIN, PENN, DART, LHIGH, COLG, RUTG, KY, MSSU, SMISS, MEMPH, MISS, GA, ECAR, LTECH, AKRON, NAVY, TOL, WMICH, MIOH, BALL, WIS, MSU, MICH, UCLA, SDSU, UTAH, USU, LBSU, SJSU, UNLV, HOUS, ASU, WSU, TCU, RICE, BAYL, ARK, TULSA, OKSU, KAS, ORSU, AZ, USC, SYR, WVA, PITT, NDAME, MIAMI, TEX, PSU, TEMP, BOST, LOUVL, BAMA, TENN, FLA, AUBN, FSU, SCAR, NCAR, MD, CLEM, ILL, IND, MO, COLO, KSU, ISU, OK, NEB, MINN, IOWA, NW, NIL, FRES, PAC, NMSU, FULL, HI, BYU, CSU, NMEX, TTECH, SMU, VAND, LSU, TA&M, SWLA, TUL, CINCI, CMICH, KENTS, OU, BOWLG, VTECH, VA, DUKE, WAKE, ARMY, USAF, OSU, PURD, WASH, CAL, OR, STAN. These two sequences have 86 games in common, although they were supposedly obtained with conflicting goals. Optimum sequences are not known.

It is interesting to note that the first sequence would never have been found by the stratified method of FOOTBALL without reversing the search direction. Consider, for example, the stratum near the middle of the sequence where Tennessee (TENN) appears. A detailed analysis, made with the help of the "verbose" option to FOOTBALL, shows that this stratum represents partial solutions in which 62 teams still remain on paths between the current team and Harvard. The accumulated score for Stanford over Tennessee at this point is only 795; but there are zillions of partial solutions that achieve much more than 795 at this stratum. When the width parameter is 6000, the cutoff value is in fact 1398. We couldn't possibly make the width high enough to retain sequences like the one shown, since it has a relatively weak start but a very strong finish.

The program for GB_GAMES gives detailed information about Associated Press and United Press International poll scores that can be used to rank the teams in various ways. By setting $ap0 = ap1 = -1$ and $upi0 = upi1 = +2$ in the parameters to *games*, you can select teams that the UPI coaches liked better than the AP journalists did. Several teams — Houston, Memphis State, Oklahoma, Oklahoma State, and (later) Florida — were ineligible for UPI ranking because they were on probation for violating NCAA rules.

The 1990 football season was especially interesting because nobody could agree afterward which team was best. "College Football's Champ? Confusion Mounts," proclaimed the International Herald Tribune on January 4, 1991: "Colorado, Georgia Tech, and Miami were awarded Wednesday what amounts to perpetual debating rights to the unofficial 1990 title." Colorado was chosen best by the Associated Press's panel of 60 writers and broadcasters, after completing the season with 11 victories, one loss, and one tie. (Its 33–31 victory over Missouri was tainted, however, by a "fifth and goal" touchdown erroneously allowed by the referees.) Georgia Tech was chosen best by United Press International's panel of 59 division 1-A coaches; with 11 victories, no losses, and one tie, it was the only major college team to be undefeated during the season. But the New York Times used a high-tech approach that may have avoided the human foibles of writers and coaches: Its computer program ranked the teams mechanically, "based on an analysis of three factors: who won, by what margin, and against what quality of opposition. The quality of an opponent is determined by examining its record against other teams. The Times's computer model collapses runaway scores and takes note of a home-field advantage. As the season progresses, results in recent games count more than results in other games." And according to that model, Miami of Florida was clearly best, despite two losses; Georgia Tech came in second, Colorado third.

A simple approximation to the Times's scheme computes a "handicap" score for each team, equal to the team's average margin of victory plus the average of the handicaps of the team's opponents, rounded to the nearest integer. Handicaps can be used to adjust the scores and make the stratified search method more effective when the width is small. This computation ranks Miami first, Washington second (two points behind), Florida State and Virginia next (five points behind); Colorado, Georgia Tech, and Nebraska then tie for 5th, 6th, and 7th place, a full seven points behind Miami.

Users of the Stanford GraphBase are invited to settle this controversy once and for all.

2.7 MILES

The data in `miles.dat` is the result of a rather elaborate sequence of computations undertaken by the author in collaboration with Sol Lederman and Brian Shriver. The starting point was Rand McNally & Company's *Standard Highway Mileage Guide*, published in 1949; but calculations of minimum spanning trees with that data gave very peculiar results. A closer look soon revealed that a dozen of the published entries were clearly wrong, since they were less than the great circle distance between cities. For example, the stated distance by road from Ravenna, Ohio, to Sarasota, Florida, was 584 miles, but the least distance as the crow flies is actually 958. On the other hand, several of the published entries were unreasonably large; the distance from Wichita Falls, Texas, to Yankton, South Dakota, was said to be 2498 miles, while a crow can fly there in 623. Four of the entries listed for Rochester, Minnesota, were found to be copied by mistake from Rochester, New York.

Of course, the existence of errors in such a compilation is not surprising, because intercity distances are difficult to get right. But even if the Rand McNally data had been 100% correct for purposes of the trucking industry, it would have needed modification for purposes of `miles.dat`, because the triangle inequality is supposed to be valid in `miles.dat`. Truckers don't always care about the triangle inequality. They might rather drive a few more miles than go through New York City or pay money to Canadian customs officials.

The author's first approach to cleaning up the data was to ask the following question: Given a (symmetric) matrix of distances between vertices, how can we make the entries satisfy the triangle inequality by changing as few of the numbers as possible? This proved to be an interesting problem, apparently not yet studied in the literature. No way was found to determine the absolute minimum number of changes in a reasonable amount of time, but a heuristic approach made it possible to satisfy the triangle inequality after changing only 379 entries. The essential idea was to first identify distances that were suspiciously low and to set those distances to ∞. Then the operation

$$d_{xy} \leftarrow \min_z(d_{xy}, d_{xz} + d_{zy})$$

was applied repeatedly for all x, y, z, until the entries converged to a steady state.

However, this minimal-alteration idea proved to be a mistake. It was the answer to the wrong question. If two cities are near each other, many triangle violations can sometimes be canceled if the distance between them is made artificially high, but the resulting minimum spanning trees will look awful.

The mileages that now appear in `miles.dat` were produced by an interactive program. This program incorporated several of the ideas in the author's earlier, fully automatic attempts, but did everything under manual supervision. A triangle violation $d_{xz} + d_{zy} < d_{xy}$ could mean that d_{xz} or d_{zy} is too small or that d_{xy} is too large; the computer assigned a score of -1 to d_{xz} and to d_{zy}, and $+2$ to d_{xy}, and summed these scores over all violations (x, y, z), in an attempt to find all the culprits. Then the scores were refined by assessing only the triangle violators who looked somewhat guilty according to the previous estimate. The refined scores were then refined again, at the user's discretion. Using the Rand McNally data, the scores converged after five rounds; 177 distances were thereby identified as suspiciously small and 158 as suspiciously large, and they were rank-ordered by probable guilt. Now the user could choose to mark a distance as definitely too big or too small. In the latter case, the distance could either be recomputed as the length of the shortest path through edges that were above suspicion, or it could be raised just enough to satisfy all triangle inequalities containing it — again under user control. The computer would display the ramifications of each decision before the user was committed to any particular choice.

Altogether 388 changes — 142 increases and 246 decreases — were made to the 8128 original mileages. About a third of these were due to serious errors in the published data; another third were caused by rerouting certain trips between cities in Michigan and upstate New York (sending them through Canada) or between southern cities and Salisbury, Maryland (taking them across Chesapeake Bay by ferry). The remaining third of the changes were small adjustments needed to prevent tiny inconsistencies from messing up the triangle inequality.

The United States government began to construct a new network of interstate highways during the 1950s, several years after the *Standard Highway Mileage Guide* was compiled. So the figures in `miles.dat` are probably obsolete even when they agree perfectly with the published data. Still, we can reasonably regard `miles.dat` as a realistic set of mileage data, suitable for testing combinatorial algorithms.

The latitude and longitude figures for U.S. cities in `miles.dat` were taken from *The National Atlas* [32]. Population data is from the 1980 census.

The TEX file `cities.texmap` places all 128 city names near their dots without overlapping other names or dots. (See Fig. 4 in Section 1.6.) The problem of finding such a layout was tackled by students in Stanford's 1989 Problem Seminar [29], and it proved to be quite challenging. The present arrangement is the author's adaptation and amalgamation of features from several partial solutions found at that time.

2.8 PLANE

A set of n points on the plane defines n *Voronoi regions*, the areas that are closer to one point than to any of the others. The *delaunay* procedure of GB_PLANE creates a planar graph in which the given points are the vertices and in which two vertices are neighbors if and only if their Voronoi regions share a common edge (see Fig. 16).

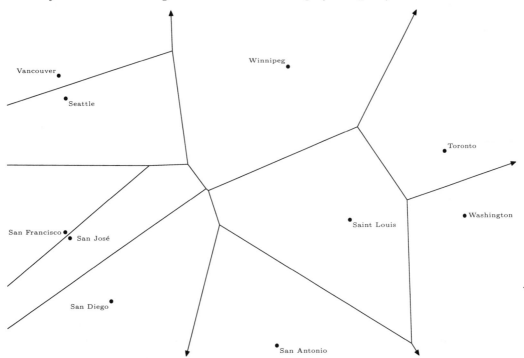

Figure 16. Voronoi regions for the ten cities whose Delaunay triangulation appeared in Section 1.7, Fig. 6. Notice that Winnipeg just barely qualifies as a "neighbor" of San Diego, with respect to these cities.

If we make slight perturbations to the coordinates, we can ensure that no two vertices are coincident, no three vertices lie on a straight line, and no four vertices lie on a circle. Then the Delaunay triangulation is the set of all triangles of vertices whose circumcircle contains no other vertices. (The circumcircle of three noncollinear points is the unique circle that passes through them.) Equivalently, the vertices are joined by an edge of a Delaunay triangle if and only if they belong to some circle that contains no other vertices. The equivalence between these two definitions can be seen by considering the family of all circles that pass through two vertices: By rotating, scaling, and shifting the coordinates we can assume that the vertices lie at positions $(0, 1)$ and $(0, -1)$, hence every circle passing through them hits the positive x-axis at a unique point $(t, 0)$. As t varies from 0 to ∞, there might be an interval of values (t_0, t_1) when the circle through $(0, 1)$, $(0, -1)$, and $(t, 0)$ contains no other vertices; then $(0, 1)$ and $(0, -1)$ belong to exactly two Delaunay triangles, whose other two vertices belong respectively to the circles through $(t_0, 0)$ and $(t_1, 0)$.

It is not difficult to prove that the Voronoi regions are polygons whose vertices are the circumcenters of the Delaunay triangles — the centers of the circumcircles. (These "Voronoi vertices" should be distinguished from the n given points, which we have also been calling vertices.) Thus the edges of the Delaunay triangulation are precisely the edges between neighbors, as defined above.

The *delaunay* procedure implements an efficient algorithm for Delaunay triangulation that was developed in 1989 and 1990 by Leo Guibas, Micha Sharir, and the author. Voronoi regions can be computed with the same procedure, slightly extended, because the data structure makes it easy to traverse the Delaunay triangles. The complete theory underlying this algorithm is described in [19], where the running time on a variety of datasets is also reported in terms of mems and determinant evaluations.

2.9   LISA

Leonardo da Vinci's *Mona Lisa* has been called "without doubt the most famous work in the entire forty-thousand-year history of the visual arts" [26]. Scholars say that it was painted about 1503–1505, and that the subject was Lisa Gherardini, the wife of Francesco di Zanobi del Giocondo. ("Mona" is short for "Madonna" and has roughly the significance of the term "Mrs." in English.) Leonardo sold the painting to King Francis I of France for 12,000 francs, and it now hangs in the Louvre Museum.

The painting's most important features have withstood the ravages of time remarkably well, but portions of the background are in poor condition — cracked and/or covered with streaks of unsightly 17th-century varnish. The white poplar wood on which Leonardo originally painted has split badly in one prominent place above Lisa's head.

The author digitized a photograph of the painting and used Adobe PhotoShop™ to retouch the defects. This process did not require great expertise, because extremely fine details do not show up when the resolution is only 5 pixels/cm. But the image was treated carefully so that Lisa's features remained sharp and so that highlights and shadows would not simply map to solid blocks of pixel values 255 and 0. The following histogram shows the actual distribution of pixel values in `lisa.dat`:

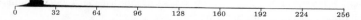

| 0 | 32 | 64 | 96 | 128 | 160 | 192 | 224 | 256 |

CHAPTER 3: *INSTALLATION AND USE*

The GraphBase programs are meant to be read, but they are also meant to be used. And there is no charge for using them. So you might want to have them in your personal computer, or in a network file system that you can access conveniently from a workstation.

3.1 OBTAINING THE FILES

The master sources for the Stanford GraphBase reside on a Stanford computer called LaBrea. If you have access to the Internet, you can obtain a copy of the sources by issuing the command 'ftp labrea.stanford.edu' and logging in as 'anonymous', typing also your email address when requested. Then you simply say 'cd pub/sgb' to connect to the public Stanford GraphBase directory, and 'prompt' to turn off unnecessary dialog, and 'mget *' to retrieve all the files — a total of approximately 1.2 megabytes. Say 'bye' to leave the ftp program.

If you don't have access to the Internet, your best bet is probably to get a copy from a friend who does, or from a friend of such a friend, etc. The files fit comfortably on a single 3.5-inch high-density floppy disk. (Your friend who has the files can perhaps make you a copy by typing the command 'make floppy' after connecting to the relevant directory.) There is no copyright restriction, except that nobody is permitted to change the GraphBase files in any way. (See the copyright page of this book.) Thus anything you obtain from another source should be just as good as the files you would get directly.

The master data files *.dat will never change. And the master program files *.w will be updated from the versions in this book only if changes are necessary to make them portable to C compilers that behave differently from the C compilers that the author has encountered in preliminary tests. Thus, the Stanford GraphBase is an essentially stable set of files. It is not going to be a moving target.

Of course, the *save_graph* and *restore_graph* routines make it possible for users to construct arbitrarily many additional graphs that might be used like those described in this book, and to share them with each other. The author hopes that one or more volunteers will come forward to maintain a growing collection of significant test problems in compatible formats. The Stanford GraphBase is intended to be a useful "fixed point" in such collections.

3.2 INSTALLING CWEB

Before you can use the Stanford GraphBase, you need to make sure that your computer supports CWEB [20, 24]. Actually, you ought to have CWEB in any case — everybody should — but it is a relatively new system, and nobody can keep up with all the announcements of significant new software these days, so you needn't feel too guilty if you haven't already acquired it. Perhaps you do have CWEB, but not an up-to-date version? The Stanford Graph-Base uses some of CWEB's latest bells and whistles, so you need CWEB version 3.0 or greater.

The installation of CWEB is a relatively painless process, unless you don't have a decent C compiler on your machine. The CWEB sources, like the Stanford GraphBase, are in the public domain and can be obtained from labrea.stanford.edu via anonymous ftp as described above. In this case, the directory is pub/cweb, and the source files (including the 11-page

CHAPTER 3: *INSTALLATION AND USE* 54

user's manual and its appendices) amount to approximately 450 kilobytes. You might also want to obtain a copy of the additional material in directory `pub/cweb/examples`, another 150 kilobytes.

When `CWEB` is installed, you will be able to run two important programs called `CTANGLE` and `CWEAVE`. The purpose of `CTANGLE` is to convert a `CWEB` file, say `foo.w`, into a C program, `foo.c`. You can then produce executable machine instructions from `foo.c` like you do with any C code, except that you never need to look at `foo.c`; all error messages from the C compiler and all communications to and from your online debugger will refer directly to the original source file `foo.w`. The other half of the `CWEB` system is `CWEAVE`, a program that converts the same file `foo.w` into `foo.tex` — another file you don't need to look at. The command `tex foo` will then produce formatted and cross-referenced documentation for program `foo`, using typographic conventions that enhance readability. (If you don't have TEX on your machine, you won't be able to use `CWEAVE`. But `CTANGLE` will still work, and programs like `foo.w` will still be easier to read than ordinary C programs.)

$$\begin{array}{ccccc}
& \text{CTANGLE} \cdots\!\!\rightarrow & \text{foo.c} & \cdots \text{cc} \cdots\!\!\rightarrow & \text{foo} \quad (\text{executable}) \\
\text{foo.w} & & & & \\
& \text{CWEAVE} \cdots\!\!\rightarrow & \text{foo.tex} & \cdots \text{T\!E\!X} \cdots\!\!\rightarrow & \text{foo.dvi} \quad (\text{printable})
\end{array}$$

Both `CTANGLE` and `CWEAVE` are written in `CWEB`, so you might wonder how the whole process can get started — it might seem that you need `CWEB` already installed before you can install it. Which could come first, the chicken or the egg? The solution, in this case, is very simple: The `CWEB` source files include not only `ctangle.w` and `cweave.w` but also `ctangle.c`, the result of applying `CTANGLE` to `ctangle.w` on another machine (or, originally, by hand). Once you have `ctangle.c`, you can make `CTANGLE` with the C compiler, and then you're in business. The `README` file that comes with `CWEB` explains how to do the job when you have an operating system something like UNIX. If you say '`make cautiously`', the `CWEB` sources will check that `CTANGLE` is indeed able to reproduce itself perfectly.

The only slightly tricky part about installing `CWEB` is the decision that needs to be made about where to put the executable programs `CTANGLE` and `CWEAVE`. You also need to install certain associated files like `cwebmac.tex` (a macro package for TEX, used by the output of `CWEAVE`). If you share your file system with other users, your system administrator should put everything in a place that's universally accessible. In any case you will need to have the executable code for `CTANGLE` and preferably also `CWEAVE` in a directory that belongs to your execution path (the list of directories searched for instructions that you type to your operating system).

3.3 INSTALLING THE GRAPHBASE

OK, you have a working `CWEB`, and you have acquired a copy of the GraphBase files. What should you do next? Answer: Read the file called `README`. It contains a brief description of all the files. It also contains any late-breaking news items that might have been added since the publication of this book.

Then turn to the file called `Makefile`. If you have a UNIX-like system, the details of GraphBase installation are encapsulated in that `Makefile` and you'll probably have to

change only a few lines that refer to your local file system. Otherwise you'll have to do the
basic steps by hand, unless someone has already prepared an automatic installation kit for
your system; basically, you need to tangle and compile the kernel modules like GB_GRAPH,
then the generator modules like GB_WORDS (see Fig. 1 at the beginning of Chapter 1), and
put the resulting compiled objects into a library that can be loaded with other C programs
like LADDERS. The following comments assume that you have a UNIX-like system, but they
also explain what the `Makefile` means so that you can simulate its actions by hand if
necessary.

Here's how the GraphBase `Makefile` begins:

```
 1 #
 2 #   Makefile for the Stanford GraphBase
 3 #

 4 #   Be sure that CWEB version 3.0 or greater is installed before proceeding!
 5 #   Skip down to "SHORTCUT" if you're going to work only from the
 6 #   current directory. (Not recommended for serious users.)

 7 #   Change SGBDIR to the directory where all GraphBase files will go:
 8 SGBDIR = /usr/local/sgb

 9 #   Change DATADIR to the directory where GraphBase data files will go:
10 DATADIR = $(SGBDIR)/data

11 #   Change INCLUDEDIR to the directory where GraphBase header files will go:
12 INCLUDEDIR = $(SGBDIR)/include

13 #   Change LIBDIR to the directory where GraphBase library routines will go:
14 LIBDIR = /usr/local/lib

15 #   Change BINDIR to the directory where installdemos will put demo programs:
16 BINDIR = /usr/local/bin

17 #   Change CWEBINPUTS to the directory where CWEB include files will go:
18 CWEBINPUTS = /usr/local/lib/cweb

19 #   SHORTCUT: Uncomment these lines, for single-directory installation:
20 #DATADIR = .
21 #INCLUDEDIR = .
22 #LIBDIR = .
23 #BINDIR = .
24 #CWEBINPUTS = .

25 #   Uncomment the next line if your C uses <string.h> but not <strings.h>:
26 #SYS = -DSYSV

27 #   If you prefer optimization to debugging, change -g to something like -O:
28 CFLAGS = -g -I$(INCLUDEDIR) $(SYS)

29 ########## You shouldn't have to change anything after this point ##########
```

These lines are the only part of the Stanford GraphBase that most users will have to
customize. (Line numbers do not actually appear in the file, of course; they are shown only
for reference. The actual file contains blank lines that aren't numbered here.)

Let's assume that you have placed the GraphBase source files in a directory called `/usr/local/src/sgb`. (Another possibility is `$(HOME)/sgb`, where `$(HOME)` is your directory after logging in; see the shortcut explained later in this section.)

Many different conventions are commonly used for computer file system nomenclature. Here we are assuming that all software not supplied by your manufacturer is generally placed in `/usr/local`; thus, for example, files needed by the GNU Emacs editor might reside in `/usr/local/emacs`. Source files are assumed to appear in the subdirectory `/usr/local/src`. These source files are regarded as expendable; that is, they may be deleted if the available disk space becomes tight. The programs themselves are installed elsewhere so that they can outlive their sources if necessary. The permanent resting place for GraphBase text files is typically a directory called `/usr/local/sgb`. On some systems another name, say `/usr/lib/sgb`, might be preferable. Line 8 of the `Makefile` should be changed if you want a nonstandard directory name. (The standard name can be used only if the parent directory `/usr/local` already exists before you say `make install`.)

Two kinds of files generally go into the directory `SGBDIR` that you have specified on line 8. First are the `*.dat` files, like `lisa.dat` and `words.dat`, which are used to create some of the graphs; they occupy about 425,000 bytes of memory. Line 10 of the `Makefile` specifies their location, which is normally a subdirectory of `SGBDIR` called `data`.

Programs that use GraphBase conventions contain `#include` statements, which tell the C compiler about standard data structures and procedure names. For example, most Graph-Base programs begin with the line `#include "gb_graph.h"`. The compiler has to know where to look for such header files; hence you must specify the `INCLUDEDIR`, as on line 12. Change it if you don't want the default, which is a subdirectory of `SGBDIR` called `include`. About 13,000 bytes of information will be placed in this directory.

The loading routine for compiled programs will combine your GraphBase executables with procedures from a library of compiled object code, so it needs to know where to find that library. The usual location is `/usr/local/lib`, but you might prefer another name; if so, change line 14. The library routines themselves, by the way, will be put into a file called `libgb.a`; you should find that file on your `LIBDIR` after the GraphBase has been successfully installed. The size of `libgb.a` varies from one machine to another, but typically is about 350,000 bytes when debugging information is included.

Line 16 of the `Makefile` defines a directory that you might not need. The Stanford GraphBase includes a dozen demonstration programs (LADDERS, FOOTBALL, GIRTH, and so on, explained in Chapter 1), and you must decide whether to install them semi-permanently in your machine or just to use them occasionally and remove them each time when you're done. In the former case, you need to specify a directory where the binary executables will be stored. This directory should be part of your normal execution path, the list of directories that the operating system searches for programs when you type a command. On a typical computer, the twelve demo programs will occupy almost 900,000 bytes on `BINDIR`, including information for debugging.

Line 18 is for communicating with `CWEB`. Most GraphBase programs begin with the `CWEB` instruction

```
@i gb_types.w
```

which tells CTANGLE and CWEAVE to include the file gb_types.w. (This file defines identifiers such as **Graph** and **Vertex** so that they will be properly typeset.) The CWEBINPUTS directory specified on line 18 should usually be the same as the macro variable of the same name in the Makefile that you used when installing CWEB. (The full name of that file may be /usr/local/src/cweb/Makefile; look at it again, if you've forgotten what CWEBINPUTS was called there.) More precisely, the value of CWEBINPUTS on line 18 should be one of the three directories in which CTANGLE and CWEAVE will look for included files when you use them with GraphBase programs. These are (1) the current working directory; (2) the directory specified in the environment variable CWEBINPUTS, if this variable has been assigned a value; (3) the directory specified as CWEBINPUTS when CWEB was installed.

Shortcut: If you don't plan to use the GraphBase regularly — maybe you just want to play with it for a day or two — you don't need to install it in separate directories all over your computer system. You can simply define DATADIR, INCLUDEDIR, LIBDIR, BINDIR, and CWEBINPUTS to be the same as the directory where the GraphBase sources already reside. Then you can run everything from the source directory. Just uncomment lines 20–24 of the Makefile, if you decide to take this shortcut. (To "uncomment" a line of the Makefile means to remove the '#' at the beginning.) However, after you start adding random programs of your own, you might find that the source directory fills up with unwanted junk, making it harder and harder for you to sort things out. Therefore separate directories, such as the suggested defaults, are recommended if you are going to do any serious experimental work with the GraphBase library.

Finally, you might also want to customize the definitions on lines 26 and/or 28. Those lines are self-explanatory.

Once you have made the proper definitions, type 'make tests' and stand back, keeping your fingers crossed. The machine should then start chugging away, tangling and compiling the kernel routines and the rest of the library, meanwhile checking that everything is hunky-dory. The triumphant conclusion to this phase will be the following message:

> Congratulations --- the tests have all been passed.

Good, you're almost done! Type 'make install', after first becoming superuser if any of the directories DATADIR, INCLUDEDIR, LIBDIR, BINDIR, or CWEBINPUTS that you specified are owned by the system rather than by you as a user. (It's best not to become superuser until after make tests has been successful, because all system defenses are down when the superuser is executing untested code. The actions done by make install, on the other hand, are not dangerous.) With any luck, you'll soon have a working GraphBase system that's rarin' to go. And you, too, can go right on to Section 3.6 below.

3.4 UNDERSTANDING THE MAKEFILE

What's in the rest of that Makefile? We've seen only lines 1–29. Its contents after line 29, which specify the exact meaning of make tests and make install, are rather cryptic, but not unfathomable. Let's list them first, then discuss them line by line.

```
30 LDFLAGS = -L. -L$(LIBDIR)
31 LDLIBS = -lgb
32 LOADLIBES = $(LDLIBS)
```

```
33  .SUFFIXES: .dvi .tex .w
34  .tex.dvi:
35          tex $*.tex
36  .w.c:
37          if test -r $*.ch; then ctangle $*.w $*.ch; else ctangle $*.w; fi
38  .w.tex:
39          if test -r $*.ch; then cweave $*.w $*.ch; else cweave $*.w; fi
40  .w.o:
41          make $*.c
42          make $*.o
43  .w:
44          make $*.c
45          make $*
46  .w.dvi:
47          make $*.tex
48          make $*.dvi
49  DATAFILES = anna.dat david.dat econ.dat games.dat homer.dat huck.dat \
50          jean.dat lisa.dat miles.dat roget.dat words.dat
51  KERNELFILES = gb_flip.w gb_graph.w gb_io.w gb_sort.w
52  GENERATORFILES = gb_basic.w gb_books.w gb_econ.w gb_games.w gb_gates.w \
53          gb_lisa.w gb_miles.w gb_plane.w gb_raman.w gb_rand.w gb_roget.w \
54          gb_words.w
55  DEMOFILES = assign_lisa.w book_components.w econ_order.w football.w \
56          girth.w ladders.w miles_span.w multiply.w queen.w roget_components.w \
57          take_risc.w word_components.w
58  MISCWEBS = boilerplate.w gb_dijk.w gb_save.w gb_types.w test_sample.w
59  CHANGEFILES = queen_wrap.ch word_giant.ch
60  MISCFILES = Makefile README abstract.plaintex cities.texmap blank.w \
61          sample.correct test.correct test.dat \#The\#Stanford\#GraphBase\#
62  ALL = $(DATAFILES) $(KERNELFILES) $(GENERATORFILES) $(DEMOFILES) \
63          $(MISCWEBS) $(CHANGEFILES) $(MISCFILES)
64  OBJS = $(KERNELFILES:.w=.o) $(GENERATORFILES:.w=.o) gb_dijk.o gb_save.o
65  HEADERS = $(OBJS:.o=.h)
66  DEMOS = $(DEMOFILES:.w=)
67  help:
68          @ echo "First 'make tests';"
69          @ echo "then (optionally) become superuser;
70          @ echo "then 'make install';
71          @ echo "then (optionally) 'make installdemos';"
72          @ echo "then (optionally) 'make clean'."
73  lib: libgb.a
74  libgb.a: $(OBJS)
75          rm -f certified
76          ar rcv libgb.a $(OBJS)
77          - ranlib libgb.a
```

```
78 gb_io.o: gb_io.c
79         $(CC) $(CFLAGS) -DDATA_DIRECTORY=\"$(DATADIR)/\" -c gb_io.c
80 test_io: gb_io.o
81         $(CC) $(CFLAGS) test_io.c gb_io.o -o test_io
82 test_graph: gb_graph.o
83         $(CC) $(CFLAGS) test_graph.c gb_graph.o -o test_graph
84 test_flip: gb_flip.o
85         $(CC) $(CFLAGS) test_flip.c gb_flip.o -o test_flip
86 tests: test_io test_graph test_flip
87         test_io
88         test_graph
89         test_flip
90         make gb_sort.o
91         make lib
92         make test_sample
93         - test_sample > sample.out
94         diff test.gb test.correct
95         diff sample.out sample.correct
96         rm test.gb sample.out test_io test_graph test_flip test_sample
97         echo "Congratulations --- the tests have all been passed."
98         touch certified
99 install: lib
100        if test ! -r certified; then echo "Please run 'make tests' first!"; fi
101        test -r certified
102        make installdata
103        - mkdir $(LIBDIR)
104        - cp libgb.a $(LIBDIR)
105        - mkdir $(CWEBINPUTS)
106        - cp -p boilerplate.w gb_types.w $(CWEBINPUTS)
107        - mkdir $(INCLUDEDIR)
108        - cp -p $(HEADERS) Makefile $(INCLUDEDIR)
109 installdata: $(DATAFILES)
110        - mkdir $(SGBDIR)
111        - mkdir $(DATADIR)
112        - cp -p $(DATAFILES) $(DATADIR)
113 installdemos: lib $(DEMOS)
114        - mkdir $(BINDIR)
115        - mv $(DEMOS) $(BINDIR)
116 uninstalldemos:
117        - cd $(BINDIR); rm -f $(DEMOS)
118 doc:
119        tex abstract.plaintex
120 clean:
121        rm -f *~ *.o *.c *.h libgb.a certified \
122                *.tex *.log *.dvi *.toc *.idx *.scn core
```

```
123 veryclean: clean
124         rm -f $(DEMOS)
125 sgb.tar: $(ALL)
126         tar cvf sgb.tar $(ALL)
127 floppy: $(ALL)
128         bar cvf /dev/rfd0 $(ALL)
129         bar tvf /dev/rfd0
130         eject
```

Lines 30–32 define parts of commands that will be passed to the system loading routine, because `Makefile` instructions are traditionally supplemented by standard default definitions that handle common cases of compiling and loading. If, for example, you have a file `foo.c` and you type 'make foo', the default rules typically convert this to

$$\texttt{\$(CC) \$(CFLAGS) \$(LDFLAGS) -o foo \$(LDLIBS)}$$

or to the same thing with `LOADLIBES` instead of `LDLIBS`.* Here `CC` denotes the C compiler, usually `cc` by default. `CFLAGS` specifies compilation options for optimization or debugging or profiling or variations of floating-point arithmetic; these options are empty by default, but we set `CFLAGS = -g` on line 28 of the `Makefile`. On that line we also added the specifications `-I$(INCLUDEDIR)` and `$(SYS)`, which tell the C preprocessor where to find header files and what special symbols are defined. (Some `make` routines have more elaborate defaults in which `CFLAGS` is split into two parts, one of which is called `CPPFLAGS` and refers to the C preprocessor. This refinement is not needed for our purposes, so both types of information have been combined in `CFLAGS`.) `LDFLAGS` specifies nonstandard directories where the loader should look for libraries of precompiled code.

If all those macro variables have their default settings, as specified in the `Makefile` that comes with GraphBase, the command 'make foo' will then be an abbreviation for the monstrous

$$\texttt{cc -g -I/usr/local/sgb/include -L. -L/usr/local/lib -o foo -lgb}$$

which will indeed make 'foo' from 'foo.c'. In detail, this command
- invokes the compiler `cc`, which will emit extra information for debugging routines because of the `-g` option;
- tells the C preprocessor to search for `#include` files on directory `/usr/local/sgb/ include`' in addition to the current working directory;
- tells the loading routine to search for libraries in the current directory '.' and in directory '/usr/local/lib', in addition to the standard system directories for libraries;
- causes the resulting object file to be called 'foo';
- says that object code for routines used but not defined in `foo.c` can be found in the supplementary library file `libgb.a`.

---

* UNIX System V make knows neither `LOADLIBES` nor `LDLIBS`, but accepts `-lgb` as part of `$(LDFLAGS)`.

(The list of standard system library directories usually includes `/usr/local/lib`, so the specification '`-L/usr/local/lib`' might be redundant; but a little unnecessary redundancy won't hurt.)

UNIX's `make` already knows the filename suffix `.c`, but CWEB users also have additional suffixes `.w`, `.tex`, and `.dvi`; these are introduced in line 33. (They are listed in the opposite order, since suffixes of derived files should be considered before suffixes of source files.) Lines 34–48 explain how to handle these new suffixes. Their main significance is that if you type '`make foo`' when both `foo.c` and `foo.w` are present, the system will look at the creation times of those files before deciding how to proceed. If `foo.w` is newer than `foo.c`, the command '`ctangle foo.w`' will be issued to create an up-to-date `foo.c` before `foo.c` is compiled by `cc`. If you type '`make foo`' when all three files `foo.w`, `foo.c`, `foo` are already present, `make` will start with the one most recently created and will recreate any others that depend on it.

Lines 34 and 35 provide a simple example of the suffix rules. If you have a file `foo.tex` and you say '`make foo.dvi`', those lines will issue the command '`tex foo.tex`'. The effect will be to invoke TEX, which will read `foo.tex` and create a device-independent file called `foo.dvi` that can be previewed or converted to hardcopy.

Line 37 is slightly more tricky. It says that a conditional command should be used when making `foo.c` from `foo.w`. If a file called `foo.ch` exists and is readable — that's the meaning of '`if test -r foo.ch`' — the command will be '`ctangle foo.w foo.ch`'; otherwise the command will be simply '`ctangle foo.w`'. The suffix `.ch` refers to a CWEB change file, which is explained later in this section. (UNIX gurus will recognize the conditional syntax on line 37 as typical of the so-called Bourne shell; `make` speaks Bourne shell language unless `SHELL` has been redefined in the `Makefile`.)

Sometimes we want to compile a C program but not load it immediately, because its object code will be loaded with other programs later. Such object code files have the suffix `.o`, and the rule for making `foo.o` from `foo.c` is typically

```
COMPILE.c = $(CC) $(CFLAGS) -c foo.c
```

by default. Lines 40–42 of the `Makefile` say that if `foo.o` is desired and `foo.w` is present, then `foo.c` should first be made from `foo.w` (unless it is already present and newer than `foo.w`).

So much for suffixes. The next several lines of the `Makefile`, lines 49–63, define several macros that list all files of the GraphBase sources, separated by spaces. `DATAFILES` represents all files named `*.dat`, except for the special file `test.dat`. Then the CWEB source files `*.w` are listed in four categories, depending on whether they belong to the kernel or generate graphs or demonstrate graphs or do none of the above. (See lines 51–58.) The remaining files, called `CHANGEFILE` and `MISCFILES`, are listed on lines 59–61, and `ALL` is defined to be the whole works (lines 62–63).

Let's skip down briefly to lines 125 and 126. According to those lines, if you type '`make sgb.tar`', the operating system will execute the command

```
tar cvf sgb.tar anna.dat david.dat ···
```

(with all the GraphBase sources listed); the effect is to create a tape archive file called `sgb.tar` containing all the sources. This single file can then be compressed and/or moved to another computer, where the original files can be restored again by saying 'tar xvf sgb.tar'. If `sgb.tar` already exists when you type 'make sgb.tar', it will not be made again unless at least one of the GraphBase source files has changed since `sgb.tar` was made. This behavior is a consequence of the '$(ALL)' on line 125, which says that `sgb.tar` depends on all the sources. Of course you will not have changed any of the GraphBase sources — that's illegal — except for `Makefile` itself; so `sgb.tar`, once made, probably won't have to be made again. On the other hand, you won't need to make `sgb.tar` at all unless you are doing something unusual like moving the GraphBase sources from one computer to another.

Lines 127–130 are similar, but they might perhaps be more relevant to your own activities. If you type 'make floppy', they specify three commands that will

- write all the GraphBase sources to device `/dev/rfd0` (typically a floppy disk drive);
- read them back to see if the write was successful;
- eject the diskette.

The diskette can then be taken to another computer and converted back to GraphBase sources, using the command 'bar xvf /dev/rfd0'. Caution: These commands are based on Sun workstation conventions and might not work on other systems. Thus you might have to modify them, in spite of what the `Makefile` says on line 29.

Now let's go back to lines 64–66, where additional macros are defined. The definition of `DEMOS` is the simplest: It says that the elements listed in `DEMOS` are the same as those listed in `DEMOFILES`, except that the suffix `.w` should be dropped. Similarly, in line 64, we get `OBJS` by replacing `.w` with `.o` in the `KERNELFILES` and `GENERATORFILES`, also appending `gb_dijk.o` and `gb_save.o`. The files named in `OBJS` are the ones that go into the library `libgb.a`; lines 92–96 create that library. (Some `make` routines are not able to understand the substitution operations specified on lines 64–66. To humor them, you'll have to expand the names of all `OBJS`, `HEADERS`, and `DEMOS` yourself. But in that case you should seriously consider replacing your `make` by `gnumake`, an excellent program that is distributed by the Free Software Foundation.)

Most of the object files are made by applying the suffix rules. For example, `gb_graph.o` is formed from the `.w.o` rule on lines 40–42, essentially

```
ctangle gb_graph.w
$(CC) $(CFLAGS) -c gb_graph.c
```

(that is, by tangling and compiling). The only exception is `gb_io.o`, for which a special rule appears in lines 78–79. After `gb_io.c` has been made (by applying `CTANGLE` to `gb_io.w`), the command is

```
$(CC) $(CFLAGS) -DDATA_DIRECTORY=\"$(DATADIR)/\" -c gb_io.c
```

The effect, when `gb_io.c` is being compiled, is to define the macro `DATA_DIRECTORY` in such a way that the C preprocessor will change it to what you called `DATADIR` on line 10, followed by `/`. (The '/' is UNIX's file-directory delimiter; on a DOS-like system you'll need to change it to '\'.)

The kernel routines GB_IO, GB_GRAPH, and GB_FLIP all include special test programs that make sure things are operating correctly. When CTANGLE is applied to gb_io.w, it produces not only the C program gb_io.c but also the header file gb_io.h and the test program test_io.c. Lines 80 and 81 of the Makefile explain how to make test_io: First make gb_io.o (thereby making test_io.c as a byproduct of tangling), then compile test_io.c and call it test_io.

But we're getting a little ahead of ourselves. The top-level work of Makefile, which comes into play if the user types 'make' with no special target, is the sequence of actions for make help, as specified on lines 67–72. These lines simply issue informal instructions about the order in which various parts of the GraphBase should be built and tested. (The '@' signs tell make not to echo the echo commands themselves; without them, the instructions would each be output twice.)

We come now to the specifications for make tests, on lines 86–98. The target tests depends on test_io, test_graph, and test_flip (see line 86), so those routines are made first. In the process, the kernel objects gb_io.o, gb_graph.o and gb_flip.o are created (lines 80–85). Then the test routines are invoked (lines 87–89), and the remaining kernel object gb_sort.o is made (line 90). (Incidentally, it is important to create gb_flip.o before gb_sort.o, because gb_sort.o has an implicit dependence on gb_flip.h; the latter file, like test_flip.c, is created as a byproduct of tangling gb_flip.w. Such dependence on header files is not mentioned in the Makefile because the rules would then become unnecessarily complex.)

Now we get to meatier stuff: Line 91 invokes make lib, which causes all the other object files to be generated. (See lines 73 and 74.) Then a file called certified is removed, if present (line 75), and the object files are collected together in a library archive file (lines 76 and 77).

At this point we have created the library libgb.a, but we haven't tested anything except the lowest-level routines (gb_io, gb_graph, and gb_flip). It is time to make the TEST_SAMPLE program, which is a major shakedown cruise involving the whole library, and to see if its output files match the correct results. (See lines 92–95.) If the results are incorrect, the diff instructions on lines 94 and 95 will stop make with an error — and you will have to read Section 3.5 below (sorry). Otherwise the test programs can be removed from memory (line 96), and congratulations are in order (line 97). Finally, in line 98, we create an empty file called certified; this file will exist only if make tests has come to a successful conclusion.

Remember that users are supposed to make tests before they become superuser and make install. This ordering is enforced by the opening lines of make install, which ensure that a valid certificate has been generated (see lines 99–101). If certified is not present, line 100 will tell the user what to do, and line 101 will abort the make install with an error message.*

---

* What if one of the object files changes after the tests have been certified? This contingency is covered by the 'lib' dependency on line 99, which will remake the library and remove the certificate in such a case, forcing another make tests before make install will run to completion. Such scenarios should, however, be rare, because the master files *.w never

Once the tests are certified, `make install` has plain sailing. Its first subproblem (line 102) is to `make installdata`, lines 109–112: Line 110 says '`- mkdir $(SGBDIR)`', which creates the directory `SGBDIR` if it isn't already present; the – sign means that `make` should keep going even if `mkdir` couldn't make a new directory. Line 111 says '`- mkdir $(DATADIR)`', thereby creating the data subdirectory in a similar fashion. And line 112, '`- cp -p $(DATAFILES) $(DATADIR)`', copies all the data files to the data directory and preserves their protection codes. (The data files are read-only.) Again, line 112 contains a minus sign; this tells `make` to plunge ahead even if `cp` complains that it cannot do its work. For example, if the user has used the shortcut and defined `DATADIR=.` on line 20, `cp` will balk at the idea of copying a file to its own directory, but no harm will be done.

The remaining steps in `make install`, lines 103–108, are similar. They simply create `LIBDIR`, `CWEBINPUTS`, and `INCLUDEDIR`, then copy the relevant files to those directories.

Let's recapitulate. The entire installation process carried out by `make tests` followed by `make install` can be summarized as follows:

- Tangle and compile `gb_io.w`, defining `DATA_DIRECTORY` to be a string containing `DATADIR` followed by the directory delimiter `/`. As a side-effect of tangling, create `gb_io.h` and `test_io.c`.
- Compile, load, and execute `test_io.c`.
- Tangle and compile `gb_graph.w`.
- Compile, load, and execute `test_graph.c`.
- Tangle and compile `gb_flip.w`.
- Compile, load, and execute `test_flip.c`.
- Tangle and compile `gb_sort.w`.
- Tangle and compile all generator files `gb_basic.w` ⋯ `gb_words.w`.
- Put all object files `*.o` into the library archive `libgb.a`.
- Tangle, compile, load, and execute `test_sample.w`, directing its standard output to file `sample.out`.
- Verify that `test.gb` matches `test.correct`.
- Verify that `sample.out` matches `sample.correct`.
- Remove files `test.gb`, `sample.out`, `test_io`, `test_graph`, `test_flip`, `test_sample`.
- Create the file `certified`. (This marks the successful completion of `make tests`; now it's safe to become superuser and `make install`.)
- Copy the data files `*.dat` to `DATADIR`.
- Copy `libgb.a` to `LIBDIR`.
- Copy `boilerplate.w` and `gb_types.w` to `CWEBINPUTS`.
- Copy the header files `*.h` and `Makefile` to `INCLUDEDIR`.

A more explicit list of these instructions appears in the `README` file.

3.5 TROUBLESHOOTING

But what if the tests fail? The GraphBase routines check themselves during installation,

---

change. An obsolescent library arises only when the user decides to modify a generator or kernel routine via a change file.

and they will complain if your computer hardware or software does not deliver the expected results.*

If this happens to you, you'll need to look inside the programs and figure out what is wrong. Comments in the documentation try to anticipate such difficulties and to explain where systems might be incompatible. The index to programs has entries called "system dependencies" and "UNIX dependencies" that list the most likely trouble spots.

The remedy for all such problems is to modify the standard GraphBase programs. But you are not allowed to change the *.w files, so what can you do? Answer: Use CWEB's handy *change files.* Both CTANGLE and CWEAVE have a built-in ability to patch the source program as it is being input, using a secondary change file. The change file, which usually has the suffix '.ch' as part of its name, contains patches of the form

```
@x
⟨verbatim copy of text lines to be replaced from the original *.w file⟩
@y
⟨new text to be substituted⟩
@z
```

in the order that the patches need to be applied. The first line after each '@x' must uniquely identify the material to be replaced.

For example, suppose you have a computer that does not use two's complement arithmetic internally. Then the GraphBase random number generator will probably fail its test, because it uses logical-and with the constant $2^{31} - 1$ to compute nonnegative remainders modulo 2^{31} (see GB_FLIP, section 7). You can correct this problem by preparing a file gb_flip.ch that contains the following lines:

```
Change file for GB_FLIP by J. H. Quick, student
@x
@d mod_diff(x,y) (((x)-(y))&0x7fffffff) /* difference modulo $2^{31}$ */
@y avoid assumption of two's complement arithmetic
@d mod_diff(x,y) ((x)>=(y)? (x)-(y):
        ((long)(((long)((x)-(y))+0x40000000))+0x40000000))
@z
```

Line 37 of the Makefile will recognize the presence of gb_flip.ch and it will tell CTANGLE to adopt this change when making gb_flip.c.

The Makefile isn't smart enough, however, to know about the dependency between gb_flip.c and gb_flip.ch. This means that you can change gb_flip.ch and it will still think gb_flip.c is up to date. The best way to remedy this is to add the line

$$gb\_flip.c: gb\_flip.w\ gb\_flip.ch$$

to the Makefile. A simpler but kludgier way is to remember to touch gb_flip.w every time you change gb_flip.ch.

---

* Are you sure you had your fingers crossed when you ran make tests?

Alternatively, if you require change files almost all the time because your system is very different from ordinary implementations of C, you can add the following lines to the Makefile:

```
%.w: %.ch
%.ch:
            touch $@
```

Then empty change files will be created whenever you use a .w file, and you won't need to insert any dependency lines like the example for gb_flip.c above.

If you have to prepare several change files because your system is rather unusual or because your computer has severe limitations (like many PCs), you might want to share your work with other people who have the same kind of system. Then you should add the new change files to the definition of $(CHANGEFILES) on line 59 of the Makefile or whatever your system has instead of a Makefile. You should also amend the README file, listing your change files and stating that you have modified the standard GraphBase sources for a particular system. Then the commands make sgb.tar or make floppy will produce copies of your special GraphBase sources for use by your cohorts. Such modifications of Makefile and README are permissible as long as you don't change any of the other files.

One failure that might be reported by TEST_SAMPLE is the so-called edge trick. But such a defect is relatively innocuous, so it does not abort the installation procedure. You can read all about it, if you're curious, by consulting *edge_trick* in the index to programs.

3.6   RUNNING THE DEMOS

Once 'make install' succeeds, you are in possession of a working GraphBase kernel and all the generators, but you still have no actual programs that use them. The fun begins when you've also compiled at least some of the demonstration programs.

You can install the demos in several ways, depending on how much disk space you're willing to devote to them. The easiest approach, but also the most expensive spacewise, is to say 'make installdemos'. Then lines 113–115 of the Makefile will take charge, causing all the demo programs mentioned in Chapter 1 (ASSIGN_LISA, BOOK_COMPONENTS, ECON_ORDER, FOOTBALL, GIRTH, LADDERS, MILES_SPAN, MULTIPLY, QUEEN, ROGET_COMPONENTS, TAKE_RISC, WORD_COMPONENTS) to be tangled, compiled, loaded, and moved to BINDIR, where they can be executed like any other software. (You might have to say something like 'rehash' before UNIX will know where they are.) After you're done playing with them, you can either remove them from BINDIR one by one, or you can remove them all at once by typing 'make uninstalldemos'.

Another approach is to make just one of the demos and play with it, while connected to the GraphBase source directory. For example, if you say 'make ladders' you can immediately interact with the word-ladder program as discussed in Section 1.1.

3.7   CONSERVING MEMORY

At any time you can type 'make clean' to get rid of files that were needed only during the installation process. If you've also made some demo programs in the GraphBase source directory, 'make veryclean' will remove them too. (You need not 'make clean' before 'make floppy' or 'make sgb.tar'; those commands don't dump the whole source directory, they write out only the files that they need.)

If you've installed the GraphBase data and library and header files in other directories, you can safely delete the entire source directory. Then the entire GraphBase library will take up less than 750,000 bytes on your system. Note, however, that online debugging routines will not be as effective when the source code for the kernel and generator modules is no longer present.

3.8 RUNNING YOUR OWN PROGRAMS

The most satisfying way to use the Stanford GraphBase is to write new programs that do interesting things with the available graphs. The demo programs, after all, are merely examples that illustrate how to get started.

When you run your own programs, you'll probably want to work in a special directory, not the directory with the GraphBase sources. On a UNIX-like system you should first copy the Makefile from the GraphBase source directory to your working directory. (You could also copy it from your INCLUDEDIR, where a duplicate has been installed for safe-keeping.) The Makefile will make life a lot simpler for you because it will automatically invoke CWEB at the appropriate times, and it will specify the relevant CFLAGS and LDFLAGS and LDLIBS to the compiler and loader, as discussed in Section 3.4 above.

It's best to start with a simple example, so begin your experiments by copying the file queen.w from the GraphBase sources to your working directory. Pretend that you have just written queen.w. Look at it with your favorite text editor, so that you can remind yourself what you wrote. (The text of queen.w appears on pages 508–509 of this book.) Also read the CWEB user manual. Then try 'make queen', and run 'queen'. Compare the standard output file produced by queen with the file queen.gb, which should have been output behind the scenes. (The two outputs represent the graph of all queen moves on a 3×4 chessboard, in two different formats: The standard output is more or less human-oriented, while queen.gb is machine-oriented.) Also try 'make queen.dvi'; this will invoke CWEAVE and TEX to produce a file queen.dvi that can be laserprinted, giving a typeset documentation of the QUEEN program. Compare the typeset output to the original source file queen.w. (Don't look at the intermediate files queen.c and queen.tex.)

Now try using an online debugger to step through the queen program one line at a time. A good debugging routine will show you exactly how control passes through corresponding lines of the source file queen.w, and you'll see how easy and pleasant it is to debug CWEB programs with a well-designed debugger.

Next, copy the file queen_wrap.ch from the GraphBase sources to your working directory. (See pages 510–511.) Type

```
ctangle queen queen_wrap queen_wrap
```

(which creates a file called queen_wrap.c) and then 'make queen_wrap'. Use your text editor to read queen_wrap.ch as an example of a CWEB change file; it illustrates how you can modify any GraphBase program without changing the program file. (A shorter example of a change file was discussed in Section 3.5.) Use the debugger to step through queen_wrap;

you should be able to see the control jump back and forth between the change file and the original file.

The GraphBase sources also include another change file, `word_giant.ch`, which you might want to try. After you've copied `word_components.w` and `word_giant.ch` into your working directory, type

```
ctangle word_components word_giant word_giant
```

and then 'make word_giant'. Running `word_giant` will now produce a file `word_giant.gb`, which represents the giant component of the *words* graph, namely the 4497 five-letter words that are reachable from each other by the rules of LADDERS and the 13,619 edges between them. The length of each edge will be the alphabetic distance between words; for example, the distance from `chaos` to `chats` will be 5, because the distance from `o` to `t` is 5. This graph can be used as input to other programs via the *restore_graph* procedure. You can also find its minimum spanning tree by issuing the UNIX command

```
miles_span -gword_giant.gb
```

(assuming that you have already made the demo program `miles_span`). To make a formatted listing of `word_giant`, type

```
cweave word_components word_giant word_giant
make word_giant.dvi
```

and print `word_giant.dvi`. This listing will show only the sections where the original program WORD-COMPONENTS has been changed.

Here's a change file that you might try in connection with the FOOTBALL program:

```
@x change to section 22 of football.w
   if (x->tot_len<=list[h]->tot_len) goto done; /* drop node |x| */
@y flip a coin in case of equality
   if (x->tot_len<list[h]->tot_len ||
   (x->tot_len==list[h]->tot_len && (gb_next_rand()&1)))
     goto done;  /* drop node |x| */
@z
```

It adds a mild amount of randomization to the method of stratified greed, so that you will often get different solutions to the problem when the same starting and ending teams are specified.

The following optional change to FOOTBALL makes it search for the worst solution instead of the best:

```
@x change to section 5 of football.w, reverses the orientation
       a->del=a->len-(a+1)->len;
@y
       a->del=(a+1)->len-a->len;
@z
@x compensating change to section 15
```

```
    printf(" (%+1d)\n",next_node->tot_len);
@y
  printf(" (%+1d)\n",-next_node->tot_len);
@z
@x compensating change to section 16
                            u->name,u->nickname,a->len-a->del);
@y
                            u->name,u->nickname,a->len+a->del);
@z
```

The worst solution for team x over team y corresponds to the best solution for team y over team x, so the effect of this change is to search backward from the goal instead of forward from the starting team. The author's best solutions for Stanford over Harvard were found in this way.

Both changes, randomization and pessimization, can be used simultaneously. But when you merge them, you must list the latter changes first, because they affect sections 5, 15, and 16 while the randomization change affects section 22. Changes must be listed in program order.

Maybe you would prefer not to use change files, because it seems easier to dive in and modify the existing code of **queen.w** directly. According to the copyright restrictions, you are allowed to do this, but only if you change the name of the file first. So rename **queen.w** to be '**knight.w**'; also, delete the lines at the top that identify the file as part of the Stanford GraphBase, since they are no longer true. Then you're free to change **knight.w** so that it creates a graph of knight moves, or whatever you want.

After these experiments you should be more than ready to write your own CWEB programs to work with GraphBase graphs. When creating your own routines, you might be pleased to discover that change files provide a convenient way to experiment with modified data structures or to design a series of related programs.

You'll soon find yourself able to compute the answers to interesting new combinatorial problems, shortly after you think of them. With practice, you will in fact be able to make an hour or two of pleasant CWEB programming work wonders. Before long you might even feel an irresistible urge to publish your programs in a book like this one.

CHAPTER 4: *HOW TO READ CWEB PROGRAMS*

The programs that make up the bulk of this book were prepared with the CWEB system of structured documentation. A CWEB program is a C program that has been cut up into pieces and rearranged into an order that is easier for a human being to understand. A C program is a CWEB program that has been rearranged into an order that is easier for a computer to understand.

In other words, CWEB programs and C programs are essentially the same kinds of things, but their parts are arranged differently. You should be able to understand a C program better when you see it in its CWEB form, if the author of the CWEB form has chosen a good order of presentation.

Before you try to read a CWEB program, you should be familiar with the C language, or at least with its most basic elements. The following paragraphs describe how a CWEB document can be translated mechanically into C code; the CWEB form serves as a high-level description that can be converted to ordinary C by a process called "tangling." The C language is thereby enriched so that its programs can be constructed, documented, and maintained in a pleasant, untangled way according to the principles of "literate programming" [20].

A CWEB program consists of numbered sections — first §1, then §2, and so on. Each section is intended to be small enough that it can be understood by itself, and the CWEB format indicates how each section relates to other sections. In this way, an entire program can be regarded as a web or network, consisting of little pieces together with interconnections between pieces. The complex whole can be understood by understanding each simple part and by understanding the simple relationships between neighboring parts.

Large collections of software are inherently complex, and there is no "royal road" to instant comprehension of their subtle features. But if you read a well-written CWEB program one section at a time, starting with §1, you will find that its ideas are not difficult to assimilate.

Every section of a CWEB program begins with commentary about the purpose of the section or about some noteworthy feature of that part of the program. Such comments are in English, and informality is the rule. Then the section concludes with a formal part that is expressed in C language; this is what actually counts, and you should be able to convince yourself that the formal C code actually does what the informal comments imply.

Between the introductory comments and the C code, there may also be one or more macro definitions, which will be explained shortly. Thus each section has three parts:

- informal commentary
- macro definitions
- C code

and these parts always occur in the stated order. In practice, the individual parts are sometimes empty: Some sections have no opening comments, some have no macro definitions, and some have no C code.

Every section is either *named* or *unnamed*. The C code in a named section begins with '⟨ Name of section ⟩ ≡' in angle brackets, and the section name is followed by its so-called replacement text. For example, the C code in §23 might be

$$\langle\,\text{Clear the arrays } 23\,\rangle \equiv$$
$$\textbf{for } (k = 1;\ k \le n;\ k\text{++})\ \{$$
$$a[k] = 0;$$
$$str[k] = \texttt{""};$$
$$\}$$

and this means that the section name is 'Clear the arrays' while the replacement text is 'for (k=1;k<=n;k++) {a[k]=0; str[k]=""; }'.

This example illustrates the fact that CWEB descriptions of replacement texts are formatted typographically in a special way. Raw C code then becomes easier to read. For example, reserved words like '**for**' are set in boldface type; identifiers like 'k' and 'str' are set in italics; the contents of strings are set in typewriter type. Program structure is shown by indentation. The following special symbols are used to enhance readability:

'$\le$'	stands for	'<=' ;
'$\ge$'	stands for	'>=' ;
'$\ne$'	stands for	'!=' ;
'$\equiv$'	stands for	'==' ;
'$\wedge$'	stands for	'&&' ;
'$\vee$'	stands for	'||' ;
'$\neg$'	stands for	'!' ;
'$\hookrightarrow$'	stands for	'->' ;
'$\oplus$'	stands for	'^' ;
'Λ'	stands for	'NULL'.

Numeric constants have their own special formatting conventions. A **long** constant, written '3L' in the C program, appears as '3_{L}' in the typeset documentation. An octal constant, written '077', appears as '$^\circ 77$'. A hexadecimal constant, written '0x7ff', appears as '$\#7ff$'. And a floating-point constant, written '1.4E6', appears as '$1.4 \cdot 10^6$'.

When the name of a section appears in another section, it stands for the corresponding replacement text. For example, suppose the C code for §20 is

$$\langle\,\text{Initialize the data structures } 20\,\rangle \equiv$$
$$sum = 0;$$
$$\langle\,\text{Clear the arrays } 23\,\rangle;$$

This means that the replacement text for 'Clear the arrays' should be inserted as part of the replacement text for 'Initialize the data structures'. The name of a section generally provides a short summary of what that section does. Therefore the full details won't distract your attention unless you need to know them. But if you do need further details, the subscript '23' in '$\langle\,$Clear the arrays $23\,\rangle$' makes it easy for you to locate the subsidiary replacement text. Incidentally, §23 in our example will contain a footnote that says 'This code is used in section 20'; thus you can move easily from a replacement text to its context or vice versa.

The process of converting a CWEB program to a C program is quite simple, at least conceptually. First you put together all the C code that appears in unnamed sections, preserving the relative order. Then you substitute the replacement texts for all named sections that appear in the resulting code. And you continue to do this until only pure C remains.

It's possible to give the same name to several different sections. Continuing our example, let's suppose that the C code of §30 is

$\langle$ Initialize the data structures $20 \rangle$ $+\equiv$
    $source\_file = fopen(source\_name, \texttt{"r"});$

The '+≡' here indicates that the name 'Initialize the data structures' has appeared before; the C code following '+≡' will be *appended* to the previous replacement text for that name. Thus, if 'Initialize the data structures' is defined only in §20 and §30, the actual replacement text for $\langle$ Initialize the data structures $20 \rangle$ will be

$sum = 0;$
$\langle$ Clear the arrays $23 \rangle;$
$source\_file = fopen(source\_name, \texttt{"r"});$

Section 20 will also contain a footnote: 'See also section 30'. This feature means that different parts of program initialization can be introduced when the reader is ready to understand them, instead of all at once when the computer has to execute them.

Macro definitions, which might appear just before the C code of any section, are collected together in a similar way. They are placed at a point called '$\langle$ Preprocessor definitions $\rangle$', which generally comes just after the #include declarations at the start of a program. This placement allows the standard C preprocessor to expand the macros in its normal fashion. (Alternatively, the program need not specifically indicate the position of $\langle$ Preprocessor definitions $\rangle$. In that case the collected macros are simply placed at the very beginning of the tangled output, before all other C language material.) Sometimes a macro definition extends over several lines; those lines are tied together appropriately, behind the scenes, in the C code that is actually created.

Sometimes the name of a section is a file name in typewriter type, such as $\langle$ **gb_save.h** $1 \rangle$. This means that the replacement text for that section is output to the named file instead of being part of the tangled C code. A single program document can therefore describe several related files.

Hundreds of identifiers generally appear in a long program, and it's hard to keep them straight. Therefore the CWEB system provides a comprehensive index that lists all sections where a given identifier appears. Entries in the index are underlined if the identifier was defined in the corresponding section. Furthermore, the right-hand pages of this book contain mini-indexes that will make it unnecessary for you to look at the big index very often. Every identifier that is used somewhere on a pair of facing pages is listed in a footnote on the right-hand page, unless it is explicitly defined or declared somewhere on the left-hand or right-hand page you are reading. These footnote entries tell you the type of the identifier or whether it is a macro, and the section in which it was declared or defined.

OK! You know enough now about CWEB conventions to understand all the programs in this book. A few additional conventions are described in [24] but not needed here; the full language contains provisions for passing verbatim text to a file, for concatenating identifiers, for defining numeric constants that depend on ASCII code, for converting 8-bit codes to strings of letters in identifiers, and for printing special identifiers in special formats. You can safely ignore such things until you happen to write a CWEB program that needs a special feature.

The programs that comprise the next sections of this book are designed to work together, as described in Chapter 1. Some of them are kernel routines, some are graph generators, some are demonstration programs. They appear in alphabetic order, with ASSIGN_LISA first and WORD_COMPONENTS last, so that cross-references are easy to find; but you shouldn't actually read them in alphabetic order unless you're the type of person who enjoys reading a dictionary from cover to cover. Each program tells you which other programs ought to be read as prerequisites. For example, you shouldn't read ASSIGN_LISA until after reading GB_LISA; and you shouldn't read GB_LISA until after reading GB_GRAPH and GB_IO.

Ready? Please turn first to the program for GB_FLIP. It's only six pages long, and it illustrates most of the rules discussed in this chapter. Then turn to the program for GB_GRAPH, which is a prerequisite for almost everything else. Then, take your pick and enjoy the rest.

ASSIGN_LISA ⸻

Important: Before reading ASSIGN_LISA, please read or at least skim the program for GB_LISA.

1. The assignment problem. This demonstration program takes a matrix of numbers constructed by the GB_LISA module and chooses at most one number from each row and column in such a way as to maximize the sum of the numbers chosen. It also reports the number of "mems" (memory references) expended during its computations, so that the algorithm it uses can be compared with alternative procedures.

The matrix has m rows and n columns. If $m \leq n$, one number will be chosen in each row; if $m \geq n$, one number will be chosen in each column. The numbers in the matrix are brightness levels (pixel values) in a digitized version of the Mona Lisa.

Of course the author does not pretend that the location of "highlights" in da Vinci's painting, one per row and one per column, has any application to art appreciation. However, this program does seem to have pedagogic value, because the relation between pixel values and shades of gray allows us to visualize the data underlying this special case of the assignment problem; ordinary matrices of numeric data are much harder to perceive. The nonrandom nature of pixels in a work of art might also have similarities to the "organic" properties of data in real-world applications.

This program is optionally able to produce an encapsulated PostScript file from which the solution can be displayed graphically, with halftone shading.

2. As explained in GB_LISA, the subroutine call $lisa(m, n, d, m0, m1, n0, n1, d0, d1, area)$ constructs an $m \times n$ matrix of integers between 0 and d, inclusive, based on the brightness levels in a rectangular region of a digitized Mona Lisa, where $m0$, $m1$, $n0$, and $n1$ define that region. The raw data is obtained as a sum of $(m1 - m0)(n1 - n0)$ pixel values between 0 and 255, then scaled in such a way that sums $\leq d0$ are mapped to zero, sums $\geq d1$ are mapped to d, and intermediate sums are mapped linearly to intermediate values. Default values $m1 = 360$, $n1 = 250$, $m = m1 - m0$, $n = n1 - n0$, $d = 255$, and $d1 = 255(m1 - m0)(n1 - n0)$ are substituted if any of the parameters m, n, d, $m1$, $n1$, or $d1$ are zero.

The user can specify the nine parameters $(m, n, d, m0, m1, n0, n1, d0, d1)$ on the command line, at least in a UNIX implementation, thereby obtaining a variety of special effects; the relevant command-line options are m=⟨number⟩, m0=⟨number⟩, and so on, with no spaces before or after the = signs that separate parameter names from parameter values. Additional options are also provided: -s (use only Mona Lisa's 16×32 "smile"); -e (use only her 20×50 eyes); -c (complement black/white); -p (print the matrix and solution); -P (produce a PostScript file **lisa.eps** for graphic output); -h (use a heuristic that applies only when $m = n$); and -v or -V (print verbose or Very verbose commentary about the algorithm's performance).

Here is the overall layout of this C program:

```
#include "gb_graph.h"    /* the GraphBase data structures */
#include "gb_lisa.h"     /* the lisa routine */
  ⟨ Preprocessor definitions ⟩
  ⟨ Global variables 3 ⟩
  main(argc, argv)
      int argc;    /* the number of command-line arguments */
```

```
        char *argv[];      /* an array of strings containing those arguments */
  { ⟨Local variables 4⟩;
    ⟨Scan the command-line options 5⟩;
    mtx = lisa(m, n, d, m0, m1, n0, n1, d0, d1, working_storage);
    if (mtx ≡ Λ) {
        fprintf(stderr, "Sorry,⎵can't⎵create⎵the⎵matrix!⎵(error⎵code⎵%ld)\n", panic_code);
        return −1;
    }
    printf("Assignment⎵problem⎵for⎵%s%s\n", lisa_id, (compl ? ",⎵complemented" : ""));
    sscanf(lisa_id, "lisa(%lu,%lu,%lu", &m, &n, &d);      /* adjust for defaults */
    if (m ≠ n) heur = 0;
    if (printing) ⟨Display the input matrix 6⟩;
    if (PostScript) ⟨Output the input matrix in PostScript format 28⟩;
    mems = 0;
    ⟨Solve the assignment problem 24⟩;
    if (printing) ⟨Display the solution 27⟩;
    if (PostScript) ⟨Output the solution in PostScript format 31⟩;
    printf("Solved⎵in⎵%ld⎵mems%s.\n", mems, (heur ? "⎵with⎵square-matrix⎵heuristic" : ""));
    return 0;      /* normal exit */
  }
```

3. ⟨Global variables 3⟩ ≡
  Area *working_storage*;   /* where to put the input data and auxiliary arrays */
  long *$*mtx$;   /* input data for the assignment problem */
  long *mems*;   /* the number of memory references counted while solving the problem */

See also section 29.

This code is used in section 2.

4. The following local variables are related to the command-line options:

⟨Local variables 4⟩ ≡
  unsigned long $m = 0$, $n = 0$;   /* number of rows and columns desired */
  unsigned long $d = 0$;   /* number of pixel values desired, minus 1 */
  unsigned long $m0 = 0$, $m1 = 0$;   /* input will be from rows $[m0 \mathinner{.\,.} m1)$ */
  unsigned long $n0 = 0$, $n1 = 0$;   /* and from columns $[n0 \mathinner{.\,.} n1)$ */
  unsigned long $d0 = 0$, $d1 = 0$;   /* lower and upper threshold of raw pixel scores */
  long *compl* $= 0$;   /* should the input values be complemented? */
  long *heur* $= 0$;   /* should the square-matrix heuristic be used? */
  long *printing* $= 0$;   /* should the input matrix and solution be printed? */
  long *PostScript* $= 0$;   /* should an encapsulated PostScript file be produced? */

See also sections 13, 14, and 26.

This code is used in section 2.

Area, GB_GRAPH §12.   *lisa_id*: **char** [], GB_LISA §4.   *sscanf*: **int** (), <stdio.h>.
fprintf: **int** (), <stdio.h>.   *panic_code*: **long**, GB_GRAPH §5.   *stderr*: **FILE** *, <stdio.h>.
lisa: **long** *(), GB_LISA §6.   *printf*: **int** (), <stdio.h>.

5. ⟨Scan the command-line options 5⟩ ≡
  while (−−*argc*) {
    if (*sscanf*(*argv*[*argc*], "m=%lu", &*m*) ≡ 1) ;
    else if (*sscanf*(*argv*[*argc*], "n=%lu", &*n*) ≡ 1) ;
    else if (*sscanf*(*argv*[*argc*], "d=%lu", &*d*) ≡ 1) ;
    else if (*sscanf*(*argv*[*argc*], "m0=%lu", &*m0*) ≡ 1) ;
    else if (*sscanf*(*argv*[*argc*], "m1=%lu", &*m1*) ≡ 1) ;
    else if (*sscanf*(*argv*[*argc*], "n0=%lu", &*n0*) ≡ 1) ;
    else if (*sscanf*(*argv*[*argc*], "n1=%lu", &*n1*) ≡ 1) ;
    else if (*sscanf*(*argv*[*argc*], "d0=%lu", &*d0*) ≡ 1) ;
    else if (*sscanf*(*argv*[*argc*], "d1=%lu", &*d1*) ≡ 1) ;
    else if (*strcmp*(*argv*[*argc*], "-s") ≡ 0) {
     smile;   /∗ sets *m0*, *m1*, *n0*, *n1* ∗/
     d1 = 100000;   /∗ makes the pixels brighter ∗/
    }
    else if (*strcmp*(*argv*[*argc*], "-e") ≡ 0) {
     eyes;
     d1 = 200000;
    }
    else if (*strcmp*(*argv*[*argc*], "-c") ≡ 0) *compl* = 1;
    else if (*strcmp*(*argv*[*argc*], "-h") ≡ 0) *heur* = 1;
    else if (*strcmp*(*argv*[*argc*], "-v") ≡ 0) *verbose* = 1;
    else if (*strcmp*(*argv*[*argc*], "-V") ≡ 0) *verbose* = 2;   /∗ terrifically verbose ∗/
    else if (*strcmp*(*argv*[*argc*], "-p") ≡ 0) *printing* = 1;
    else if (*strcmp*(*argv*[*argc*], "-P") ≡ 0) *PostScript* = 1;
    else {
     fprintf(*stderr*, "Usage:␣%s␣[param=value]␣[-s]␣[-c]␣[-h]␣[-v]␣[-p]␣[-P]\n", *argv*[0]);
     return −2;
    }
  }
This code is used in section 2.

6. ⟨Display the input matrix 6⟩ ≡
  for (*k* = 0; *k* < *m*; *k*++) {
    for (*l* = 0; *l* < *n*; *l*++) *printf*("%␣4ld", *compl* ? *d* − ∗(*mtx* + *k* ∗ *n* + *l*) : ∗(*mtx* + *k* ∗ *n* + *l*));
    printf("\n");
  }
This code is used in section 2.

7. We obtain a crude but useful estimate of the computation time by counting mem units, as explained in the MILES_SPAN program.

#define *o* *mems*++
#define *oo* *mems* += 2
#define *ooo* *mems* += 3

8. Algorithmic overview. The assignment problem is the classical problem of weighted bipartite matching: to choose a maximum-weight set of disjoint edges in a bipartite graph. We will consider only the case of complete bipartite graphs, when the weights are specified by an $m \times n$ matrix.

An algorithm is most easily developed if we begin with the assumption that the matrix is square (i.e., that $m = n$), and if we change from maximization to minimization. Then the assignment problem is the task of finding a permutation $\pi[0] \ldots \pi[n-1]$ of $\{0, \ldots, n-1\}$ such that $\sum_{k=0}^{n-1} a_{k\pi[k]}$ is minimized, where $A = (a_{kl})$ is a given matrix of numbers a_{kl} for $0 \le k, l < n$. The algorithm below works for arbitrary real numbers a_{kl}, but we will assume in our implementation that the matrix entries are integers.

One way to approach the assignment problem is to make three simple observations: (a) Adding a constant to any row of the matrix does not change the solution $\pi[0] \ldots \pi[n-1]$. (b) Adding a constant to any column of the matrix does not change the solution. (c) If $a_{kl} \ge 0$ for all k and l, and if $\pi[0] \ldots \pi[n-1]$ is a permutation with the property that $a_{k\pi[k]} = 0$ for all k, then $\pi[0] \ldots \pi[n-1]$ solves the assignment problem.

The remarkable fact is that these three observations actually suffice. In other words, there is always a sequence of constants $(\sigma_0, \ldots, \sigma_{n-1})$ and $(\tau_0, \ldots, \tau_{n-1})$ and a permutation $\pi[0] \ldots \pi[n-1]$ such that

$$a_{kl} - \sigma_k + \tau_l \ge 0, \qquad \text{for } 0 \le k < n \text{ and } 0 \le l < n;$$
$$a_{k\pi[k]} - \sigma_k + \tau_{\pi[k]} = 0, \quad \text{for } 0 \le k < n.$$

---

argc: **int**, §2.	*k*: **register long**, §13.	*n1*: **unsigned long**, §4.
argv: **char** *[], §2.	*l*: **register long**, §13.	*PostScript*: **long**, §4.
compl: **long**, §4.	*m*: **unsigned long**, §4.	*printf*: **int** ( ), <stdio.h>.
d: **unsigned long**, §4.	*m0*: **unsigned long**, §4.	*printing*: **long**, §4.
d0: **unsigned long**, §4.	*m1*: **unsigned long**, §4.	*smile* = macro, GB_LISA §3.
d1: **unsigned long**, §4.	*mems*: **long**, §3.	*sscanf*: **int** ( ), <stdio.h>.
eyes = macro, GB_LISA §3.	*mtx*: **long** *, §3.	*stderr*: **FILE** *, <stdio.h>.
fprintf: **int** ( ), <stdio.h>.	*n*: **unsigned long**, §4.	*strcmp*: **int** ( ), <string.h>.
heur: **long**, §4.	*n0*: **unsigned long**, §4.	*verbose*: **long**, GB_GRAPH §5.

9. To prove the remarkable fact just stated, we start by reviewing the theory of *unweighted* bipartite matching. Any $m \times n$ matrix $A = (a_{kl})$ defines a bipartite graph on the vertices $(r_0, \ldots, r_{m-1})$ and $(c_0, \ldots, c_{n-1})$ if we say that r_k — c_l whenever $a_{kl} = 0$; in other words, the edges of the bipartite graph are the zeroes of the matrix. Two zeroes of A are called *independent* if they appear in different rows and columns; this means that the corresponding edges have no vertices in common. A set of mutually independent zeroes of the matrix therefore corresponds to a set of mutually disjoint edges, also called a *matching* between rows and columns.

The Hungarian mathematicians Egerváry and König proved [*Matematikai és Fizikai Lapok* **38** (1931), 16–28, 116–119] that the maximum number of independent zeroes in a matrix is equal to the minimum number of rows and/or columns that are needed to "cover" every zero. In other words, if we can find p independent zeroes but not $p+1$, then there is a way to choose p lines in such a way that every zero of the matrix is included in at least one of the chosen lines, where a "line" is either a row or a column.

Their proof was constructive, and it leads to a useful computer algorithm. Given a set of p independent zeroes of a matrix, let us write $r_k = c_l$ or $c_l = r_k$ and say that r_k is matched with c_l if a_{kl} is one of these p special zeroes, while we continue to write r_k — c_l or c_l — r_k if a_{kl} is one of the nonspecial zeroes. A given set of p special zeroes defines a choice of p lines in the following way: Column c is chosen if and only if it is reachable by a path of the form

$$r_0 - c_1 = r_1 - c_2 = \cdots - c_q = r_q, \qquad (*)$$

where r_0 is unmatched, $q \geq 1$, and $c = c_q$. Row r is chosen if and only if it is matched with a column that is not chosen. Thus exactly p lines are chosen. We can now prove that the chosen lines cover all the zeroes, unless there is a way to find $p+1$ independent zeroes.

For if $c = r$, either c or r has been chosen. And if c — r, one of the following cases must arise. (1) If r and c are both unmatched, we can increase p by matching them to each other. (2) If r is unmatched and $c = r'$, then c has been chosen, so the zero has been covered. (3) If r is matched to $c' \neq c$, then either r has been chosen or c' has been chosen. In the latter case, there is a path of the form

$$r_0 - c_1 = r_1 - c_2 = \cdots = r_{q-1} - c' = r - c,$$

where r_0 is unmatched and $q \geq 1$. If c is matched, it has therefore been chosen; otherwise we can increase p by redefining the matching to include

$$r_0 = c_1 - r_1 = c_2 - \cdots - r_{q-1} = c' - r = c.$$

10. Now suppose A is a *nonnegative* matrix. Cover the zeroes of A with a minimum number of lines, p, using the algorithm of Egerváry and König. If $p < n$, some elements are still uncovered, so those elements are positive. Suppose the minimum uncovered value is $\delta > 0$. Then we can subtract δ from each unchosen row and add δ to each chosen column. The net effect is to subtract δ from all uncovered elements and to add δ to all doubly covered elements, while leaving all singly covered elements unchanged. This transformation causes a new zero to appear, while preserving p independent zeroes of the previous matrix (since they were each covered only once).

If we repeat the Egerváry-König construction with the same p independent zeroes, we find that either p is no longer maximum or at least one more column has been chosen. (The new zero $r - c$ occurs in a row r that was either unmatched or matched to a previously chosen column, because row r was not chosen.) Therefore if we repeat the process, we must eventually be able to increase p until finally $p = n$. This will solve the assignment problem, proving the remarkable claim made earlier.

11. If the given matrix A has m rows and $n > m$ columns, we can extend it artificially until it is square, by setting $a_{kl} = 0$ for all $m \leq k < n$ and $0 \leq l < n$. The construction above will then apply. But we need not waste time making such an extension, because it suffices to run the algorithm on the original $m \times n$ matrix until m independent zeroes have been found. The reason is that the set of matched vertices always grows monotonically in the Egerváry-König construction: If a column is matched at some stage, it will remain matched from that time on, although it might well change partners. The $n - m$ dummy rows at the bottom of A are always chosen to be part of the covering; so the dummy entries become nonzero only in the columns that are part of some covering. Such columns are part of some matching, so they are part of the final matching. Therefore at most m columns of the dummy entries become nonzero during the procedure. We can always find $n - m$ independent zeroes in the $n - m$ dummy rows of the matrix, so we need not deal with the dummy elements explicitly.

12. It has been convenient to describe the algorithm by saying that we add and subtract constants to and from the columns and rows of A. But all those additions and subtractions can take a lot of time. So we will merely pretend to make the adjustments that the method calls for; we will represent them implicitly by two vectors $(\sigma_0, \ldots, \sigma_{m-1})$ and $(\tau_0, \ldots, \tau_{n-1})$. Then the current value of each matrix entry will be $a_{kl} - \sigma_k + \tau_l$, instead of a_{kl}. The "zeroes" will be positions such that $a_{kl} = \sigma_k - \tau_l$.

Initially we will set $\tau_l = 0$ for $0 \leq l < n$ and $\sigma_k = \min\{a_{k0}, \ldots, a_{k(n-1)}\}$ for $0 \leq k < m$. If $m = n$ we can also make sure that there's a zero in every column by subtracting $\min\{a_{0l}, \ldots, a_{(n-1)l}\}$ from a_{kl} for all k and l. (This initial adjustment can conveniently be made to the original matrix entries, instead of indirectly via the τ's.) Users can discover if such a transformation is worthwhile by trying the program both with and without the **-h** option.

We have been saying a lot of things and proving a bunch of theorems, without writing any code. Let's get back into programming mode by writing the routine that is called into action when the **-h** option has been specified:

#define $aa(k, l)$ $*(mtx + k * n + l)$ /* a macro to access the matrix elements */

$\langle$ Subtract column minima in order to start with lots of zeroes 12 $\rangle \equiv$

```
{
   for (l = 0; l < n; l++) {
      o, s = aa(0, l);        /* the o macro counts one mem */
      for (k = 1; k < n; k++)
         if (o, aa(k, l) < s) s = aa(k, l);
      if (s ≠ 0)
         for (k = 0; k < n; k++) oo, aa(k, l) -= s;    /* oo counts two mems */
   }
   if (verbose) printf("␣The␣heuristic␣has␣cost␣%ld␣mems.\n", mems);
}
```

This code is used in section 24.

13. $\langle$ Local variables 4 $\rangle \mathrel{+}\equiv$

   register long k;   /* the current row of interest */
   register long l;   /* the current column of interest */
   register long j;   /* another interesting column */
   register long s;   /* the current matrix element of interest */

14. Algorithmic details.   The algorithm sketched above is quite simple, except that we did not discuss how to determine the chosen columns c_q that are reachable by paths of the stated form ($*$). It is easy to find all such columns by constructing an unordered forest whose nodes are rows, beginning with all unmatched rows r_0 and adding a row r for which $c = r$ when c is adjacent to a row already in the forest.

Our data structure, which is based on suggestions of Papadimitriou and Steiglitz [*Combinatorial Optimization* (Prentice-Hall, 1982), §11.1], will use several arrays. If row r is matched with column c, we will have $col\_mate[r] = c$ and $row\_mate[c] = r$; if row r is unmatched, $col\_mate[r]$ will be -1, and if column c is unmatched, $row\_mate[c]$ will be -1. If column c has a mate and is also reachable in a path of the form ($*$), we will have $parent\_row[c] = r'$ for some r' in the forest. Otherwise column c is not chosen, and we will have $parent\_row[c] = -1$. The rows in the current forest will be called $unchosen\_row[0]$ through $unchosen\_row[t-1]$, where t is the current total number of nodes.

The amount σ_k subtracted from row k is called $row\_dec[k]$; the amount τ_l added to row l is called $col\_inc[l]$. To compute the minimum uncovered element efficiently, we maintain a quantity called $slack[l]$, which represents the minimum uncovered element in each column. More precisely, if column l is not chosen, $slack[l]$ is the minimum of $a_{kl} - \sigma_k + \tau_l$ for $k \in \{unchosen\_row[0], \ldots,$ $unchosen\_row[q-1]\}$, where $q \le t$ is the number of rows in the forest that we have explored so far. We also remember $slack\_row[l]$, the number of a row where the stated minimum occurs.

Column l is chosen if and only if $parent\_row[l] \ge 0$. We will arrange things so that we also have $slack[l] = 0$ in every chosen column.

⟨ Local variables 4 ⟩ $+\equiv$
    long $*col\_mate$;   /$*$ the column matching a given row, or -1 $*$/
    long $*row\_mate$;   /$*$ the row matching a given column, or -1 $*$/
    long $*parent\_row$;   /$*$ ancestor of a given column's mate, or -1 $*$/
    long $*unchosen\_row$;   /$*$ node in the forest $*$/
    long t;   /$*$ total number of nodes in the forest $*$/
    long q;   /$*$ total number of explored nodes in the forest $*$/
    long $*row\_dec$;   /$*$ σ_k, the amount subtracted from a given row $*$/
    long $*col\_inc$;   /$*$ τ_l, the amount added to a given column $*$/
    long $*slack$;   /$*$ minimum uncovered entry seen in a given column $*$/
    long $*slack\_row$;   /$*$ where the *slack* in a given column can be found $*$/
    long $unmatched$;   /$*$ this many rows have yet to be matched $*$/

---

mems: **long**, §3. $o = $ macro, §7.   *printf*: **int** ( ), <stdio.h>.
mtx: **long** $*$, §3. $oo = $ macro, §7.   *verbose*: **long**, GB_GRAPH §5.
n: **unsigned long**, §4.

15. ⟨ Allocate the intermediate data structures 15 ⟩ ≡
  $col\_mate = gb\_typed\_alloc(m, \textbf{long}, working\_storage);$
  $row\_mate = gb\_typed\_alloc(n, \textbf{long}, working\_storage);$
  $parent\_row = gb\_typed\_alloc(n, \textbf{long}, working\_storage);$
  $unchosen\_row = gb\_typed\_alloc(m, \textbf{long}, working\_storage);$
  $row\_dec = gb\_typed\_alloc(m, \textbf{long}, working\_storage);$
  $col\_inc = gb\_typed\_alloc(n, \textbf{long}, working\_storage);$
  $slack = gb\_typed\_alloc(n, \textbf{long}, working\_storage);$
  $slack\_row = gb\_typed\_alloc(n, \textbf{long}, working\_storage);$
  if $(gb\_trouble\_code)$ {
    $fprintf(stderr, \texttt{"Sorry,}_{\sqcup}\texttt{out}_{\sqcup}\texttt{of}_{\sqcup}\texttt{memory!\\n"});$
    return $-3;$
  }

This code is used in section 24.

16. The algorithm operates in stages, where each stage terminates when we are able to increase the number of matched elements.

The first stage is different from the others; it simply goes through the matrix and looks for zeroes, matching as many rows and columns as it can. This stage also initializes table entries that will be useful in later stages.

#define INF #7fffffff   /* infinity (or darn near) */
⟨ Do the initial stage 16 ⟩ ≡
  $t = 0;$ /* the forest starts out empty */
  for $(l = 0; \ l < n; \ l\text{++})$ {
    $o, row\_mate[l] = -1;$
    $o, parent\_row[l] = -1;$
    $o, col\_inc[l] = 0;$
    $o, slack[l] = \texttt{INF};$
  }
  for $(k = 0; \ k < m; \ k\text{++})$ {
    $o, s = aa(k, 0);$ /* get ready to calculate the minimum entry of row k */
    for $(l = 1; \ l < n; \ l\text{++})$
      if $(o, aa(k, l) < s) \ s = aa(k, l);$
    $o, row\_dec[k] = s;$
    for $(l = 0; \ l < n; \ l\text{++})$
      if $((o, s \equiv aa(k, l)) \wedge (o, row\_mate[l] < 0))$ {
        $o, col\_mate[k] = l;$
        $o, row\_mate[l] = k;$
        if $(verbose > 1) \ printf(\texttt{"}_{\sqcup}\texttt{matching}_{\sqcup}\texttt{col}_{\sqcup}\texttt{\%ld==row}_{\sqcup}\texttt{\%ld\\n"}, l, k);$
        goto $row\_done;$
      }
    $o, col\_mate[k] = -1;$
    if $(verbose > 1) \ printf(\texttt{"}_{\sqcup\sqcup}\texttt{node}_{\sqcup}\texttt{\%ld:}_{\sqcup}\texttt{unmatched}_{\sqcup}\texttt{row}_{\sqcup}\texttt{\%ld\\n"}, t, k);$
    $o, unchosen\_row[t\text{++}] = k;$
  $row\_done:$;
  }

This code is used in section 18.

17. If a subsequent stage has not succeeded in matching every row, we prepare for a new stage by reinitializing the forest as follows.

⟨ Get ready for another stage 17 ⟩ ≡
    $t = 0$;
    for $(l = 0;\ l < n;\ l\texttt{++})$ {
        $o, parent\_row[l] = -1$;
        $o, slack[l] = \texttt{INF}$;
    }
    for $(k = 0;\ k < m;\ k\texttt{++})$
        if $(o, col\_mate[k] < 0)$ {
            if $(verbose > 1)$ $printf(\texttt{"\_\_node\_\%ld:\_unmatched\_row\_\%ld\\n"}, t, k)$;
            $o, unchosen\_row[t\texttt{++}] = k$;
        }

This code is used in section 18.

18. Here, then, is the algorithm's overall control structure. There are at most m stages, and each stage does $O(mn)$ operations, so the total running time is $O(m^2n)$.

⟨ Do the Hungarian algorithm 18 ⟩ ≡
    ⟨ Do the initial stage 16 ⟩;
    if $(t \equiv 0)$ **goto** *done*;
    $unmatched = t$;
    while (1) {
        if $(verbose)$ $printf(\texttt{"\_After\_\%ld\_mems\_I've\_matched\_\%ld\_rows.\\n"}, mems, m - t)$;
        $q = 0$;
        while (1) {
            while $(q < t)$ {
                ⟨ Explore node q of the forest; if the matching can be increased, **goto** *breakthru* 19 ⟩;
                $q\texttt{++}$;
            }
            ⟨ Introduce a new zero into the matrix by modifying *row_dec* and *col_inc*; if the matching can be
                increased, **goto** *breakthru* 21 ⟩;
        }
    breakthru: ⟨ Update the matching by pairing row k with column l 20 ⟩;
        if $(\texttt{--}unmatched \equiv 0)$ **goto** *done*;
        ⟨ Get ready for another stage 17 ⟩;
    }
done: ⟨ Doublecheck the solution 23 ⟩;

This code is used in section 24.

$aa = $ macro ( ), §12.
col_inc: **long** *, §14.
col_mate: **long** *, §14.
fprintf: **int** ( ), `<stdio.h>`.
gb_trouble_code: **long**,
    GB_GRAPH §14.
gb_typed_alloc = macro ( ),
    GB_GRAPH §11.
k: **register long**, §13.
l: **register long**, §13.

m: **unsigned long**, §4.
mems: **long**, §3.
n: **unsigned long**, §4.
o = macro, §7.
parent_row: **long** *, §14.
printf: **int** ( ), `<stdio.h>`.
q: **long**, §14.
row_dec: **long** *, §14.
row_mate: **long** *, §14.

s: **register long**, §13.
slack: **long** *, §14.
slack_row: **long** *, §14.
stderr: **FILE** *, `<stdio.h>`.
t: **long**, §14.
unchosen_row: **long** *, §14.
unmatched: **long**, §14.
verbose: **long**, GB_GRAPH §5.
working_storage: **Area**, §3.

19. $\langle$ Explore node q of the forest; if the matching can be increased, **goto** *breakthru* $19\,\rangle \equiv$

```
{
  o, k = unchosen_row[q];
  o, s = row_dec[k];
  for (l = 0; l < n; l++)
    if (o, slack[l]) { register long del;
      oo, del = aa(k, l) - s + col_inc[l];
      if (del < slack[l]) {
        if (del ≡ 0) {      /* we found a new zero */
          if (o, row_mate[l] < 0) goto breakthru;
          o, slack[l] = 0;    /* this column will now be chosen */
          o, parent_row[l] = k;
          if (verbose > 1)
            printf("␣␣node␣%ld:␣row␣%ld==col␣%ld--row␣%ld\n", t, row_mate[l], l, k);
          oo, unchosen_row[t++] = row_mate[l];
        } else {
          o, slack[l] = del;
          o, slack_row[l] = k;
        }
      }
    }
}
```

This code is used in section 18.

20. At this point, column l is unmatched, and row k is in the forest. By following parent links in the forest, we can rematch rows and columns so that a previously unmatched row r_0 gets a mate.

$\langle$ Update the matching by pairing row k with column l $20\,\rangle \equiv$

```
if (verbose) printf("␣Breakthrough␣at␣node␣%ld␣of␣%ld!\n", q, t);
while (1) {
  o, j = col_mate[k];
  o, col_mate[k] = l;
  o, row_mate[l] = k;
  if (verbose > 1) printf("␣rematching␣col␣%ld==row␣%ld\n", l, k);
  if (j < 0) break;
  o, k = parent_row[j];
  l = j;
}
```

This code is used in section 18.

21. If we get to this point, we have explored the entire forest; none of the unchosen rows has led to a breakthrough. An unchosen column with smallest *slack* will allow us to make further progress.

$\langle$ Introduce a new zero into the matrix by modifying *row_dec* and *col_inc*; if the matching can be increased, **goto** *breakthru* $21\,\rangle \equiv$

```
s = INF;
for (l = 0; l < n; l++)
```

```
    if (o, slack[l] ∧ slack[l] < s)  s = slack[l];
    for (q = 0;  q < t;  q++)  ooo, row_dec[unchosen_row[q]] += s;
    for (l = 0;  l < n;  l++)
      if (o, slack[l]) {        /* column l is not chosen */
        o, slack[l] -= s;
        if (slack[l] ≡ 0) ⟨Look at a new zero, and goto breakthru with col_inc up to date if there's a
                    breakthrough 22⟩;
      } else  oo, col_inc[l] += s;
```
This code is used in section 18.

22.   There might be several columns tied for smallest slack. If any of them leads to a breakthrough, we are very happy; but we must finish the loop on l before going to *breakthru*, because
the *col_inc* variables need to be maintained for the next stage.

Within column l, there might be several rows that produce the same slack; we have remembered
only one of them, *slack_row*[l]. Fortunately, one is sufficient for our purposes. Either we have
a breakthrough or we choose column l, regardless of which row or rows led us to consider that
column.

⟨Look at a new zero, and **goto** *breakthru* with *col_inc* up to date if there's a breakthrough 22⟩ ≡
```
    {
      o, k = slack_row[l];
      if (verbose > 1)
        printf("␣Decreasing␣uncovered␣elements␣by␣%ld␣produces␣zero␣at␣[%ld,%ld]\n", s, k, l);
      if (o, row_mate[l] < 0) {
        for (j = l + 1;  j < n;  j++)
          if (o, slack[j] ≡ 0)  oo, col_inc[j] += s;
        goto breakthru;
      } else {        /* not a breakthrough, but the forest continues to grow */
        o, parent_row[l] = k;
        if (verbose > 1)  printf("␣␣node␣%ld:␣row␣%ld==col␣%ld--row␣%ld\n", t, row_mate[l], l, k);
        oo, unchosen_row[t++] = row_mate[l];
      }
    }
```
This code is used in section 21.

---

aa = macro ( ), §12.
breakthru: label, §18.
col_inc: **long** *, §14.
col_mate: **long** *, §14.
INF = #7fffffff, §16.
j: **register long**, §13.
k: **register long**, §13.
l: **register long**, §13.

n: **unsigned long**, §4.
o = macro, §7.
oo = macro, §7.
ooo = macro, §7.
parent_row: **long** *, §14.
printf: **int** ( ), <stdio.h>.
q: **long**, §14.
row_dec: **long** *, §14.

row_mate: **long** *, §14.
s: **register long**, §13.
slack: **long** *, §14.
slack_row: **long** *, §14.
t: **long**, §14.
unchosen_row: **long** *, §14.
verbose: **long**, GB_GRAPH §5.

23. The code in the present section is redundant, unless cosmic radiation has caused the hardware to malfunction. But there is some reassurance whenever we find that mathematics still appears to be consistent, so the author could not resist writing these few unnecessary lines, which verify that the assignment problem has indeed been solved optimally. (We don't count the mems.)

⟨ Doublecheck the solution 23 ⟩ ≡

```
for (k = 0; k < m; k++)
    for (l = 0; l < n; l++)
        if (aa(k, l) < row_dec[k] − col_inc[l]) {
            fprintf(stderr, "Oops,␣I␣made␣a␣mistake!\n");
            return −6;      /* can't happen */
        }
for (k = 0; k < m; k++) {
    l = col_mate[k];
    if (l < 0 ∨ aa(k, l) ≠ row_dec[k] − col_inc[l]) {
        fprintf(stderr, "Oops,␣I␣blew␣it!\n");
        return −66;      /* can't happen */
    }
}
k = 0;
for (l = 0; l < n; l++)
    if (col_inc[l]) k++;
if (k > m) {
    fprintf(stderr, "Oops,␣I␣adjusted␣too␣many␣columns!\n");
    return −666;      /* can't happen */
}
```

This code is used in section 18.

24. Interfacing. A few nitty-gritty details still need to be handled: Our algorithm is not symmetric between rows and columns, and it works only for $m \leq n$; so we will transpose the matrix when $m > n$. Furthermore, our algorithm minimizes, but we actually want it to maximize (except when *compl* is nonzero).

Hence, we want to make the following transformations to the data before processing it with the algorithm developed above.

⟨ Solve the assignment problem 24 ⟩ ≡
    if $(m > n)$ ⟨ Transpose the matrix 25 ⟩ **else** *transposed* = 0;
    ⟨ Allocate the intermediate data structures 15 ⟩;
    if $(compl \equiv 0)$
        for $(k = 0;\ k < m;\ k{+}{+})$
            for $(l = 0;\ l < n;\ l{+}{+})\ \ aa(k, l) = d - aa(k, l);$
    if (*heur*) ⟨ Subtract column minima in order to start with lots of zeroes 12 ⟩;
    ⟨ Do the Hungarian algorithm 18 ⟩;
This code is used in section 2.

25. ⟨ Transpose the matrix 25 ⟩ ≡
    {
        if (*verbose* > 1) *printf* ("Temporarily␣transposing␣rows␣and␣columns...\n");
        tmtx = *gb_typed_alloc*$(m * n, $ **long**$, $ *working_storage*$);$
        if (*tmtx* ≡ Λ) {
            fprintf (*stderr*, "Sorry,␣out␣of␣memory!\n"); **return** $-4;$
        }
        for $(k = 0;\ k < m;\ k{+}{+})$
            for $(l = 0;\ l < n;\ l{+}{+})\ *(tmtx + l * m + k) = *(mtx + k * n + l);$
        $m = n;\ n = k;$ /* k holds the former value of m */
        $mtx = tmtx;\ transposed = 1;$
    }
This code is used in section 24.

26. ⟨ Local variables 4 ⟩ +≡
    long $*tmtx;$ /* the transpose of *mtx* */
    long *transposed*;   /* has the data been transposed? */

27. ⟨ Display the solution 27 ⟩ ≡
    {
        printf ("The␣following␣entries␣produce␣an␣optimum␣assignment:\n");
        for $(k = 0;\ k < m;\ k{+}{+})$
            printf ("␣[%ld,%ld]\n", *transposed* ? *col_mate*[k] : k, *transposed* ? k : *col_mate*[k]);
    }
This code is used in section 2.

---

aa = macro ( ), §12.
col_inc: **long** *, §14.
col_mate: **long** *, §14.
compl: **long**, §4.
d: **unsigned long**, §4.
fprintf: **int** ( ), <stdio.h>.
gb_typed_alloc = macro ( ),

GB_GRAPH §11.
heur: **long**, §4.
k: **register long**, §13.
l: **register long**, §13.
m: **unsigned long**, §4.
mtx: **long** *, §3.

n: **unsigned long**, §4.
printf: **int** ( ), <stdio.h>.
row_dec: **long** *, §14.
stderr: **FILE** *, <stdio.h>.
verbose: **long**, GB_GRAPH §5.
working_storage: **Area**, §3.

28. Encapsulated PostScript. A special output file called `lisa.eps` is written if the user has selected the `-P` option. This file contains a sequence of PostScript commands that can be used to generate an illustration within many kinds of documents. For example, if TEX is being used with the `dvips` output driver from Radical Eye Software and with the associated `epsf.tex` macros, one can say

$$\text{\epsfxsize=10cm \epsfbox\{lisa.eps\}}$$

within a TEX document and the illustration will be typeset in a box that is 10 centimeters wide.

The conventions of PostScript allow the illustration to be scaled to any size. Best results are probably obtained if each pixel is at least one millimeter wide (about 1/25 inch) when printed.

The illustration is formed by first "painting" the input data as a rectangle of pixels, with up to 256 shades of gray. Then the solution pixels are framed in black, with a white trim just inside the black edges to help make the frame visible in already-dark places. The frames are created by painting over the original image; the center of each solution pixel retains its original color.

Encapsulated PostScript files have a simple format that is recognized by many software packages and printing devices. We use a subset of PostScript that should be easy to convert to other languages if necessary.

⟨ Output the input matrix in PostScript format 28 ⟩ ≡
```
{
    eps_file = fopen("lisa.eps", "w");
    if (¬eps_file) {
        fprintf(stderr, "Sorry,␣I␣can't␣open␣the␣file␣'lisa.eps'!\n");
        PostScript = 0;
    } else {
        fprintf(eps_file, "%%!PS-Adobe-3.0␣EPSF-3.0\n");      /* 1.0 and 2.0 also OK */
        fprintf(eps_file, "%%%%BoundingBox:␣-1␣-1␣%ld␣%ld\n", n + 1, m + 1);
        fprintf(eps_file, "/buffer␣%ld␣string␣def\n", n);
        fprintf(eps_file, "%ld␣%ld␣8␣[%ld␣0␣0␣-%ld␣0␣%ld]\n", n, m, n, m, m);
        fprintf(eps_file, "{currentfile␣buffer␣readhexstring␣pop}␣bind\n");
        fprintf(eps_file, "gsave␣%ld␣%ld␣scale␣image\n", n, m);
        for (k = 0; k < m; k++) ⟨ Output row k as a hexadecimal string 30 ⟩;
        fprintf(eps_file, "grestore\n");
    }
}
```
This code is used in section 2.

29. ⟨ Global variables 3 ⟩ +≡
FILE *eps_file;   /* file for encapsulated PostScript output */

30. This program need not produce machine-independent output, so we can safely use floating-point arithmetic here. At most 64 characters (32 pixel-bytes) are output on each line.

⟨ Output row k as a hexadecimal string 30 ⟩ ≡
```
{ register float conv = 255.0/(float) d;
  register long x;
  for (l = 0; l < n; l++) {
      x = (long) (conv * (float) (compl ? d - aa(k, l) : aa(k, l)));
      fprintf(eps_file, "%02lx", x > 255 ? 255 L : x);
```

        if $((l \mathrel{\&} {}^{\#}\mathtt{1f}) \equiv {}^{\#}\mathtt{1f})$ *fprintf* $(eps\_file, \texttt{"\textbackslash n"})$;
      }
      if $(n \mathrel{\&} {}^{\#}\mathtt{1f})$ *fprintf* $(eps\_file, \texttt{"\textbackslash n"})$;
   }

This code is used in section 28.

31.   ⟨ Output the solution in PostScript format 31 ⟩ ≡
   {
      fprintf $(eps\_file, \texttt{"/bx\_\{moveto\_0\_1\_rlineto\_1\_0\_rlineto\_0\_-1\_rlineto\_closepath\textbackslash n"})$;
      fprintf $(eps\_file, \texttt{"\_gsave\_.3\_setlinewidth\_1\_setgray\_clip\_stroke"})$;
      fprintf $(eps\_file, \texttt{"\_grestore\_stroke\}\_bind\_def\textbackslash n"})$;
      fprintf $(eps\_file, \texttt{"\_.1\_setlinewidth\textbackslash n"})$;
      for $(k = 0;\ k < m;\ k{+}{+})$ *fprintf* $(eps\_file, \texttt{"\_\%ld\_\%ld\_bx\textbackslash n"},$
         $transposed\ ?\ k : col\_mate[k],$
         $transposed\ ?\ n - 1 - col\_mate[k] : m - 1 - k)$;
      fclose (eps_file);
   }

This code is used in section 2.

$aa = $ macro ( ), §12.
col_mate: **long** *, §14.
compl: **long**, §4.
d: **unsigned long**, §4.
fclose: **int** ( ), <stdio.h>.

FILE, <stdio.h>.
fopen: **FILE** *( ), <stdio.h>.
fprintf: **int** ( ), <stdio.h>.
k: **register long**, §13.
l: **register long**, §13.

m: **unsigned long**, §4.
n: **unsigned long**, §4.
PostScript: **long**, §4.
stderr: **FILE** *, <stdio.h>.
transposed: **long**, §26.

BOOK_COMPONENTS _____

Important: Before reading BOOK_COMPONENTS, please read or at least skim the program for GB_BOOKS.

1. Bicomponents. This demonstration program computes the biconnected components of GraphBase graphs derived from world literature, using a variant of Hopcroft and Tarjan's algorithm [R. E. Tarjan, "Depth-first search and linear graph algorithms," *SIAM Journal on Computing* **1** (1972), 146–160]. Articulation points and ordinary (connected) components are also obtained as byproducts of the computation.

Two edges belong to the same biconnected component—or "bicomponent" for short—if and only if they are identical or both belong to a simple cycle. This defines an equivalence relation on edges. The bicomponents of a connected graph form a free tree, if we say that two bicomponents are adjacent when they have a common vertex (i.e., when there is a vertex belonging to at least one edge in each of the bicomponents). Such a vertex is called an *articulation point*; there is a unique articulation point between any two adjacent bicomponents. If we choose one bicomponent to be the "root" of the free tree, the other bicomponents can be represented conveniently as lists of vertices, with the articulation point that leads toward the root listed last. This program displays the bicomponents in exactly that way.

2. We permit command-line options in typical UNIX style so that a variety of graphs can be studied: The user can say '-t⟨title⟩', '-n⟨number⟩', '-x⟨number⟩', '-f⟨number⟩', '-l⟨number⟩', '-i⟨number⟩', '-o⟨number⟩', and/or '-s⟨number⟩' to change the default values of the parameters in the graph generated by $book(t, n, x, f, l, i, o, s)$.

When the bicomponents are listed, each character in the book is identified by a two-letter code, as found in the associated data file. An explanation of these codes will appear first if the -v or -V option is specified. The -V option prints a fuller explanation than -v; it also shows each character's weighted number of appearances.

The special command-line option -g⟨filename⟩ overrides all others. It substitutes an external graph previously saved by *save_graph* for the graphs produced by *book*.

```
#include "gb_graph.h"    /* the GraphBase data structures */
#include "gb_books.h"    /* the book routine */
#include "gb_io.h"    /* the imap_chr routine */
#include "gb_save.h"    /* restore_graph */
⟨Preprocessor definitions⟩
⟨Global variables 7⟩;
⟨Subroutines 4⟩;

main(argc, argv)
    int argc;    /* the number of command-line arguments */
    char *argv[];    /* an array of strings containing those arguments */
{ Graph *g;    /* the graph we will work on */
  register Vertex *v;    /* the current vertex of interest */
  char *t = "anna";    /* the book to use */
  unsigned long n = 0;    /* the desired number of vertices (0 means infinity) */
  unsigned long x = 0;    /* the number of major characters to exclude */
```

```
unsigned long f = 0;      /* the first chapter to include */
unsigned long l = 0;      /* the last chapter to include (0 means infinity) */
long i = 1;      /* the weight for appearances in selected chapters */
long o = 1;      /* the weight for appearances in unselected chapters */
long s = 0;      /* the random number seed */
⟨ Scan the command-line options 3 ⟩;
if (filename) g = restore_graph(filename);
else g = book(t, n, x, f, l, i, o, s);
if (g ≡ Λ) {
  fprintf(stderr, "Sorry,␣can't␣create␣the␣graph!␣(error␣code␣%ld)\n", panic_code);
  return −1;
}
printf("Biconnectivity␣analysis␣of␣%s\n\n", g⃗id);
if (verbose) ⟨ Print the cast of selected characters 5 ⟩;
⟨ Perform the Hopcroft-Tarjan algorithm on g 12 ⟩;
return 0;      /* normal exit */
}
```

3. ⟨ Scan the command-line options 3 ⟩ ≡

```
while (−−argc) {
  if (strncmp(argv[argc], "-t", 2) ≡ 0)  t = argv[argc] + 2;
  else if (sscanf(argv[argc], "-n%lu", &n) ≡ 1)  ;
  else if (sscanf(argv[argc], "-x%lu", &x) ≡ 1)  ;
  else if (sscanf(argv[argc], "-f%lu", &f) ≡ 1)  ;
  else if (sscanf(argv[argc], "-l%lu", &l) ≡ 1)  ;
  else if (sscanf(argv[argc], "-i%ld", &i) ≡ 1)  ;
  else if (sscanf(argv[argc], "-o%ld", &o) ≡ 1)  ;
  else if (sscanf(argv[argc], "-s%ld", &s) ≡ 1)  ;
  else if (strcmp(argv[argc], "-v") ≡ 0)  verbose = 1;
  else if (strcmp(argv[argc], "-V") ≡ 0)  verbose = 2;
  else if (strncmp(argv[argc], "-g", 2) ≡ 0)  filename = argv[argc] + 2;
  else {
    fprintf(stderr, "Usage:␣%s␣[-ttitle] [-nN] [-xN] [-fN] [-lN] [-iN] [-oN] [-sN] [-v] [-gfoo]\n",
        argv[0]);
    return −2;
  }
}
if (filename)  verbose = 0;
```

This code is used in section 2.

---

book: **Graph** *( ), GB_BOOKS §8.
filename: **char** *, §4.
fprintf: **int** ( ), <stdio.h>.
Graph = **struct**, GB_GRAPH §20.
id: **char** [ ], GB_GRAPH §20.
imap_chr: **char** ( ), GB_IO §12.

panic_code: **long**, GB_GRAPH §5.
printf: **int** ( ), <stdio.h>.
restore_graph: **Graph** *( ),
  GB_SAVE §4.
save_graph: **long** ( ), GB_SAVE §20.
sscanf: **int** ( ), <stdio.h>.

stderr: **FILE** *, <stdio.h>.
strcmp: **int** ( ), <string.h>.
strncmp: **int** ( ), <string.h>.
verbose: **long**, GB_GRAPH §5.
Vertex = **struct**, GB_GRAPH §9.

4. ⟨ Subroutines 4 ⟩ ≡

    char *filename = Λ;   /* external graph to be restored */

    char *code_name*[3][3];

    char *vertex_name(*v*, *i*)   /* return (as a string) the name of vertex *v* */

        Vertex *v;

        char *i*;   /* *i* should be 0, 1, or 2 to avoid clash in *code_name* array */

  {

    if (*filename*) **return** *v→name*;   /* not a *book* graph */

    code_name[*i*][0] = *imap_chr*(*v→short_code* /36);

    code_name[*i*][1] = *imap_chr*(*v→short_code* % 36);

    return *code_name*[*i*];

  }

This code is used in section 2.

5. ⟨ Print the cast of selected characters 5 ⟩ ≡

  {

    for (*v* = *g→vertices*; *v* < *g→vertices* + *g→n*; *v*++) {

      if (*verbose* ≡ 1) *printf*("%s=%s\n", *vertex_name*(*v*, 0), *v→name*);

      else *printf*("%s=%s,␣%s␣[weight␣%ld]\n", *vertex_name*(*v*, 0), *v→name*, *v→desc*,

          i * *v→in_count* + *o* * *v→out_count*);

    }

    printf("\n");

  }

This code is used in section 2.

6.   The algorithm.   The Hopcroft-Tarjan algorithm is inherently recursive. We will implement the recursion explicitly via linked lists, instead of using C's runtime stack, because some
computer systems bog down in the presence of deeply nested recursion.

Each vertex goes through three stages during the algorithm. First it is "unseen"; then it is
"active"; finally it becomes "settled," when it has been assigned to a bicomponent.

The data structures that represent the current state of the algorithm are implemented by using
five of the utility fields in each vertex: *rank*, *parent*, *untagged*, *link*, and *min*. We will consider
each of these in turn.

7.   First is the integer *rank* field, which is zero when a vertex is unseen. As soon as the vertex
is first examined, it becomes active and its *rank* becomes and remains nonzero. Indeed, the kth
vertex to become active will receive rank k.

It's convenient to think of the Hopcroft-Tarjan algorithm as a simple adventure game in which
we want to explore all the rooms of a cave. Passageways between the rooms allow two-way travel.
When we come into a room for the first time, we assign a new number to that room; this is its
rank. Later on we might happen to come into the same room again, and we will notice that it
has nonzero rank. Then we'll be able to make a quick exit, saying "we've already been here."
(The extra complexities of computer games, like dragons that might need to be vanquished, do
not arise.)

#define *rank*   z.I   /* the *rank* of a vertex is stored in utility field z */

⟨ Global variables 7 ⟩ ≡
    long *nn*;   /* the number of vertices that have been seen */

See also sections 8, 10, 13, and 20.

This code is used in section 2.

8.   The active vertices will always form an oriented tree, whose arcs are a subset of the arcs in
the original graph. A tree arc from u to v will be represented by $v\text{-}parent \equiv u$. Every active
vertex has a parent, which is usually another active vertex; the only exception is the root of the
tree, whose *parent* is a dummy vertex called *dummy*. The dummy vertex has rank zero.

In the cave analogy, the "parent" of room v is the room we were in immediately before entering
v the first time. By following parent pointers, we will be able to leave the cave whenever we want.

#define *parent*   y.V   /* the *parent* of a vertex is stored in utility field y */

⟨ Global variables 7 ⟩ +≡
    Vertex *dummy*;   /* imaginary parent of the root vertex */

---

book: **Graph** *( ), GB_BOOKS §8.
desc = *z.S*, GB_BOOKS §17.
g: **Graph** *, §2.
i: **long**, §2.
I: **long**, GB_GRAPH §8.
imap_chr: **char** ( ), GB_IO §12.
in_count = *y.I*, GB_BOOKS §17.
link = *w.V*, §10.

min = *v.V*, §11.
n: **long**, GB_GRAPH §20.
name: **char** *, GB_GRAPH §9.
o: **long**, §2.
out_count = *x.I*, GB_BOOKS §17.
printf: **int** ( ), <stdio.h>.
short_code = *u.I*, GB_BOOKS §17.
untagged = *x.A*, §9.

v: **register Vertex** *, §2.
V: **Vertex** *, GB_GRAPH §8.
verbose: **long**, GB_GRAPH §5.
Vertex = **struct**, GB_GRAPH §9.
vertices: **Vertex** *, GB_GRAPH §20.
y: **util**, GB_GRAPH §9.
z: **util**, GB_GRAPH §9.

9. All edges in the original undirected graph are explored systematically during a depth-first search. Whenever we look at an edge, we tag it so that we won't need to explore it again. In a cave, for example, we might mark each passageway between rooms once we've tried to go through it.

In a GraphBase graph, undirected edges are represented as a pair of directed arcs. Each of these arcs will be examined and eventually tagged.

The algorithm doesn't actually place a tag on its **Arc** records; instead, each vertex v has a pointer $v \rightarrow untagged$ that leads to all hitherto-unexplored arcs from v. The arcs of the list that appear between $v \rightarrow arcs$ and $v \rightarrow untagged$ are the ones already examined.

#define *untagged* *x.A*   /* the *untagged* field points to an **Arc** record, or Λ */

10. The algorithm maintains a special stack, the *active_stack*, which contains all the currently active vertices. Each vertex has a *link* field that points to the vertex that is next lower on its stack, or to Λ if the vertex is at the bottom. The vertices on *active_stack* always appear in increasing order of rank from bottom to top.

#define *link* *w.V*   /* the *link* field of a vertex occupies utility field w */
⟨ Global variables 7 ⟩ +≡
  Vertex **active_stack*;   /* the top of the stack of active vertices */

11. Finally there's a *min* field, which is the tricky part that makes everything work. If vertex v is unseen or settled, its *min* field is irrelevant. Otherwise $v \rightarrow min$ points to the active vertex u of smallest rank having the following property: There is a directed path from v to u consisting of zero or more mature tree arcs followed by a single non-tree arc.

What is a tree arc, you ask. And what is a mature arc? Good questions. At the moment when arcs of the graph are tagged, we classify them either as tree arcs (if they correspond to a new *parent* link in the tree of active nodes) or non-tree arcs (otherwise). The tree arcs therefore correspond to passageways that have led us to new territory. A tree arc becomes mature when it is no longer on the path from the root to the current vertex being explored. We also say that a vertex becomes mature when it is no longer on that path. All arcs from a mature vertex have been tagged.

We said before that every vertex is initially unseen, then active, and finally settled. With our new definitions, we see further that every arc starts out untagged, then it becomes either a non-tree arc or a tree arc. In the latter case, the arc begins as an immature tree arc and eventually matures.

The dummy vertex is considered to be active, and we assume that there is a non-tree arc from the root vertex back to *dummy*. Thus there is a non-tree arc from v to $v \rightarrow parent$ for all v, and $v \rightarrow min$ will always point to a vertex whose rank is less than or equal to $v \rightarrow parent \rightarrow rank$. It will turn out that $v \rightarrow min$ is always an ancestor of v.

Just believe these definitions, for now. All will become clear soon.

#define *min* *v.V*   /* the *min* field of a vertex occupies utility field v */

12. Depth-first search explores a graph by systematically visiting all vertices and seeing what they can lead to. In the Hopcroft-Tarjan algorithm, as we have said, the active vertices form an oriented tree. One of these vertices is called the current vertex.

If the current vertex still has an arc that hasn't been tagged, we tag one such arc and there are two cases: Either the arc leads to an unseen vertex, or it doesn't. If it does, the arc becomes a tree arc; the previously unseen vertex becomes active, and it becomes the new current vertex. On the other hand if the arc leads to a vertex that has already been seen, the arc becomes a non-tree arc and the current vertex doesn't change.

Finally there will come a time when the current vertex v has no untagged arcs. At this point, the algorithm might decide that v and all its descendants form a bicomponent, together with $v\text{-}min \equiv v\text{-}parent$. Indeed, this condition turns out to be true if and only if $v\text{-}min \equiv v\text{-}parent$; a proof appears below. If so, v and all its descendants become settled, and they leave the tree. If not, the tree arc from v's parent u to v becomes mature, so the value of $v\text{-}min$ is used to update the value of $u\text{-}min$. In both cases, v becomes mature and the new current vertex will be the parent of v. Notice that only the value of $u\text{-}min$ needs to be updated, when the arc from u to v matures; all other values $w\text{-}min$ stay the same, because a newly mature arc has no mature predecessors.

The cave analogy helps to clarify the situation: Suppose we enter room v from room u. If there's no way out of the subcave starting at v unless we come back through u, and if we can get to u from all of v's descendants without passing through v, then room v and its descendants will become a bicomponent together with u. Once such a bicomponent is identified, we close it off and don't explore that subcave any further.

If v is the root of the tree, it always has $v\text{-}min \equiv dummy$, so it will always define a new bicomponent at the moment it matures. Then the depth-first search will terminate, since v has no real parent. But the Hopcroft-Tarjan algorithm will press on, trying to find a vertex u that is still unseen. If such a vertex exists, a new depth-first search will begin with u as the root. This process keeps on going until at last all vertices are happily settled.

The beauty of this algorithm is that it all works very efficiently when we organize it as follows:

⟨ Perform the Hopcroft-Tarjan algorithm on g 12 ⟩ ≡
  ⟨ Make all vertices unseen and all arcs untagged 14 ⟩;
  for $(vv = g\text{-}vertices;\ vv < g\text{-}vertices + g\text{-}n;\ vv{+}{+})$
    if $(vv\text{-}rank \equiv 0)$ /∗ vv is still unseen ∗/
      ⟨ Perform a depth-first search with vv as the root, finding the bicomponents of all unseen vertices
         reachable from vv 15 ⟩;

This code is used in section 2.

13.   ⟨ Global variables 7 ⟩ +≡
  Vertex ∗vv;   /∗ sweeps over all vertices, making sure none is left unseen ∗/

A: **Arc** ∗, GB_GRAPH §8.
Arc = **struct**, GB_GRAPH §10.
arcs: **Arc** ∗, GB_GRAPH §9.
dummy: **Vertex**, §8.
g: **Graph** ∗, §2.

n: **long**, GB_GRAPH §20.
parent = $y.V$, §8.
rank = $z.I$, §7.
V: **Vertex** ∗, GB_GRAPH §8.
v: **util**, GB_GRAPH §9.

Vertex = **struct**, GB_GRAPH §9.
vertices: **Vertex** ∗, GB_GRAPH §20.
w: **util**, GB_GRAPH §9.
x: **util**, GB_GRAPH §9.

14. It's easy to get the data structures started, according to the conventions stipulated above.

⟨ Make all vertices unseen and all arcs untagged 14 ⟩ ≡
    for $(v = g\text{-}vertices;\ v < g\text{-}vertices + g\text{-}n;\ v\text{++})$ {
        $v\text{-}rank = 0$;
        $v\text{-}untagged = v\text{-}arcs$;
    }
    $nn = 0$;
    $active\_stack = \Lambda$;
    $dummy.rank = 0$;

This code is used in section 12.

15. The task of starting a depth-first search isn't too bad either. Throughout this part of the algorithm, variable v will point to the current vertex.

⟨ Perform a depth-first search with vv as the root, finding the bicomponents of all unseen vertices
      reachable from vv 15 ⟩ ≡
    {
      $v = vv$;
      $v\text{-}parent = \&\,dummy$;
      ⟨ Make vertex v active 16 ⟩;
      do ⟨ Explore one step from the current vertex v, possibly moving to another current vertex and
          calling it v 17 ⟩ **while** $(v \neq \&\,dummy)$;
    }

This code is used in section 12.

16. ⟨ Make vertex v active 16 ⟩ ≡
    $v\text{-}rank = \text{++}nn$;
    $v\text{-}link = active\_stack$;
    $active\_stack = v$;
    $v\text{-}min = v\text{-}parent$;

This code is used in sections 15 and 17.

17. Now things get interesting. But we're just doing what any well-organized spelunker would do when calmly exploring a cave. There are three main cases, depending on whether the current vertex stays where it is, moves to a new child, or backtracks to a parent.

⟨ Explore one step from the current vertex v, possibly moving to another current vertex and
      calling it v 17 ⟩ ≡
    { **register Vertex** *u;   /* a vertex adjacent to v */
      register Arc *$a = v\text{-}untagged$;   /* v's first remaining untagged arc, if any */
      if (a) {
        $u = a\text{-}tip$;
        $v\text{-}untagged = a\text{-}next$;   /* tag the arc from v to u */
        if $(u\text{-}rank)$ {   /* we've seen u already */
          if $(u\text{-}rank < v\text{-}min\text{-}rank)\ v\text{-}min = u$;   /* non-tree arc, just update $v\text{-}min$ */
        } **else** {   /* u is presently unseen */
          $u\text{-}parent = v$;   /* the arc from v to u is a new tree arc */
          $v = u$;   /* u will now be the current vertex */
          ⟨ Make vertex v active 16 ⟩;

```
      }
    } else {       /* all arcs from v are tagged, so v matures */
      u = v⃗parent;       /* prepare to backtrack in the tree */
      if (v⃗min ≡ u) ⟨Remove v and all its successors on the active stack from the tree, and report
            them as a bicomponent of the graph together with u 19⟩
      else       /* the arc from u to v has just matured, making v⃗min visible from u */
        if (v⃗min⃗rank < u⃗min⃗rank) u⃗min = v⃗min;
      v = u;       /* the former parent of v is the new current vertex v */
    }
  }
```

This code is used in section 15.

18.   The elements of the active stack are always in order by rank, and all children of a vertex v in the tree have rank higher than v. The Hopcroft-Tarjan algorithm relies on a converse property: *All active nodes whose rank exceeds that of the current vertex v are descendants of v.* (This property holds because the algorithm has constructed the tree by assigning ranks in preorder, "the order of succession to the throne." First come v's firstborn and descendants, then the nextborn, and so on.) Therefore the descendants of the current vertex always appear consecutively at the top of the stack.

Suppose v is a mature, active vertex with $v⃗min ≡ v⃗parent$, and let $u = v⃗parent$. We want to prove that v and its descendants, together with u and all edges between these vertices, form a biconnected graph. Call this subgraph H. The parent links define a subtree of H, rooted at u, and v is the only vertex having u as a parent (because all other vertices are descendants of v). Let x be any vertex of H different from u and v. Then there is a path from x to $x⃗min$ that does not touch $x⃗parent$, and $x⃗min$ is a proper ancestor of $x⃗parent$. This property is sufficient to establish the biconnectedness of H. (A proof appears at the conclusion of this program.) Moreover, we cannot add any more vertices to H without losing biconnectivity. If w is another vertex, either w has been output already as a non-articulation point of a previous biconnected component, or we can prove that there is no path from w to v that avoids the vertex u.

---

active_stack: **Vertex** *, §10.
Arc = **struct**, GB_GRAPH §10.
arcs: **Arc** *, GB_GRAPH §9.
dummy: **Vertex** *, §8.
g: **Graph** *, §2.
link = *w.V*, §10.
min = *v.V*, §11.

n: **long**, GB_GRAPH §20.
next: **Arc** *, GB_GRAPH §10.
nn: **long**, §7.
parent = *y.V*, §8.
rank = *z.I*, §7.
tip: **Vertex** *, GB_GRAPH §10.
untagged = *x.A*, §9.

v: **register Vertex** *, §2.
Vertex = **struct**, GB_GRAPH §9.
vertices: **Vertex** *, GB_GRAPH §20.
vv: **Vertex** *, §13.
w: **util**, GB_GRAPH §9.
x: **unsigned long**, §2.

19. Therefore we are justified in settling v and its active descendants now. Removing them from the tree of active vertices does not remove any vertex from which there is a path to a vertex of rank less than $u\rightarrow rank$. Hence their removal does not affect the validity of the $w\rightarrow min$ value for any vertex w that remains active.

A slight technicality arises with respect to whether or not the parent of v, vertex u, is part of the present bicomponent. When u is the dummy vertex, we have already printed the final bicomponent of a connected component of the original graph, unless v was an isolated vertex. Otherwise u is an articulation point that will occur in subsequent bicomponents, unless the new bicomponent is the final bicomponent of a connected component. (This aspect of the algorithm is probably its most subtle point; consideration of an example or two should clarify everything.)

We print out enough information for a reader to verify the biconnectedness of the claimed component easily.

⟨ Remove v and all its successors on the active stack from the tree, and report them as a bicomponent of the graph together with u 19 ⟩ ≡

```
if (u ≡ &dummy) {       /* active_stack contains just v */
    if (artic_pt) printf("␣and␣%s␣(this␣ends␣a␣connected␣component␣of␣the␣graph)\n",
            vertex_name(artic_pt, 0));
    else printf("Isolated␣vertex␣%s\n", vertex_name(v, 0));
    active_stack = artic_pt = Λ;
} else { register Vertex *t;       /* runs through the vertices of the new bicomponent */
    if (artic_pt) printf("␣and␣articulation␣point␣%s\n", vertex_name(artic_pt, 0));
    t = active_stack;
    active_stack = v→link;
    printf("Bicomponent␣%s", vertex_name(v, 0));
    if (t ≡ v) putchar('\n');       /* single vertex */
    else {
        printf("␣also␣includes:\n");
        while (t ≠ v) {
            printf("␣%s␣(from␣%s;␣..to␣%s)\n", vertex_name(t, 0), vertex_name(t→parent, 1),
                    vertex_name(t→min, 2));
            t = t→link;
        }
    }
    artic_pt = u;       /* the printout will be finished later */
}
```

This code is used in section 17.

20. Like all global variables, *artic_pt* is initially zero (Λ).

⟨ Global variables 7 ⟩ +≡

Vertex **artic_pt*;   /* articulation point to be printed if the current bicomponent isn't the last in its connected component */

21.   Proofs.   The program is done, but we still should prove that it works. First we want to clarify the informal definition by verifying that the cycle relation between edges, as stated in the introduction, is indeed an equivalence relation.

Suppose $u - v$ and $w - x$ are edges of a simple cycle C, while $w - x$ and $y - z$ are edges of a simple cycle D. We want to show that there is a simple cycle containing the edges $u - v$ and $x - y$. There are vertices $a, b \in C$ such that $a -^* y - z -^* b$ is a subpath of D containing no other vertices of C besides a and b. Join this subpath to the subpath in C that runs from b to a through the edge $u - v$.

Therefore the stated relation between edges is transitive, and it is an equivalence relation. A graph is biconnected if it contains a single vertex, or if each of its vertices is adjacent to at least one other vertex and any two edges are equivalent.

22.   Next we prove the well-known fact that a graph is biconnected if and only if it is connected and, for any three distinct vertices x, y, z, it contains a path from x to y that does not touch z. Call the latter condition property P.

Suppose G is biconnected, and let x, y be distinct vertices of G. Then there exist edges $u - x$ and $v - y$, which are either identical (hence x and y are adjacent) or part of a simple cycle (hence there are two paths from x to y, having no other vertices in common). Thus G has property P.

Suppose, conversely, that G has property P, and let $u - v, w - x$ be distinct edges of G. We want to show that these edges belong to some simple cycle. The proof is by induction on $k = \min\big(d(u,w), d(u,x), d(v,w), d(v,x)\big)$, where d denotes distance. If $k = 0$, property P gives the result directly. If $k > 0$, we can assume by symmetry that $k = d(u, w)$; so there's a vertex y with $u - y$ and $d(y, w) = k - 1$. And we have $u - v$ equivalent to $u - y$ by property P, $u - y$ equivalent to $w - x$ by induction, hence $u - v$ is equivalent to $w - x$ by transitivity.

23.   Finally, we prove that G satisfies property P if it has the following properties: (1) There are two distinguished vertices u and v. (2) Some of the edges of G form a subtree rooted at u, and v is the only vertex whose parent in this tree is u. (3) Every vertex x other than u or v has a path to its grandparent that does not go through its parent.

If property P doesn't hold, there are distinct vertices x, y, z such that every path from x to y goes through z. In particular, z must be between x and y in the unique path π that joins them in the subtree. It follows that $z \neq u$ is the parent of some node z' in that path; hence $z' \neq u$ and $z' \neq v$. But we can avoid z by going from z' to the grandparent of z', which is also part of path π unless z is also the parent of another node z'' in π. In the latter case, however, we can avoid z by going from z' to the grandparent of z' and from there to z'', since z' and z'' have the same grandparent.

$active\_stack$: **Vertex** $*$, §10. $parent = y.V$, §8. u: **register Vertex** $*$, §17.
$dummy$: **Vertex**, §8. $printf$: **int** ( ), <stdio.h>. v: **register Vertex** $*$, §2.
$link = w.V$, §10. $putchar =$ macro ( ), <stdio.h>.   **Vertex** = **struct**, GB_GRAPH §9.
$min = v.V$, §11. $rank = z.I$, §7. $vertex\_name$: **char** $*$( ), §4.

ECON_ORDER _____

Important: Before reading ECON_ORDER, please read or at least skim the program for GB_ECON.

1. Near-triangular ordering. This demonstration program takes a matrix of data constructed by the GB_ECON module and permutes the economic sectors so that the first sectors of the ordering tend to be producers of primary materials for other industries, while the last sectors tend to be final-product industries that deliver their output mostly to end users.

More precisely, suppose the rows of the matrix represent the outputs of a sector and the columns represent the inputs. This program attempts to find a permutation of rows and columns that minimizes the sum of the elements below the main diagonal. (If this sum were zero, the matrix would be upper triangular; each supplier of a sector would precede it in the ordering, while each customer of that sector would follow it.)

The general problem of finding a minimizing permutation is NP-complete; it includes, as a very special case, the FEEDBACK ARC SET problem discussed in Karp's classic paper [*Complexity of Computer Computations* (Plenum Press, 1972), 85–103]. But sophisticated "branch and cut" methods have been developed that work well in practice on problems of reasonable size. Here we use a simple heuristic downhill method to find a permutation that is locally optimum, in the sense that the below-diagonal sum does not decrease if any individual sector is moved to another position while preserving the relative order of the other sectors. We start with a random permutation and repeatedly improve it, choosing the improvement that gives the least positive gain at each step. A primary motive for the present implementation was to get further experience with this method of cautious descent, which was proposed by A. M. Gleason in *AMS Proceedings of Symposia in Applied Mathematics* **10** (1958), 175–178. (See the comments following the program below.)

2. As explained in GB_ECON, the subroutine call $econ(n, 2, 0, s)$ constructs a graph whose $n \leq 79$ vertices represent sectors of the U.S. economy and whose arcs $u \to v$ are assigned numbers corresponding to the flow of products from sector u to sector v. When $n < 79$, the n sectors are obtained from a basic set of 79 sectors by combining related commodities. If $s = 0$, the combination is done in a way that tends to equalize the row sums, while if $s > 0$, the combination is done by choosing a random subtree of a given 79-leaf tree; the "randomness" is fully determined by the value of s.

This program uses two random number seeds, one for *econ* and one for choosing the random initial permutation. The former is called s and the latter is called t. A further parameter, r, governs the number of repetitions to be made; the machine will try r different starting permutations on the same matrix. When $r > 1$, new solutions are displayed only when they improve on the previous best.

By default, $n = 79$, $r = 1$, and $s = t = 0$. The user can change these default parameters by specifying options on the command line, at least in a UNIX implementation, thereby obtaining a variety of special effects. The relevant command-line options are -n⟨number⟩, -r⟨number⟩, -s⟨number⟩, and/or -t⟨number⟩. Additional options -v (verbose), -V (extreme verbosity), and -g (greedy or steepest descent instead of cautious descent) are also provided.

Here is the overall layout of this C program:

```
#include "gb_graph.h"      /* the GraphBase data structures */
#include "gb_flip.h"       /* the random number generator */
#include "gb_econ.h"       /* the econ routine */
⟨ Preprocessor definitions ⟩
⟨ Global variables 3 ⟩
main(argc, argv)
     int argc;      /* the number of command-line arguments */
     char *argv[];      /* an array of strings containing those arguments */
{ unsigned long n = 79;      /* the desired number of sectors */
  long s = 0;      /* random seed for econ */
  long t = 0;      /* random seed for initial permutation */
  unsigned long r = 1;      /* the number of repetitions */
  long greedy = 0;      /* should we use steepest descent? */
  register long j, k;      /* all-purpose indices */
  ⟨ Scan the command-line options 4 ⟩;
  g = econ(n, 2_L, 0_L, s);
  if (g ≡ Λ) {
    fprintf(stderr, "Sorry,␣can't␣create␣the␣matrix!␣(error␣code␣%ld)\n", panic_code);
    return −1;
  }
  printf("Ordering␣the␣sectors␣of␣%s,␣using␣seed␣%ld:\n", g→id, t);
  printf("␣(%s␣descent␣method)\n", greedy ? "Steepest" : "Cautious");
  ⟨ Put the graph data into matrix form 5 ⟩;
  ⟨ Print an obvious lower bound 6 ⟩;
  gb_init_rand(t);
  while (r−−) ⟨ Find a locally optimum permutation and report the below-diagonal sum 8 ⟩;
  return 0;      /* normal exit */
}
```

3. Besides the matrix M of input/output coefficients, we will find it convenient to use the matrix Δ, where $\Delta_{jk} = M_{jk} - M_{kj}$.

```
#define INF  #7fffffff      /* infinity (or darn near) */
⟨ Global variables 3 ⟩ ≡
  Graph *g;      /* the graph we will work on */
  long mat[79][79];      /* the corresponding matrix */
  long del[79][79];      /* skew-symmetric differences */
  long best_score = INF;      /* the smallest below-diagonal sum we've seen so far */
```

See also sections 7 and 12.

This code is used in section 2.

---

econ: **Graph** *(), GB_ECON §7.   **Graph** = **struct**, GB_GRAPH §20.   printf: **int** (), <stdio.h>.
fprintf: **int** (), <stdio.h>.   id: **char** [], GB_GRAPH §20.   stderr: **FILE** *, <stdio.h>.
gb_init_rand: **void** (), GB_FLIP §8.   panic_code: **long**, GB_GRAPH §5.

4. ⟨ Scan the command-line options 4 ⟩ ≡

```
while (-- argc) {
    if (sscanf (argv[argc], "-n%lu", &n) ≡ 1) ;
    else if (sscanf (argv[argc], "-r%lu", &r) ≡ 1) ;
    else if (sscanf (argv[argc], "-s%ld", &s) ≡ 1) ;
    else if (sscanf (argv[argc], "-t%ld", &t) ≡ 1) ;
    else if (strcmp (argv[argc], "-v") ≡ 0) verbose = 1;
    else if (strcmp (argv[argc], "-V") ≡ 0) verbose = 2;
    else if (strcmp (argv[argc], "-g") ≡ 0) greedy = 1;
    else {
        fprintf (stderr, "Usage:␣%s␣[-nN] [-rN] [-sN] [-tN] [-g] [-v] [-V]\n", argv[0]);
        return -2;
    }
}
```

This code is used in section 2.

5. The optimum permutation is a function only of the Δ matrix, because we can subtract any constant from both M_{jk} and M_{kj} without changing the basic problem.

⟨ Put the graph data into matrix form 5 ⟩ ≡

```
{ register Vertex *v; register Arc *a;
    n = g→n;
    for (v = g→vertices; v < g→vertices + n; v++)
        for (a = v→arcs; a; a = a→next) mat[v - g→vertices][a→tip - g→vertices] = a→flow;
    for (j = 0; j < n; j++) for (k = 0; k < n; k++) del[j][k] = mat[j][k] - mat[k][j];
}
```

This code is used in section 2.

6. Nontrivial lower bounds that can be made strong enough to find provably optimum solutions to the ordering problem can be based on linear programming, as shown for example by Grötschel, Jünger, and Reinelt [*Operations Research* **32** (1984), 1195–1220]. The basic idea is to formulate the problem as the task of minimizing $\sum M_{jk}x_{jk}$ for integer variables $x_{jk} \geq 0$, subject to the conditions $x_{jk} + x_{kj} = 1$ and $x_{ik} \leq x_{ij} + x_{jk}$ for all triples (i, j, k) of distinct subscripts; these conditions are necessary and sufficient. Relaxing the integrality constraints gives a lower bound, and we can also add additional inequalities such as $x_{14} + x_{25} + x_{36} + x_{42} + x_{43} + x_{51} + x_{53} + x_{61} + x_{62} \leq 7$. The interesting story of inequalities like this has been surveyed by P. C. Fishburn [*Mathematical Social Sciences* **23** (1992), 67–80].

However, our goal is more modest—we just want to study two of the simplest heuristics. So we will be happy with a trivial bound based only on the constraints $x_{jk} + x_{kj} = 1$.

⟨ Print an obvious lower bound 6 ⟩ ≡

```
{ register long sum = 0;
    for (j = 1; j < n; j++) for (k = 0; k < j; k++)
        if (mat[j][k] ≤ mat[k][j]) sum += mat[j][k];
        else sum += mat[k][j];
    printf ("(The␣amount␣of␣feed-forward␣must␣be␣at␣least␣%ld.)\n", sum);
}
```

This code is used in section 2.

7. Descent. At each stage in our search, *mapping* will be the current permutation; in other words, the sector in row and column k will be g-*vertices* + *mapping*[k]. The current below-diagonal sum will be the value of *score*. We will not actually have to permute anything inside of *mat*.

#define *sec_name*(k) (g-*vertices* + *mapping*[k])-*name*

⟨ Global variables 3 ⟩ +≡
   long *mapping*[79];   /* current permutation */
   long *score*;   /* current sum of elements above main diagonal */
   long *steps*;   /* the number of iterations so far */

8. ⟨ Find a locally optimum permutation and report the below-diagonal sum 8 ⟩ ≡
```
{
    ⟨ Initialize mapping to a random permutation 9 ⟩;
    while (1) {
        ⟨ Figure out the next move to make; break if at local optimum 10 ⟩;
        if (verbose) printf("%8ld␣after␣step␣%ld\n", score, steps);
        else if (steps % 1000 ≡ 0 ∧ steps > 0) {
            putchar('.');
            fflush(stdout);      /* progress report */
        }
        ⟨ Take the next step 13 ⟩;
    }
    printf("\n%s␣is␣%ld,␣found␣after␣%ld␣step%s.\n",
        best_score ≡ INF ? "Local␣minimum␣feed-forward" : "Another␣local␣minimum",
        score, steps, steps ≡ 1 ? "" : "s");
    if (verbose ∨ score < best_score) {
        printf("The␣corresponding␣economic␣order␣is:\n");
        for (k = 0; k < n; k++) printf("␣%s\n", sec_name(k));
        if (score < best_score) best_score = score;
    }
}
```
This code is used in section 2.

Arc = **struct**, GB_GRAPH §10.
arcs: *Arc* ∗, GB_GRAPH §9.
argc: **int**, §2.
argv: **char** ∗[], §2.
best_score: **long**, §3.
del: **long** [][][], §3.
fflush: **int** ( ), <stdio.h>.
flow = *a.I*, GB_ECON §2.
fprintf: **int** ( ), <stdio.h>.
g: **Graph** ∗, §3.
greedy: **long**, §2.

INF = #7fffffff, §3.
j: **register long**, §2.
k: **register long**, §2.
mat: **long** [][][], §3.
n: **unsigned long**, §2.
name: **char** ∗, GB_GRAPH §9.
next: **Arc** ∗, GB_GRAPH §10.
printf: **int** ( ), <stdio.h>.
putchar = **macro** ( ), <stdio.h>.
r: **unsigned long**, §2.

s: **long**, §2.
sscanf: **int** ( ), <stdio.h>.
stderr: **FILE** ∗, <stdio.h>.
stdout: **FILE** ∗, <stdio.h>.
strcmp: **int** ( ), <string.h>.
t: **long**, §2.
tip: **Vertex** ∗, GB_GRAPH §10.
verbose: **long**, GB_GRAPH §5.
Vertex = **struct**, GB_GRAPH §9.
vertices: **Vertex** ∗, GB_GRAPH §20.

9. ⟨Initialize *mapping* to a random permutation 9⟩ ≡
```
  steps = score = 0;
  for (k = 0; k < n; k++) {
    j = gb_unif_rand(k + 1);
    mapping[k] = mapping[j];
    mapping[j] = k;
  }
  for (j = 1; j < n; j++)
    for (k = 0; k < j; k++) score += mat[mapping[j]][mapping[k]];
  if (verbose > 1) {
    printf("\nInitial␣permutation:\n");
    for (k = 0; k < n; k++) printf("␣%s\n", sec_name(k));
  }
```
This code is used in section 8.

10. If we move, say, *mapping*[5] to *mapping*[3] and shift the previous entries *mapping*[3] and *mapping*[4] right one, the score decreases by

$$del[mapping[5]][mapping[3]] + del[mapping[5]][mapping[4]].$$

Similarly, if we move *mapping*[5] to *mapping*[7] and shift the previous entries *mapping*[6] and *mapping*[7] left one, the score decreases by

$$del[mapping[6]][mapping[5]] + del[mapping[7]][mapping[5]].$$

The number of possible moves is $(n-1)^2$. Our job is to find the one that makes the score decrease, but by as little as possible (or, if *greedy* ≠ 0, to make the score decrease as much as possible).

⟨Figure out the next move to make; **break** if at local optimum 10⟩ ≡
```
  best_d = greedy ? 0 : INF;
  best_k = −1;
  for (k = 0; k < n; k++) { register long d = 0;
    for (j = k − 1; j ≥ 0; j−−) {
      d += del[mapping[k]][mapping[j]];
      ⟨Record the move from k to j, if d is better than best_d 11⟩;
    }
    d = 0;
    for (j = k + 1; j < n; j++) {
      d += del[mapping[j]][mapping[k]];
      ⟨Record the move from k to j, if d is better than best_d 11⟩;
    }
  }
  if (best_k < 0) break;
```
This code is used in section 8.

11. ⟨Record the move from k to j, if d is better than best_d 11⟩ ≡
```
  if (d > 0 ∧ (greedy ? d > best_d : d < best_d)) {
    best_k = k; best_j = j; best_d = d;
  }
```
This code is used in section 10 (twice).

12. ⟨ Global variables 3 ⟩ +≡

  long $best\_d$;   /∗ best improvement seen so far on this step ∗/

  long $best\_k$, $best\_j$;   /∗ moving $best\_k$ to $best\_j$ improves by $best\_d$ ∗/

13. ⟨ Take the next step 13 ⟩ ≡

  if ($verbose > 1$)

    $printf$ ("Now␣move␣%s␣to␣the␣%s,␣past\n", sec_name ($best\_k$), $best\_j < best\_k$? "left" : "right");

  $j = best\_k$;

  $k = mapping[j]$;

  do {

    if ($best\_j < best\_k$) $mapping[j] = mapping[j-1], j--$;

    else $mapping[j] = mapping[j+1], j++$;

    if ($verbose > 1$) $printf$ ("␣␣␣␣%s␣(%ld)\n", sec_name (j),

        $best\_j < best\_k$? $del[mapping[j+1]][k]$: $del[k][mapping[j-1]]$);

  } **while** ($j \neq best\_j$);

  $mapping[j] = k$;

  $score\ -= best\_d$;

  $steps\ ++$;

This code is used in section 8.

14. How well does cautious descent work? In this application, it is definitely too cautious. For example, after lots of computation with the default settings, it comes up with a pretty good value (457342), but only after taking 39,418 steps! Then (if $r > 1$) it tries again and stops with 461584 after 47,634 steps. The greedy algorithm with the same starting permutations obtains the local minimum 457408 after only 93 steps, then 460411 after 83 steps. The greedy algorithm tends to find solutions that are a bit inferior, but it is so much faster that it allows us to run many more experiments. After 20 trials with the default settings, it finds a permutation with only 456315 below the diagonal, and after about 250 more it reduces this upper bound to 456295. (Gerhard Reinelt has proved, via branch-and-cut, that 456295 is in fact optimum.)

The method of stratified greed, which is illustrated in the FOOTBALL module, should do better than the ordinary greedy algorithm; and interesting results can be expected when stratified greed is compared also to other methods like simulated annealing and genetic breeding. Comparisons should be made by seeing which method can come up with the best upper bound after calculating for a given number of mems (see MILES_SPAN). The upper bound obtained in any run is a random variable, so several independent trials of each method should be made.

Question: Suppose we divide the vertices into two subsets and prescribe a fixed permutation on each subset. Is it NP-complete to find the optimum way to merge these two permutations—i.e., to find a permutation, extending the given ones, that has the smallest below-diagonal sum?

---

del: **long** [ ][ ], §3.

gb_unif_rand: **long** ( ), GB_FLIP §12.

$greedy$: **long**, §2.

INF = #7fffffff, §3.

j: **register long**, §2.

k: **register long**, §2.

$mapping$: **long** [ ], §7.

mat: **long** [ ][ ], §3.

n: **unsigned long**, §2.

$printf$: **int** ( ), <stdio.h>.

r: **unsigned long**, §2.

$score$: **long**, §7.

sec_name = **macro** ( ), §7.

$steps$: **long**, §7.

$verbose$: **long**, GB_GRAPH §5.

FOOTBALL ────────────────────────────────

Important: Before reading FOOTBALL, please read or at least skim the program for GB_GAMES.

1. Introduction. This demonstration program uses graphs constructed by the GB_GAMES module to produce an interactive program called `football`, which finds preposterously long chains of scores to "prove" that one given team might outrank another by a huge margin.

The program prompts you for a starting team. If you simply type ⟨return⟩, it exits; otherwise you should enter a team name (e.g., '`Stanford`') before typing ⟨return⟩.

Then the program prompts you for another team. If you simply type ⟨return⟩ at this point, it will go back and ask for a new starting team; otherwise you should specify another name (e.g., '`Harvard`').

Then the program finds and displays a chain from the starting team to the other one. For example, you might see the following:

```
Oct 06:   Stanford Cardinal 36, Notre Dame Fighting Irish 31 (+5)
Oct 20:   Notre Dame Fighting Irish 29, Miami Hurricanes 20 (+14)
Jan 01:   Miami Hurricanes 46, Texas Longhorns 3 (+57)
Nov 03:   Texas Longhorns 41, Texas Tech Red Raiders 22 (+76)
Nov 17:   Texas Tech Red Raiders 62, Southern Methodist Mustangs 7 (+131)
Sep 08:   Southern Methodist Mustangs 44, Vanderbilt Commodores 7 (+168)
            ⋮
Nov 10:   Cornell Big Red 41, Columbia Lions 0 (+2188)
Sep 15:   Columbia Lions 6, Harvard Crimson 9 (+2185)
```

The chain isn't necessarily optimal; it's just this particular program's best guess. Another chain, which establishes a victory margin of +2279 points, can in fact be produced by modifying this program to search back from Harvard instead of forward from Stanford. Algorithms that find even better chains should be fun to invent.

Actually this program has two variants. If you invoke it by saying simply '`football`', you get chains found by a simple "greedy algorithm." But if you invoke it by saying '`football` ⟨number⟩', assuming UNIX command-line conventions, the program works harder. Higher values of ⟨number⟩ do more calculation and tend to find better chains. For example, the simple greedy algorithm favors Stanford over Harvard by only 781; `football 10` raises this to 1895; the example above corresponds to `football 4000`.

2. Here is the general program layout, as seen by the C compiler:

```
#include "gb_graph.h"      /* the standard GraphBase data structures */
#include "gb_games.h"      /* the routine that sets up the graph of scores */
#include "gb_flip.h"       /* random number generator */
```
⟨ Preprocessor definitions ⟩

⟨ Type declarations 10 ⟩
⟨ Global variables 4 ⟩
⟨ Subroutines 7 ⟩

main(*argc*, *argv*)
      int *argc*;   /* the number of command-line arguments */
      char *argv*[];   /* an array of strings containing those arguments */
{
    ⟨ Scan the command-line options 3 ⟩;
    ⟨ Set up the graph 5 ⟩;
    while (1) {
      ⟨ Prompt for starting team and goal team; **break** if none given 6 ⟩;
      ⟨ Find a chain from *start* to *goal*, and print it 9 ⟩;
    }
    return 0;   /* normal exit */
}
```

**3.** Let's deal with UNIX-dependent stuff first. The rest of this program should work without change on any operating system.

⟨ Scan the command-line options 3 ⟩ ≡
    **if** (*argc* ≡ 3 ∧ *strcmp*(*argv*[2], "-v") ≡ 0) *verbose* = *argc* = 2;    /* secret option */
    **if** (*argc* ≡ 1) *width* = 0;
    **else if** (*argc* ≡ 2 ∧ *sscanf*(*argv*[1], "%ld", &*width*) ≡ 1) {
      **if** (*width* < 0) *width* = −*width*;    /* a UNIX user might have used a hyphen */
    } **else** {
      *fprintf*(*stderr*, "Usage:␣%s␣[searchwidth]\n", *argv*[0]);
      **return** −2;
    }

This code is used in section 2.

**4.** ⟨ Global variables 4 ⟩ ≡
    **long** *width*;    /* number of cases examined per stratum */
    **Graph** *\*g*;    /* the graph containing score information */
    **Vertex** *\*u*, *\*v*;    /* vertices of current interest */
    **Arc** *\*a*;    /* arc of current interest */
    **Vertex** *\*start*, *\*goal*;    /* teams specified by the user */
    **long** *mm*;    /* counter used only in *verbose* mode */

See also sections 11, 20, and 29.

This code is used in section 2.

---

**Arc** = **struct**, GB_GRAPH §10.    *sscanf*: **int** ( ), <stdio.h>.    *verbose*: **long**, GB_GRAPH §5.
*fprintf*: **int** ( ), <stdio.h>.    *stderr*: **FILE** *\**, <stdio.h>.    **Vertex** = **struct**, GB_GRAPH §9.
**Graph** = **struct**, GB_GRAPH §20.    *strcmp*: **int** ( ), <string.h>.
```

5. An arc from u to v in the graph generated by *games* has a *len* field equal to the number of points scored by u against v. For our purposes we want also a *del* field, which gives the difference between the number of points scored by u and the number of points scored by v in that game.

#define *del*   *a.I*   /* *del* info appears in utility field *a* of an **Arc** record */

⟨ Set up the graph 5 ⟩ ≡

```
  g = games(0 L, 0 L, 0 L, 0 L, 0 L, 0 L, 0 L, 0 L);
     /* this default graph has the data for the entire 1990 season */
  if (g ≡ Λ) {
    fprintf(stderr, "Sorry,⎵can't⎵create⎵the⎵graph!⎵(error⎵code⎵%ld)\n", panic_code);
    return −1;
  }
  for (v = g→vertices; v < g→vertices + g→n; v++)
    for (a = v→arcs; a; a = a→next)
      if (a→tip > v) {      /* arc a + 1 is the mate of arc a iff a→tip > v */
        a→del = a→len − (a + 1)→len;
        (a + 1)→del = −a→del;
      }
```

This code is used in section 2.

6. Terminal interaction. While we're getting trivialities out of the way, we might as well take care of the simple dialog that transpires between this program and the user.

⟨ Prompt for starting team and goal team; **break** if none given 6 ⟩ ≡
> $putchar$ (`'\n'`);   /∗ make a blank line for visual punctuation ∗/
> *restart*:   /∗ if we avoid this label, the **break** command will be broken ∗/
>   **if** (($start$ = $prompt\_for\_team$ (`"Starting"`)) ≡ Λ) **break**;
>   **if** (($goal$ = $prompt\_for\_team$ (`"␣␣␣Other"`)) ≡ Λ) **goto** *restart*;
>   **if** ($start$ ≡ $goal$) { $printf$ (`"␣(Um,␣please␣give␣me␣the␣names␣of␣two␣DISTINCT␣teams.)\n"`);
>   **goto** *restart*;
>   }

This code is used in section 2.

7. The user must spell team names exactly as they appear in the file **games.dat**. Thus, for example, 'Berkeley' and 'Cal' don't work; it has to be '**California**'. Similarly, a person must type '**Pennsylvania**' instead of '**Penn**', '**Nevada-Las Vegas**' instead of '**UNLV**'. A backslash is necessary in '**Texas A\&M**'.

⟨ Subroutines 7 ⟩ ≡
>   **Vertex** ∗*prompt_for_team*(s)
>   **char** ∗s;   /∗ string used in prompt message ∗/
>   { **register char** ∗q;   /∗ current position in *buffer* ∗/
>   **register Vertex** ∗v;   /∗ current vertex being examined in sequential search ∗/
>   **char** $buffer$[30];   /∗ a line of input ∗/
>   **while** (1) { $printf$ (`"%s␣team:␣"`, s);
> $fflush$ ($stdout$);   /∗ make sure the user sees the prompt ∗/
> $fgets$ ($buffer$, 30, $stdin$);
>   **if** ($buffer$[0] ≡ `'\n'`) **return** Λ;   /∗ the user just hit ⟨return⟩ ∗/
> $buffer$[29] = `'\n'`;
>   **for** (q = $buffer$; ∗q ≠ `'\n'`; q++) ;   /∗ scan to end of input ∗/
>   ∗q = `'\0'`;
>   **for** (v = g⃗$vertices$; v < g⃗$vertices$ + g⃗n; v++)
>   **if** ($strcmp$ ($buffer$, v⃗$name$) ≡ 0) **return** v;   /∗ aha, we found it ∗/
> $printf$ (`"␣(Sorry,␣I␣don't␣know␣any␣team␣by␣that␣name.)\n"`);
> $printf$ (`"␣(One␣team␣I␣do␣know␣is␣%s...)\n"`, (g⃗$vertices$ + gb_unif_rand (g⃗n))⃗$name$);
>   }
>   }

See also section 13.

This code is used in section 2.

---

a: **Arc** ∗, §4.
a: **util**, GB_GRAPH §10.
Arc = **struct**, GB_GRAPH §10.
arcs: **Arc** ∗, GB_GRAPH §9.
fflush: **int** ( ), <stdio.h>.
fgets: **char** ∗( ), <stdio.h>.
fprintf: **int** ( ), <stdio.h>.
g: **Graph** ∗, §4.
games: **Graph** ∗( ), GB_GAMES §7.
gb_unif_rand: **long** ( ), GB_FLIP §12.

goal: **Vertex** ∗, §4.
l: **long**, GB_GRAPH §8.
len: **long**, GB_GRAPH §10.
n: **long**, GB_GRAPH §20.
name: **char** ∗, GB_GRAPH §9.
next: **Arc** ∗, GB_GRAPH §10.
panic_code: **long**, GB_GRAPH §5.
printf: **int** ( ), <stdio.h>.
putchar = macro ( ), <stdio.h>.

start: **Vertex** ∗, §4.
stderr: **FILE** ∗, <stdio.h>.
stdin: **FILE** ∗, <stdio.h>.
stdout: **FILE** ∗, <stdio.h>.
strcmp: **int** ( ), <string.h>.
tip: **Vertex** ∗, GB_GRAPH §10.
v: **Vertex** ∗, §4.
Vertex = **struct**, GB_GRAPH §9.
vertices: **Vertex** ∗, GB_GRAPH §20.

8. Greed. This program's primary task is to find the longest possible simple path from *start* to *goal*, using *del* as the length of each arc in the path. This is an NP-complete problem, and the number of possibilities is pretty huge, so the present program is content to use heuristics that are reasonably easy to compute. (Researchers are hereby challenged to come up with better heuristics. Does simulated annealing give good results? How about genetic algorithms?)

Perhaps the first approach that comes to mind is a simple "greedy" approach in which each step takes the largest possible *del* that doesn't prevent us from eventually getting to *goal*. So that's the method we will implement first.

9. ⟨Find a chain from *start* to *goal*, and print it 9⟩ ≡
  ⟨Initialize the allocation of auxiliary memory 12⟩;
  if (*width* ≡ 0) ⟨Use a simple-minded greedy algorithm to find a chain from *start* to *goal* 17⟩
  else ⟨Use a stratified heuristic to find a chain from *start* to *goal* 19⟩;
  ⟨Print the solution corresponding to *cur_node* 15⟩;
  ⟨Recycle the auxiliary memory used 14⟩;

This code is used in section 2.

10. We might as well use data structures that are more general than we need, in anticipation of a more complex heuristic that will be implemented later. The set of all possible solutions can be viewed as a backtrack tree in which the branches from each node are the games that can possibly follow that node. We will examine a small part of that gigantic tree.

⟨Type declarations 10⟩ ≡
  typedef struct node_struct {
      Arc **game*;   /* game from the current team to the next team */
      long *tot_len*;   /* accumulated length from *start* to here */
      struct node_struct **prev*;   /* node that gave us the current team */
      struct node_struct **next*;   /* list pointer to node in same stratum (see below) */
  } **node**;

This code is used in section 2.

11. ⟨Global variables 4⟩ +≡
  Area *node_storage*;   /* working storage for heuristic calculations */
  node **next_node*;   /* where the next node is slated to go */
  node **bad_node*;   /* end of current allocation block */
  node **cur_node*;   /* current node of particular interest */

12. ⟨Initialize the allocation of auxiliary memory 12⟩ ≡
  next_node = *bad_node* = Λ;

This code is used in section 9.

13. ⟨Subroutines 7⟩ +≡
  node **new_node*(*x*, *d*)
      node **x*;   /* an old node that the new node will call *prev* */
      long *d*;   /* incremental change to *tot_len* */
  {
    if (*next_node* ≡ *bad_node*) {
      next_node = *gb_typed_alloc*(1000, **node**, *node_storage*);
      if (*next_node* ≡ Λ) **return** Λ;   /* we're out of space */

$$bad\_node = next\_node + 1000;$$
    }
    $next\_node \rightarrow prev = x;$
    $next\_node \rightarrow tot\_len = (x\ ?\ x \rightarrow tot\_len : 0) + d;$
    return $next\_node\mathbin{++};$
}

14. ⟨ Recycle the auxiliary memory used 14 ⟩ ≡
    $gb\_free(node\_storage);$

This code is used in section 9.

15. When we're done, $cur\_node \rightarrow game \rightarrow tip$ will be the *goal* vertex, and we can get back to the *start* vertex by following *prev* links from cur_node. It looks better to print the answers from *start* to *goal*, so maybe we should have changed our algorithm to go the other way.

But let's not worry over trifles. It's easy to change the order of a linked list. The secret is simply to think of the list as a stack, from which we pop all the elements off to another stack; the new stack has the elements in reverse order.

⟨ Print the solution corresponding to cur_node 15 ⟩ ≡
    $next\_node = \Lambda;$ /∗ now we'll use $next\_node$ as top of temporary stack ∗/
    do { **register node** ∗$t;$
      $t = cur\_node;$
      $cur\_node = t \rightarrow prev;$ /∗ pop ∗/
      $t \rightarrow prev = next\_node;$
      $next\_node = t;$ /∗ push ∗/
    } **while** $(cur\_node);$
    for $(v = start;\ v \neq goal;\ v = u, next\_node = next\_node \rightarrow prev)$ {
      $a = next\_node \rightarrow game;$
      $u = a \rightarrow tip;$
      ⟨ Print the score of game a between v and u 16 ⟩;
      $printf(\texttt{"\textvisiblespace(\%+1d)\textbackslash n"}, next\_node \rightarrow tot\_len);$
    }

This code is used in section 9.

---

a: **Arc** ∗, §4.
Arc = **struct**, GB_GRAPH §10.
Area, GB_GRAPH §12.
$del = a.I$, §5.
gb_free: **void** ( ), GB_GRAPH §16.

gb_typed_alloc = macro ( ),
    GB_GRAPH §11.
$goal$: **Vertex** ∗, §4.
$printf$: **int** ( ), <stdio.h>.
$start$: **Vertex** ∗, §4.

tip: **Vertex** ∗, GB_GRAPH §10.
u: **Vertex** ∗, §4.
v: **Vertex** ∗, §4.
$width$: **long**, §4.

16. ⟨ Print the score of game a between v and u 16 ⟩ ≡
  { **register long** $d = a\text{→}date$;   /∗ date of the game, 0 means Aug 26 ∗/
    if $(d \leq 5)$ *printf* ("␣Aug␣%02ld", $d + 26$);
    else if $(d \leq 35)$ *printf* ("␣Sep␣%02ld", $d - 5$);
    else if $(d \leq 66)$ *printf* ("␣Oct␣%02ld", $d - 35$);
    else if $(d \leq 96)$ *printf* ("␣Nov␣%02ld", $d - 66$);
    else if $(d \leq 127)$ *printf* ("␣Dec␣%02ld", $d - 96$);
    else *printf* ("␣Jan␣01");   /∗ $d = 128$ ∗/
    printf (":␣%s␣%s␣%ld,␣%s␣%s␣%ld", $v\text{→}name$, $v\text{→}nickname$, $a\text{→}len$, $u\text{→}name$, $u\text{→}nickname$, $a\text{→}len - a\text{→}del$);
  }

This code is used in section 15.

17. We can't just move from v to any adjacent vertex; we can go only to a vertex from which *goal* can be reached without touching v or any other vertex already used on the path from *start*.

Furthermore, if the locally best move from v is directly to *goal*, we don't want to make that move unless it's our last chance; we can probably do better by making the chain longer. Otherwise, for example, a chain between a team and its worst opponent would consist of only a single game.

To keep track of untouchable vertices, we use a utility field called *blocked* in each vertex record. Another utility field, *valid*, will be set to a validation code in each vertex that still leads to the goal.

#define *blocked* *u.I*
#define *valid* *v.V*

⟨ Use a simple-minded greedy algorithm to find a chain from *start* to *goal* 17 ⟩ ≡
  {
    for $(v = g\text{→}vertices;\ v < g\text{→}vertices + g\text{→}n;\ v\text{++})$ $v\text{→}blocked = 0, v\text{→}valid = \Lambda$;
    $cur\_node = \Lambda$;
    for $(v = start;\ v \neq goal;\ v = cur\_node\text{→}game\text{→}tip)$ { **register long** $d = -10000$;
      register Arc ∗*best_arc*;   /∗ arc that achieves $del = d$ ∗/
      register Arc ∗*last_arc*;   /∗ arc that goes directly to *goal* ∗/
      $v\text{→}blocked = 1$;
      $cur\_node = new\_node(cur\_node, 0_{\text{L}})$;
      if $(cur\_node \equiv \Lambda)$ {
        fprintf (*stderr*, "Oops,␣there␣isn't␣enough␣memory!\n"); **return** -2;
      }
      ⟨ Set $u\text{→}valid = v$ for all u to which v might now move 18 ⟩;
      for $(a = v\text{→}arcs;\ a;\ a = a\text{→}next)$
        if $(a\text{→}del > d \wedge a\text{→}tip\text{→}valid \equiv v)$
          if $(a\text{→}tip \equiv goal)$ $last\_arc = a$;
          else $best\_arc = a, d = a\text{→}del$;
      $cur\_node\text{→}game = (d \equiv -10000\ ?\ last\_arc : best\_arc)$;   /∗ use *last_arc* as a last resort ∗/
      $cur\_node\text{→}tot\_len\ += cur\_node\text{→}game\text{→}del$;
    }
  }

This code is used in section 9.

18. A standard marking algorithm supplies the final missing link in our algorithm.
#define *link* *w.V*

⟨ Set u→$valid = v$ for all u to which v might now move 18 ⟩ ≡

```
u = goal;      /* u will be the top of a stack of nodes to be explored */
u→link = Λ;
u→valid = v;
do {
    for (a = u→arcs, u = u→link;  a;  a = a→next)
        if (a→tip→blocked ≡ 0 ∧ a→tip→valid ≠ v) {
            a→tip→valid = v;      /* mark a→tip reachable from goal */
            a→tip→link = u;
            u = a→tip;      /* push it on the stack, so that its successors will be marked too */
        }
} while (u);
```

This code is used in section 17.

---

a: **Arc** *, §4.

Arc = **struct**, GB_GRAPH §10.

arcs: **Arc** *, GB_GRAPH §9.

cur_node: **node** *, §11.

date = *b.I*, GB_GAMES §5.

del = *a.I*, §5.

fprintf: **int** ( ), <stdio.h>.

g: **Graph** *, §4.

game: **Arc** *, §10.

goal: **Vertex** *, §4.

I: **long**, GB_GRAPH §8.

len: **long**, GB_GRAPH §10.

n: **long**, GB_GRAPH §20.

name: **char** *, GB_GRAPH §9.

new_node: **node** *( ), §13.

next: **Arc** *, GB_GRAPH §10.

nickname = *y.S*, GB_GAMES §4.

printf: **int** ( ), <stdio.h>.

start: **Vertex** *, §4.

stderr: **FILE** *, <stdio.h>.

tip: **Vertex** *, GB_GRAPH §10.

tot_len: **long**, §10.

u: **Vertex** *, §4.

u: **util**, GB_GRAPH §9.

v: **Vertex** *, §4.

v: **util**, GB_GRAPH §9.

V: **Vertex** *, GB_GRAPH §8.

vertices: **Vertex** *, GB_GRAPH §20.

w: **util**, GB_GRAPH §9.

19. Stratified greed.   One approach to better chains is the following algorithm, motivated by similar ideas of Pang Chen [Ph.D. thesis, Stanford University, 1989]: Suppose the nodes of a (possibly huge) backtrack tree are classified into a (fairly small) number of strata, by a function h with the property that $h(\text{child}) < h(\text{parent})$ and $h(\text{goal}) = 0$. Suppose further that we want to find a node x that maximizes a given function $f(x)$, where it is reasonable to believe that $f(\text{child})$ will be relatively large among nodes in a child's stratum only if $f(\text{parent})$ is relatively large in the parent's stratum. Then it makes sense to restrict backtracking to, say, the top w nodes of each stratum, ranked by their f values.

The greedy algorithm already described is a special case of this general approach, with $w = 1$ and with $h(x) = -(\text{length of chain leading to } x)$. The refined algorithm we are about to describe uses a general value of w and a somewhat more relevant stratification function: Given a node x of the backtrack tree for longest paths, corresponding to a path from *start* to a certain vertex $u = u(x)$, we will let $h(x)$ be the number of vertices that lie between u and *goal* (in the sense that the simple path from *start* to u can be extended until it passes through such a vertex and then all the way to *goal*).

Here is the top level of the stratified greedy algorithm. We maintain a linked list of nodes for each stratum, that is, for each possible value of h. The number of nodes required is bounded by w times the number of strata.

⟨ Use a stratified heuristic to find a chain from *start* to *goal* 19 ⟩ ≡
```
    {
      ⟨ Make list[0] through list[n − 1] empty 21 ⟩;
      cur_node = Λ;      /* Λ represents the root of the backtrack tree */
      m = g⁀n − 1;       /* the highest stratum not yet fully explored */
      do {
        ⟨ Place each child x of cur_node into list[h(x)], retaining at most width nodes of maximum
            tot_len on each list 27 ⟩;
        while (list[m] ≡ Λ) ⟨ Decrease m and get ready to explore another list 23 ⟩;
        cur_node = list[m];
        list[m] = cur_node⁀next;      /* remove a node from highest remaining stratum */
        if (verbose) ⟨ Print "verbose" info about cur_node 24 ⟩;
      } while (m > 0);      /* exactly one node should be in list[0] (see below) */
    }
```
This code is used in section 9.

20.   The calculation of $h(x)$ is somewhat delicate, and we will defer it for a moment. But the list manipulation is easy, so we can finish it quickly while it's fresh in our minds.

#define MAX_N 120   /* the number of teams in `games.dat` */

⟨ Global variables 4 ⟩ +≡
```
    node *list[MAX_N];     /* the best nodes known in given strata */
    long size[MAX_N];      /* the number of elements in a given list */
    long m, h;      /* current lists of interest */
    node *x;      /* a child of cur_node */
```

21.   ⟨ Make list[0] through list[n − 1] empty 21 ⟩ ≡
```
    for (m = 0; m < g⁀n; m++) list[m] = Λ, size[m] = 0;
```
This code is used in section 19.

22. The lists are maintained in order by *tot_len*, with the largest *tot_len* value at the end so that we can easily delete the smallest.

When $h = 0$, we retain only one node instead of *width* different nodes, because we are interested in only one solution.

⟨ Place node x into *list*[h], retaining at most *width* nodes of maximum *tot_len* 22 ⟩ ≡
```
    if ((h > 0 ∧ size[h] ≡ width) ∨ (h ≡ 0 ∧ size[0] > 0)) {
        if (x→tot_len ≤ list[h]→tot_len) goto done;      /* drop node x */
        list[h] = list[h]→next;        /* drop one node from list[h] */
    } else  size[h]++;
    { register node *p, *q;      /* node in list and its predecessor */
        for (p = list[h], q = Λ;  p;  q = p, p = p→next)  if (x→tot_len ≤ p→tot_len)  break;
        x→next = p;
        if (q)  q→next = x;  else  list[h] = x;
    }
done: ;
```
This code is used in section 27.

23. We reverse the list so that large entries will tend to go in first.

⟨ Decrease m and get ready to explore another list 23 ⟩ ≡
```
    { register node *r = Λ, *s = list[--m], *t;
        while (s)  t = s→next, s→next = r, r = s, s = t;
        list[m] = r;
        mm = 0;      /* mm is an index for "verbose" printing */
    }
```
This code is used in section 19.

24. ⟨ Print "verbose" info about *cur_node* 24 ⟩ ≡
```
    {
        cur_node→next = (node *) ((++mm ≪ 8) + m);      /* pack an ID for this node */
        printf("[%lu,%lu]=[%lu,%lu]&%s␣(%+ld)\n", m, mm,
            cur_node→prev ? ((unsigned long) cur_node→prev→next) & #ff : 0L,
            cur_node→prev ? ((unsigned long) cur_node→prev→next) ≫ 8 : 0L,
            cur_node→game→tip→name, cur_node→tot_len);
    }
```
This code is used in section 19.

25. Incidentally, it is plausible to conjecture that the stratified algorithm always beats the simple greedy algorithm, but that conjecture is false. For example, the greedy algorithm is able to rank Harvard over Stanford by 1529, while the stratified algorithm achieves only 1527 when *width* $= 1$. On the other hand, the greedy algorithm often fails miserably; when comparing two Ivy League teams, it doesn't find a way to break out of the Ivy and Patriot Leagues.

---

cur_node: **node** *, §11.
g: **Graph** *, §4.
game: **Arc** *, §10.
goal: **Vertex** *, §4.
mm: **long**, §4.
n: **long**, GB_GRAPH §20.

name: **char** *, GB_GRAPH §9.
next: **struct node_struct** *, §10.
node = **struct node_struct**, §10.
prev: **struct node_struct** *, §10.
printf: **int** (), <stdio.h>.

start: **Vertex** *, §4.
tip: **Vertex** *, GB_GRAPH §10.
tot_len: **long**, §10.
verbose: **long**, GB_GRAPH §5.
width: **long**, §4.

26. Bicomponents revisited. How difficult is it to compute the function h? Given a connected graph G with two distinguished vertices u and v, we want to count the number of vertices that might appear on a simple path from u to v. (This is *not* the same as the number of vertices reachable from both u and v. For example, consider a "claw" graph with four vertices $\{u, v, w, x\}$ and with edges only from x to the other three vertices; in this graph w is reachable from u and v, but it is not on any simple path between them.)

The best way to solve this problem is probably to compute the bicomponents of G, or least to compute some of them. Another demo program, BOOK_COMPONENTS, explains the relevant theory in some detail, and we will assume familiarity with that algorithm in the present discussion.

Let us imagine extending G to a slightly larger graph G^+ by adding a dummy vertex o that is adjacent only to v. Suppose we determine the bicomponents of G^+ by depth-first search starting at o. These bicomponents form a tree rooted at the bicomponent that contains just o and v. The number of vertices on paths between u and v, not counting v itself, is then the number of vertices in the bicomponent containing u and in any other bicomponents between that one and the root.

Strictly speaking, each articulation point belongs to two or more bicomponents. But we will assign each articulation point to its bicomponent that is nearest the root of the tree; then the vertices of each bicomponent are precisely the vertices output in bursts by the depth-first procedure. The bicomponents we want to enumerate are B_1, B_2, ..., B_k, where B_1 is the bicomponent containing u and B_{j+1} is the bicomponent containing the articulation point associated with B_j; we stop at B_k when its associated articulation point is v. (Often $k = 1$.)

The "children" of a given graph G are obtained by removing vertex u and by considering paths from u' to v, where u' is a vertex formerly adjacent to u; thus u' is either in B_1 or it is B_1's associated articulation point. Removing u will, in general, split B_1 into a tree of smaller bicomponents, but $B_2, \ldots, B_k$ will be unaffected. The implementation below does not take full advantage of this observation, because the amount of memory required to avoid recomputation would probably be prohibitive.

27. The following program is copied almost verbatim from BOOK_COMPONENTS. Instead of repeating the commentary that appears there, we will mention only the significant differences. One difference is that we start the depth-first search at a definite place, the *goal*.

⟨ Place each child x of *cur_node* into *list*$[h(x)]$, retaining at most *width* nodes of maximum *tot_len* on each list 27 ⟩ ≡
  ⟨ Make all vertices unseen and all arcs untagged, except for vertices that have already been used in steps leading up to *cur_node* 28 ⟩;
  ⟨ Perform a depth-first search with *goal* as the root, finding bicomponents and determining the number of vertices accessible between any given vertex and *goal* 30 ⟩;
  for $(a = (cur\_node \ ? \ cur\_node{\rightarrow}game{\rightarrow}tip : start){\rightarrow}arcs; \ a; \ a = a{\rightarrow}next)$
    if $((u = a{\rightarrow}tip){\rightarrow}untagged \equiv \Lambda)$ {   /* *goal* is reachable from u */
      $x = new\_node(cur\_node, a{\rightarrow}del)$;
      if $(x \equiv \Lambda)$ { *fprintf*(*stderr*, "Oops,␣there␣isn't␣enough␣memory!\n"); **return** -3; }
      $x{\rightarrow}game = a$;
      ⟨ Set h to the number of vertices on paths between u and *goal* 35 ⟩;
      ⟨ Place node x into *list*$[h]$, retaining at most *width* nodes of maximum *tot_len* 22 ⟩;
    }

This code is used in section 19.

28.   Setting the *rank* field of a vertex to infinity before beginning a depth-first search is tantamount to removing that vertex from the graph, because it tells the algorithm not to look further at such a vertex.

#define *rank*   *z.I*   /* when was this vertex first seen? */
#define *parent* *u.V*   /* who told me about this vertex? */
#define *untagged* *x.A*   /* what is its first untagged arc? */
#define *min* *v.V*   /* how low in the tree can we jump from its mature descendants? */

$\langle$ Make all vertices unseen and all arcs untagged, except for vertices that have already been used in steps
        leading up to *cur_node* 28 $\rangle \equiv$

  for $(v = g\text{-}vertices; \ v < g\text{-}vertices + g\text{-}n; \ v\text{++})$ {
    $v\text{-}rank = 0;$
    $v\text{-}untagged = v\text{-}arcs;$
  }
  for $(x = cur\_node; \ x; \ x = x\text{-}prev) \ x\text{-}game\text{-}tip\text{-}rank = g\text{-}n;$
    /* "infinite" rank (or close enough) */
  $start\text{-}rank = g\text{-}n;$
  $nn = 0;$
  $active\_stack = settled\_stack = \Lambda;$

This code is used in section 27.

29. $\langle$ Global variables 4 $\rangle$ $+\equiv$
  Vertex *$*active\_stack$*;   /* the top of the stack of active vertices */
  Vertex *$*settled\_stack$*;   /* the top of the stack of bicomponents found */
  long *nn*;   /* the number of vertices that have been seen */
  Vertex *dummy*;   /* imaginary parent of *goal*; its *rank* is zero */

---

a: **Arc** *, §4.
A: **Arc** *, GB_GRAPH §8.
arcs: **Arc** *, GB_GRAPH §9.
cur_node: **node** *, §11.
del = *a.I*, §5.
fprintf: **int** ( ), <stdio.h>.
g: **Graph** *, §4.
game: **Arc** *, §10.
goal: **Vertex** *, §4.
h: **long**, §20.
I: **long**, GB_GRAPH §8.

list: **node** *[], §20.
n: **long**, GB_GRAPH §20.
new_node: **node** *( ), §13.
next: **Arc** *, GB_GRAPH §10.
prev: **struct node_struct** *, §10.
start: **Vertex** *, §4.
stderr: **FILE** *, <stdio.h>.
tip: **Vertex** *, GB_GRAPH §10.
tot_len: **long**, §10.
u: **Vertex** *, §4.

u: **util**, GB_GRAPH §9.
v: **Vertex** *, §4.
v: **util**, GB_GRAPH §9.
V: **Vertex** *, GB_GRAPH §8.
Vertex = **struct**, GB_GRAPH §9.
vertices: **Vertex** *, GB_GRAPH §20.
width: **long**, §4.
x: **node** *, §20.
x: **util**, GB_GRAPH §9.
z: **util**, GB_GRAPH §9.

30. The *settled_stack* will contain a list of all bicomponents in the opposite order from which they are discovered. This is the order we'll need later for computing the h function in each bicomponent.

⟨ Perform a depth-first search with *goal* as the root, finding bicomponents and determining the number of vertices accessible between any given vertex and *goal* 30 ⟩ ≡
```
{
    v = goal;
    v→parent = &dummy;
    ⟨ Make vertex v active 31 ⟩;
    do ⟨ Explore one step from the current vertex v, possibly moving to another current vertex and
        calling it v 32 ⟩ while (v ≠ &dummy);
    ⟨ Use settled_stack to put the mutual reachability count for each vertex u in u→parent→rank 34 ⟩;
}
```
This code is used in section 27.

31. ⟨ Make vertex v active 31 ⟩ ≡
```
    v→rank = ++nn;
    v→link = active_stack;
    active_stack = v;
    v→min = v→parent;
```
This code is used in sections 30 and 32.

32. ⟨ Explore one step from the current vertex v, possibly moving to another current vertex and calling it v 32 ⟩ ≡
```
{ register Vertex *u;      /* a vertex adjacent to v */
  register Arc *a = v→untagged;      /* v's first remaining untagged arc, if any */
  if (a) {
      u = a→tip;
      v→untagged = a→next;      /* tag the arc from v to u */
      if (u→rank) {      /* we've seen u already */
          if (u→rank < v→min→rank) v→min = u;      /* non-tree arc, just update v→min */
      } else {      /* u is presently unseen */
          u→parent = v;      /* the arc from v to u is a new tree arc */
          v = u;      /* u will now be the current vertex */
          ⟨ Make vertex v active 31 ⟩;
      }
  } else {      /* all arcs from v are tagged, so v matures */
      u = v→parent;      /* prepare to backtrack in the tree */
      if (v→min ≡ u) ⟨ Remove v and all its successors on the active stack from the tree, and report
              them as a bicomponent of the graph together with u 33 ⟩
      else      /* the arc from u to v has just matured, making v→min visible from u */
          if (v→min→rank < u→min→rank) u→min = v→min;
      v = u;      /* the former parent of v is the new current vertex v */
  }
}
```
This code is used in section 30.

33. When a bicomponent is found, we reset the *parent* field of each vertex so that, afterwards, two vertices will belong to the same bicomponent if and only if they have the same *parent*. (This trick was not used in BOOK_COMPONENTS, but it does appear in the similar algorithm of ROGET_COMPONENTS.) The new parent, v, will represent that bicomponent in subsequent computation; we put it onto *settled_stack*. We also reset $v \rightarrow rank$ to be the bicomponent's size, plus a constant large enough to keep the algorithm from getting confused. (Vertex u might still have untagged arcs leading into this bicomponent; we need to keep the ranks at least as big as the rank of $u \rightarrow min$.) Notice that $v \rightarrow min$ is u, the articulation point associated with this bicomponent. Later the *rank* field will contain the sum of all counts between here and the root.

We don't have to do anything when $v \equiv goal$; the trivial root bicomponent always comes out last.

⟨ Remove v and all its successors on the active stack from the tree, and report them as a bicomponent of the graph together with u 33 ⟩ ≡

```
{ if (v ≠ goal) { register Vertex *t;      /* runs through the vertices of the new bicomponent */
    long c = 0;      /* the number of vertices removed */
    t = active_stack;
    while (t ≠ v) {
      c++; t→parent = v; t = t→link;
    }
    active_stack = v→link;
    v→parent = v;
    v→rank = c + g→n;      /* the true component size is c + 1 */
    v→link = settled_stack; settled_stack = v;
  }
}
```

This code is used in section 32.

34. So here's how we sum the ranks. When we get to this step, the *settled* stack contains all bicomponent representatives except *goal* itself.

⟨ Use *settled_stack* to put the mutual reachability count for each vertex u in $u \rightarrow parent \rightarrow rank$ 34 ⟩ ≡

```
while (settled_stack) {
  v = settled_stack; settled_stack = v→link;
  v→rank += v→min→parent→rank + 1 - g→n;
}      /* note that goal→parent→rank = 0 */
```

This code is used in section 30.

35. And here's the last piece of the puzzle.

⟨ Set h to the number of vertices on paths between u and *goal* 35 ⟩ ≡

```
h = u→parent→rank;
```

This code is used in section 27.

---

active_stack: **Vertex** *, §29.
Arc = **struct**, GB_GRAPH §10.
dummy: **Vertex**, §29.
g: **Graph** *, §4.
goal: **Vertex** *, §4.
h: **long**, §20.

link = $w.V$, §18.
min = $v.V$, §28.
n: **long**, GB_GRAPH §20.
next: **Arc** *, GB_GRAPH §10.
nn: **long**, §29.
parent = $u.V$, §28.

rank = $z.I$, §28.
settled_stack: **Vertex** *, §29.
tip: **Vertex** *, GB_GRAPH §10.
untagged = $x.A$, §28.
v: **Vertex** *, §4.
Vertex = **struct**, GB_GRAPH §9.

GB_BASIC

Important: Before reading GB_BASIC, please read or at least skim the program for GB_GRAPH.

1. Introduction. This GraphBase module contains six subroutines that generate standard graphs of various types, together with six routines that combine or transform existing graphs.

Simple examples of the use of these routines can be found in the demonstration programs QUEEN and QUEEN_WRAP.

⟨ gb_basic.h 1 ⟩ ≡
  extern Graph *board*( );   /* moves on generalized chessboards */
  extern Graph *simplex*( );   /* generalized triangular configurations */
  extern Graph *subsets*( );   /* patterns of subset intersection */
  extern Graph *perms*( );   /* permutations of a multiset */
  extern Graph *parts*( );   /* partitions of an integer */
  extern Graph *binary*( );   /* binary trees */

  extern Graph *complement*( );   /* the complement of a graph */
  extern Graph *gunion*( );   /* the union of two graphs */
  extern Graph *intersection*( );   /* the intersection of two graphs */
  extern Graph *lines*( );   /* the line graph of a graph */
  extern Graph *product*( );   /* the product of two graphs */
  extern Graph *induced*( );   /* a graph induced from another */

See also sections 7, 36, 41, 54, 63, 94, 100, 102, and 104.

2. The C file gb_basic.c has the following overall shape:

#include "gb_graph.h"   /* we use the GB_GRAPH data structures */
  ⟨ Preprocessor definitions ⟩
  ⟨ Private variables 3 ⟩
  ⟨ Basic subroutines 8 ⟩
  ⟨ Applications of basic subroutines 101 ⟩

3. Several of the programs below allocate arrays that will be freed again before the routine is finished.

⟨ Private variables 3 ⟩ ≡
  static Area *working_storage*;

See also sections 5, 10, and 51.

This code is used in section 2.

4. If a graph-generating subroutine encounters a problem, it returns Λ (that is, NULL), after putting a code number into the external variable *panic_code*. This code number identifies the type of failure. Otherwise the routine returns a pointer to the newly created graph, which will be represented with the data structures explained in GB_GRAPH. (The external variable *panic_code* is itself defined in GB_GRAPH.)

#define *panic*(c)
        { *panic_code* = c;
          gb_free(*working_storage*);
          gb_trouble_code = 0;
          return Λ;
        }

5. The names of vertices are sometimes formed from the names of other vertices, or from potentially long sequences of numbers. We assemble them in the *buffer* array, which is sufficiently long that the vast majority of applications will be unconstrained by size limitations. The programs do always make sure that `BUF_SIZE` is not exceeded, but they assume that it is rather large.

#**define** `BUF_SIZE` 4096

⟨ Private variables 3 ⟩ +≡
  static char *buffer*[`BUF_SIZE`];

---

6.   Grids and game boards.   The subroutine call *board*($n1$, $n2$, $n3$, $n4$, *piece*, *wrap*, *directed*) constructs a graph based on the moves of generalized chesspieces on a generalized rectangular board. Each vertex of the graph corresponds to a position on the board. Each arc of the graph corresponds to a move from one position to another.

The first parameters, $n1$ through $n4$, specify the size of the board. If, for example, a two-dimensional board with n_1 rows and n_2 columns is desired, you set $n1 = n_1$, $n2 = n_2$, and $n3 = 0$; the resulting graph will have $n_1 n_2$ vertices. If you want a three-dimensional board with n_3 layers, set $n3 = n_3$ and $n4 = 0$. If you want a 4-D board, put the number of 4th coordinates in $n4$. If you want a d-dimensional board with 2^d positions, set $n1 = 2$ and $n2 = -d$.

In general, the *board* subroutine determines the dimensions by scanning the sequence ($n1$, $n2$, $n3$, $n4$, 0) = ($n_1, n_2, n_3, n_4, 0$) from left to right until coming to the first nonpositive parameter n_{k+1}. If $k = 0$ (i.e., if $n1 \le 0$), the default size 8×8 will be used; this is an ordinary chessboard with 8 rows and 8 columns. Otherwise if $n_{k+1} = 0$, the board will have k dimensions $n_1, \ldots, n_k$. Otherwise we must have $n_{k+1} < 0$; in this case, the board will have $d = |n_{k+1}|$ dimensions, chosen as the first d elements of the infinite periodic sequence ($n_1, \ldots, n_k, n_1, \ldots, n_k, n_1, \ldots$). For example, the specification ($n1$, $n2$, $n3$, $n4$) = $(2, 3, 5, -7)$ is about as tricky as you can get. It produces a seven-dimensional board with dimensions ($n_1, \ldots, n_7$) = $(2, 3, 5, 2, 3, 5, 2)$, hence a graph with $2 \cdot 3 \cdot 5 \cdot 2 \cdot 3 \cdot 5 \cdot 2 = 1800$ vertices.

The *piece* parameter specifies the legal moves of a generalized chesspiece. If *piece* > 0, a move from position u to position v is considered legal if and only if the Euclidean distance between points u and v is equal to $\sqrt{piece}$. For example, if *piece* $= 1$ and if we have a two-dimensional board, the legal moves from (x, y) are to $(x, y \pm 1)$ and $(x \pm 1, y)$; these are the moves of a so-called wazir, the only moves that a king and a rook can both make. If *piece* $= 2$, the legal moves from (x, y) are to $(x \pm 1, y \pm 1)$; these are the four moves that a king and a bishop can both make. (A piece that can make only these moves was called a "fers" in ancient Muslim chess.) If *piece* $= 5$, the legal moves are those of a knight, from (x, y) to $(x \pm 1, y \pm 2)$ or to $(x \pm 2, y \pm 1)$. If *piece* $= 3$, there are no legal moves on a two-dimensional board; but moves from (x, y, z) to $(x \pm 1, y \pm 1, z \pm 1)$ would be legal in three dimensions. If *piece* $= 0$, it is changed to the default value *piece* $= 1$.

If the value of *piece* is negative, arbitrary multiples of the basic moves for $|piece|$ are permitted. For example, *piece* $= -1$ defines the moves of a rook, from (x, y) to $(x \pm a, y)$ or to $(x, y \pm a)$ for all $a > 0$; *piece* $= -2$ defines the moves of a bishop, from (x, y) to $(x \pm a, y \pm a)$. The literature of "fairy chess" assigns standard names to the following *piece* values: wazir $= 1$, fers $= 2$, dabbaba $= 4$, knight $= 5$, alfil $= 8$, camel $= 10$, zebra $= 13$, giraffe $= 17$, fiveleaper $= 25$, root-50-leaper $= 50$, etc.; rook $= -1$, bishop $= -2$, unicorn $= -3$, dabbabarider $= -4$, nightrider $= -5$, alfilrider $= -8$, camelrider $= -10$, etc.

To generate a board with the moves of a king, you can use the *gunion* subroutine below to take the union of boards with *piece* $= 1$ and *piece* $= 2$. Similarly, you can get queen moves by taking the union of boards with *piece* $= -1$ and *piece* $= -2$.

If *piece* > 0, all arcs of the graph will have length 1. If *piece* < 0, the length of each arc will be the number of multiples of a basic move that produced the arc.

7.   If the *wrap* parameter is nonzero, it specifies a subset of coordinates in which values are computed modulo the corresponding size. For example, the coordinates (x, y) for vertices on a two-dimensional board are restricted to the range $0 \le x < n_1$, $0 \le y < n_2$; therefore when $wrap = 0$, a move from (x, y) to $(x+\delta_1, y+\delta_2)$ is legal only if $0 \le x+\delta_1 < n_1$ and $0 \le y+\delta_2 < n_2$. But when $wrap = 1$, the x coordinates are allowed to "wrap around"; the move would then be made to $((x + \delta_1) \bmod n_1, y + \delta_2)$, provided that $0 \le y + \delta_2 < n_2$. Setting $wrap = 1$ effectively makes the board into a cylinder instead of a rectangle. Similarly, the y coordinates are allowed to wrap around when $wrap = 2$. Both x and y coordinates are treated modulo their corresponding sizes when $wrap = 3$; the board is then effectively a torus. In general, coordinates $k_1, k_2, \ldots$ will wrap around when $wrap = 2^{k_1-1} + 2^{k_2-1} + \cdots$. Setting $wrap = -1$ causes all coordinates to be computed modulo their size.

The graph constructed by *board* will be undirected unless *directed* $\ne 0$. Directed *board* graphs will be acyclic when $wrap = 0$, but they may have cycles when $wrap \ne 0$. Precise rules defining the directed arcs are given below.

Several important special cases are worth noting. To get the complete graph on n vertices, you can say $board(n, 0, 0, 0, -1, 0, 0)$. To get the transitive tournament on n vertices, i.e., the directed graph with arcs from u to v when $u < v$, you can say $board(n, 0, 0, 0, -1, 0, 1)$. To get the empty graph on n vertices, you can say $board(n, 0, 0, 0, 2, 0, 0)$. To get a circuit (undirected) or a cycle (directed) of length n, you can say $board(n, 0, 0, 0, 1, 1, 0)$ and $board(n, 0, 0, 0, 1, 1, 1)$, respectively.

⟨ gb_basic.h 1 ⟩ +≡
```
#define complete(n)   board((long) (n), 0_L, 0_L, 0_L, -1_L, 0_L, 0_L)
#define transitive(n)   board((long) (n), 0_L, 0_L, 0_L, -1_L, 0_L, 1_L)
#define empty(n)   board((long) (n), 0_L, 0_L, 0_L, 2_L, 0_L, 0_L)
#define circuit(n)   board((long) (n), 0_L, 0_L, 0_L, 1_L, 1_L, 0_L)
#define cycle(n)   board((long) (n), 0_L, 0_L, 0_L, 1_L, 1_L, 1_L)
```

---

board: **Graph** *( )**, §8.	*n1*: **long**, §8.	*n4*: **long**, §8.
directed: **long**, §8.	*n2*: **long**, §8.	*piece*: **long**, §8.
gunion: **Graph** *( )**, §78.	*n3*: **long**, §8.	*wrap*: **long**, §8.

8. ⟨ Basic subroutines 8 ⟩ ≡

  Graph *∗board*(*n1*, *n2*, *n3*, *n4*, *piece*, *wrap*, *directed*)
    long *n1*, *n2*, *n3*, *n4*; /∗ size of board desired ∗/
    long *piece*; /∗ type of moves desired ∗/
    long *wrap*; /∗ mask for coordinate positions that wrap around ∗/
    long *directed*; /∗ should the graph be directed? ∗/
  { ⟨ Vanilla local variables 9 ⟩

   long *n*; /∗ total number of vertices ∗/
   long *p*; /∗ |*piece*| ∗/
   long *l*; /∗ length of current arc ∗/

   ⟨ Normalize the board-size parameters 11 ⟩;
   ⟨ Set up a graph with *n* vertices 13 ⟩;
   ⟨ Insert arcs or edges for all legal moves 15 ⟩;
   if (*gb_trouble_code*) {
    gb_recycle(*new_graph*);
    panic(*alloc_fault*); /∗ alas, we ran out of memory somewhere back there ∗/
   }
   return *new_graph*;
  }

See also sections 26, 37, 43, 55, 64, 74, 78, 81, 87, 95, and 105.

This code is used in section 2.

9. Most of the subroutines in GB_BASIC use the following local variables.

⟨ Vanilla local variables 9 ⟩ ≡

  Graph *∗new_graph*; /∗ the graph being constructed ∗/
  register long *i*, *j*, *k*; /∗ all-purpose indices ∗/
  register long *d*; /∗ the number of dimensions ∗/
  register Vertex *∗v*; /∗ the current vertex of interest ∗/
  register long *s*; /∗ accumulator ∗/

This code is used in sections 8, 26, 37, 43, 55, 64, 74, 78, 81, 87, 95, and 105.

10. Several arrays will facilitate the calculations that *board* needs to make. The number of distinct values in coordinate position k will be $nn[k]$; this coordinate position will wrap around if and only if $wr[k] \neq 0$. The current moves under consideration will be from $(x_1, \ldots, x_d)$ to $(x_1 + \delta_1, \ldots, x_k + \delta_k)$, where δ_k is stored in $del[k]$. An auxiliary array *sig* holds the sums $\sigma_k = \delta_1^2 + \cdots + \delta_{k-1}^2$. Additional arrays xx and yy hold coordinates of vertices before and after a move is made.

Some of these arrays are also used for other purposes by other programs besides *board*; we will meet those programs later.

We limit the number of dimensions to 91 or less. This is hardly a limitation, since the number of vertices would be astronomical even if the dimensionality were only half this big. But some of our later programs will be able to make good use of 40 or 50 dimensions and perhaps more; the number 91 is an upper limit imposed by the number of standard printable characters (see the convention for vertex names in the *perms* routine).

#define MAX_D 91

⟨ Private variables 3 ⟩ +≡
    static long $nn[\texttt{MAX\_D} + 1]$;   /* component sizes */
    static long $wr[\texttt{MAX\_D} + 1]$;   /* does this component wrap around? */
    static long $del[\texttt{MAX\_D} + 1]$;   /* displacements for the current move */
    static long $sig[\texttt{MAX\_D} + 2]$;   /* partial sums of squares of displacements */
    static long $xx[\texttt{MAX\_D} + 1]$, $yy[\texttt{MAX\_D} + 1]$;   /* coordinate values */

11.　⟨ Normalize the board-size parameters 11 ⟩ ≡
  if $(piece \equiv 0)$ $piece = 1$;
  if $(n1 \leq 0)$ { $n1 = n2 = 8$; $n3 = 0$; }
  $nn[1] = n1$;
  if $(n2 \leq 0)$ { $k = 2$; $d = -n2$; $n3 = n4 = 0$; }
  else {
    $nn[2] = n2$;
    if $(n3 \leq 0)$ { $k = 3$; $d = -n3$; $n4 = 0$; }
    else {
      $nn[3] = n3$;
      if $(n4 \leq 0)$ { $k = 4$; $d = -n4$; }
      else { $nn[4] = n4$; $d = 4$; **goto** *done*; }
    }
  }
  if $(d \equiv 0)$ { $d = k - 1$; **goto** *done*; }
  ⟨ Compute component sizes periodically for d dimensions 12 ⟩;
  done: /* now $nn[1]$ through $nn[d]$ are set up */
This code is used in section 8.

12.　At this point, $nn[1]$ through $nn[k - 1]$ are the component sizes that should be replicated periodically. In unusual cases, the number of dimensions might not be as large as the number of specifications.

⟨ Compute component sizes periodically for d dimensions 12 ⟩ ≡
  if $(d > \texttt{MAX\_D})$ $panic(bad\_specs)$;   /* too many dimensions */
  for $(j = 1$; $k \leq d$; $j{+}{+}, k{+}{+})$ $nn[k] = nn[j]$;
This code is used in sections 11 and 27.

$alloc\_fault = -1$, GB_GRAPH §7.
$bad\_specs = 30$, GB_GRAPH §7.
$gb\_recycle$: **void** ( ), GB_GRAPH §40.
$gb\_trouble\_code$: **long**,
  GB_GRAPH §14.
Graph = **struct**, GB_GRAPH §20.
$panic$ = macro ( ), §4.
$perms$: **Graph** *( ), §43.
Vertex = **struct**, GB_GRAPH §9.

13. We want to make the subroutine idiot-proof, so we use floating-point arithmetic to make sure that boards with more than a billion cells have not been specified.

#define MAX_NNN 1000000000.0

⟨ Set up a graph with n vertices 13 ⟩ ≡
```
  { float nnn;      /* approximate size */
    for (n = 1, nnn = 1.0, j = 1; j ≤ d; j++) {
      nnn *= (float) nn[j];
      if (nnn > MAX_NNN) panic(very_bad_specs);      /* way too big */
      n *= nn[j];      /* this multiplication cannot cause integer overflow */
    }
    new_graph = gb_new_graph(n);
    if (new_graph ≡ Λ) panic(no_room);      /* out of memory before we're even started */
    sprintf(new_graph⃗id, "board(%ld,%ld,%ld,%ld,%ld,%ld,%d)", n1, n2, n3, n4, piece, wrap,
        directed ? 1 : 0);
    strcpy(new_graph⃗util_types, "ZZZIIIZZZZZZZ");
    ⟨ Give names to the vertices 14 ⟩;
  }
```
This code is used in section 8.

14. The symbolic name of a board position like $(3, 1)$ will be the string '3.1'. The first three coordinates are also stored as integers, in utility fields $x.I$, $y.I$, and $z.I$, because immediate access to those values will be helpful in certain applications. (The coordinates can, of course, always be recovered in a slower fashion from the vertex name, via *sscanf*.)

The process of assigning coordinate values and names is equivalent to adding unity in a mixed-radix number system. Vertex $(x_1, \ldots, x_d)$ will be in position $x_1 n_2 \ldots n_d + \cdots + x_{d-1} n_d + x_d$ relative to the first vertex of the new graph; therefore it is also possible to deduce the coordinates of a vertex from its address.

⟨ Give names to the vertices 14 ⟩ ≡
```
  { register char *q;      /* string pointer */
    nn[0] = xx[0] = xx[1] = xx[2] = xx[3] = 0;
    for (k = 4; k ≤ d; k++) xx[k] = 0;
    for (v = new_graph⃗vertices; ; v++) {
      q = buffer;
      for (k = 1; k ≤ d; k++) {
        sprintf(q, ".%ld", xx[k]);
        while (*q) q++;
      }
      v⃗name = gb_save_string(&buffer[1]);      /* omit buffer[0], which is '.' */
      v⃗x.I = xx[1]; v⃗y.I = xx[2]; v⃗z.I = xx[3];
      for (k = d; xx[k] + 1 ≡ nn[k]; k--) xx[k] = 0;
      if (k ≡ 0) break;      /* a "carry" has occurred all the way to the left */
      xx[k]++;      /* increase coordinate k */
    }
  }
```
This code is used in section 13.

15. Now we come to a slightly tricky part of the routine: the move generator. Let $p = |piece|$. The outer loop of this procedure runs through all solutions of the equation $\delta_1^2 + \cdots + \delta_d^2 = p$, where the δ's are nonnegative integers. Within that loop, we attach signs to the δ's, but we always leave δ_k positive if $\delta_1 = \cdots = \delta_{k-1} = 0$. For every such vector δ, we generate moves from v to $v + \delta$ for every vertex v. When *directed* $= 0$, we use *gb_new_edge* instead of *gb_new_arc*, so that the reverse arc from $v + \delta$ to v is also generated.

⟨ Insert arcs or edges for all legal moves 15 ⟩ ≡
  ⟨ Initialize the *wr*, *sig*, and *del* tables 16 ⟩;
  $p = piece$;
  if $(p < 0)$ $p = -p$;
  while (1) {
    ⟨ Advance to the next nonnegative *del* vector, or **break** if done 17 ⟩;
    while (1) {
      ⟨ Generate moves for the current *del* vector 19 ⟩;
      ⟨ Advance to the next signed *del* vector, or restore *del* to nonnegative values and **break** 18 ⟩;
    }
  }
This code is used in section 8.

16. The C language does not define $\gg$ unambiguously. If w is negative, the assignment '$w \gg= 1$' here should keep w negative. (However, this technicality doesn't matter except in highly unusual cases when there are more than 32 dimensions.)

⟨ Initialize the *wr*, *sig*, and *del* tables 16 ⟩ ≡
  { **register long** $w = wrap$;
    for $(k = 1;\ k \leq d;\ k{+}{+}, w \gg= 1)$ {
      $wr[k] = w\ \&\ 1$;
      $del[k] = sig[k] = 0$;
    }
    $sig[0] = del[0] = sig[d + 1] = 0$;
  }
This code is used in section 15.

---

buffer: **static char** [], §5.
d: **register long**, §9.
del: **static long** [], §10.
directed: **long**, §8.
gb_new_arc: **void** ( ),
  GB_GRAPH §30.
gb_new_edge: **void** ( ),
  GB_GRAPH §31.
gb_new_graph: **Graph** *( ),
  GB_GRAPH §23.
gb_save_string: **char** *( ),
  GB_GRAPH §35.
I: **long**, GB_GRAPH §8.
id: **char** [], GB_GRAPH §20.

j: **register long**, §9.
k: **register long**, §9.
n: **long**, §8.
n1: **long**, §8.
n2: **long**, §8.
n3: **long**, §8.
n4: **long**, §8.
name: **char** *, GB_GRAPH §9.
new_graph: **Graph** *, §9.
nn: **static long** [], §10.
no_room = 1, GB_GRAPH §7.
p: **long**, §8.
panic = macro ( ), §4.
piece: **long**, §8.

sig: **static long** [], §10.
sprintf: **int** ( ), <stdio.h>.
sscanf: **int** ( ), <stdio.h>.
strcpy: **char** *( ), <string.h>.
util_types: **char** [], GB_GRAPH §20.
v: **register Vertex** *, §9.
vertices: **Vertex** *, GB_GRAPH §20.
very_bad_specs = 40, GB_GRAPH §7.
wr: **static long** [], §10.
wrap: **long**, §8.
x: **util**, GB_GRAPH §9.
xx: **static long** [], §10.
y: **util**, GB_GRAPH §9.
z: **util**, GB_GRAPH §9.

17. The *sig* array makes it easy to backtrack through all partitions of p into an ordered sum of squares.

⟨ Advance to the next nonnegative *del* vector, or **break** if done 17 ⟩ ≡
```
    for (k = d; sig[k] + (del[k] + 1) * (del[k] + 1) > p; k--)  del[k] = 0;
    if (k ≡ 0) break;
    del[k]++;
    sig[k + 1] = sig[k] + del[k] * del[k];
    for (k++; k ≤ d; k++)  sig[k + 1] = sig[k];
    if (sig[d + 1] < p) continue;
```
This code is used in section 15.

18. ⟨ Advance to the next signed *del* vector, or restore *del* to nonnegative values and **break** 18 ⟩ ≡
```
    for (k = d; del[k] ≤ 0; k--)  del[k] = −del[k];
    if (sig[k] ≡ 0) break;        /* all but del[k] were negative or zero */
    del[k] = −del[k];        /* some entry preceding del[k] is positive */
```
This code is used in section 15.

19. We use the mixed-radix addition technique again when generating moves.

⟨ Generate moves for the current *del* vector 19 ⟩ ≡
```
    for (k = 1; k ≤ d; k++)  xx[k] = 0;
    for (v = new_graph→vertices; ; v++) {
        ⟨ Generate moves from v corresponding to del 20 ⟩;
        for (k = d; xx[k] + 1 ≡ nn[k]; k--)  xx[k] = 0;
        if (k ≡ 0) break;        /* a "carry" has occurred all the way to the left */
        xx[k]++;        /* increase coordinate k */
    }
```
This code is used in section 15.

20. The legal moves when *piece* is negative are derived as follows, in the presence of possible wraparound: Starting at $(x_1, \ldots, x_d)$, we move to $(x_1+\delta_1, \ldots, x_d+\delta_d)$, $(x_1+2\delta_1, \ldots, x_d+2\delta_d)$, $\ldots$, until either coming to a position with a nonwrapped coordinate out of range or coming back to the original point.

A subtle technicality should be noted: When coordinates are wrapped and *piece* > 0, self-loops are possible—for example, in $board(1, 0, 0, 0, 1, 1, 1)$. But self-loops never arise when *piece* < 0.

⟨ Generate moves from v corresponding to del 20 ⟩ ≡
```
    for (k = 1; k ≤ d; k++)  yy[k] = xx[k] + del[k];
    for (l = 1; ; l++) {
        ⟨ Correct for wraparound, or goto no_more if off the board 22 ⟩;
        if (piece < 0) ⟨ Go to no_more if yy = xx 21 ⟩;
        ⟨ Record a legal move from xx to yy 23 ⟩;
        if (piece > 0) goto no_more;
        for (k = 1; k ≤ d; k++)  yy[k] += del[k];
    }
    no_more:
```
This code is used in section 19.

21. ⟨ Go to *no_more* if $yy = xx$ 21 ⟩ ≡
```
{
    for (k = 1; k ≤ d; k++)
        if (yy[k] ≠ xx[k]) goto unequal;
    goto no_more;
unequal: ;
}
```
This code is used in section 20.

22. ⟨ Correct for wraparound, or **goto** *no_more* if off the board 22 ⟩ ≡
```
for (k = 1; k ≤ d; k++) {
    if (yy[k] < 0) {
        if (¬wr[k]) goto no_more;
        do yy[k] += nn[k];   while (yy[k] < 0);
    } else if (yy[k] ≥ nn[k]) {
        if (¬wr[k]) goto no_more;
        do yy[k] −= nn[k];   while (yy[k] ≥ nn[k]);
    }
}
```
This code is used in section 20.

23. ⟨ Record a legal move from *xx* to *yy* 23 ⟩ ≡
```
for (k = 2, j = yy[1]; k ≤ d; k++) j = nn[k] * j + yy[k];
if (directed) gb_new_arc(v, new_graph→vertices + j, l);
else gb_new_edge(v, new_graph→vertices + j, l);
```
This code is used in section 20.

---

board: **Graph** *( ), §8.
d: **register long**, §9.
del: **static long** [], §10.
directed: **long**, §8.
gb_new_arc: **void** ( ),
  GB_GRAPH §30.
gb_new_edge: **void** ( ),

GB_GRAPH §31.
j: **register long**, §9.
k: **register long**, §9.
l: **long**, §8.
new_graph: **Graph** *, §9.
nn: **static long** [], §10.
p: **long**, §8.

piece: **long**, §8.
sig: **static long** [], §10.
v: **register Vertex** *, §9.
vertices: **Vertex** *, GB_GRAPH §20.
wr: **static long** [], §10.
xx: **static long** [], §10.
yy: **static long** [], §10.

24. Generalized triangular boards. The subroutine call $simplex(n, n0, n1, n2, n3, n4,$ *directed*) creates a graph based on generalized triangular or tetrahedral configurations. Such graphs are similar in spirit to the game boards created by *board*, but they pertain to nonrectangular grids like those in Chinese checkers. As with *board* in the case *piece* = 1, the vertices represent board positions and the arcs run from board positions to their nearest neighbors. Each arc has length 1.

More formally, the vertices can be defined as sequences of nonnegative integers $(x_0, x_1, \ldots, x_d)$ whose sum is n, where two sequences are considered adjacent if and only if they differ by ± 1 in exactly two components—equivalently, if the Euclidean distance between them is $\sqrt{2}$. When $d = 2$, for example, the vertices can be visualized as a triangular array

$$(0,0,3)$$
$$(0,1,2) \quad (1,0,2)$$
$$(0,2,1) \quad (1,1,1) \quad (2,0,1)$$
$$(0,3,0) \quad (1,2,0) \quad (2,1,0) \quad (3,0,0)$$

containing $(n+1)(n+2)/2$ elements, illustrated here when $n = 3$; each vertex of the array has up to 6 neighbors. When $d = 3$ the vertices form a tetrahedral array, a stack of triangular layers, and they can have as many as 12 neighbors. In general, a vertex in a d-simplicial array will have up to $d(d+1)$ neighbors.

If the *directed* parameter is nonzero, arcs run only from vertices to neighbors that are lexicographically greater—for example, downward or to the right in the triangular array shown. The directed graph is therefore acyclic, and a vertex of a d-simplicial array has out-degree at most $d(d+1)/2$.

25. The first parameter, n, specifies the sum of the coordinates $(x_0, x_1, \ldots, x_d)$. The following parameters $n0$ through $n4$ specify upper bounds on those coordinates, and they also specify the dimensionality d.

If, for example, $n0$, $n1$, and $n2$ are positive while $n3 = 0$, the value of d will be 2 and the coordinates will be constrained to satisfy $0 \le x_0 \le n0$, $0 \le x_1 \le n1$, $0 \le x_2 \le n2$. These upper bounds essentially lop off the corners of the triangular array. We obtain a hexagonal board with $6m$ boundary cells by asking for $simplex(3m, 2m, 2m, 2m, 0, 0, 0)$. We obtain the diamond-shaped board used in the game of Hex [Martin Gardner, *The Scientific American Book of Mathematical Puzzles & Diversions* (Simon & Schuster, 1959), Chapter 8] by calling $simplex(20, 10, 20, 10, 0, 0, 0)$.

In general, *simplex* determines d and upper bounds $(n_0, n_1, \ldots, n_d)$ in the following way: Let the first nonpositive entry of the sequence $(n0, n1, n2, n3, n4, 0) = (n_0, n_1, n_2, n_3, n_4, 0)$ be n_k. If $k > 0$ and $n_k = 0$, the value of d will be $k - 1$ and the coordinates will be bounded by the given numbers $(n_0, \ldots, n_d)$. If $k > 0$ and $n_k < 0$, the value of d will be $|n_k|$ and the coordinates will be bounded by the first $d + 1$ elements of the infinite periodic sequence $(n_0, \ldots, n_{k-1}, n_0, \ldots, n_{k-1}, n_0, \ldots)$. If $k = 0$ and $n_0 < 0$, the value of d will be $|n_0|$ and the coordinates will be unbounded; equivalently, we may set $n_0 = \cdots = n_d = n$. In this case the number of vertices will be $\binom{n+d}{d}$. Finally, if $k = 0$ and $n_0 = 0$, we have the default case of a triangular array with $3n$ boundary cells, exactly as if $n_0 = -2$.

For example, the specification $n0 = 3$, $n1 = -5$ will produce all vertices $(x_0, x_1, \ldots, x_5)$ such that $x_0 + x_1 + \cdots + x_5 = n$ and $0 \le x_j \le 3$. The specification $n0 = 1$, $n1 = -d$ will essentially produce all n-element subsets of the $(d+1)$-element set $\{0, 1, \ldots, d\}$, because we can regard an element k as being present in the set if $x_k = 1$ and absent if $x_k = 0$. In that case two subsets are adjacent if and only if they have exactly $n - 1$ elements in common.

26.   ⟨ Basic subroutines 8 ⟩ +≡

> **Graph** *$simplex(n, n0, n1, n2, n3, n4, directed)$
>> **unsigned long** n;   /* the constant sum of all coordinates */
>> **long** $n0$, $n1$, $n2$, $n3$, $n4$;   /* constraints on coordinates */
>> **long** $directed$;   /* should the graph be directed? */
>
> { ⟨ Vanilla local variables 9 ⟩
> ⟨ Normalize the simplex parameters 27 ⟩;
> ⟨ Create a graph with one vertex for each point 28 ⟩;
> ⟨ Name the points and create the arcs or edges 31 ⟩;
> **if** ($gb\_trouble\_code$) {
>> $gb\_recycle(new\_graph)$;
>> $panic(alloc\_fault)$;   /* darn, we ran out of memory somewhere back there */
>
> }
> **return** new_graph;
> }

$alloc\_fault = -1$, GB_GRAPH §7. $gb\_trouble\_code$: **long**, new_graph: **Graph** *, §9.
$board$: **Graph** *( ), §8.   GB_GRAPH §14. $panic = $ macro ( ), §4.
$gb\_recycle$: **void** ( ), GB_GRAPH §40.   **Graph** = **struct**, GB_GRAPH §20. $piece$: **long**, §8.

27. ⟨ Normalize the simplex parameters 27 ⟩ ≡
```
if (n0 ≡ 0) n0 = −2;
if (n0 < 0) { k = 2; nn[0] = n; d = −n0; n1 = n2 = n3 = n4 = 0; }
else {
  if (n0 > n) n0 = n;
  nn[0] = n0;
  if (n1 ≤ 0) { k = 2; d = −n1; n2 = n3 = n4 = 0; }
  else {
    if (n1 > n) n1 = n;
    nn[1] = n1;
    if (n2 ≤ 0) { k = 3; d = −n2; n3 = n4 = 0; }
    else {
      if (n2 > n) n2 = n;
      nn[2] = n2;
      if (n3 ≤ 0) { k = 4; d = −n3; n4 = 0; }
      else {
        if (n3 > n) n3 = n;
        nn[3] = n3;
        if (n4 ≤ 0) { k = 5; d = −n4; }
        else { if (n4 > n) n4 = n;
          nn[4] = n4; d = 4; goto done; }
      }
    }
  }
}
if (d ≡ 0) { d = k − 2; goto done; }
nn[k − 1] = nn[0];
⟨ Compute component sizes periodically for d dimensions 12 ⟩;
done:     /* now nn[0] through nn[d] are set up */
```
This code is used in sections 26, 37, and 44.

28. ⟨ Create a graph with one vertex for each point 28 ⟩ ≡
⟨ Determine the number of feasible $(x_0, \ldots, x_d)$, and allocate the graph 29 ⟩;
$sprintf(new\_graph{\to}id,$ `"simplex(%lu,%ld,%ld,%ld,%ld,%ld,%d)"`$, n, n0, n1, n2, n3, n4,$
    $directed\ ?\ 1:0);$
$strcpy(new\_graph{\to}util\_types,$ `"VVZIIIZZZZZZZZ"`$);$ /* hash table will be used */
This code is used in section 26.

29. We determine the number of vertices by determining the coefficient of z^n in the power series

$$(1 + z + \cdots + z^{n_0})(1 + z + \cdots + z^{n_1}) \ldots (1 + z + \cdots + z^{n_d}).$$

⟨ Determine the number of feasible $(x_0, \ldots, x_d)$, and allocate the graph 29 ⟩ ≡
```
{ long nverts;       /* the number of vertices */
  register long *coef = gb_typed_alloc(n + 1, long, working_storage);
  if (gb_trouble_code) panic(no_room + 1);     /* can't allocate coef array */
  for (k = 0; k ≤ nn[0]; k++) coef[k] = 1;
      /* now coef represents the coefficients of 1 + z + ⋯ + z^{n0} */
```

```
    for (j = 1; j ≤ d; j++) ⟨Multiply the power series coefficients by 1 + z + · · · + z^{n_j} 30⟩;
    nverts = coef[n];
    gb_free(working_storage);        /* recycle the coef array */
    new_graph = gb_new_graph(nverts);
    if (new_graph ≡ Λ) panic(no_room);        /* out of memory before we're even started */
  }
```

This code is used in sections 28 and 38.

30.   There's a neat way to multiply by $1 + z + \cdots + z^{n_j}$: We multiply first by $1 - z^{n_j+1}$, then sum the coefficients.

We want to detect impossibly large specifications without risking integer overflow. It is easy to do this because multiplication is being done via addition.

⟨Multiply the power series coefficients by $1 + z + \cdots + z^{n_j}$ 30⟩ ≡

```
  {
    for (k = n, i = n − nn[j] − 1; i ≥ 0; k−−, i−−) coef[k] −= coef[i];
    s = 1;
    for (k = 1; k ≤ n; k++) {
      s += coef[k];
      if (s > 1000000000) panic(very_bad_specs);        /* way too big */
      coef[k] = s;
    }
  }
```

This code is used in section 29.

---

d: **register long**, §9.
directed: **long**, §26.
gb_free: **void** ( ), GB_GRAPH §16.
gb_new_graph: **Graph** *( ),
    GB_GRAPH §23.
gb_trouble_code: **long**,
    GB_GRAPH §14.
gb_typed_alloc = macro ( ),
    GB_GRAPH §11.
i: **register long**, §9.
id: **char** [ ], GB_GRAPH §20.

j: **register long**, §9.
k: **register long**, §9.
n: **unsigned long**, §26.
n: **unsigned long**, §37.
n: **register long**, §43.
n0: **long**, §26.
n1: **long**, §26.
n2: **long**, §26.
n3: **long**, §26.
n4: **long**, §26.

new_graph: **Graph** *, §9.
nn: **static long** [ ], §10.
no_room = 1, GB_GRAPH §7.
panic = macro ( ), §4.
s: **register long**, §9.
sprintf: **int** ( ), <stdio.h>.
strcpy: **char** *( ), <string.h>.
util_types: **char** [ ], GB_GRAPH §20.
very_bad_specs = 40, GB_GRAPH §7.
working_storage: **static Area**, §3.

31. As we generate the vertices, it proves convenient to precompute an array containing the numbers $y_j = n_j + \cdots + n_d$, which represent the largest possible sums $x_j + \cdots + x_d$. We also want to maintain the numbers $\sigma_j = n - (x_0 + \cdots + x_{j-1}) = x_j + \cdots + x_d$. The conditions

$$0 \le x_j \le n_j, \qquad \sigma_j - y_{j+1} \le x_j \le \sigma_j$$

are necessary and sufficient, in the sense that we can find at least one way to complete a partial solution $(x_0, \ldots, x_k)$ to a full solution $(x_0, \ldots, x_d)$ if and only if the conditions hold for all $j \le k$.

There is at least one solution if and only if $n \le y_0$.

We enter the name string into a hash table, using the *hash_in* routine of GB_GRAPH, because there is no simple way to compute the location of a vertex from its coordinates.

⟨ Name the points and create the arcs or edges 31 ⟩ ≡
```
v = new_graph→vertices;
yy[d + 1] = 0;  sig[0] = n;
for (k = d;  k ≥ 0;  k−−)  yy[k] = yy[k + 1] + nn[k];
if (yy[0] ≥ n) {
    k = 0;  xx[0] = (yy[1] ≥ n ? 0 : n − yy[1]);
    while (1) {
        ⟨ Complete the partial solution (x₀, . . . , xₖ) 32 ⟩;
        ⟨ Assign a symbolic name for (x₀, . . . , x_d) to vertex v 34 ⟩;
        hash_in(v);      /* enter v→name into the hash table (via utility fields u, v) */
        ⟨ Create arcs or edges from previous points to v 35 ⟩;
        v++;
        ⟨ Advance to the next partial solution (x₀, . . . , xₖ), where k is as large as possible; goto last if
            there are no more solutions 33 ⟩;
    }
}
last: if (v ≠ new_graph→vertices + new_graph→n) panic(impossible);      /* can't happen */
```
This code is used in section 26.

32. ⟨ Complete the partial solution $(x_0, \ldots, x_k)$ 32 ⟩ ≡
```
for (s = sig[k] − xx[k], k++;  k ≤ d;  s −= xx[k], k++) {
    sig[k] = s;
    if (s ≤ yy[k + 1]) xx[k] = 0;
    else  xx[k] = s − yy[k + 1];
}
if (s ≠ 0) panic(impossible + 1)      /* can't happen */
```
This code is used in sections 31 and 39.

33. Here we seek the largest k such that x_k can be increased without violating the necessary and sufficient conditions stated earlier.

⟨ Advance to the next partial solution $(x_0, \ldots, x_k)$, where k is as large as possible; **goto** *last* if there are no more solutions 33 ⟩ ≡
```
for (k = d − 1;  ;  k−−) {
    if (xx[k] < sig[k] ∧ xx[k] < nn[k]) break;
    if (k ≡ 0) goto last;
}
xx[k]++;
```
This code is used in sections 31 and 39.

34. As in the *board* routine, we represent the sequence of coordinates $(2, 0, 1)$ by the string '2.0.1'. The string won't exceed BUF_SIZE, because the ratio BUF_SIZE/MAX_D is plenty big.

The first three coordinate values, (x_0, x_1, x_2), are placed into utility fields x, y, and z, so that they can be accessed immediately if an application needs them.

⟨ Assign a symbolic name for $(x_0, \ldots, x_d)$ to vertex v 34 ⟩ ≡
 { **register char** $*p = buffer$;　　/* string pointer */
  for $(k = 0; \ k \leq d; \ k{+}{+})$ { $sprintf(p, \texttt{".\%1d"}, xx[k])$; **while** $(*p)$ $p{+}{+}$; }
  $v{\rightarrow}name = gb\_save\_string(\&buffer[1])$;　　/* omit $buffer[0]$, which is '.' */
  $v{\rightarrow}x.I = xx[0]$; $v{\rightarrow}y.I = xx[1]$; $v{\rightarrow}z.I = xx[2]$;
 }

This code is used in sections 31 and 39.

35. Since we are generating the vertices in lexicographic order of their coordinates, it is easy to identify all adjacent vertices that precede the current setting of $(x_0, x_1, \ldots, x_d)$. We locate them via their symbolic names.

⟨ Create arcs or edges from previous points to v 35 ⟩ ≡
 for $(j = 0; \ j < d; \ j{+}{+})$
  if $(xx[j])$ { **register Vertex** $*u$;　　/* previous vertex adjacent to v */
   $xx[j]{-}{-}$;
   for $(k = j + 1; \ k \leq d; \ k{+}{+})$
    if $(xx[k] < nn[k])$ { **register char** $*p = buffer$;　　/* string pointer */
     $xx[k]{+}{+}$;
     for $(i = 0; \ i \leq d; \ i{+}{+})$ { $sprintf(p, \texttt{".\%1d"}, xx[i])$; **while** $(*p)$ $p{+}{+}$; }
     $u = hash\_out(\&buffer[1])$;
     if $(u \equiv \Lambda)$ $panic(impossible + 2)$;　　/* can't happen */
     if $(directed)$ $gb\_new\_arc(u, v, 1\,\text{L})$; **else** $gb\_new\_edge(u, v, 1\,\text{L})$;
     $xx[k]{-}{-}$;
    }
   $xx[j]{+}{+}$;
  }

This code is used in section 31.

---

board: **Graph** *( ), §8.
BUF_SIZE = 4096, §5.
buffer: **static char** [ ], §5.
d: **register long**, §9.
directed: **long**, §26.
gb_new_arc: **void** ( ),
 GB_GRAPH §30.
gb_new_edge: **void** ( ),
 GB_GRAPH §31.
gb_save_string: **char** *( ),
 GB_GRAPH §35.
hash_in: **void** ( ), GB_GRAPH §44.
hash_out: **Vertex** *( ),
 GB_GRAPH §46.

i: **register long**, §9.
I: **long**, GB_GRAPH §8.
impossible = 90, GB_GRAPH §7.
j: **register long**, §9.
k: **register long**, §9.
last: **label**, §39.
MAX_D = 91, §10.
n: **unsigned long**, §26.
n: **long**, GB_GRAPH §20.
name: **char** *, GB_GRAPH §9.
new_graph: **Graph** *, §9.
nn: **static long** [ ], §10.
panic = macro ( ), §4.

s: **register long**, §9.
sig: **static long** [ ], §10.
sprintf: **int** ( ), <stdio.h>.
u: **util**, GB_GRAPH §9.
v: **register Vertex** *, §9.
v: **util**, GB_GRAPH §9.
Vertex = **struct**, GB_GRAPH §9.
vertices: **Vertex** *, GB_GRAPH §20.
x: **util**, GB_GRAPH §9.
xx: **static long** [ ], §10.
y: **util**, GB_GRAPH §9.
yy: **static long** [ ], §10.
z: **util**, GB_GRAPH §9.

36. **Subset graphs.** The subroutine call $subsets(n, n0, n1, n2, n3, n4, size\_bits, directed)$ creates a graph having the same vertices as $simplex(n, n0, n1, n2, n3, n4, directed)$ but with a quite different notion of adjacency. In this we interpret a solution $(x_0, x_1, \ldots, x_d)$ to the conditions $x_0 + x_1 + \cdots + x_d = n$ and $0 \le x_j \le n_j$ not as a position on a game board but as an n-element submultiset of the multiset $\{n_0 \cdot 0, n_1 \cdot 1, \ldots, n_d \cdot d\}$ that has x_j elements equal to j. (If each $n_j = 1$, the multiset is a set; this is an important special case.) Two vertices are adjacent if and only if their intersection has a cardinality that matches one of the bits in $size\_bits$, which is an unsigned integer. Each arc has length 1.

For example, suppose $n = 3$ and $(n0, n1, n2, n3) = (2, 2, 2, 0)$. Then the vertices are the 3-element submultisets of $\{0, 0, 1, 1, 2, 2\}$, namely

$$\{0,0,1\}, \quad \{0,0,2\}, \quad \{0,1,2\}, \quad \{0,2,2\}, \quad \{1,1,2\}, \quad \{1,2,2\},$$

which are represented by the respective vectors

$$(2,1,0), \quad (2,0,1), \quad (1,1,1), \quad (1,0,2), \quad (0,2,1), \quad (0,1,2).$$

The intersection of multisets represented by $(x_0, x_1, \ldots, x_d)$ and $(y_0, y_1, \ldots, y_d)$ is

$$\big(\min(x_0, y_0), \min(x_1, y_1), \ldots, \min(x_d, y_d)\big);$$

each element occurs as often as it occurs in both multisets being intersected. If now $size\_bits = 3$, the multisets will be considered adjacent whenever their intersection contains exactly 0 or 1 elements, because $3 = 2^0 + 2^1$. The vertices adjacent to $\{0,0,1\}$, for example, will be $\{0,2,2\}$ and $\{1,2,2\}$. In this case, every pair of submultisets has a nonempty intersection, so the same graph would be obtained if $size\_bits = 2$.

If $directed$ is nonzero, the graph will have directed arcs, from u to v only if $u \le v$. Notice that the graph will have self-loops if and only if the binary representation of $size\_bits$ contains the term 2^n, in which case there will be a loop from every vertex to itself. (In an undirected graph, such loops are represented by two arcs.)

We define a macro $disjoint\_subsets(n, k)$ for the case of $\binom{n}{k}$ vertices, adjacent if and only if they represent disjoint k-subsets of an n-set. One important special case is the Petersen graph, whose vertices are the 2-element subsets of $\{0, 1, 2, 3, 4\}$, adjacent when they are disjoint. This graph is remarkable because it contains 10 vertices, each of degree 3, but it has no circuits of length less than 5.

$\langle$ gb_basic.h 1 $\rangle$ $+\equiv$
#define $disjoint\_subsets(n, k)$ $subsets(($**long**$)\,(k), 1_L, ($**long**$)\,(1 - (n)), 0_L, 0_L, 0_L, 1_L, 0_L)$
#define $petersen()$ $disjoint\_subsets(5, 2)$

37. $\langle$ Basic subroutines 8 $\rangle$ $+\equiv$
  Graph $*subsets(n, n0, n1, n2, n3, n4, size\_bits, directed)$
      unsigned long n;   /∗ the number of elements in the multiset ∗/
      long $n0$, $n1$, $n2$, $n3$, $n4$;   /∗ multiplicities of elements ∗/
      unsigned long $size\_bits$;   /∗ intersection sizes that trigger arcs ∗/
      long $directed$;   /∗ should the graph be directed? ∗/

```
{  ⟨ Vanilla local variables 9 ⟩
   ⟨ Normalize the simplex parameters 27 ⟩;
   ⟨ Create a graph with one vertex for each subset 38 ⟩;
   ⟨ Name the subsets and create the arcs or edges 39 ⟩;
   if (gb_trouble_code) {
      gb_recycle(new_graph);
      panic(alloc_fault);       /* rats, we ran out of memory somewhere back there */
   }
   return new_graph;
}
```

38. ⟨ Create a graph with one vertex for each subset 38 ⟩ ≡
  ⟨ Determine the number of feasible $(x_0, \ldots, x_d)$, and allocate the graph 29 ⟩;
  sprintf(*new_graph→id*, "subsets(%lu,%ld,%ld,%ld,%ld,%ld,0x%lx,%d)", $n, n0, n1, n2, n3, n4$,
      size_bits, *directed* ? 1 : 0);
  strcpy(*new_graph→util_types*, "ZZZIIIZZZZZZZ");   /* hash table will not be used */
This code is used in section 37.

39. We generate the vertices with exactly the logic used in *simplex*.

⟨ Name the subsets and create the arcs or edges 39 ⟩ ≡
  $v = new\_graph→vertices$;
  $yy[d+1] = 0$; $sig[0] = n$;
  for $(k = d;\ k \geq 0;\ k\!-\!-)$ $yy[k] = yy[k+1] + nn[k]$;
  if $(yy[0] \geq n)$ {
     $k = 0$; $xx[0] = (yy[1] \geq n\ ?\ 0 : n - yy[1])$;
     while (1) {
        ⟨ Complete the partial solution $(x_0, \ldots, x_k)$ 32 ⟩;
        ⟨ Assign a symbolic name for $(x_0, \ldots, x_d)$ to vertex v 34 ⟩;
        ⟨ Create arcs or edges from previous subsets to v 40 ⟩;
        $v{+}{+}$;
        ⟨ Advance to the next partial solution $(x_0, \ldots, x_k)$, where k is as large as possible; **goto** *last* if
            there are no more solutions 33 ⟩;
     }
  }
last: **if** $(v \neq new\_graph→vertices + new\_graph→n)$ *panic*(*impossible*);   /* can't happen */
This code is used in section 37.

---

alloc_fault = −1, GB_GRAPH §7.
d: **register long**, §9.
gb_recycle: **void** ( ), GB_GRAPH §40.
gb_trouble_code: **long**,
   GB_GRAPH §14.
Graph = **struct**, GB_GRAPH §20.
id: **char** [ ], GB_GRAPH §20.
impossible = 90, GB_GRAPH §7.

k: **register long**, §9.
n: **long**, GB_GRAPH §20.
new_graph: **Graph** *, §9.
nn: **static long** [ ], §10.
panic = **macro** ( ), §4.
sig: **static long** [ ], §10.
simplex: **Graph** *( ), §26.

sprintf: **int** ( ), <stdio.h>.
strcpy: **char** *( ), <string.h>.
util_types: **char** [ ], GB_GRAPH §20.
v: **register Vertex** *, §9.
vertices: **Vertex** *, GB_GRAPH §20.
xx: **static long** [ ], §10.
yy: **static long** [ ], §10.

40. The only difference is that we generate the arcs or edges by brute force, examining each pair of vertices to see if they are adjacent or not.

The code here is character-set dependent: It assumes that '.' and null have a character code less than '0', as in ASCII. It also assumes that characters occupy exactly eight bits.

#define UL_BITS $8 * \mathbf{sizeof}(\mathbf{unsigned\ long})$ /* the number of bits in *size_bits* */

⟨ Create arcs or edges from previous subsets to v 40 ⟩ ≡

```
{ register Vertex *u;
  for (u = new_graph⁻vertices; u ≤ v; u++) { register char *p = u⁻name;
    long ss = 0;       /* the number of elements common to u and v */
    for (j = 0; j ≤ d; j++, p++) {
      for (s = (*p++) − '0'; *p ≥ '0'; p++) s = 10 * s + *p − '0';     /* sscanf (p, "%ld", &s) */
      if (xx[j] < s) ss += xx[j];
      else ss += s;
    }
    if ((size_bits & (((unsigned long) 1) ≪ ss)) ∧ ss < UL_BITS) {
      if (directed) gb_new_arc(u, v, 1_L);
      else gb_new_edge(u, v, 1_L);
    }
  }
}
```

This code is used in section 39.

41. Permutation graphs. The subroutine call $perms(n0, n1, n2, n3, n4, max\_inv, directed)$ creates a graph whose vertices represent the permutations of a multiset that have at most max_inv inversions. Two permutations are adjacent in the graph if one is obtained from the other by interchanging two adjacent elements. Each arc has length 1.

For example, the multiset $\{0, 0, 1, 2\}$ has the following twelve permutations:

$$0012, \quad 0021, \quad 0102, \quad 0120, \quad 0201, \quad 0210, \quad 1002, \quad 1020, \quad 1200, \quad 2001, \quad 2010, \quad 2100.$$

The first of these, 0012, has two neighbors, 0021 and 0102.

The number of inversions is the number of pairs of elements xy such that $x > y$ and x precedes y from left to right, counting multiplicity. For example, 2010 has four inversions, corresponding to $xy \in \{20, 21, 20, 10\}$. It is not difficult to verify that the number of inversions of a permutation is the distance in the graph from that permutation to the lexicographically first permutation.

Parameters $n0$ through $n4$ specify the composition of the multiset, just as in the *subsets* routine. Roughly speaking, there are $n0$ elements equal to 0, $n1$ elements equal to 1, and so on. The multiset $\{0, 0, 1, 2, 3, 3\}$, for example, would be represented by $(n0, n1, n2, n3, n4) = (2, 1, 1, 2, 0)$.

Of course, we sometimes want to have multisets with more than five distinct elements; when there are $d + 1$ distinct elements, the multiset should have n_k elements equal to k and $n = n_0 + n_1 + \cdots + n_d$ elements in all. Larger values of d can be specified by using $-d$ as a parameter: If $n0 = -d$, each multiplicity n_k is taken to be 1; if $n0 > 0$ and $n1 = -d$, each multiplicity n_k is taken to be equal to $n0$; if $n0 > 0$, $n1 > 0$, and $n2 = -d$, the multiplicities are alternately $(n0, n1, n0, n1, n0, \ldots)$; if $n0 > 0$, $n1 > 0$, $n2 > 0$, and $n3 = -d$, the multiplicities are the first $d + 1$ elements of the periodic sequence $(n0, n1, n2, n0, n1, \ldots)$; and if all but $n4$ are positive, while $n4 = -d$, the multiplicities again are periodic.

An example like $(n0, n1, n2, n3, n4) = (1, 2, 3, 4, -8)$ is about as tricky as you can get. It specifies the multiset $\{0, 1, 1, 2, 2, 2, 3, 3, 3, 3, 4, 5, 5, 6, 6, 6, 7, 7, 7, 7, 8\}$.

If any of the multiplicity parameters is negative or zero, the remaining multiplicities are ignored. For example, if $n2 \leq 0$, the subroutine does not look at $n3$ or $n4$.

You probably don't want to try $perms(n0, 0, 0, 0, 0, max\_inv, directed)$ when $n0 > 0$, because a multiset with $n0$ identical elements has only one permutation.

The special case when you want all $n!$ permutations of an n-element set can be obtained by calling $all\_perms(n, directed)$.

$\langle$ gb_basic.h 1 $\rangle$ $+\equiv$
#**define** $all\_perms(n, directed)$ $perms((\textbf{long})\ (1 - (n)), 0_{\text{L}}, 0_{\text{L}}, 0_{\text{L}}, 0_{\text{L}}, 0_{\text{L}}, (\textbf{long})\ (directed))$

---

d: **register long**, §9.	$n0$: **long**, §43.	s: **register long**, §9.
directed: **long**, §37.	$n1$: **long**, §43.	*size_bits*: **unsigned long**, §37.
directed: **long**, §43.	$n2$: **long**, §43.	*sscanf*: **int** ( ), <stdio.h>.
gb_new_arc: **void** ( ),	$n3$: **long**, §43.	*subsets*: **Graph** *( ), §37.
GB_GRAPH §30.	$n4$: **long**, §43.	v: **register Vertex** *, §9.
gb_new_edge: **void** ( ),	*name*: **char** *, GB_GRAPH §9.	**Vertex** = **struct**, GB_GRAPH §9.
GB_GRAPH §31.	*new_graph*: **Graph** *, §9.	*vertices*: **Vertex** *, GB_GRAPH §20.
j: **register long**, §9.	*perms*: **Graph** *( ), §43.	xx: **static long** [ ], §10.
max_inv: **unsigned long**, §43.		

42. If $max\_inv = 0$, all permutations will be considered, regardless of the number of inversions. In that case the total number of vertices in the graph will be the multinomial coefficient

$$\binom{n}{n_0, n_1, \ldots, n_d}, \qquad n = n_0 + n_1 + \cdots + n_d.$$

The maximum number of inversions in general is the number of inversions of the lexicographically last permutation, namely $\binom{n}{2} - \binom{n_0}{2} - \binom{n_1}{2} - \cdots - \binom{n_d}{2} = \sum_{0 \le j < k \le d} n_j n_k$.

Notice that in the case $d = 1$, we essentially obtain all combinations of $n0 + n1$ elements taken $n1$ at a time. The positions of the 1's correspond to the elements of a subset or sample.

If *directed* is nonzero, the graph will contain only directed arcs from permutations to neighboring permutations that have exactly one more inversion. In this case the graph corresponds to a partial ordering that is a lattice with interesting properties; see the article by Bennett and Birkhoff in *Algebra Universalis* (1994), to appear.

43. The program for *perms* is very similar in structure to the program for *simplex* already considered.

⟨ Basic subroutines 8 ⟩ +≡
  Graph *perms*($n0$, $n1$, $n2$, $n3$, $n4$, *max_inv*, *directed*)
    long $n0$, $n1$, $n2$, $n3$, $n4$;   /* composition of the multiset */
    unsigned long *max_inv*;   /* maximum number of inversions */
    long *directed*;   /* should the graph be directed? */
  { ⟨ Vanilla local variables 9 ⟩
    register long n;   /* total number of elements in multiset */
    ⟨ Normalize the permutation parameters 44 ⟩;
    ⟨ Create a graph with one vertex for each permutation 46 ⟩;
    ⟨ Name the permutations and create the arcs or edges 48 ⟩;
    if (*gb_trouble_code*) {
      gb_recycle(*new_graph*);
      panic(*alloc_fault*);   /* shucks, we ran out of memory somewhere back there */
    }
    return *new_graph*;
  }

44. ⟨ Normalize the permutation parameters 44 ⟩ ≡
  if ($n0 \equiv 0$) { $n0 = 1$; $n1 = 0$; }   /* convert the empty set into {0} */
  else if ($n0 < 0$) { $n1 = n0$; $n0 = 1$; }
  $n = $ BUF_SIZE;   /* this allows us to borrow code from *simplex*, already written */
  ⟨ Normalize the simplex parameters 27 ⟩;
  ⟨ Determine n and the maximum possible number of inversions 45 ⟩;
This code is used in section 43.

45. Here we want to set *max_inv* to the maximum possible number of inversions, if the given value of *max_inv* is zero or if it exceeds that maximum number.

⟨ Determine n and the maximum possible number of inversions 45 ⟩ ≡
  { **register long** ss;   /* max inversions known to be possible */
    for ($k = 0, s = ss = 0$; $k \le d$; $ss \mathrel{+}= s * nn[k], s \mathrel{+}= nn[k], k{+}{+}$)

      if $(nn[k] \geq$ BUF_SIZE$)$ *panic*(bad_specs);   /* too many elements in the multiset */
      if $(s \geq$ BUF_SIZE$)$ *panic*($bad\_specs + 1$);   /* too many elements in the multiset */
      $n = s$;
      if $(max\_inv \equiv 0 \vee max\_inv > ss)$ $max\_inv = ss$;
   }

This code is used in section 44.

46.   To determine the number of vertices, we sum the first $max\_inv + 1$ coefficients of a power series in which the coefficient of z^j is the number of permutations having j inversions. It is known [*Sorting and Searching*, exercise 5.1.2–16] that this power series is the "z-multinomial coefficient"

$$\binom{n}{n_0, \ldots, n_d}_z = \frac{n!_z}{n_0!_z \ldots n_d!_z}, \qquad \text{where} \qquad m!_z = \prod_{k=1}^{m} \frac{1 - z^k}{1 - z}.$$

⟨ Create a graph with one vertex for each permutation 46 ⟩ ≡
   { **long** *nverts*;   /* the number of vertices */
   register long *$*coef$* = *gb_typed_alloc*($max\_inv + 1$, **long**, *working_storage*);
   if (*gb_trouble_code*) *panic*($no\_room + 1$);   /* can't allocate *coef* array */
   $coef[0] = 1$;
   for $(j = 1, s = nn[0];\ j \leq d;\ s\ {+}{=}\ nn[j], j{+}{+})$
     ⟨ Multiply the power series coefficients by $\prod_{1 \leq k \leq n_j}(1 - z^{s+k})/(1 - z^k)$ 47 ⟩;
   for $(k = 1, nverts = 1;\ k \leq max\_inv;\ k{+}{+})$ {
     $nverts\ {+}{=}\ coef[k]$;
     if $(nverts > 1000000000)$ *panic*($very\_bad\_specs$);   /* way too big */
   }
   gb_free(*working_storage*);   /* recycle the *coef* array */
   new_graph = *gb_new_graph*(*nverts*);
   if (*new_graph* $\equiv \Lambda$) *panic*(no_room);   /* out of memory before we're even started */
   sprintf(*new_graph*→*id*, "`perms(%ld,%ld,%ld,%ld,%ld,%lu,%d)`", *n0*, *n1*, *n2*, *n3*, *n4*, *max_inv*,
     directed ? 1 : 0);
   strcpy(*new_graph*→*util_types*, "`VVZZZZZZZZZZZZ`");   /* hash table will be used */
   }

This code is used in section 43.

---

47. After multiplication by $(1 - z^{k+s})/(1 - z^k)$, the coefficients of the power series will be nonnegative, because they are the coefficients of a z-multinomial coefficient.

$\langle$ Multiply the power series coefficients by $\prod_{1 \le k \le n_j} (1 - z^{s+k})/(1 - z^k)$ 47 $\rangle \equiv$

```
for (k = 1; k ≤ nn[j]; k++) {  register long ii;
    for (i = max_inv, ii = i − k − s; ii ≥ 0; ii−−, i−−)  coef[i] −= coef[ii];
    for (i = k, ii = 0; i ≤ max_inv; i++, ii++) {
        coef[i] += coef[ii];
        if (coef[i] > 1000000000) panic(very_bad_specs + 1);       /* way too big */
    }
}
```

This code is used in section 46.

48. As we generate the permutations, we maintain a table $(y_1, \ldots, y_n)$, where y_k is the number of inversions whose first element is the kth element of the multiset. For example, if the multiset is $\{0, 0, 1, 2\}$ and the current permutation is $(2, 0, 1, 0)$, the inversion table is $(y_1, y_2, y_3, y_4) = (0, 0, 1, 3)$. Clearly $0 \le y_k < k$, and $y_k \le y_{k-1}$ when the kth element of the multiset is the same as the $(k-1)$st element. These conditions are necessary and sufficient to define a valid inversion table. We will generate permutations in lexicographic order of their inversion tables.

For convenience, we set up another array z, which holds the initial inversion-free permutation.

$\langle$ Name the permutations and create the arcs or edges 48 $\rangle \equiv$

```
{ register long *xtab, *ytab, *ztab;     /* permutations and their inversions */
  long m = 0;       /* current number of inversions */
  ⟨Initialize xtab, ytab, and ztab 49⟩;
  v = new_graph→vertices;
  while (1) {
      ⟨Assign a symbolic name for (x₁,…,xₙ) to vertex v 52⟩;
      ⟨Create arcs or edges from previous permutations to v 53⟩;
      v++;
      ⟨Advance to the next perm; goto last if there are no more solutions 50⟩;
  }
last: if (v ≠ new_graph→vertices + new_graph→n) panic(impossible);     /* can't happen */
  gb_free(working_storage);
}
```

This code is used in section 43.

49. $\langle$ Initialize xtab, ytab, and ztab 49 $\rangle \equiv$

```
xtab = gb_typed_alloc(3 * n + 3, long, working_storage);
if (gb_trouble_code) {      /* can't allocate xtab */
    gb_recycle(new_graph); panic(no_room + 2); }
ytab = xtab + (n + 1); ztab = ytab + (n + 1);
for (j = 0, k = 1, s = nn[0]; ; k++) {
    xtab[k] = ztab[k] = j;     /* ytab[k] = 0 */
    if (k ≡ s) {
        if (++j > d) break; else s += nn[j];
    }
}
```

This code is used in section 48.

50. Here is the heart of the permutation logic. We find the largest k such that y_k can legitimately be increased by 1. When we encounter a k for which y_k cannot be increased, we set $y_k = 0$ and adjust the x's accordingly. If no y_k can be increased, we are done.

$\langle$ Advance to the next perm; **goto** *last* if there are no more solutions 50 $\rangle \equiv$

```
for (k = n; k; k--) {
    if (m < max_inv ∧ ytab[k] < k - 1)
        if (ytab[k] < ytab[k - 1] ∨ ztab[k] > ztab[k - 1]) goto move;
    if (ytab[k]) {
        for (j = k - ytab[k]; j < k; j++) xtab[j] = xtab[j + 1];
        m -= ytab[k];
        ytab[k] = 0;
        xtab[k] = ztab[k];
    }
}
goto last;
move: j = k - ytab[k];        /* the current location of the kth element, z_k */
    xtab[j] = xtab[j - 1];  xtab[j - 1] = ztab[k];
    ytab[k]++;  m++;
```

This code is used in section 48.

51. A permutation is encoded as a sequence of nonblank characters, using an abbreviated copy of the *imap* code from GB_IO and omitting the characters that need to be quoted within strings. If the number of distinct elements in the multiset is at most 62, only digits and letters will appear in the vertex name.

$\langle$ Private variables 3 $\rangle$ $+\equiv$

```
static char *short_imap = "0123456789ABCDEFGHIJKLMNOPQRSTUVWXYZabcdefghijklmnopqrstuvw\
    xyz_^~&@,;.:?!%#$+-*/|<=>()[]{}'\"";
```

52. $\langle$ Assign a symbolic name for $(x_1, \ldots, x_n)$ to vertex v 52 $\rangle \equiv$

```
{ register char *p;
  register long *q;

  for (p = &buffer[n - 1], q = &xtab[n]; q > xtab; p--, q--) *p = short_imap[*q];
  v→name = gb_save_string(buffer);
  hash_in(v);     /* enter v→name into the hash table (via utility fields u, v) */
}
```

This code is used in section 48.

---

buffer: **static char** [], §5.
coef: **register long** *, §46.
d: **register long**, §9.
gb_free: **void** ( ), GB_GRAPH §16.
gb_recycle: **void** ( ), GB_GRAPH §40.
gb_save_string: **char** *( ),
   GB_GRAPH §35.
gb_trouble_code: **long**,
   GB_GRAPH §14.
gb_typed_alloc = macro ( ),
   GB_GRAPH §11.

hash_in: **void** ( ), GB_GRAPH §44.
i: **register long**, §9.
imap: **static char** *, GB_IO §11.
impossible = 90, GB_GRAPH §7.
j: **register long**, §9.
k: **register long**, §9.
max_inv: **unsigned long**, §43.
n: **register long**, §43.
n: **long**, GB_GRAPH §20.
name: **char** *, GB_GRAPH §9.
new_graph: **Graph** *, §9.

nn: **static long** [], §10.
no_room = 1, GB_GRAPH §7.
panic = macro ( ), §4.
s: **register long**, §9.
u: util, GB_GRAPH §9.
v: **register Vertex** *, §9.
v: util, GB_GRAPH §9.
vertices: **Vertex** *, GB_GRAPH §20.
very_bad_specs = 40, GB_GRAPH §7.
working_storage: **static Area**, §3.

53. Since we are generating the vertices in lexicographic order of their inversions, it is easy to identify all adjacent vertices that precede the current setting of $(x_1, \ldots, x_n)$. We locate them via their symbolic names.

⟨ Create arcs or edges from previous permutations to v 53 ⟩ ≡

    for $(j = 1; \ j < n; \ j\texttt{++})$

        if $(xtab[j] > xtab[j+1])$ { **register Vertex** $*u$;   /* previous vertex adjacent to v */

            $buffer[j-1] = short\_imap[xtab[j+1]]$; $buffer[j] = short\_imap[xtab[j]]$;

            $u = hash\_out(buffer)$;

            if $(u \equiv \Lambda)$ $panic(impossible + 2)$;   /* can't happen */

            if $(directed)$ $gb\_new\_arc(u, v, 1_{\,\mathrm{L}})$;

            else $gb\_new\_edge(u, v, 1_{\,\mathrm{L}})$;

            $buffer[j-1] = short\_imap[xtab[j]]$; $buffer[j] = short\_imap[xtab[j+1]]$;

        }

This code is used in section 48.

54.   Partition graphs.   The subroutine call $parts(n, max\_parts, max\_size, directed)$ creates a graph whose vertices represent the different ways to partition the integer n into at most max_parts parts, where each part is at most max_size. Two partitions are adjacent in the graph if one can be obtained from the other by combining two parts. Each arc has length 1.

For example, the partitions of 5 are

$$5, \quad 4+1, \quad 3+2, \quad 3+1+1, \quad 2+2+1, \quad 2+1+1+1, \quad 1+1+1+1+1.$$

Here 5 is adjacent to $4+1$ and to $3+2$; $4+1$ is adjacent also to $3+1+1$ and to $2+2+1$; $3+2$ is adjacent also to $3+1+1$ and to $2+2+1$; etc. If max_size is 3, the partitions 5 and $4+1$ would not be included in the graph. If max_parts is 3, the partitions $2+1+1+1$ and $1+1+1+1+1$ would not be included.

If max_parts or max_size are zero, they are reset to be equal to n so that they make no restriction on the partitions.

If *directed* is nonzero, the graph will contain only directed arcs from partitions to their neighbors that have exactly one more part.

The special case when we want to generate all $p(n)$ partitions of the integer n can be obtained by calling $all\_parts(n, directed)$.

⟨ gb_basic.h   1 ⟩ +≡
  #**define** $all\_parts(n, directed)$ $parts(($**long**$)(n), 0_L, 0_L, ($**long**$)(directed))$

---

buffer: **static char** [], §5.
directed: **long**, §43.
directed: **long**, §55.
gb_new_arc: **void** ( ),
   GB_GRAPH §30.
gb_new_edge: **void** ( ),
   GB_GRAPH §31.

hash_out: **Vertex** *( ),
   GB_GRAPH §46.
impossible = 90, GB_GRAPH §7.
j: **register long**, §9.
max_parts: **unsigned long**, §55.
max_size: **unsigned long**, §55.
n: **register long**, §43.

n: **unsigned long**, §55.
panic = macro ( ), §4.
parts: **Graph** *( ), §55.
short_imap: **static char** *, §51.
v: **register Vertex** *, §9.
Vertex = **struct**, GB_GRAPH §9.
xtab: **register long** *, §48.

55. The program for *parts* is very similar in structure to the program for *perms* already considered.

⟨ Basic subroutines 8 ⟩ +≡
  Graph **parts*(n, max_parts, max_size, directed)
      unsigned long n;   /* the number being partitioned */
      unsigned long max_parts;   /* maximum number of parts */
      unsigned long max_size;   /* maximum size of each part */
      long directed;   /* should the graph be directed? */
  { ⟨ Vanilla local variables 9 ⟩
    if (max_parts ≡ 0 ∨ max_parts > n) max_parts = n;
    if (max_size ≡ 0 ∨ max_size > n) max_size = n;
    if (max_parts > MAX_D) panic(bad_specs);   /* too many parts allowed */
    ⟨ Create a graph with one vertex for each partition 56 ⟩;
    ⟨ Name the partitions and create the arcs or edges 57 ⟩;
    if (gb_trouble_code) {
      gb_recycle(new_graph);
      panic(alloc_fault);   /* doggone it, we ran out of memory somewhere back there */
    }
    return new_graph;
  }

56. The number of vertices is the coefficient of z^n in the z-binomial coefficient $\binom{m+p}{m}_z$, where $m = max\_parts$ and $p = max\_size$. This coefficient is calculated as in the *perms* routine.

⟨ Create a graph with one vertex for each partition 56 ⟩ ≡
  { **long** nverts;   /* the number of vertices */
    register long **coef* = gb_typed_alloc(n + 1, **long**, working_storage);

    if (gb_trouble_code) panic(no_room + 1);   /* can't allocate coef array */
    coef[0] = 1;
    for (k = 1; k ≤ max_parts; k++) {
      for (j = n, i = n − k − max_size; i ≥ 0; i−−, j−−) coef[j] −= coef[i];
      for (j = k, i = 0; j ≤ n; i++, j++) {
        coef[j] += coef[i];
        if (coef[j] > 1000000000) panic(very_bad_specs);   /* way too big */
      }
    }
    nverts = coef[n];
    gb_free(working_storage);   /* recycle the coef array */
    new_graph = gb_new_graph(nverts);
    if (new_graph ≡ Λ) panic(no_room);   /* out of memory before we're even started */
    sprintf(new_graph→id, "parts(%lu,%lu,%lu,%d)", n, max_parts, max_size, directed ? 1 : 0);
    strcpy(new_graph→util_types, "VVZZZZZZZZZZZZ");   /* hash table will be used */
  }

This code is used in section 55.

57. As we generate the partitions, we maintain the numbers $\sigma_j = n - (x_1 + \cdots + x_{j-1}) = x_j + x_{j+1} + \cdots$, somewhat as we did in the *simplex* routine. We set $x_0 = max\_size$ and $y_j = max\_parts + 1 - j$; then when values $(x_1, \ldots, x_{j-1})$ are given, the conditions

$$\sigma_j / y_j \leq x_j \leq \sigma_j, \qquad x_j \leq x_{j-1}$$

characterize the legal values of x_j.

⟨ Name the partitions and create the arcs or edges 57 ⟩ ≡

```
v = new_graph→vertices;
xx[0] = max_size;  sig[1] = n;
for (k = max_parts, s = 1;  k > 0;  k−−, s++)  yy[k] = s;
if (max_size * max_parts ≥ n) {
    k = 1;  xx[1] = (n − 1)/max_parts + 1;      /* ⌈n/max_parts⌉ */
    while (1) {
        ⟨ Complete the partial solution (x₁,…,x_k) 58 ⟩;
        ⟨ Assign the name x₁ + ⋯ + x_d to vertex v 60 ⟩;
        ⟨ Create arcs or edges from v to previous partitions 61 ⟩;
        v++;
        ⟨ Advance to the next partial solution (x₁,…,x_k), where k is as large as possible; goto last if
            there are no more solutions 59 ⟩;
    }
}
last: if (v ≠ new_graph→vertices + new_graph→n)  panic(impossible);      /* can't happen */
```

This code is used in section 55.

58. ⟨ Complete the partial solution $(x_1, \ldots, x_k)$ 58 ⟩ ≡

```
for (s = sig[k] − xx[k], k++;  s;  k++) {
    sig[k] = s;
    xx[k] = (s − 1)/yy[k] + 1;
    s −= xx[k];
}
d = k − 1;      /* the smallest part is x_d */
```

This code is used in section 57.

---

alloc_fault = −1, GB_GRAPH §7.
bad_specs = 30, GB_GRAPH §7.
d: **register long**, §9.
gb_free: **void** ( ), GB_GRAPH §16.
gb_new_graph: **Graph** *( ),
   GB_GRAPH §23.
gb_recycle: **void** ( ), GB_GRAPH §40.
gb_trouble_code: **long**,
   GB_GRAPH §14.
gb_typed_alloc = macro ( ),
   GB_GRAPH §11.
Graph = **struct**, GB_GRAPH §20.

i: **register long**, §9.
id: **char** [], GB_GRAPH §20.
impossible = 90, GB_GRAPH §7.
j: **register long**, §9.
k: **register long**, §9.
MAX_D = 91, §10.
n: **long**, GB_GRAPH §20.
new_graph: **Graph** *, §9.
no_room = 1, GB_GRAPH §7.
panic = macro ( ), §4.
perms: **Graph** *( ), §43.
s: **register long**, §9.

sig: **static long** [], §10.
simplex: **Graph** *( ), §26.
sprintf: **int** ( ), <stdio.h>.
strcpy: **char** *( ), <string.h>.
util_types: **char** [], GB_GRAPH §20.
v: **register Vertex** *, §9.
vertices: **Vertex** *, GB_GRAPH §20.
very_bad_specs = 40, GB_GRAPH §7.
working_storage: **static Area**, §3.
xx: **static long** [], §10.
yy: **static long** [], §10.

59. Here we seek the largest k such that x_k can be increased without violating the necessary and sufficient conditions stated earlier.

⟨ Advance to the next partial solution $(x_1, \ldots, x_k)$, where k is as large as possible; **goto** *last* if there are no more solutions 59 ⟩ ≡

    if $(d \equiv 1)$ **goto** *last*;
    for $(k = d - 1;\ ;\ k\!-\!-)$ {
        if $(xx[k] < sig[k] \wedge xx[k] < xx[k-1])$ **break**;
        if $(k \equiv 1)$ **goto** *last*;
    }
    $xx[k]\!+\!+;$

This code is used in section 57.

60.   ⟨ Assign the name $x_1 + \cdots + x_d$ to vertex v 60 ⟩ ≡
    { **register char** $*p = buffer;$ /* string pointer */
      for $(k = 1;\ k \le d;\ k\!+\!+)$ {
        sprintf $(p,$ "+%1d", $xx[k]);$
        while $(*p)\ p\!+\!+;$
      }
      $v{\to}name = gb\_save\_string(\&buffer[1]);$ /* omit $buffer[0]$, which is '+' */
      $hash\_in(v);$ /* enter $v{\to}name$ into the hash table (via utility fields u, v) */
    }

This code is used in section 57.

61. Since we are generating the partitions in lexicographic order of their parts, it is reasonably easy to identify all adjacent vertices that precede the current setting of $(x_1, \ldots, x_d)$, by splitting x_j into two parts when $x_j \ne x_{j+1}$. We locate previous partitions via their symbolic names.

⟨ Create arcs or edges from v to previous partitions 61 ⟩ ≡

    if $(d < max\_parts)$ {
      $xx[d + 1] = 0;$
      for $(j = 1;\ j \le d;\ j\!+\!+)$ {
        if $(xx[j] \ne xx[j+1])$ { **long** $a, b;$
          for $(b = xx[j]/2, a = xx[j] - b;\ b;\ a\!+\!+, b\!-\!-)$ ⟨ Generate a subpartition $(n_1, \ldots, n_{d+1})$ by
             splitting x_j into $a + b$, and make that subpartition adjacent to v 62 ⟩;
        }
        $nn[j] = xx[j];$
      }
    }

This code is used in section 57.

62. The values of $(x_1, \ldots, x_{j-1})$ have already been copied into $(n_1, \ldots, n_{j-1})$. Our job is to copy the smaller parts $(x_{j+1}, \ldots, x_d)$ while inserting a and b in their proper places, knowing that $a \ge b$.

⟨ Generate a subpartition $(n_1, \ldots, n_{d+1})$ by splitting x_j into $a + b$, and make that subpartition adjacent to v 62 ⟩ ≡

    { **register Vertex** $*u;$ /* previous vertex adjacent to v */
      register char $*p = buffer;$
      for $(k = j + 1;\ xx[k] > a;\ k\!+\!+)\ nn[k-1] = xx[k];$

$nn[k-1] = a;$
for (; $xx[k] > b$; k++) $nn[k] = xx[k]$;
$nn[k] = b;$
for (; $k \leq d$; k++) $nn[k+1] = xx[k]$;
for ($k = 1$; $k \leq d+1$; k++) {
    $sprintf(p, \texttt{"+\%ld"}, nn[k])$;
    while ($*p$) p++;
}
$u = hash\_out(\&buffer[1]);$
if ($u \equiv \Lambda$) $panic(impossible + 2)$;   /* can't happen */
if ($directed$) $gb\_new\_arc(v, u, 1_L)$;
else $gb\_new\_edge(v, u, 1_L)$;
}

This code is used in section 61.

---

buffer: **static char** [ ], §5.
d: **register long**, §9.
directed: **long**, §55.
gb_new_arc: **void** ( ),
    GB_GRAPH §30.
gb_new_edge: **void** ( ),
    GB_GRAPH §31.
gb_save_string: **char** *( ),
    GB_GRAPH §35.

hash_in: **void** ( ), GB_GRAPH §44.
hash_out: **Vertex** *( ),
    GB_GRAPH §46.
impossible = 90, GB_GRAPH §7.
j: **register long**, §9.
k: **register long**, §9.
last: label, §57.
max_parts: **unsigned long**, §55.
name: **char** *, GB_GRAPH §9.

nn: **static long** [ ], §10.
panic = macro ( ), §4.
sig: **static long** [ ], §10.
sprintf: **int** ( ), <stdio.h>.
u: **util**, GB_GRAPH §9.
v: **register Vertex** *, §9.
v: **util**, GB_GRAPH §9.
Vertex = **struct**, GB_GRAPH §9.
xx: **static long** [ ], §10.

63. Binary tree graphs. The subroutine call $binary(n, max\_height, directed)$ creates a graph whose vertices represent the binary trees with n internal nodes and with all leaves at a distance that is at most max_height from the root. Two binary trees are adjacent in the graph if one can be obtained from the other by a single application of the associative law for binary operations, i.e., by replacing some subtree of the form $(\alpha \cdot \beta) \cdot \gamma$ by the subtree $\alpha \cdot (\beta \cdot \gamma)$. (This transformation on binary trees is often called a "rotation.") If the *directed* parameter is nonzero, the directed arcs go from a tree containing $(\alpha \cdot \beta) \cdot \gamma$ to a tree containing $\alpha \cdot (\beta \cdot \gamma)$ in its place; otherwise the graph is undirected. Each arc has length 1.

For example, the binary trees with three internal nodes form a circuit of length 5. They are

$$(a \cdot b) \cdot (c \cdot d), \quad a \cdot (b \cdot (c \cdot d)), \quad a \cdot ((b \cdot c) \cdot d), \quad (a \cdot (b \cdot c)) \cdot d, \quad ((a \cdot b) \cdot c) \cdot d,$$

if we use infix notation and name the leaves (a, b, c, d) from left to right. Here each tree is related to its two neighbors by associativity. The first and last trees are also related in the same way.

If $max\_height = 0$, it is changed to n, which means there is no restriction on the height of a leaf. In this case the graph will have exactly $\binom{2n+1}{n}/(2n + 1)$ vertices; furthermore, each vertex will have exactly $n - 1$ neighbors, because a rotation will be possible just above every internal node except the root. The graph in this case can also be interpreted geometrically: The vertices are in one-to-one correspondence with the triangulations of a regular $(n+2)$-gon; two triangulations are adjacent if and only if one is obtained from the other by replacing the pair of adjacent triangles ABC, DCB by the pair ADC, BDA.

The partial ordering corresponding to the directed graph on $\binom{2n+1}{n}/(2n + 1)$ vertices created by $all\_trees(n, 1)$ is a lattice with interesting properties. See Huang and Tamari, *Journal of Combinatorial Theory* **A13** (1972), 7–13.

⟨ gb_basic.h   1 ⟩ +≡
#**define** $all\_trees(n, directed)$ $binary((\textbf{long}) (n), 0_L, (\textbf{long}) (directed))$

64. The program for *binary* is very similar in structure to the program for *parts* already considered. But the details are more exciting.

⟨ Basic subroutines 8 ⟩ +≡
  Graph $*binary(n, max\_height, directed)$
    unsigned long n;   /* the number of internal nodes */
    unsigned long max_height;   /* maximum height of a leaf */
    long $directed$;   /* should the graph be directed? */
  { ⟨ Vanilla local variables 9 ⟩
    if $(2 * n + 2 > \texttt{BUF\_SIZE})$ $panic(bad\_specs)$;   /* n is too huge for us */
    if $(max\_height \equiv 0 \lor max\_height > n)$ $max\_height = n$;
    if $(max\_height > 30)$ $panic(very\_bad\_specs)$;   /* more than a billion vertices */
    ⟨ Create a graph with one vertex for each binary tree 65 ⟩;
    ⟨ Name the trees and create the arcs or edges 67 ⟩;
    if $(gb\_trouble\_code)$ {
      $gb\_recycle(new\_graph)$;
      $panic(alloc\_fault)$;   /* uff da, we ran out of memory somewhere back there */
    }
    return new_graph;
  }

65.   The number of vertices is the coefficient of z^n in the power series G_h, where $h = max\_height$ and G_h satisfies the recurrence

$$G_0 = 1, \qquad G_{h+1} = 1 + zG_h^2.$$

The coefficients of G_5 are ≤ 55308, but the coefficients of G_6 are much larger; they exceed one billion when $28 \leq n \leq 49$, and they exceed one million when $17 \leq n \leq 56$. In order to avoid overflow during this calculation, we use a special method when $h \geq 6$ and $n \geq 20$: In such cases, graphs of reasonable size arise only if $n \geq 2^h - 7$, and we look at the coefficient of $z^{-(2^h-1-n)}$ in $R_h = G_h/z^{2^h-1}$, which is a power series in z^{-1} defined by the recurrence

$$R_0 = 1, \qquad R_{h+1} = R_h^2 + z^{1-2^{h+1}}.$$

⟨ Create a graph with one vertex for each binary tree 65 ⟩ ≡
```
{ long nverts;      /* the number of vertices */
  if (n ≥ 20 ∧ max_height ≥ 6) ⟨Compute nverts using the R series 66⟩
  else {
    nn[0] = nn[1] = 1;
    for (k = 2; k ≤ n; k++) nn[k] = 0;
    for (j = 2; j ≤ max_height; j++)
      for (k = n − 1; k; k−−) {
        for (s = 0, i = k; i ≥ 0; i−−) s += nn[i] * nn[k − i];      /* overflow impossible */
        nn[k + 1] = s;
      }
    nverts = nn[n];
  }
  new_graph = gb_new_graph(nverts);
  if (new_graph ≡ Λ) panic(no_room);      /* out of memory before we're even started */
  sprintf(new_graph⃗id, "binary(%lu,%lu,%d)", n, max_height, directed ? 1 : 0);
  strcpy(new_graph⃗util_types, "VVZZZZZZZZZZZZ");      /* hash table will be used */
}
```
This code is used in section 64.

---

66. The smallest nontrivial graph that is unilaterally disallowed by this procedure on the grounds of size limitations occurs when *max_height* = 6 and $n = 20$; it has 14,162,220 vertices.

⟨ Compute *nverts* using the R series 66 ⟩ ≡

```
{ register float ss;
    d = (1 L ≪ max_height) − 1 − n;
    if (d > 8) panic(bad_specs + 1);        /* too many vertices */
    if (d < 0) nverts = 0;
    else {
        nn[0] = nn[1] = 1;
        for (k = 2; k ≤ d; k++) nn[k] = 0;
        for (j = 2; j ≤ max_height; j++) {
            for (k = d; k; k−−) {
                for (ss = 0.0, i = k; i ≥ 0; i−−) ss += ((float) nn[i]) * ((float) nn[k − i]);
                if (ss > MAX_NNN) panic(very_bad_specs + 1);      /* way too big */
                for (s = 0, i = k; i ≥ 0; i−−) s += nn[i] * nn[k − i];      /* overflow impossible */
                nn[k] = s;
            }
            i = (1 L ≪ j) − 1;
            if (i ≤ d) nn[i]++;      /* add z^{1−2^j} */
        }
        nverts = nn[d];
    }
}
```

This code is used in section 65.

67. We generate the trees in lexicographic order of their Polish prefix notation, encoded in binary notation as $x_0 x_1 \ldots x_{2n}$, using '1' for an internal node and '0' for a leaf. For example, the five trees when $n = 3$ are

$$1010100, \quad 1011000, \quad 1100100, \quad 1101000, \quad 1110000,$$

in lexicographic order. The algorithm for lexicographic generation maintains three auxiliary arrays l_j, y_j, and σ_j, where

$$\sigma_j = n - j + \sum_{i=0}^{j-1} x_i = -1 + \sum_{i=j}^{2n} (1 - x_i)$$

is one less than the number of 0's (leaves) in $(x_j, \ldots, x_{2n})$. The values of l_j and y_j are harder to describe formally; l_j is 2^{h-l} when $h = max\_height$ and when x_j represents a node at level l of the tree, based on the values of $(x_0, \ldots, x_{j-1})$. The value of y_j is a binary encoding of tree levels in which an internal node has not yet received a right child; y_j is also the maximum number of future leaves that can be produced by previously specified internal nodes without exceeding the maximum height. The number of 1-bits in y_j is the minimum number of future leaves, based on previous specifications.

Therefore if $\sigma_j > y_j$, x_j is forced to be 1. If $l_j = 1$, x_j is forced to be 0. If the number of 1-bits of y_j is equal to σ_j, x_j is forced to be 0. Otherwise x_j can be either 0 or 1, and it will be possible to complete the partial solution $x_0 \ldots x_j$ to a full Polish prefix code $x_0 \ldots x_{2n}$.

For example, here are the arrays for one of the binary trees that is generated when $n = h = 3$:

$$
\begin{array}{rccccccccc}
j & = & 0 & 1 & 2 & 3 & 4 & 5 & 6 \\
l_j & = & 8 & 4 & 2 & 2 & 1 & 1 & 4 \\
y_j & = & 0 & 4 & 6 & 4 & 5 & 4 & 0 \\
\sigma_j & = & 3 & 3 & 3 & 2 & 2 & 1 & 0 \\
x_j & = & 1 & 1 & 0 & 1 & 0 & 0 & 0
\end{array}
$$

If $x_j = 1$ and $j < 2n$, we have $l_{j+1} = l_j/2$, $y_{j+1} = y_j + l_{j+1}$, and $\sigma_{j+1} = \sigma_j$. If $x_j = 0$ and $j < 2n$, we have $l_{j+1} = 2^t$, $y_{j+1} = y_j - 2^t$, and $\sigma_{j+1} = \sigma_j - 1$, where 2^t is the least power of 2 in the binary representation of y_j. It is not difficult to prove by induction that $\sigma_j < y_j + l_j$, assuming that $n < 2^h$.

⟨ Name the trees and create the arcs or edges 67 ⟩ ≡
  { **register long** *xtab, *ytab, *ltab, *stab;
    ⟨ Initialize *xtab*, *ytab*, *ltab*, and *stab*; also set $d = 2n$ 68 ⟩;
    $v = new\_graph{\rightarrow}vertices$;
    if $(ltab[0] > n)$ {
      $k = 0$; $xtab[0] = n$? 1 : 0;
      while (1) {
        ⟨ Complete the partial tree $x_0 \ldots x_k$ 69 ⟩;
        ⟨ Assign a Polish prefix code name to vertex v 71 ⟩;
        ⟨ Create arcs or edges from v to previous trees 72 ⟩;
        $v{+}{+}$;
        ⟨ Advance to the next partial tree $x_0 \ldots x_k$, where k is as large as possible; **goto** *last* if there
            are no more solutions 70 ⟩;
      }
    }
  }
last: **if** $(v \neq new\_graph{\rightarrow}vertices + new\_graph{\rightarrow}n)$ *panic*(*impossible*);   /* can't happen */
  $gb\_free(working\_storage)$;

This code is used in section 64.

---

68. ⟨ Initialize *xtab*, *ytab*, *ltab*, and *stab*; also set $d = 2n$ 68 ⟩ ≡
$xtab = gb\_typed\_alloc(8 * n + 4, \textbf{long}, working\_storage);$
if $(gb\_trouble\_code)$ {   /* no room for *xtab* */
  $gb\_recycle(new\_graph);$ $panic(no\_room + 2);$ }
$d = n + n;$
$ytab = xtab + (d + 1);$
$ltab = ytab + (d + 1);$
$stab = ltab + (d + 1);$
$ltab[0] = 1_\text{L} \ll max\_height;$
$stab[0] = n;$ /* $ytab[0] = 0$ */

This code is used in section 67.

69. ⟨ Complete the partial tree $x_0 \ldots x_k$ 69 ⟩ ≡
for $(j = k + 1;\ j \leq d;\ j{+}{+})$ {
  if $(xtab[j - 1])$ {
    $ltab[j] = ltab[j - 1] \gg 1;$
    $ytab[j] = ytab[j - 1] + ltab[j];$
    $stab[j] = stab[j - 1];$
  } **else** {
    $ytab[j] = ytab[j - 1]\ \&\ (ytab[j - 1] - 1);$ /* remove least significant 1-bit */
    $ltab[j] = ytab[j - 1] - ytab[j];$
    $stab[j] = stab[j - 1] - 1;$
  }
  if $(stab[j] \leq ytab[j])$ $xtab[j] = 0;$
  else $xtab[j] = 1;$ /* this is the lexicographically smallest completion */
}

This code is used in section 67.

70. As in previous routines, we seek the largest k such that x_k can be increased without violating the necessary and sufficient conditions stated earlier.

⟨ Advance to the next partial tree $x_0 \ldots x_k$, where k is as large as possible; **goto** *last* if there are no
    more solutions 70 ⟩ ≡
  for $(k = d - 1;\ ;\ k{-}{-})$ {
    if $(k \leq 0)$ **goto** *last*;   /* this happens only when $n \leq 1$ */
    if $(xtab[k])$ **break**;   /* find rightmost 1 */
  }
  for $(k{-}{-};\ ;\ k{-}{-})$ {
    if $(xtab[k] \equiv 0 \wedge ltab[k] > 1)$ **break**;
    if $(k \equiv 0)$ **goto** *last*;
  }
  $xtab[k]{+}{+};$

This code is used in section 67.

71. In the *name* field, we encode internal nodes of the binary tree by '.' and leaves by 'x'. Thus the five trees shown above in binary code will be named

.x.x.xx,   .x..xxx,   ..xx.xx,   ..x.xxx,   ...xxxx,

respectively.

⟨ Assign a Polish prefix code name to vertex v 71 ⟩ ≡
  { **register char** $*p = buffer$;   /∗ string pointer ∗/
    for $(k = 0;\ k \le d;\ k{+}{+}, p{+}{+})$ $*p = (xtab[k]\ ?\ \text{'.'} : \text{'x'})$;
    $v{\to}name = gb\_save\_string(buffer)$;
    $hash\_in(v)$;   /∗ enter $v{\to}name$ into the hash table (via utility fields u, v) ∗/
  }

This code is used in section 67.

72. Since we are generating the trees in lexicographic order of their Polish prefix notation, it is relatively easy to find all pairs of trees that are adjacent via one application of the associative law: We simply replace a substring of the form $..\alpha\beta$ by $.\alpha.\beta$, when α and β are Polish prefix strings. The result comes earlier in lexicographic order, so it will be an existing vertex unless it violates the *max_height* restriction.

⟨ Create arcs or edges from v to previous trees 72 ⟩ ≡
  for $(j = 0;\ j < d;\ j{+}{+})$
    if $(xtab[j] \equiv 1 \wedge xtab[j + 1] \equiv 1)$ {
      for $(i = j + 1, s = 0;\ s \ge 0;\ s\mathrel{+}= (xtab[i + 1] \ll 1) - 1, i{+}{+})$ $xtab[i] = xtab[i + 1]$;
      $xtab[i] = 1$;
      { **register char** $*p = buffer$;   /∗ string pointer ∗/
      register Vertex $*u$;
        for $(k = 0;\ k \le d;\ k{+}{+}, p{+}{+})$ $*p = (xtab[k]\ ?\ \text{'.'} : \text{'x'})$;
        $u = hash\_out(buffer)$;
        if (u) {
          if $(directed)$ $gb\_new\_arc(v, u, 1_\text{L})$;
          else $gb\_new\_edge(v, u, 1_\text{L})$;
        }
      }
      for $(i{-}{-};\ i > j;\ i{-}{-})$ $xtab[i + 1] = xtab[i]$;   /∗ restore *xtab* ∗/
      $xtab[i + 1] = 1$;
    }

This code is used in section 67.

---

buffer: **static char** [ ], §5.
d: **register long**, §9.
directed: **long**, §64.
gb_new_arc: **void** ( ),
  GB_GRAPH §30.
gb_new_edge: **void** ( ),
  GB_GRAPH §31.
gb_recycle: **void** ( ), GB_GRAPH §40.
gb_save_string: **char** ∗( ),
  GB_GRAPH §35.
gb_trouble_code: **long**,
  GB_GRAPH §14.
gb_typed_alloc = macro ( ),

GB_GRAPH §11.
hash_in: **void** ( ), GB_GRAPH §44.
hash_out: **Vertex** ∗( ),
  GB_GRAPH §46.
i: **register long**, §9.
j: **register long**, §9.
k: **register long**, §9.
last: **label**, §67.
ltab: **register long** ∗, §67.
max_height: **unsigned long**, §64.
n: **unsigned long**, §64.
name: **char** ∗, GB_GRAPH §9.

new_graph: **Graph** ∗, §9.
no_room = 1, GB_GRAPH §7.
panic = macro ( ), §4.
s: **register long**, §9.
stab: **register long** ∗, §67.
u: **util**, GB_GRAPH §9.
v: **register Vertex** ∗, §9.
v: **util**, GB_GRAPH §9.
Vertex = **struct**, GB_GRAPH §9.
working_storage: **static Area**, §3.
xtab: **register long** ∗, §67.
ytab: **register long** ∗, §67.

73. Complementing and copying.   We have seen how to create a wide variety of basic graphs with the *board*, *simplex*, *subsets*, *perms*, *parts*, and *binary* procedures. The remaining routines of GB_BASIC are somewhat different. They transform existing graphs into new ones, thereby presenting us with an almost mind-boggling array of further possibilities.

The first of these transformations is perhaps the simplest. It complements a·given graph, making vertices adjacent if and only if they were previously nonadjacent. More precisely, the subroutine call *complement*(*g*, *copy*, *self*, *directed*) returns a graph with the same vertices as *g*, but with complemented arcs. If *self* $\neq$ 0, the new graph will have a self-loop from a vertex *v* to itself when the original graph did not; if *self* = 0, the new graph will have no self-loops. If *directed* $\neq$ 0, the new graph will have an arc from *u* to *v* when the original graph did not; if *directed* = 0, the new graph will be undirected, and it will have an edge between *u* and *v* when the original graph did not. In the latter case, the original graph should also be undirected (that is, its arcs should come in pairs, as described in the *gb_new_edge* routine of GB_GRAPH).

If *copy* $\neq$ 0, a double complement will actually be done. This means that the new graph will essentially be a copy of the old, except that duplicate arcs (and possibly self-loops) will be removed. Information that might have been in the utility fields is not copied, and arc lengths are all set to 1.

One possibly useful feature of the graphs returned by *complement* is worth noting. The vertices adjacent to *v*, namely the list

$$v{\rightarrow}arcs{\rightarrow}tip, \quad v{\rightarrow}arcs{\rightarrow}next{\rightarrow}tip, \quad v{\rightarrow}arcs{\rightarrow}next{\rightarrow}next{\rightarrow}tip, \quad \ldots ,$$

will be in strictly decreasing order (except in the case of an undirected self-loop, when *v* itself will appear twice in succession).

74.   ⟨ Basic subroutines 8 ⟩ +≡
  Graph **complement*(*g*, *copy*, *self*, *directed*)
      Graph **g*;   /* graph to be complemented */
      long *copy*;   /* should we double-complement? */
      long *self*;   /* should we produce self-loops? */
      long *directed*;   /* should the graph be directed? */
  { ⟨ Vanilla local variables 9 ⟩

  register long *n*;
  register Vertex **u*;
  register siz_t *delta*;   /* difference in memory addresses */
  if (*g* ≡ Λ) *panic*(*missing_operand*);   /* where's *g*? */
  ⟨ Set up a graph with the vertices of *g* 75 ⟩;
  sprintf(*buffer*, ",%d,%d,%d", *copy* ? 1 : 0, *self* ? 1 : 0, *directed* ? 1 : 0);
  make_compound_id(*new_graph*, "complement(", *g*, *buffer*);
  ⟨ Insert complementary arcs or edges 76 ⟩;
  if (*gb_trouble_code*) {
    gb_recycle(*new_graph*);
    panic(*alloc_fault*);   /* worse luck, we ran out of memory somewhere back there */
  }
  return *new_graph*;
  }

75. In several of the following routines, it is efficient to circumvent C's normal rules for pointer arithmetic, and to use the fact that the vertices of a graph being copied are a constant distance away in memory from the vertices of its clone.

#define *vert_offset*($v, delta$) (($\textbf{Vertex}$ *) ((($\textbf{siz\_t}$) v) + $delta$))

⟨ Set up a graph with the vertices of g 75 ⟩ ≡
   $n = g{\rightarrow}n$; *new_graph* = *gb_new_graph*(n);
   if (*new_graph* ≡ Λ) *panic*(*no_room*);   /* out of memory before we're even started */
   $delta = (($\textbf{siz_t}$) (new\_graph{\rightarrow}vertices)) - (($\textbf{siz_t}$) (g{\rightarrow}vertices))$;
   for ($u = new\_graph{\rightarrow}vertices, v = g{\rightarrow}vertices$; $v < g{\rightarrow}vertices + n$; $u{+}{+}, v{+}{+}$)
     $u{\rightarrow}name$ = *gb_save_string*($v{\rightarrow}name$);

This code is used in sections 74, 78, and 81.

76. A temporary utility field in the new graph is used to remember which vertices are adjacent to a given vertex in the old one. We stamp the *tmp* field of v with a pointer to u when there's an arc from u to v.

#define *tmp* *u.V*   /* utility field u for temporary use as a vertex pointer */

⟨ Insert complementary arcs or edges 76 ⟩ ≡
   for ($v = g{\rightarrow}vertices$; $v < g{\rightarrow}vertices + n$; $v{+}{+}$) { **register Vertex** *vv;
     u = *vert_offset*($v, delta$);   /* vertex in *new_graph* corresponding to v in g */
     { **register Arc** *a; **for** ($a = v{\rightarrow}arcs$; a; $a = a{\rightarrow}next$) *vert_offset*($a{\rightarrow}tip, delta$)${\rightarrow}tmp = u$; }
     if (*directed*) {
       for ($vv = new\_graph{\rightarrow}vertices$; $vv < new\_graph{\rightarrow}vertices + n$; $vv{+}{+}$)
         if (($vv{\rightarrow}tmp \equiv u \wedge copy$) ∨ ($vv{\rightarrow}tmp \neq u \wedge \neg copy$))
           if ($vv \neq u \vee self$) *gb_new_arc*($u, vv, 1_\text{L}$);
     } **else** {
       for ($vv = (self\ ?\ u : u + 1)$; $vv < new\_graph{\rightarrow}vertices + n$; $vv{+}{+}$)
         if (($vv{\rightarrow}tmp \equiv u \wedge copy$) ∨ ($vv{\rightarrow}tmp \neq u \wedge \neg copy$)) *gb_new_edge*($u, vv, 1_\text{L}$);
     }
   }
   for ($v = new\_graph{\rightarrow}vertices$; $v < new\_graph{\rightarrow}vertices + n$; $v{+}{+}$) $v{\rightarrow}tmp = \Lambda$;

This code is used in section 74.

---

alloc_fault = −1, GB_GRAPH §7.
Arc = **struct**, GB_GRAPH §10.
arcs: **Arc** *, GB_GRAPH §9.
binary: **Graph** *( ), §64.
board: **Graph** *( ), §8.
buffer: **static char** [ ], §5.
delta: **register siz_t**, §78.
delta: **register siz_t**, §81.
g: **Graph** *, §78.
g: **Graph** *, §81.
gb_new_arc: **void** ( ),
  GB_GRAPH §30.
gb_new_edge: **void** ( ),
  GB_GRAPH §31.
gb_new_graph: **Graph** *( ),
  GB_GRAPH §23.
gb_recycle: **void** ( ), GB_GRAPH §40.

gb_save_string: **char** *( ),
  GB_GRAPH §35.
gb_trouble_code: **long**,
  GB_GRAPH §14.
Graph = **struct**, GB_GRAPH §20.
make_compound_id: **void** ( ),
  GB_GRAPH §26.
missing_operand = 50,
  GB_GRAPH §7.
n: **register long**, §78.
n: **register long**, §81.
n: **long**, GB_GRAPH §20.
name: **char** *, GB_GRAPH §9.
new_graph: **Graph** *, §9.
next: **Arc** *, GB_GRAPH §10.
no_room = 1, GB_GRAPH §7.

panic = macro ( ), §4.
parts: **Graph** *( ), §55.
perms: **Graph** *( ), §43.
simplex: **Graph** *( ), §26.
siz_t = **unsigned long**,
  GB_GRAPH §34.
sprintf: **int** ( ), <stdio.h>.
subsets: **Graph** *( ), §37.
tip: **Vertex** *, GB_GRAPH §10.
u: **register Vertex** *, §78.
u: **register Vertex** *, §81.
u: **util**, GB_GRAPH §9.
v: **register Vertex** *, §9.
V: **Vertex** *, GB_GRAPH §8.
Vertex = **struct**, GB_GRAPH §9.
vertices: **Vertex** *, GB_GRAPH §20.

77. Graph union and intersection. Another simple way to get new graphs from old ones is to take the union or intersection of their sets of arcs. The subroutine call *gunion*(*g*, *gg*, *multi*, *directed*) produces a graph with the vertices and arcs of *g* together with the arcs of another graph *gg*. The subroutine call *intersection*(*g*, *gg*, *multi*, *directed*) produces a graph with the vertices of *g* but with only the arcs that appear in both *g* and *gg*. In both cases we assume that *gg* has the same vertices as *g*, in the sense that vertices in the same relative position from the beginning of the vertex array are considered identical. If the actual number of vertices in *gg* exceeds the number in *g*, the extra vertices and all arcs touching them in *gg* are suppressed.

The input graphs are assumed to be undirected, unless the *directed* parameter is nonzero. Peculiar results might occur if you mix directed and undirected graphs, but the subroutines will not "crash" when they are asked to produce undirected output from directed input.

If *multi* is nonzero, the new graph may have multiple edges: Suppose there are k_1 arcs from *u* to *v* in *g* and k_2 such arcs in *gg*. Then there will be $k_1 + k_2$ in the union and $\min(k_1, k_2)$ in the intersection when $multi \neq 0$, but at most one in the union or intersection when $multi = 0$.

The lengths of arcs are copied to the union graph when $multi \neq 0$; the minimum length of multiple arcs is retained in the union when $multi = 0$.

The lengths of arcs in the intersection graph are a bit trickier. If multiple arcs occur in *g*, their minimum length, *l*, is computed. Then we compute the maximum of *l* and the lengths of corresponding arcs in *gg*. If $multi = 0$, only the minimum of those maxima will survive.

78. ⟨ Basic subroutines 8 ⟩ +≡
  Graph *∗gunion*(*g*, *gg*, *multi*, *directed*)
      Graph *∗g*, *∗gg*;   /∗ graphs to be united ∗/
      long *multi*;   /∗ should we reproduce multiple arcs? ∗/
      long *directed*;   /∗ should the graph be directed? ∗/
  { ⟨ Vanilla local variables 9 ⟩
    register long *n*;
    register Vertex *∗u*;
    register siz_t *delta*, *ddelta*;   /∗ differences in memory addresses ∗/
    if $(g \equiv \Lambda \vee gg \equiv \Lambda)$ *panic*(*missing_operand*);   /∗ where are *g* and *gg*? ∗/
    ⟨ Set up a graph with the vertices of *g* 75 ⟩;
    sprintf(*buffer*, ",%d,%d", *multi* ? 1 : 0, *directed* ? 1 : 0);
    make_double_compound_id(*new_graph*, "gunion(", *g*, ",", *gg*, *buffer*);
    ddelta = ((**siz_t**) (*new_graph→vertices*)) − ((**siz_t**) (*gg→vertices*));
    ⟨ Insert arcs or edges present in either *g* or *gg* 79 ⟩;
    if (*gb_trouble_code*) {
      gb_recycle(*new_graph*);
      panic(*alloc_fault*);   /∗ uh oh, we ran out of memory somewhere back there ∗/
    }
    return *new_graph*;
  }

79. ⟨ Insert arcs or edges present in either *g* or *gg* 79 ⟩ ≡
  for (*v* = *g→vertices*; *v* < *g→vertices* + *n*; *v*++) {
    register Arc *∗a*;
    register Vertex *∗vv* = *vert_offset*(*v*, *delta*);   /∗ vertex in *new_graph* corresponding to *v* in *g* ∗/
    register Vertex *∗vvv* = *vert_offset*(*vv*, −*ddelta*);   /∗ vertex in *gg* corresponding to *v* in *g* ∗/

```
  for (a = v→arcs; a; a = a→next) {
    u = vert_offset(a→tip, delta);
    ⟨Insert a union arc or edge from vv to u, if appropriate 80⟩;
  }
  if (vvv < gg→vertices + gg→n)
    for (a = vvv→arcs; a; a = a→next) {
      u = vert_offset(a→tip, ddelta);
      if (u < new_graph→vertices + n) ⟨Insert a union arc or edge from vv to u, if appropriate 80⟩;
    }
}
for (v = new_graph→vertices; v < new_graph→vertices + n; v++) v→tmp = Λ, v→tlen = Λ;
```

This code is used in section 78.

80. We use the *tmp* trick of *complement* to remember which arcs have already been recorded from u, and we extend it so that we can maintain minimum lengths. Namely, $uu\text{-}tmp$ will equal u if and only if we have already seen an arc from u to uu; and if so, $uu\text{-}tlen$ will be one such arc. In the undirected case, $uu\text{-}tlen$ will point to the first arc of an edge pair that touches u.

The only thing slightly nontrivial here is the way we keep undirected edges grouped in pairs. We generate a new edge from vv to u only if $vv \le u$, and if equality holds we advance a so that we don't see the self-loop in both directions. Similar logic will be repeated in many of the programs below.

#define *tlen*   *z.A*   /* utility field z regarded as a pointer to an arc */

⟨ Insert a union arc or edge from vv to u, if appropriate 80 ⟩ ≡
```
{ register Arc *b;
   if (directed) {
      if (multi ∨ u→tmp ≠ vv)  gb_new_arc(vv, u, a→len);
      else {
         b = u→tlen;
         if (a→len < b→len)  b→len = a→len;
      }
      u→tmp = vv;      /* remember that we've seen this */
      u→tlen = vv→arcs;
   } else if (u ≥ vv) {
      if (multi ∨ u→tmp ≠ vv)  gb_new_edge(vv, u, a→len);
      else {
         b = u→tlen;
         if (a→len < b→len)  b→len = (b + 1)→len = a→len;
      }
      u→tmp = vv;
      u→tlen = vv→arcs;
      if (u ≡ vv ∧ a→next ≡ a + 1)  a++;      /* bypass second half of self-loop */
   }
}
```
This code is used in section 79 (twice).

81. ⟨ Basic subroutines 8 ⟩ +≡
```
Graph *intersection(g, gg, multi, directed)
      Graph *g, *gg;      /* graphs to be intersected */
      long multi;      /* should we reproduce multiple arcs? */
      long directed;      /* should the graph be directed? */
{ ⟨ Vanilla local variables 9 ⟩
   register long n;
   register Vertex *u;
   register siz_t delta, ddelta;      /* differences in memory addresses */
   if (g ≡ Λ ∨ gg ≡ Λ)  panic(no_room + 1);      /* where are g and gg? */
   ⟨ Set up a graph with the vertices of g 75 ⟩;
   sprintf(buffer, ",%d,%d)", multi ? 1 : 0, directed ? 1 : 0);
   make_double_compound_id(new_graph, "intersection(", g, ",", gg, buffer);
   ddelta = ((siz_t) (new_graph→vertices)) − ((siz_t) (gg→vertices));
```

⟨ Insert arcs or edges present in both g and gg 82 ⟩;
if (*gb_trouble_code*) {
    gb_recycle(*new_graph*);
    panic(*alloc_fault*);   /* whoops, we ran out of memory somewhere back there */
}
return *new_graph*;
}

82.   Two more temporary utility fields are needed here.

#define *mult*   *v.I*   /* utility field v, counts multiplicity of arcs */
#define *minlen*   *w.I*   /* utility field w, records the smallest length */

⟨ Insert arcs or edges present in both g and gg 82 ⟩ ≡
   for ($v = g\text{→}vertices$; $v < g\text{→}vertices + n$; $v{+}{+}$) { **register Arc** $*a$;
     register Vertex $*vv = vert\_offset(v, delta)$;   /* vertex in *new_graph* corresponding to v in g */
     register Vertex $*vvv = vert\_offset(vv, -ddelta)$;   /* vertex in gg corresponding to v in g */

     if ($vvv \geq gg\text{→}vertices + gg\text{→}n$) **continue**;
     ⟨ Take note of all arcs from v 85 ⟩;
     for ($a = vvv\text{→}arcs$; a; $a = a\text{→}next$) {
       $u = vert\_offset(a\text{→}tip, ddelta)$;
       if ($u \geq new\_graph\text{→}vertices + n$) **continue**;
       if ($u\text{→}tmp \equiv vv$) { **long** $l = u\text{→}minlen$;
         if ($a\text{→}len > l$) $l = a\text{→}len$;   /* maximum */
         if ($u\text{→}mult < 0$) ⟨ Update minimum of multiple maxima 84 ⟩
         else   ⟨ Generate a new arc or edge for the intersection, and reduce the multiplicity 83 ⟩;
       }
     }
   }
⟨ Clear out the temporary utility fields 86 ⟩;
This code is used in section 81.

---

a: **register Arc** *, §79.
A: **Arc** *, GB_GRAPH §8.
alloc_fault = −1, GB_GRAPH §7.
Arc = **struct**, GB_GRAPH §10.
arcs: **Arc** *, GB_GRAPH §9.
buffer: **static char** [], §5.
complement: **Graph** *( ), §74.
directed: **long**, §78.
gb_new_arc: **void** ( ),
   GB_GRAPH §30.
gb_new_edge: **void** ( ),
   GB_GRAPH §31.
gb_recycle: **void** ( ), GB_GRAPH §40.
gb_trouble_code: **long**,

GB_GRAPH §14.
Graph = **struct**, GB_GRAPH §20.
I: **long**, GB_GRAPH §8.
len: **long**, GB_GRAPH §10.
make_double_compound_id: **void**
   ( ), GB_GRAPH §27.
multi: **long**, §78.
n: **long**, GB_GRAPH §20.
new_graph: **Graph** *, §9.
next: **Arc** *, GB_GRAPH §10.
no_room = 1, GB_GRAPH §7.
panic = **macro** ( ), §4.
siz_t = **unsigned long**,

GB_GRAPH §34.
sprintf: **int** ( ), <stdio.h>.
tip: **Vertex** *, GB_GRAPH §10.
tmp = *u.V*, §76.
u: **register Vertex** *, §78.
v: **register Vertex** *, §9.
v: **util**, GB_GRAPH §9.
vert_offset = **macro** ( ), §75.
Vertex = **struct**, GB_GRAPH §9.
vertices: **Vertex** *, GB_GRAPH §20.
vv: **register Vertex** *, §79.
w: **util**, GB_GRAPH §9.
z: **util**, GB_GRAPH §9.

83. ⟨ Generate a new arc or edge for the intersection, and reduce the multiplicity 83 ⟩ ≡
```
{
    if (directed) gb_new_arc(vv, u, l);
    else {
        if (vv ≤ u) gb_new_edge(vv, u, l);
        if (vv ≡ u ∧ a⃗next ≡ a + 1) a++;      /* skip second half of self-loop */
    }
    if (¬multi) {
        u⃗tlen = vv⃗arcs;
        u⃗mult = −1;
    } else if (u⃗mult ≡ 0) u⃗tmp = Λ;
    else u⃗mult −−;
}
```
This code is used in section 82.

84. We get here if and only if *multi* = 0 and *gg* has more than one arc from *vv* to *u* and *g* has at least one arc from *vv* to *u*.

⟨ Update minimum of multiple maxima 84 ⟩ ≡
```
{ register Arc *b = u⃗tlen;     /* previous arc or edge from vv to u */
    if (l < b⃗len) {
        b⃗len = l;
        if (¬directed) (b + 1)⃗len = l;
    }
}
```
This code is used in section 82.

85. ⟨ Take note of all arcs from *v* 85 ⟩ ≡
```
for (a = v⃗arcs; a; a = a⃗next) {
    u = vert_offset(a⃗tip, delta);
    if (u⃗tmp ≡ vv) {
        u⃗mult ++;
        if (a⃗len < u⃗minlen) u⃗minlen = a⃗len;
    } else u⃗tmp = vv, u⃗mult = 0, u⃗minlen = a⃗len;
    if (u ≡ vv ∧ ¬directed ∧ a⃗next ≡ a + 1) a++;     /* skip second half of self-loop */
}
```
This code is used in section 82.

86. ⟨ Clear out the temporary utility fields 86 ⟩ ≡
```
for (v = new_graph⃗vertices; v < new_graph⃗vertices + n; v++) {
    v⃗tmp = Λ;
    v⃗tlen = Λ;
    v⃗mult = 0;
    v⃗minlen = 0;
}
```
This code is used in section 82.

87. Line graphs. The next operation in GB_BASIC's repertoire constructs the so-called line graph of a given graph g. The subroutine that does this is invoked by calling '$lines(g, directed)$'.

If $directed = 0$, the line graph has one vertex for each edge of g; two vertices are adjacent if and only if the corresponding edges have a common vertex.

If $directed \neq 0$, the line graph has one vertex for each arc of g; there is an arc from vertex u to vertex v if and only if the arc corresponding to u ends at the vertex that begins the arc corresponding to v.

All arcs of the line graph will have length 1.

Utility fields $u.V$ and $v.V$ of each vertex in the line graph will point to the vertices of g that define the corresponding arc or edge, and $w.A$ will point to the arc from $u.V$ to $v.V$ in g. In the undirected case we will have $u.V \leq v.V$.

⟨ Basic subroutines 8 ⟩ +≡
  Graph *$lines(g, directed)$
      Graph *g;   /* graph whose lines will become vertices */
      long $directed$;   /* should the graph be directed? */
  { ⟨ Vanilla local variables 9 ⟩

  register long m;   /* the number of lines */
  register Vertex *u;

  if $(g \equiv \Lambda)$ $panic(no\_room + 1)$;   /* where is g? */
  ⟨ Set up a graph whose vertices are the lines of g 89 ⟩;
  if $(directed)$ ⟨ Insert arcs of a directed line graph 92 ⟩
  else ⟨ Insert edges of an undirected line graph 93 ⟩;
  ⟨ Restore g to its pristine original condition 88 ⟩;
  if $(gb\_trouble\_code)$ {
    $gb\_recycle(new\_graph)$;
    $panic(alloc\_fault)$;   /* (sigh) we ran out of memory somewhere back there */
  }
  return new_graph;
  $near\_panic$: ⟨ Recover from potential disaster due to bad data 90 ⟩;
  }

---

a: **register Arc** *, §82.
A: **Arc** *, GB_GRAPH §8.
$alloc\_fault = -1$, GB_GRAPH §7.
Arc = **struct**, GB_GRAPH §10.
$arcs$: **Arc** *, GB_GRAPH §9.
$delta$: **register siz_t**, §81.
$directed$: **long**, §81.
g: **Graph** *, §81.
gb_new_arc: **void** ( ),
  GB_GRAPH §30.
gb_new_edge: **void** ( ),
  GB_GRAPH §31.
$gb\_recycle$: **void** ( ), GB_GRAPH §40.
$gb\_trouble\_code$: **long**,

GB_GRAPH §14.
gg: **Graph** *, §81.
Graph = **struct**, GB_GRAPH §20.
l: **long**, §82.
len: **long**, GB_GRAPH §10.
$minlen = w.I$, §82.
$mult = v.I$, §82.
$multi$: **long**, §81.
n: **register long**, §81.
new_graph: **Graph** *, §9.
$next$: **Arc** *, GB_GRAPH §10.
$no\_room = 1$, GB_GRAPH §7.
$panic$ = macro ( ), §4.

tip: **Vertex** *, GB_GRAPH §10.
$tlen = z.A$, §80.
$tmp = u.V$, §76.
u: **register Vertex** *, §81.
u: **util**, GB_GRAPH §9.
v: **register Vertex** *, §9.
v: **util**, GB_GRAPH §9.
V: **Vertex** *, GB_GRAPH §8.
$vert\_offset$ = macro ( ), §75.
Vertex = **struct**, GB_GRAPH §9.
$vertices$: **Vertex** *, GB_GRAPH §20.
vv: **register Vertex** *, §82.
w: **util**, GB_GRAPH §9.

88. We want to add a data structure to g so that the line graph can be built efficiently. But we also want to preserve g so that it exhibits no traces of occupation when *lines* has finished its work. To do this, we will move utility field $v \rightarrow z$ temporarily into a utility field $u \rightarrow z$ of the line graph, where u is the first vertex having $u \rightarrow u.V \equiv v$, whenever such a u exists. Then we'll set $v \rightarrow map = u$. We will then be able to find u when v is given, and we'll be able to cover our tracks later.

In the undirected case, further structure is needed. We will temporarily change the *tip* field in the second arc of each edge pair so that it points to the line-graph vertex that points to the first arc of the pair.

The *util_types* field of the graph does not indicate the fact that utility fields $u.V$, $v.V$, and $w.A$ of each vertex will be set, because those utility fields are pointers from the new graph to the original graph. The *save_graph* procedure does not deal with pointers between graphs.

#define *map* $z.V$ /* the z field treated as a vertex pointer */
$\langle$ Restore g to its pristine original condition 88 $\rangle \equiv$
  for $(u = new\_graph \rightarrow vertices, v = \Lambda;\ u < new\_graph \rightarrow vertices + m;\ u{++})\ \{$
    if $(u \rightarrow u.V \neq v)\ \{$
      $v = u \rightarrow u.V;$ /* original vertex of g */
      $v \rightarrow map = u \rightarrow map;$ /* restore original value of $v \rightarrow z$ */
      $u \rightarrow map = \Lambda;$
    $\}$
    if $(\neg directed)\ ((u \rightarrow w.A) + 1) \rightarrow tip = v;$
  $\}$
This code is used in sections 87 and 90.

89. Special care must be taken to avoid chaos when the user is trying to construct the undirected line graph of a directed graph. Otherwise we might trash the memory, or we might leave the original graph in a garbled state with pointers leading into supposedly free space.

$\langle$ Set up a graph whose vertices are the lines of g 89 $\rangle \equiv$
  $m = (directed\ ?\ g \rightarrow m : (g \rightarrow m)/2);$
  $new\_graph = gb\_new\_graph(m);$
  if $(new\_graph \equiv \Lambda)\ panic(no\_room);$ /* out of memory before we're even started */
  $make\_compound\_id(new\_graph, \texttt{"lines("}, g, directed\ ?\ \texttt{",1)"} : \texttt{",0)"});$
  $u = new\_graph \rightarrow vertices;$
  for $(v = g \rightarrow vertices + g \rightarrow n - 1;\ v \geq g \rightarrow vertices;\ v{-}{-})\ \{$ **register Arc** $*a;$
    register long $mapped = 0;$ /* has $v \rightarrow map$ been set? */
    for $(a = v \rightarrow arcs;\ a;\ a = a \rightarrow next)\ \{$ **register Vertex** $*vv = a \rightarrow tip;$
      if $(\neg directed)\ \{$
        if $(vv < v)$ **continue;**
        if $(vv \geq g \rightarrow vertices + g \rightarrow n)$ **goto** *near_panic;*   /* original graph not undirected */
      $\}$
      $\langle$ Make u a vertex representing the arc a from v to vv 91 $\rangle;$
      if $(\neg mapped)\ \{$
        $u \rightarrow map = v \rightarrow map;$ /* $z.V = map$ incorporates all bits of utility field z, whatever its type */
        $v \rightarrow map = u;$
        $mapped = 1;$
      $\}$

```
        u++;
      }
    }
if (u ≠ new_graph→vertices + m) goto near_panic;
```
This code is used in section 87.

90. ⟨ Recover from potential disaster due to bad data 90 ⟩ ≡
```
m = u − new_graph→vertices;
⟨ Restore g to its pristine original condition 88 ⟩;
gb_recycle(new_graph);
panic(invalid_operand);        /* g did not obey the conventions for an undirected graph */
```
This code is used in section 87.

91. The vertex names in the line graph are pairs of original vertex names, separated by '`--`' when undirected, '`->`' when directed. If either of the original names is horrendously long, the villainous Procrustes chops it off arbitrarily so that it fills at most half of the name buffer.

⟨ Make u a vertex representing the arc a from v to vv 91 ⟩ ≡
```
u→u.V = v;
u→v.V = vv;
u→w.A = a;
if (¬directed) {
    if (u ≥ new_graph→vertices + m ∨ (a + 1)→tip ≠ v) goto near_panic;
    if (v ≡ vv ∧ a→next ≡ a + 1) a++;        /* skip second half of self-loop */
    else (a + 1)→tip = u;
}
sprintf(buffer, "%.*s-%c%.*s", (BUF_SIZE − 3)/2, v→name,
        directed ? '>' : '-', BUF_SIZE/2 − 1, vv→name);
u→name = gb_save_string(buffer);
```
This code is used in section 89.

A: **Arc** ∗, GB_GRAPH §8.
Arc = **struct**, GB_GRAPH §10.
arcs: **Arc** ∗, GB_GRAPH §9.
BUF_SIZE = 4096, §5.
buffer: **static char** [ ], §5.
directed: **long**, §87.
g: **Graph** ∗, §87.
gb_new_graph: **Graph** ∗( ),
    GB_GRAPH §23.
gb_recycle: **void** ( ), GB_GRAPH §40.
gb_save_string: **char** ∗( ),
    GB_GRAPH §35.
invalid_operand = 60, GB_GRAPH §7.

lines: **Graph** ∗( ), §87.
m: **register long**, §87.
m: **long**, GB_GRAPH §20.
make_compound_id: **void** ( ),
    GB_GRAPH §26.
n: **long**, GB_GRAPH §20.
name: **char** ∗, GB_GRAPH §9.
near_panic: **label**, §87.
new_graph: **Graph** ∗, §9.
next: **Arc** ∗, GB_GRAPH §10.
no_room = 1, GB_GRAPH §7.
panic = **macro** ( ), §4.
save_graph: **long** ( ), GB_SAVE §20.

sprintf: **int** ( ), <stdio.h>.
tip: **Vertex** ∗, GB_GRAPH §10.
u: **register Vertex** ∗, §87.
u: **util**, GB_GRAPH §9.
util_types: **char** [ ], GB_GRAPH §20.
v: **register Vertex** ∗, §9.
v: **util**, GB_GRAPH §9.
V: **Vertex** ∗, GB_GRAPH §8.
Vertex = **struct**, GB_GRAPH §9.
vertices: **Vertex** ∗, GB_GRAPH §20.
w: **util**, GB_GRAPH §9.
z: **util**, GB_GRAPH §9.

92. ⟨ Insert arcs of a directed line graph 92 ⟩ ≡

```
for (u = new_graph⃗vertices; u < new_graph⃗vertices + m; u++) {
  v = u⃗v.V;
  if (v⃗arcs) {      /* v⃗map has been set up */
    v = v⃗map;
    do {
      gb_new_arc(u, v, 1 L);
      v++;
    } while (v⃗u.V ≡ u⃗v.V);
  }
}
```

This code is used in section 87.

93. An undirected line graph will contain no self-loops. It contains multiple edges only if the original graph did; in that case, there are two edges joining a line to each of its parallel mates, because each mate hits both of its endpoints.

The details of this section deserve careful study. We use the fact that the first vertices of the lines occur in nonincreasing order.

⟨ Insert edges of an undirected line graph 93 ⟩ ≡

```
for (u = new_graph⃗vertices; u < new_graph⃗vertices + m; u++) { register Vertex *vv;
  register Arc *a; register long mapped = 0;
  v = u⃗u.V;      /* we look first for prior lines that touch the first vertex */
  for (vv = v⃗map; vv < u; vv++) gb_new_edge(u, vv, 1 L);
  v = u⃗v.V;      /* then we look for prior lines that touch the other one */
  for (a = v⃗arcs; a; a = a⃗next) {
    vv = a⃗tip;
    if (vv < u ∧ vv ≥ new_graph⃗vertices) gb_new_edge(u, vv, 1 L);
    else if (vv ≥ v ∧ vv < g⃗vertices + g⃗n) mapped = 1;
  }
  if (mapped ∧ v > u⃗u.V)
    for (vv = v⃗map; vv⃗u.V ≡ v; vv++) gb_new_edge(u, vv, 1 L);
}
```

This code is used in section 87.

94. Graph products.   Three ways have traditionally been used to define the product of two graphs. In all three cases the vertices of the product graph are ordered pairs (v, v'), where v and v' are vertices of the original graphs; the difference occurs in the definition of arcs. Suppose g has m arcs and n vertices, while g' has m' arcs and n' vertices. The *cartesian product* of g and g' has $mn' + m'n$ arcs, namely from (u, u') to (v, u') whenever there's an arc from u to v in g, and from (u, u') to (u, v') whenever there's an arc from u' to v' in g'. The *direct product* has mm' arcs, namely from (u, u') to (v, v') in the same circumstances. The *strong product* has both the arcs of the cartesian product and the direct product.

Notice that an undirected graph with m edges has $2m$ arcs. Thus the number of edges in the direct product of two undirected graphs is twice the product of the number of edges in the individual graphs. A self-loop in g will combine with an edge in g' to make two parallel edges in the direct product.

The subroutine call *product*$(g, gg, type, directed)$ produces the product graph of one of these three types, where *type* $= 0$ for cartesian product, *type* $= 1$ for direct product, and *type* $= 2$ for strong product. The length of an arc in the cartesian product is copied from the length of the original arc that it replicates; the length of an arc in the direct product is the minimum of the two arc lengths that induce it. If *directed* $= 0$, the product graph will be an undirected graph with edges consisting of consecutive arc pairs according to the standard GraphBase conventions, and the input graphs should adhere to the same conventions.

⟨ gb_basic.h   1 ⟩ +≡
#**define** *cartesian*   0
#**define** *direct*   1
#**define** *strong*   2

Arc = **struct**, GB_GRAPH §10.
arcs: **Arc** ∗, GB_GRAPH §9.
directed: **long**, §95.
g: **Graph** ∗, §87.
g: **Graph** ∗, §95.
gb_new_arc: **void** ( ),
   GB_GRAPH §30.
gb_new_edge: **void** ( ),
   GB_GRAPH §31.

gg: **Graph** ∗, §95.
m: **register long**, §87.
map = *z.V*, §88.
n: **long**, GB_GRAPH §20.
new_graph: **Graph** ∗, §9.
next: **Arc** ∗, GB_GRAPH §10.
product: **Graph** ∗( ), §95.
tip: **Vertex** ∗, GB_GRAPH §10.

type: **long**, §95.
u: **register Vertex** ∗, §87.
u: **util**, GB_GRAPH §9.
v: **register Vertex** ∗, §9.
v: **util**, GB_GRAPH §9.
V: **Vertex** ∗, GB_GRAPH §8.
Vertex = **struct**, GB_GRAPH §9.
vertices: **Vertex** ∗, GB_GRAPH §20.

95. ⟨ Basic subroutines 8 ⟩ +≡

```
  Graph *product(g, gg, type, directed)
      Graph *g, *gg;       /* graphs to be multiplied */
      long type;        /* cartesian, direct, or strong */
      long directed;       /* should the graph be directed? */
  { ⟨ Vanilla local variables 9 ⟩
    register Vertex *u, *vv;
    register long n;       /* the number of vertices in the product graph */
    if (g ≡ Λ ∨ gg ≡ Λ) panic(no_room + 1);      /* where are g and gg? */
    ⟨ Set up a graph with ordered pairs of vertices 96 ⟩;
    if ((type & 1) ≡ 0) ⟨ Insert arcs or edges for cartesian product 97 ⟩;
    if (type) ⟨ Insert arcs or edges for direct product 99 ⟩;
    if (gb_trouble_code) {
      gb_recycle(new_graph);
      panic(alloc_fault);      /* @¿*#!, we ran out of memory somewhere back there */
    }
    return new_graph;
  }
```

96. We must be constantly on guard against running out of memory, especially when multiplying information.

The vertex names in the product are pairs of original vertex names separated by commas. Thus, for example, if you cross an *econ* graph with a *roget* graph, you can get vertices like `"Financial␣services,␣mediocrity"`.

⟨ Set up a graph with ordered pairs of vertices 96 ⟩ ≡

```
  { float test_product = ((float) (g→n)) * ((float) (gg→n));
    if (test_product > MAX_NNN) panic(very_bad_specs);      /* way too many vertices */
  }
  n = (g→n) * (gg→n);
  new_graph = gb_new_graph(n);
  if (new_graph ≡ Λ) panic(no_room);      /* out of memory before we're even started */
  for (u = new_graph→vertices, v = g→vertices, vv = gg→vertices;
        u < new_graph→vertices + n; u++) {
    sprintf(buffer, "%.*s,%.*s", BUF_SIZE/2 − 1, v→name, (BUF_SIZE − 1)/2, vv→name);
    u→name = gb_save_string(buffer);
    if (++vv ≡ gg→vertices + gg→n) vv = gg→vertices, v++;      /* "carry" */
  }
  sprintf(buffer, ",%d,%d", (type ? 2 : 0) − (int) (type & 1), directed ? 1 : 0);
  make_double_compound_id(new_graph, "product(", g, ",", gg, buffer);
```

This code is used in section 95.

97. ⟨ Insert arcs or edges for cartesian product 97 ⟩ ≡

```
  { register Vertex *uu, *uuu;
    register Arc *a;
    register siz_t delta;      /* difference in memory addresses */
    delta = ((siz_t) (new_graph→vertices)) − ((siz_t) (gg→vertices));
    for (u = gg→vertices; u < gg→vertices + gg→n; u++)
```

```
    for (a = u⃗arcs; a; a = a⃗next) {
        v = a⃗tip;
        if (¬directed) {
            if (u > v) continue;
            if (u ≡ v ∧ a⃗next ≡ a + 1) a++;        /* skip second half of self-loop */
        }
        for (uu = vert_offset(u, delta), vv = vert_offset(v, delta);
                uu < new_graph⃗vertices + n; uu += gg⃗n, vv += gg⃗n)
            if (directed) gb_new_arc(uu, vv, a⃗len);
            else gb_new_edge(uu, vv, a⃗len);
    }
    ⟨ Insert arcs or edges for first component of cartesian product 98 ⟩;
}
```
This code is used in section 95.

98. ⟨ Insert arcs or edges for first component of cartesian product 98 ⟩ ≡
```
    for (u = g⃗vertices, uu = new_graph⃗vertices; uu < new_graph⃗vertices + n; u++, uu += gg⃗n)
        for (a = u⃗arcs; a; a = a⃗next) {
            v = a⃗tip;
            if (¬directed) {
                if (u > v) continue;
                if (u ≡ v ∧ a⃗next ≡ a + 1) a++;        /* skip second half of self-loop */
            }
            vv = new_graph⃗vertices + ((gg⃗n) * (v − g⃗vertices));
            for (uuu = uu; uuu < uu + gg⃗n; uuu++, vv++)
                if (directed) gb_new_arc(uuu, vv, a⃗len);
                else gb_new_edge(uuu, vv, a⃗len);
        }
```
This code is used in section 97.

---

alloc_fault = −1, GB_GRAPH §7.
Arc = **struct**, GB_GRAPH §10.
arcs: **Arc** *, GB_GRAPH §9.
BUF_SIZE = 4096, §5.
buffer: **static char** [], §5.
cartesian = 0, §94.
direct = 1, §94.
econ: **Graph** *( ), GB_ECON §7.
gb_new_arc: **void** ( ),
    GB_GRAPH §30.
gb_new_edge: **void** ( ),
    GB_GRAPH §31.
gb_new_graph: **Graph** *( ),
    GB_GRAPH §23.

gb_recycle: **void** ( ), GB_GRAPH §40.
gb_save_string: **char** *( ),
    GB_GRAPH §35.
gb_trouble_code: **long**,
    GB_GRAPH §14.
Graph = **struct**, GB_GRAPH §20.
len: **long**, GB_GRAPH §10.
make_double_compound_id: **void**
    ( ), GB_GRAPH §27.
MAX_NNN = 1000000000.0, §13.
n: **long**, GB_GRAPH §20.
name: **char** *, GB_GRAPH §9.
new_graph: **Graph** *, §9.
next: **Arc** *, GB_GRAPH §10.

no_room = 1, GB_GRAPH §7.
panic = macro ( ), §4.
roget: **Graph** *( ), GB_ROGET §4.
siz_t = **unsigned long**,
    GB_GRAPH §34.
sprintf: **int** ( ), <stdio.h>.
strong = 2, §94.
tip: **Vertex** *, GB_GRAPH §10.
v: **register Vertex** *, §9.
vert_offset = macro ( ), §75.
Vertex = **struct**, GB_GRAPH §9.
vertices: **Vertex** *, GB_GRAPH §20.
very_bad_specs = 40, GB_GRAPH §7.

99. ⟨Insert arcs or edges for direct product 99⟩ ≡

```
{ Vertex *uu; Arc *a;
  siz_t delta0 = ((siz_t) (new_graph→vertices)) − ((siz_t) (gg→vertices));
  siz_t del = (gg→n) * sizeof(Vertex);
  register siz_t delta, ddelta;
  for (uu = g→vertices, delta = delta0; uu < g→vertices + g→n; uu++, delta += del)
    for (a = uu→arcs; a; a = a→next) {
      vv = a→tip;
      if (¬directed) {
        if (uu > vv) continue;
        if (uu ≡ vv ∧ a→next ≡ a + 1) a++;        /* skip second half of self-loop */
      }
      ddelta = delta0 + del * (vv − g→vertices);
      for (u = gg→vertices; u < gg→vertices + gg→n; u++) { register Arc *aa;
        for (aa = u→arcs; aa; aa = aa→next) { long length = a→len;
          if (length > aa→len) length = aa→len;
          v = aa→tip;
          if (directed) gb_new_arc(vert_offset(u, delta), vert_offset(v, ddelta), length);
          else gb_new_edge(vert_offset(u, delta), vert_offset(v, ddelta), length);
        }
      }
    }
}
```

This code is used in section 95.

100. Induced graphs.   Another important way to transform a graph is to remove, identify, or split some of its vertices. All of these operations are performed by the *induced* routine, which users can invoke by calling '*induced*(g, *description*, *self*, *multi*, *directed*)'.

Each vertex v of g should first be assigned an "induction code" in its field $v \rightarrow ind$, which is actually utility field z. The induction code is 0 if v is to be eliminated; it is 1 if v is to be retained; it is $k > 1$ if v is to be split into k nonadjacent vertices having the same neighbors as v did; and it is $k < 0$ if v is to be identified with all other vertices having the same value of k.

For example, suppose g is a circuit with vertices $\{0, 1, \ldots, 9\}$, where j is adjacent to k if and only if $k = (j \pm 1) \bmod 10$. If we set

$$0 \rightarrow ind = 0, \quad 1 \rightarrow ind = 5 \rightarrow ind = 9 \rightarrow ind = -1, \quad 2 \rightarrow ind = 3 \rightarrow ind = -2,$$
$$4 \rightarrow ind = 6 \rightarrow ind = 8 \rightarrow ind = 1, \quad \text{and } 7 \rightarrow ind = 3,$$

the induced graph will have vertices $\{-1, -2, 4, 6, 7, 7', 7'', 8\}$. The vertices adjacent to 6, say, will be -1 (formerly 5), 7, 7', and 7''. The vertices adjacent to -1 will be those formerly adjacent to 1, 5, or 9, namely -2 (formerly 2), 4, 6, and 8. The vertices adjacent to -2 will be those formerly adjacent to 2 or 3, namely -1 (formerly 1), -2 (formerly 3), -2 (formerly 2), and 4. Duplicate edges will be discarded if *multi* $\equiv 0$, and self-loops will be discarded if *self* $\equiv 0$.

The total number of vertices in the induced graph will be the sum of the positive *ind* fields plus the absolute value of the most negative *ind* field. This rule implies, for example, that if at least one vertex has *ind* $= -5$, the induced graph will always have a vertex -4, even though no *ind* field has been set to -4.

The *description* parameter is a string that will appear as part of the name of the induced graph; if *description* $= 0$, this string will be empty. In the latter case, users are encouraged to assign a suitable name to the *id* field of the induced graph themselves, characterizing the method by which the *ind* codes were set.

If the *directed* parameter is zero, the input graph will be assumed to be undirected, and the output graph will be undirected.

When *multi* $= 0$, the length of an arc that represents multiple arcs will be the minimum of the multiple arc lengths.

#**define** *ind* $z.I$

⟨ `gb_basic.h` 1 ⟩ +≡
  #**define** *ind* $z.I$ /* utility field z when used to induce a graph */

Arc = **struct**, GB_GRAPH §10.
arcs: **Arc** *, GB_GRAPH §9.
description: **char** *, §105.
directed: **long**, §95.
directed: **long**, §105.
g: **Graph** *, §95.
g: **Graph** *, §105.
gb_new_arc: **void** ( ),
  GB_GRAPH §30.
gb_new_edge: **void** ( ),
  GB_GRAPH §31.

gg: **Graph** *, §95.
I: **long**, GB_GRAPH §8.
id: **char** [], GB_GRAPH §20.
induced: **Graph** *( ), §105.
len: **long**, GB_GRAPH §10.
multi: **long**, §105.
n: **long**, GB_GRAPH §20.
new_graph: **Graph** *, §9.
next: **Arc** *, GB_GRAPH §10.
self: **long**, §105.

siz_t = **unsigned long**,
  GB_GRAPH §34.
tip: **Vertex** *, GB_GRAPH §10.
u: **register Vertex** *, §95.
v: **register Vertex** *, §9.
vert_offset = macro ( ), §75.
Vertex = **struct**, GB_GRAPH §9.
vertices: **Vertex** *, GB_GRAPH §20.
vv: **register Vertex** *, §95.
z: **util**, GB_GRAPH §9.

101. Here's a simple example: To get a complete bipartite graph with parts of sizes *n1* and *n2*, we can start with a trivial two-point graph and split its vertices into *n1* and *n2* parts.

⟨ Applications of basic subroutines 101 ⟩ ≡

```
Graph *bi_complete(n1, n2, directed)
    unsigned long n1;      /* size of first part */
    unsigned long n2;      /* size of second part */
    long directed;         /* should all arcs go from first part to second? */
{ Graph *new_graph = board(2 L, 0 L, 0 L, 0 L, 1 L, 0 L, directed);
  if (new_graph) {
    new_graph→vertices→ind = n1;
    (new_graph→vertices + 1)→ind = n2;
    new_graph = induced(new_graph, Λ, 0 L, 0 L, directed);
    if (new_graph) {
      sprintf(new_graph→id, "bi_complete(%lu,%lu,%d)",
          n1, n2, directed ? 1 : 0);
      mark_bipartite(new_graph, n1);
    }
  }
  return new_graph;
}
```

See also section 103.

This code is used in section 2.

102. The *induced* routine also provides a special feature not mentioned above: If the *ind* field of any vertex v is `IND_GRAPH` or greater (where `IND_GRAPH` is a large constant, much larger than the number of vertices that would fit in computer memory), then utility field v→*subst* should point to a graph. A copy of the vertices of that graph will then be substituted for v in the induced graph.

This feature extends the ordinary case when v→*ind* > 0, which essentially substitutes an empty graph for v.

If substitution is being used to replace all of g's vertices by disjoint copies of some other graph g', the induced graph will be somewhat similar to a product graph. But it will not be the same as any of the three types of output produced by *product*, because the relation between g and g' is not symmetrical. Assuming that no self-loops are present, and that graphs (g, g') have respectively (m, m') arcs and (n, n') vertices, the result of substituting g' for all vertices of g has $m'n + mn'^2$ arcs.

```
#define IND_GRAPH  1000000000    /* when ind is a billion or more, */
#define subst  y.G    /* we'll look at the subst field */
```

⟨ gb_basic.h 1 ⟩ +≡

```
#define IND_GRAPH  1000000000
#define subst  y.G
```

103. For example, we can use the `IND_GRAPH` feature to create a "wheel" of n vertices arranged cyclically, all connected to one or more center points. In the directed case, the arcs will run from the center(s) to a cycle; in the undirected case, the edges will join the center(s) to a circuit.

⟨ Applications of basic subroutines 101 ⟩ +≡

Graph $*wheel(n, n1, directed)$
  unsigned long n; /* size of the rim */
  unsigned long $n1$; /* number of center points */
  long *directed*; /* should all arcs go from center to rim and around? */
$\{$ **Graph** $*new\_graph = board(2_L, 0_L, 0_L, 0_L, 1_L, 0_L, directed);$ /* trivial 2-vertex graph */

 if (new_graph) $\{$
  $new\_graph{\to}vertices{\to}ind = n1$;
  $(new\_graph{\to}vertices + 1){\to}ind = \text{IND\_GRAPH}$;
  $(new\_graph{\to}vertices + 1){\to}subst = board(n, 0_L, 0_L, 0_L, 1_L, 1_L, directed);$ /* cycle or circuit */
  $new\_graph = induced(new\_graph, \Lambda, 0_L, 0_L, directed);$
  if (new_graph) $\{$
   $sprintf(new\_graph{\to}id, \texttt{"wheel(\%lu,\%lu,\%d)"},$
    $n, n1, directed\ ?\ 1 : 0);$
  $\}$
 $\}$
 return new_graph;
$\}$

104. $\langle$ `gb_basic.h` 1 $\rangle$ $+\equiv$
 extern Graph $*bi\_complete(\ )$;
 extern Graph $*wheel(\ )$; /* standard applications of *induced* */

---

board: **Graph** *( ), §8.
G: **Graph** *, GB_GRAPH §8.
Graph = **struct**, GB_GRAPH §20.
id: **char** [ ], GB_GRAPH §20.
ind = $z.I$, §100.

induced: **Graph** *( ), §105.
mark_bipartite = macro ( ),
 GB_GRAPH §22.
product: **Graph** *( ), §95.

sprintf: **int** ( ), <stdio.h>.
v: **register Vertex** *, §9.
vertices: **Vertex** *, GB_GRAPH §20.
y: **util**, GB_GRAPH §9.

105. ⟨ Basic subroutines 8 ⟩ +≡

    Graph *$*induced(g, description, self, multi, directed)$
        Graph *$*g$;   /* graph marked for induction in its *ind* fields */
        char *$*description$;   /* string to be mentioned in *new_graph⃗id* */
        long *self*;   /* should self-loops be permitted? */
        long *multi*;   /* should multiple arcs be permitted? */
        long *directed*;   /* should the graph be directed? */
  { ⟨ Vanilla local variables 9 ⟩

    register Vertex *$*u$;
    register long $n = 0$;   /* total number of vertices in induced graph */
    register long $nn = 0$;   /* number of negative vertices in induced graph */
    if $(g \equiv \Lambda)$ *panic*(*missing_operand*);   /* where is *g*? */
    ⟨ Set up a graph with the induced vertices 106 ⟩;
    ⟨ Insert arcs or edges for induced vertices 110 ⟩;
    ⟨ Restore *g* to its original state 109 ⟩;
    if (*gb_trouble_code*) {
      gb_recycle(*new_graph*);
      panic(*alloc_fault*);   /* aargh, we ran out of memory somewhere back there */
    }
    return *new_graph*;
  }

106. ⟨ Set up a graph with the induced vertices 106 ⟩ ≡
  ⟨ Determine *n* and *nn* 107 ⟩;
  new_graph = *gb_new_graph*(n);
  if (*new_graph* ≡ Λ) *panic*(*no_room*);   /* out of memory before we're even started */
  ⟨ Assign names to the new vertices, and create a map from *g* to *new_graph* 108 ⟩;
  sprintf(*buffer*, ",%s,%d,%d,%d)",
      description ? *description* : *null_string*,
      self ? 1 : 0, *multi* ? 1 : 0, *directed* ? 1 : 0);
  make_compound_id(*new_graph*, "induced(", g, *buffer*);

This code is used in section 105.

107. ⟨ Determine *n* and *nn* 107 ⟩ ≡
  for ($v = g⃗vertices$; $v < g⃗vertices + g⃗n$; $v{+}{+}$)
    if ($v⃗ind > 0$) {
      if ($n >$ IND_GRAPH) *panic*(*very_bad_specs*);   /* way too big */
      if ($v⃗ind \geq$ IND_GRAPH) {
        if ($v⃗subst \equiv \Lambda$) *panic*(*missing_operand* + 1);   /* substitute graph is missing */
        $n \mathrel{+}= v⃗subst⃗n$;
      } **else** $n \mathrel{+}= v⃗ind$;
    } **else if** ($v⃗ind < -nn$) $nn = -(v⃗ind)$;
  if ($n >$ IND_GRAPH $\lor$ $nn >$ IND_GRAPH) *panic*(*very_bad_specs* + 1);   /* gigantic */
  $n \mathrel{+}= nn$;

This code is used in section 106.

108.   The negative vertices get the negative number as their name. Split vertices get names with an optional prime appended, if the *ind* field is 2; otherwise split vertex names are obtained by appending a colon and an index number between 0 and $ind - 1$. The name of a vertex within a graph $v\text{-}subst$ is composed of the name of v followed by a colon and the name within that graph.

We store the original ind field in the $mult$ field of the first corresponding vertex in the new graph, and change ind to point to that vertex. This convention makes it easy to determine the location of each vertex's clone or clones. Of course, if the original ind field is zero, we leave it zero (Λ), because it has no corresponding vertex in the new graph.

⟨ Assign names to the new vertices, and create a map from g to new_graph 108 ⟩ ≡

```
for (k = 1, u = new_graph→vertices; k ≤ nn; k++, u++) {
    u→mult = −k;
    sprintf (buffer, "%ld", −k);  u→name = gb_save_string (buffer);
}
for (v = g→vertices; v < g→vertices + g→n; v++)
    if ((k = v→ind) < 0)  v→map = (new_graph→vertices) − (k + 1);
    else if (k > 0) {
        u→mult = k;
        v→map = u;
        if (k ≤ 2) {
            u→name = gb_save_string (v→name);
            u++;
            if (k ≡ 2) {
                sprintf (buffer, "%s'", v→name);  u→name = gb_save_string (buffer);
                u++;
            }
        } else if (k ≥ IND_GRAPH)  ⟨ Make names and arcs for a substituted graph 114 ⟩
        else
            for (j = 0; j < k; j++, u++) {
                sprintf (buffer, "%.*s:%ld", BUF_SIZE − 12, v→name, j);
                u→name = gb_save_string (buffer);
            }
    }
```

This code is used in section 106.

$alloc\_fault = -1$, GB_GRAPH §7.
BUF_SIZE $= 4096$, §5.
$buffer$: **static char** [ ], §5.
gb_new_graph: **Graph** $*()$,
 GB_GRAPH §23.
$gb\_recycle$: **void** ( ), GB_GRAPH §40.
gb_save_string: **char** $*()$,
 GB_GRAPH §35.
$gb\_trouble\_code$: **long**,
 GB_GRAPH §14.
Graph $=$ **struct**, GB_GRAPH §20.
id: **char** [ ], GB_GRAPH §20.

$ind = z.I$, §100.
IND_GRAPH $= 1000000000$, §102.
j: **register long**, §9.
k: **register long**, §9.
$make\_compound\_id$: **void** ( ),
 GB_GRAPH §26.
$map = z.V$, §88.
$missing\_operand = 50$,
 GB_GRAPH §7.
$mult = v.I$, §82.
n: **long**, GB_GRAPH §20.

$name$: **char** $*$, GB_GRAPH §9.
new_graph: **Graph** $*$, §9.
$no\_room = 1$, GB_GRAPH §7.
$null\_string$: **char** [ ], GB_GRAPH §24.
$panic =$ macro ( ), §4.
$sprintf$: **int** ( ), <stdio.h>.
$subst = y.G$, §102.
v: **register Vertex** $*$, §9.
Vertex $=$ **struct**, GB_GRAPH §9.
$vertices$: **Vertex** $*$, GB_GRAPH §20.
$very\_bad\_specs = 40$, GB_GRAPH §7.

109. ⟨Restore g to its original state 109⟩ ≡

    for $(v = g\text{→}vertices;\ v < g\text{→}vertices + g\text{→}n;\ v\text{++})$

      if $(v\text{→}map)\ v\text{→}ind = v\text{→}map\text{→}mult;$

    for $(v = new\_graph\text{→}vertices;\ v < new\_graph\text{→}vertices + n;\ v\text{++})\ v\text{→}u.I = v\text{→}v.I = v\text{→}z.I = 0;$

      /* clear *tmp*, *mult*, *tlen* */

This code is used in section 105.

110. The heart of the procedure to construct an induced graph is, of course, the part where we map the arcs of g into arcs of *new_graph*.

Notice that if v has a self-loop in the original graph and if v is being split into several vertices, it will produce arcs between different clones of itself, but it will not produce self-loops unless *self* $\neq 0$. In an undirected graph, a loop from a vertex to itself will not produce multiple edges among its clones, even if *multi* $\neq 0$.

More precisely, if v has k clones u through $u + k - 1$, an original directed arc from v to v will generate all k^2 possible arcs between them, except that the k self-loops will be eliminated when *self* $\equiv 0$. An original undirected edge from v to v will generate $\binom{k}{2}$ edges between distinct clones, together with k undirected self-loops if *self* $\neq 0$.

⟨Insert arcs or edges for induced vertices 110⟩ ≡

    for $(v = g\text{→}vertices;\ v < g\text{→}vertices + g\text{→}n;\ v\text{++})$ {

      $u = v\text{→}map;$

      if (u) { **register Arc** $*a;$ **register Vertex** $*uu, *vv;$

        $k = u\text{→}mult;$

        if $(k < 0)\ k = 1;$ /* k is the number of clones of v */

        else if $(k \geq$ IND_GRAPH$)\ k = v\text{→}subst\text{→}n;$

        for (; $k;\ k\text{--}, u\text{++})$ {

          if $(\neg multi)$ ⟨Take note of existing edges that touch u 111⟩;

          for $(a = v\text{→}arcs;\ a;\ a = a\text{→}next)$ {

            $vv = a\text{→}tip;$

            $uu = vv\text{→}map;$

            if $(uu \equiv \Lambda)$ **continue**;

            $j = uu\text{→}mult;$

            if $(j < 0)\ j = 1;$ /* j is the number of clones of vv */

            else if $(j \geq$ IND_GRAPH$)\ j = vv\text{→}subst\text{→}n;$

            if $(\neg directed)$ {

              if $(vv < v)$ **continue**;

              if $(vv \equiv v)$ {

                if $(a\text{→}next \equiv a + 1)\ a\text{++};$ /* skip second half of self-loop */

                $j = k, uu = u;$ /* also skip duplicate edges generated by self-loop */

              }

            }

            ⟨Insert arcs or edges from vertex u to vertices uu through $uu + j - 1$ 112⟩;

          }

        }

      }

    }

This code is used in section 105.

111. Again we use the *tmp* and *tlen* trick of *gunion* to handle multiple arcs. (This trick explains why the code in the previous section tries to generate as many arcs as possible from a single vertex u, before changing u.)

⟨ Take note of existing edges that touch u 111 ⟩ ≡
   for $(a = u{\to}arcs;\ a;\ a = a{\to}next)$ {
     $a{\to}tip{\to}tmp = u;$
     if $(directed \lor a{\to}tip > u \lor a{\to}next \equiv a + 1)$ $a{\to}tip{\to}tlen = a;$
     else $a{\to}tip{\to}tlen = a + 1;$
   }

This code is used in section 110.

112. ⟨ Insert arcs or edges from vertex u to vertices uu through $uu + j - 1$ 112 ⟩ ≡
   for $(\ ;\ j;\ j{-}{-},uu{+}{+})$ {
     if $(u \equiv uu \land \lnot self)$ **continue**;
     if $(uu{\to}tmp \equiv u \land \lnot multi)$ ⟨ Update the minimum arc length from u to uu, then **continue** 113 ⟩;
     if $(directed)$ $gb\_new\_arc(u, uu, a{\to}len)$;
     else $gb\_new\_edge(u, uu, a{\to}len)$;
     $uu{\to}tmp = u;$
     $uu{\to}tlen = ((directed \lor u \le uu)\ ?\ u{\to}arcs : uu{\to}arcs);$
   }

This code is used in section 110.

113. ⟨ Update the minimum arc length from u to uu, then **continue** 113 ⟩ ≡
   { **register Arc** $*b = uu{\to}tlen;$ /* existing arc or edge from u to uu */
     if $(a{\to}len < b{\to}len)$ {
       $b{\to}len = a{\to}len;$ /* remember the minimum length */
       if $(\lnot directed)$ $(b + 1){\to}len = a{\to}len;$
     }
     continue;
   }

This code is used in sections 112 and 114.

---

a: **register Arc** *, §114.
Arc = **struct**, GB_GRAPH §10.
arcs: **Arc** *, GB_GRAPH §9.
directed: **long**, §105.
g: **Graph** *, §105.
gb_new_arc: **void** ( ),
   GB_GRAPH §30.
gb_new_edge: **void** ( ),
   GB_GRAPH §31.
gunion: **Graph** *( ), §78.
I: **long**, GB_GRAPH §8.
ind = *z.I*, §100.

IND_GRAPH = 1000000000, §102.
j: **register long**, §9.
k: **register long**, §9.
len: **long**, GB_GRAPH §10.
map = *z.V*, §88.
mult = *v.I*, §82.
multi: **long**, §105.
n: **long**, GB_GRAPH §20.
new_graph: **Graph** *, §9.
next: **Arc** *, GB_GRAPH §10.
self: **long**, §105.
subst = *y.G*, §102.

tip: **Vertex** *, GB_GRAPH §10.
tlen = *z.A*, §80.
tmp = *u.V*, §76.
u: **register Vertex** *, §105.
u: **util**, GB_GRAPH §9.
uu: **Vertex** *, §114.
v: **register Vertex** *, §9.
v: **util**, GB_GRAPH §9.
Vertex = **struct**, GB_GRAPH §9.
vertices: **Vertex** *, GB_GRAPH §20.
z: **util**, GB_GRAPH §9.

114. We have now accumulated enough experience to finish off the one remaining piece of program with ease.

⟨ Make names and arcs for a substituted graph 114 ⟩ ≡

```
{ register Graph *gg = v→subst;
  register Vertex *vv = gg→vertices;
  register Arc *a;
  siz_t delta = ((siz_t) u) − ((siz_t) vv);
  for (j = 0; j < v→subst→n; j++, u++, vv++) {
    sprintf(buffer, "%.*s:%.*s", BUF_SIZE/2 − 1, v→name, (BUF_SIZE − 1)/2, vv→name);
    u→name = gb_save_string(buffer);
    for (a = vv→arcs; a; a = a→next) { register Vertex *vvv = a→tip;
      Vertex *uu = vert_offset(vvv, delta);
      if (vvv ≡ vv ∧ ¬self) continue;
      if (uu→tmp ≡ u ∧ ¬multi)
        ⟨ Update the minimum arc length from u to uu, then continue 113 ⟩;
      if (¬directed) {
        if (vvv < vv) continue;
        if (vvv ≡ vv ∧ a→next ≡ a + 1) a++;      /* skip second half of self-loop */
        gb_new_edge(u, uu, a→len);
      } else gb_new_arc(u, uu, a→len);
      uu→tmp = u;
      uu→tlen = ((directed ∨ u ≤ uu) ? u→arcs : uu→arcs);
    }
  }
}
```

This code is used in section 108.

Arc = **struct**, GB_GRAPH §10.
arcs: **Arc** *, GB_GRAPH §9.
BUF_SIZE = 4096, §5.
buffer: **static char** [], §5.
directed: **long**, §105.
gb_new_arc: **void** ( ),
    GB_GRAPH §30.
gb_new_edge: **void** ( ),
    GB_GRAPH §31.
gb_save_string: **char** *( ),
    GB_GRAPH §35.

Graph = **struct**, GB_GRAPH §20.
j: **register long**, §9.
len: **long**, GB_GRAPH §10.
multi: **long**, §105.
n: **long**, GB_GRAPH §20.
name: **char** *, GB_GRAPH §9.
next: **Arc** *, GB_GRAPH §10.
self: **long**, §105.
siz_t = **unsigned long**,
    GB_GRAPH §34.

sprintf: **int** ( ), <stdio.h>.
subst = *y.G*, §102.
tip: **Vertex** *, GB_GRAPH §10.
tlen = *z.A*, §80.
tmp = *u.V*, §76.
u: **register Vertex** *, §105.
v: **register Vertex** *, §9.
vert_offset = macro ( ), §75.
Vertex = **struct**, GB_GRAPH §9.
vertices: **Vertex** *, GB_GRAPH §20.

GB_BOOKS

Important: Before reading GB_BOOKS, please read or at least skim the programs for GB_GRAPH and GB_IO.

1. Introduction. This GraphBase module contains the *book* subroutine, which creates a family of undirected graphs that are based on classic works of literature. It also contains the *bi_book* subroutine, which creates a related family of bipartite graphs. An example of the use of *book* can be found in the demonstration program BOOK_COMPONENTS.

⟨ gb_books.h 1 ⟩ ≡
  extern Graph *book( );
  extern Graph *bi_book( );

See also sections 6, 18, and 23.

2. The subroutine call *book*(⟨title⟩, n, x, *first_chapter*, *last_chapter*, *in_weight*, *out_weight*, *seed*) constructs a graph based on the information in ⟨title⟩.dat, where ⟨title⟩ is either "anna" (for *Anna Karenina*), "david" (for *David Copperfield*), "jean" (for *Les Misérables*), "huck" (for *Huckleberry Finn*), or "homer" (for *The Iliad*). Each vertex of the graph corresponds to one of the characters in the selected book. Edges between vertices correspond to encounters between those characters. The length of each edge is 1.

Subsets of the book can be selected by specifying that the edge data should be restricted to chapters between *first_chapter* and *last_chapter*, inclusive. If *first_chapter* = 0, the result is the same as if *first_chapter* = 1. If *last_chapter* = 0 or if *last_chapter* exceeds the total number of chapters in the book, the result is the same as if *last_chapter* were the number of the book's final chapter.

The constructed graph will have $\min(n, N) - x$ vertices, where N is the total number of characters in the selected book. However, if n is zero, n is automatically made equal to the maximum possible value, N. If n is less than N, the $n - x$ characters will be selected by assigning a weight to each character and choosing the n with largest weight, then excluding the largest x of these, using random numbers to break ties in case of equal weights. Weights are computed by the formula

$$in\_weight \cdot chapters\_in + out\_weight \cdot chapters\_out,$$

where *chapters_in* is the number of chapters between *first_chapter* and *last_chapter* in which a particular character appears, and *chapters_out* is the number of other chapters in which that character appears. Both *in_weight* and *out_weight* must be at most 1,000,000 in absolute value.

Vertices of the graph will appear in order of decreasing weight. The *seed* parameter defines the pseudo-random numbers used wherever a "random" choice between equal-weight vertices needs to be made. As usual with GraphBase routines, different choices of *seed* will in general produce different selections, but in a system-independent manner; identical results will be obtained on all computers when identical parameters have been specified. Any *seed* value between 0 and $2^{31} - 1$ is permissible.

3. Examples: The call *book*("anna", 0, 0, 0, 0, 0, 0, 0) will construct a graph on 138 vertices that represent all 138 characters of Tolstoy's *Anna Karenina*, as recorded in anna.dat. Two vertices will be adjacent if the corresponding characters encounter each other anywhere in the book. The call *book*("anna", 50, 0, 0, 0, 1, 1, 0) is similar, but it is restricted to the 50 characters that occur most frequently, i.e., in the most chapters. The call *book*("anna", 50, 0, 10, 120, 1, 1, 0) has the same vertices, but it has edges only for encounters that take place between chapter 10 and chapter 120, inclusive. The call *book*("anna", 50, 0, 10, 120, 1, 0, 0) is similar, but its vertices are the 50 characters that occur most often in chapters 10 through 120, without regard to how often they occur in the rest of the book. The call *book*("anna", 50, 0, 10, 120, 0, 0, 0) is also similar, but it chooses 50 characters completely at random (possibly from those that don't occur in the selected chapters at all).

Parameter x, which causes the x vertices of highest weight to be excluded, is usually either 0 or 1. It is provided primarily so that users can set $x = 1$ with respect to *David Copperfield* and *Huckleberry Finn*; those novels are narrated by their principal character, so they have edges between the principal character and almost everybody else. (Characters cannot get into the action of a first-person account unless they encounter the narrator or unless the narrator is quoting some other person's story.) The corresponding graphs tend to have more interesting connectivity properties if we leave the narrator out by setting $x = 1$. For example, there are 87 characters in *David Copperfield*; the call *book*("david", 0, 1, 0, 0, 1, 1, 0) produces a graph with 86 vertices, one for every character except David Copperfield himself.

4.   The subroutine call *bi_book*(⟨title⟩, n, x, *first_chapter*, *last_chapter*, *in_weight*, *out_weight*, *seed*) produces a bipartite graph in which the vertices of the first part are exactly the same as the vertices of the graph returned by *book*, while the vertices of the second part are the selected chapters. For example, *bi_book*("anna", 50, 0, 10, 120, 1, 1, 0) creates a bipartite graph with 50 + 111 vertices. There is an edge between each character and the chapters in which that character appears.

5.   Chapter numbering needs further explanation. *Anna Karenina* has 239 chapters, which are numbered 1.1 through 8.19 in the work itself but renumbered 1 through 239 as far as the *book* routine is concerned. Thus, setting *first_chapter* = 10 and *last_chapter* = 120 turns out to be equivalent to selecting chapters 1.10 through 4.19 (more precisely, chapter 10 of book 1 through chapter 19 of book 4). *Les Misérables* has an even more involved scheme; its 356 chapters range from 1.1.1 (part 1, book 1, chapter 1) to 5.9.6 (part 5, book 9, chapter 6). After *book* or *bi_book* has created a graph, the external integer variable *chapters* will contain the total number of chapters, and *chap_name* will be an array of strings containing the structured chapter numbers. For example, after *book*("jean", ...), we will have *chapters* = 356, *chap_name*[1] = "1.1.1", ..., *chap_name*[356] = "5.9.6"; *chap_name*[0] will be "".

#**define** MAX_CHAPS 360   /* no book will have this many chapters */

⟨ External variables 5 ⟩ ≡
    long *chapters*;   /* the total number of chapters in the selected book */
    char **chap_name*[MAX_CHAPS] = {""};   /* string names of those chapters */
This code is used in section 8.

---

bi_book: **Graph** *( ), §8.
book: **Graph** *( ), §8.
first_chapter: **unsigned long**, §8.
Graph = **struct**, GB_GRAPH §20.

in_weight: **long**, §8.
last_chapter: **unsigned long**, §8.
n: **unsigned long**, §8.

out_weight: **long**, §8.
seed: **long**, §8.
x: **unsigned long**, §8.

6. As usual, we put declarations of the external variables into the header file for users to **include**.

⟨ gb_books.h 1 ⟩ +≡
    extern long *chapters*;   /∗ the total number of chapters in the selected book ∗/
    extern char ∗*chap_name*[ ];   /∗ string names of those chapters ∗/

7. If the *book* or *bi_book* routine encounters a problem, it returns Λ (NULL), after putting a code number into the external variable *panic_code*. This code number identifies the type of failure. Otherwise *book* returns a pointer to the newly created graph, which will be represented with the data structures explained in GB_GRAPH. (The external variable *panic_code* is itself defined in GB_GRAPH.)

    #define *panic*(c) { *panic_code* = c; *gb_trouble_code* = 0; **return** Λ; }

8. The C file **gb_books.c** has the overall shape shown here. It makes use of an internal subroutine called *bgraph*, which combines the work of *book* and *bi_book*.

```
#include "gb_io.h"    /* we will use the GB_IO routines for input */
#include "gb_flip.h"     /* we will use the GB_FLIP routines for random numbers */
#include "gb_graph.h"     /* we will use the GB_GRAPH data structures */
#include "gb_sort.h"     /* and the gb_linksort routine */
```
  ⟨ Preprocessor definitions ⟩

  ⟨ Type declarations 13 ⟩
  ⟨ Private variables 11 ⟩
  ⟨ External variables 5 ⟩

  static Graph ∗*bgraph*(*bipartite*, *title*, *n*, *x*, *first_chapter*, *last_chapter*, *in_weight*, *out_weight*, *seed*)
      long *bipartite*;   /∗ should we make the graph bipartite? ∗/
      char ∗*title*;   /∗ identification of the selected book ∗/
      unsigned long *n*;   /∗ number of vertices desired before exclusion ∗/
      unsigned long *x*;   /∗ number of vertices to exclude ∗/
      unsigned long *first_chapter*, *last_chapter*;   /∗ interval of chapters leading to edges ∗/
      long *in_weight*;   /∗ weight coefficient pertaining to chapters in that interval ∗/
      long *out_weight*;   /∗ weight coefficient pertaining to chapters not in that interval ∗/
      long *seed*;   /∗ random number seed ∗/
  { ⟨ Local variables 9 ⟩
    gb_init_rand(*seed*);
    ⟨ Check that the parameters are valid 10 ⟩;
    ⟨ Skim the data file, recording the characters and computing their weights 15 ⟩;
    ⟨ Choose the vertices and put them into an empty graph 27 ⟩;
    ⟨ Read the data file more carefully and fill the graph as instructed 29 ⟩;
    if (*gb_trouble_code*) {
      gb_recycle(*new_graph*);
      panic(*alloc_fault*);   /∗ (expletive deleted) we ran out of memory somewhere back there ∗/
    }
    return *new_graph*;
  }

Graph $*book\,(title, n, x, first\_chapter, last\_chapter, in\_weight, out\_weight, seed\,)$
    char $*title$;
    unsigned long n, x, $first\_chapter$, $last\_chapter$;
    long in_weight, out_weight, $seed$;
  { **return** $bgraph\,(0\,{\rm L}, title, n, x, first\_chapter, last\_chapter, in\_weight, out\_weight, seed\,)$; }
Graph $*bi\_book\,(title, n, x, first\_chapter, last\_chapter, in\_weight, out\_weight, seed\,)$
    char $*title$;
    unsigned long n, x, $first\_chapter$, $last\_chapter$;
    long in_weight, out_weight, $seed$;
  { **return** $bgraph\,(1\,{\rm L}, title, n, x, first\_chapter, last\_chapter, in\_weight, out\_weight, seed\,)$; }

9. ⟨ Local variables 9 ⟩ ≡
  Graph $*new\_graph$;   /* the graph constructed by *book* or *bi_book* */
  register long j, k;   /* all-purpose indices */
  long $characters$;   /* the total number of characters in the selected book */
  register node $*p$;   /* information about the current character */
See also section 21.

This code is used in section 8.

10. **#define** MAX_CHARS 600   /* there won't be more characters than this */
⟨ Check that the parameters are valid 10 ⟩ ≡
  if $(n \equiv 0)$ $n =$ MAX_CHARS;
  if $(first\_chapter \equiv 0)$ $first\_chapter = 1$;
  if $(last\_chapter \equiv 0)$ $last\_chapter =$ MAX_CHAPS;
  if $(in\_weight > 1000000 \lor in\_weight < -1000000 \lor out\_weight > 1000000 \lor out\_weight < -1000000)$
    $panic\,(bad\_specs)$;   /* the magnitude of at least one weight is too big */
  $sprintf\,(file\_name, \texttt{"\%.6s.dat"}, title)$;
  if $(gb\_open\,(file\_name) \neq 0)$ $panic\,(early\_data\_fault)$;
    /* couldn't open the file; *io_errors* tells why */
This code is used in section 8.

11. ⟨ Private variables 11 ⟩ ≡
  static char $file\_name[\,] = \texttt{"xxxxxx.dat"}$;
See also sections 14 and 24.

This code is used in section 8.

$alloc\_fault = -1$, GB_GRAPH §7.
$bad\_specs = 30$, GB_GRAPH §7.
$chap\_name$: **char** $*[\,]$, §5.
$chapters$: **long**, §5.
$early\_data\_fault = 10$, GB_GRAPH §7.
gb_init_rand: **void** ( ), GB_FLIP §8.
$gb\_linksort$: **void** ( ), GB_SORT §5.
gb_open: **long** ( ), GB_IO §32.
$gb\_recycle$: **void** ( ), GB_GRAPH §40.
$gb\_trouble\_code$: **long**,
  GB_GRAPH §14.
Graph = **struct**, GB_GRAPH §20.
io_errors: **long**, GB_IO §5.
MAX_CHAPS $= 360$, §5.
$panic\_code$: **long**, GB_GRAPH §5.
$sprintf$: **int** ( ), <stdio.h>.

12. Vertices. Each character in a book has been given a two-letter code name for internal use. The code names are explained at the beginning of each data file by a number of lines that look like this:

$$\text{XX } \langle name \rangle, \langle description \rangle$$

For example, here's one of the lines near the beginning of `"anna.dat"`:

```
AL Alexey Alexandrovitch Karenin, minister of state
```

The $\langle name \rangle$ does not contain a comma; the $\langle description \rangle$ might.

A blank line follows the cast of characters.

Internally, we will think of the two-letter code as a radix-36 integer. Thus **AA** will be the number $10 \times 36 + 10$, and **ZZ** will be $35 \times 36 + 35$. The *gb_number* routine in GB_IO is set up to input radix-36 integers just as it does hexadecimal ones. In *The Iliad*, many of the minor characters have numeric digits in their code names because the total number of characters is too large to permit mnemonic codes for everybody.

#define `MAX_CODE` 1296   /* 36×36, the number of two-digit codes in radix 36 */

13. In order to choose the vertices, we want to represent each character as a node whose key corresponds to its weight; then the *gb_linksort* routine of GB_SORT will provide the desired rank-ordering. We will find it convenient to use these nodes for all the data processing that *bgraph* has to do.

⟨ Type declarations 13 ⟩ ≡
  typedef struct node_struct {   /* records to be sorted by *gb_linksort* */
    long *key*;   /* the nonnegative sort key (weight plus 2^{30}) */
    struct node_struct **link*;   /* pointer to next record */
    long *code*;   /* code number of this character */
    long *in*;   /* number of occurrences in selected chapters */
    long *out*;   /* number of occurrences in unselected chapters */
    long *chap*;   /* seen most recently in this chapter */
    Vertex **vert*;   /* vertex corresponding to this character */
  } **node**;

This code is used in section 8.

14. Not only do nodes point to codes, we also want codes to point to nodes.

⟨ Private variables 11 ⟩ +≡
  static node *node_block*[`MAX_CHARS`];   /* array of nodes for working storage */
  static node **xnode*[`MAX_CODE`];   /* the node, if any, having a given code */

15. We will read the data file twice, once quickly (to collect statistics) and once more thoroughly (to record detailed information). Here is the quick version.

⟨ Skim the data file, recording the characters and computing their weights 15 ⟩ ≡
  ⟨ Read the character codes at the beginning of the data file, and prepare a node for each one 16 ⟩;
  ⟨ Skim the chapter information, counting the number of chapters in which each character appears 19 ⟩;
  if (*gb_close*( ) ≠ 0) *panic*(*late_data_fault*);
      /* check sum or other failure in data file; see *io_errors* */

This code is used in section 8.

16. ⟨ Read the character codes at the beginning of the data file, and prepare a node for each
       one 16 ⟩ ≡
  for ($k = 0$; $k <$ MAX_CODE; $k{+}{+}$) $xnode[k] = \Lambda$;
  { **register long** c;   /∗ current code entering the system ∗/
    $p = node\_block$;   /∗ current node entering the system ∗/
    while (($c = gb\_number\,(36)) \neq 0$) {   /∗ note that 00 is not a legal code ∗/
      if ($c \geq$ MAX_CODE $\lor$ $gb\_char\,( ) \neq$ '␣') $panic(syntax\_error)$;   /∗ unreadable line in data file ∗/
      if ($p \geq \&node\_block[$MAX_CHARS$]$) $panic(syntax\_error + 1)$;
          /∗ data has too many characters ∗/
      $p{\rightarrow}link = (p \equiv node\_block\,?\,\Lambda : p - 1)$;
      $p{\rightarrow}code = c$;
      $xnode[c] = p$;
      $p{\rightarrow}in = p{\rightarrow}out = p{\rightarrow}chap = 0$;
      $p{\rightarrow}vert = \Lambda$;
      $p{+}{+}$;
      $gb\_newline\,( )$;
    }
    $characters = p - node\_block$;
    $gb\_newline\,( )$;   /∗ bypass the blank line that terminates the character data ∗/
  }
This code is used in section 15.

---

bgraph: **static Graph** ∗( ), §8.
characters: **long**, §9.
gb_char: **char** ( ), GB_IO §22.
gb_close: **long** ( ), GB_IO §39.
gb_linksort: **void** ( ), GB_SORT §5.
gb_newline: **void** ( ), GB_IO §18.

gb_number: **unsigned long** ( ),
   GB_IO §24.
io_errors: **long**, GB_IO §5.
k: **register long**, §9.
late_data_fault = 11, GB_GRAPH §7.

MAX_CHARS = 600, §10.
p: **register node** ∗, §9.
panic = macro ( ), §7.
syntax_error = 20, GB_GRAPH §7.
Vertex = **struct**, GB_GRAPH §9.

17.   Later we will read through this part of the file again, extracting additional information if it turns out to be relevant. The ⟨description⟩ string is provided to users in a *desc* field, in case anybody cares to look at it. The *in* and *out* statistics are also made available in utility fields called *in_count* and *out_count*. The code value is placed in the *short_code* field.

#define *desc* *z.S*   /* utility field *z* points to the ⟨description⟩ string */
#define *in_count* *y.I*   /* utility field *y* counts appearances in selected chapters */
#define *out_count* *x.I*   /* utility field *x* counts appearances in other chapters */
#define *short_code* *u.I*   /* utility field *u* contains a radix-36 number */

⟨ Read the data about characters again, noting vertex names and the associated descriptions 17 ⟩ ≡
 { **register long** *c*;   /* current code entering the system a second time */
  while ((*c* = *gb_number*(36)) ≠ 0) { **register Vertex** **v* = *xnode*[*c*]→*vert*;
   if (*v*) {
    if (*gb_char*( ) ≠ '␣') *panic*(*impossible*);   /* can't happen */
    gb_string(*str_buf*, ',');   /* scan the ⟨name⟩ part */
    v→*name* = *gb_save_string*(*str_buf*);
    if (*gb_char*( ) ≠ ',') *panic*(*syntax_error* + 2);   /* missing comma after ⟨name⟩ */
    if (*gb_char*( ) ≠ '␣') *panic*(*syntax_error* + 3);   /* missing space after comma */
    gb_string(*str_buf*, '\n');   /* scan the ⟨description⟩ part */
    v→*desc* = *gb_save_string*(*str_buf*);
    v→*in_count* = *xnode*[*c*]→*in*;
    v→*out_count* = *xnode*[*c*]→*out*;
    v→*short_code* = *c*;
   }
   gb_newline( );
  }
 gb_newline( );   /* bypass the blank line that terminates the character data */
}

This code is used in section 29.

18.   ⟨ gb_books.h 1 ⟩ +≡
#define *desc* *z.S*   /* utility field definitions for the header file */
#define *in_count* *y.I*
#define *out_count* *x.I*
#define *short_code* *u.I*

19. Edges. The second part of the data file has a line for each chapter, containing "cliques of encounters." For example, the line

$$\texttt{3.22:AA,BB,CC,DD;CC,DD,EE;AA,FF}$$

means that, in chapter 22 of book 3, there were encounters between the pairs

$$\texttt{AA-BB, AA-CC, AA-DD, BB-CC, BB-DD, CC-DD, CC-EE, DD-EE, and AA-FF.}$$

(The encounter `CC-DD` is specified twice, once in the clique `AA,BB,CC,DD` and once in `CC,DD,EE`; this does not imply anything about the actual number of encounters between `CC` and `DD` in the chapter.)

A clique might involve one character only, when that character is featured in sort of a soliloquy.

A chapter might contain no references to characters at all. In such a case the ':' following the chapter number is omitted.

There might be more encounters than will fit on a single line. In such cases, continuation lines begin with '`&:`'. This convention turns out to be needed only in `homer.dat`; chapters in *The Iliad* are substantially more complex than the chapters in other GraphBase books.

On our first pass over the data, we simply want to compute statistics about who appears in what chapters, so we ignore the distinction between commas and semicolons.

$\langle$ Skim the chapter information, counting the number of chapters in which each character appears 19 $\rangle$ $\equiv$

```
for (k = 1; k < MAX_CHAPS ∧ ¬gb_eof( ); k++) {
    gb_string(str_buf, ':');      /* read past the chapter number */
    if (str_buf[0] ≡ '&') k--;    /* continuation of previous chapter */
    while (gb_char( ) ≠ '\n') { register long c = gb_number(36);
        if (c ≥ MAX_CODE) panic(syntax_error + 4);   /* missing punctuation between characters */
        p = xnode[c];
        if (p ≡ Λ) panic(syntax_error + 5);    /* unknown character */
        if (p⁀chap ≠ k) { p⁀chap = k;
            if (k ≥ first_chapter ∧ k ≤ last_chapter) p⁀in++; else p⁀out++;
        }
    }
    gb_newline( );
}
if (k ≡ MAX_CHAPS) panic(syntax_error + 6);    /* too many chapters */
chapters = k - 1;
```

This code is used in section 15.

---

chap: **long**, §13.	*I*: **long**, GB_GRAPH §8.	*S*: **char** *, GB_GRAPH §8.
chapters: **long**, §5.	*impossible* = 90, GB_GRAPH §7.	*str_buf*: **char** [ ], GB_IO §26.
first_chapter: **unsigned long**, §8.	*in*: **long**, §13.	*syntax_error* = 20, GB_GRAPH §7.
gb_char: **char** ( ), GB_IO §22.	*k*: **register long**, §9.	*u*: **util**, GB_GRAPH §9.
gb_eof: **long** ( ), GB_IO §20.	*last_chapter*: **unsigned long**, §8.	*vert*: **Vertex** *, §13.
gb_newline: **void** ( ), GB_IO §18.	MAX_CHAPS = 360, §5.	**Vertex** = **struct**, GB_GRAPH §9.
gb_number: **unsigned long** ( ),	MAX_CODE = 1296, §12.	*x*: **util**, GB_GRAPH §9.
GB_IO §24.	*name*: **char** *, GB_GRAPH §9.	*xnode*: **static node** *[ ], §14.
gb_save_string: **char** *( ),	*out*: **long**, §13.	*y*: **util**, GB_GRAPH §9.
GB_GRAPH §35.	*p*: **register node** *, §9.	*z*: **util**, GB_GRAPH §9.
gb_string: **char** *( ), GB_IO §26.	*panic* = **macro** ( ), §7.	

20. Our second pass over the data is very similar to the first, if we are simply computing a bipartite graph. In that case we add an edge to the graph between each selected chapter and each selected character in that chapter. Local variable *chap_base* will point to a vertex such that *chap_base* + k is the vertex corresponding to chapter k.

The *in_count* of a chapter vertex is the degree of that vertex, i.e., the number of selected characters that appear in the corresponding chapter. The *out_count* is the number of characters that appear in the chapter but were omitted from the graph. Thus the *in_count* and *out_count* for chapters are analogous to the *in_count* and *out_count* for characters.

⟨ Read the chapter information a second time and create the appropriate bipartite edges 20 ⟩ ≡

```
{
    for (p = node_block; p < node_block + characters; p++) p⁻chap = 0;
    for (k = 1; ¬gb_eof(); k++) {
        gb_string(str_buf, ':');        /* read the chapter number */
        if (str_buf[0] ≡ '&') k--;
        else chap_name[k] = gb_save_string(str_buf);
        if (k ≥ first_chapter ∧ k ≤ last_chapter) { register Vertex *u = chap_base + k;
            if (str_buf[0] ≠ '&') {
                u⁻name = chap_name[k];
                u⁻desc = null_string;
                u⁻in_count = u⁻out_count = 0;
            }
            while (gb_char() ≠ '\n') { register long c = gb_number(36);
                p = xnode[c];
                if (p⁻chap ≠ k) { register Vertex *v = p⁻vert;
                    p⁻chap = k;
                    if (v) { gb_new_edge(v, u, 1 L);
                        u⁻in_count++;
                    } else u⁻out_count++;
                }
            }
        }
        gb_newline();
    }
}
```

This code is used in section 29.

21. ⟨ Local variables 9 ⟩ +≡
    Vertex *chap_base;   /* the bipartite vertex for chapter k is *chap_base* + k */

22. The second pass has to work a little harder when we are recording encounters from cliques, but the logic isn't difficult really. We insert a reference to the first chapter that generated each edge, in utility field *chap_no* of the corresponding **Arc** record.

#define *chap_no* a.I   /* utility field *a* holds a chapter number */

⟨ Read the chapter information a second time and create the appropriate edges for encounters 22 ⟩ ≡
    for (k = 1; ¬gb_eof(); k++) { **char** *s;
      s = gb_string(str_buf, ':');   /* read the chapter number */

```
        if (str_buf [0] ≡ '&') k−−;
        else { if (*(s − 2) ≡ '\n') *(s − 2) = '\0';
            chap_name[k] = gb_save_string (str_buf );
        }
        if (k ≥ first_chapter ∧ k ≤ last_chapter) { register long c = gb_char ( );
            while (c ≠ '\n') { register Vertex **pp = clique_table;
                register Vertex **qq, **rr;        /* pointers within the clique table */
                do {  c = gb_number (36);          /* set c to code for next character of clique */
                    if (xnode[c]→vert)     /* is that character a selected vertex? */
                        *pp ++ = xnode[c]→vert;    /* if so, that vertex joins the current clique */
                    c = gb_char ( );
                } while (c ≡ ',');          /* repeat until end of the clique */
                for (qq = clique_table;  qq + 1 < pp;  qq ++)
                    for (rr = qq + 1;  rr < pp;  rr ++)
                        ⟨Make the vertices *qq and *rr adjacent, if they aren't already 25⟩;
            }
        }
        gb_newline ( );
    }
```

This code is used in section 29.

23. ⟨gb_books.h 1⟩ +≡
#define *chap_no* *a.I*   /* utility field definition in the header file */

24. ⟨Private variables 11⟩ +≡
static Vertex *∗clique_table*[30];   /* pointers to vertices in the current clique */

25. ⟨Make the vertices *qq and *rr adjacent, if they aren't already 25⟩ ≡
```
    { register Vertex *u = *qq, *v = *rr;
        register Arc *a;
        for (a = u→arcs;  a;  a = a→next) if (a→tip ≡ v) goto found;
        gb_new_edge(u, v, 1 L);        /* not found, so they weren't already adjacent */
        if (u < v) a = u→arcs;
        else  a = v→arcs;        /* the new edge consists of arcs a and a + 1 */
        a→chap_no = (a + 1)→chap_no = k;
    found: ;
    }
```

This code is used in section 22.

---

a: util, GB_GRAPH §10.
Arc = struct, GB_GRAPH §10.
arcs: **Arc** ∗, GB_GRAPH §9.
chap: **long**, §13.
chap_name: **char** ∗[], §5.
characters: **long**, §9.
desc = *z.S*, §17.
first_chapter: **unsigned long**, §8.
gb_char: **char** ( ), GB_IO §22.
gb_eof: **long** ( ), GB_IO §20.
gb_new_edge: **void** ( ),

GB_GRAPH §31.
gb_newline: **void** ( ), GB_IO §18.
gb_number: **unsigned long** ( ),
    GB_IO §24.
gb_save_string: **char** ∗( ),
    GB_GRAPH §35.
gb_string: **char** ∗( ), GB_IO §26.
I: **long**, GB_GRAPH §8.
in_count = *y.I*, §17.
k: **register long**, §9.
last_chapter: **unsigned long**, §8.

name: **char** ∗, GB_GRAPH §9.
next: **Arc** ∗, GB_GRAPH §10.
node_block: **static node** [], §14.
null_string: **char** [], GB_GRAPH §24.
out_count = *x.I*, §17.
p: **register node** ∗, §9.
str_buf: **char** [], GB_IO §26.
tip: **Vertex** ∗, GB_GRAPH §10.
vert: **Vertex** ∗, §13.
Vertex = struct, GB_GRAPH §9.
xnode: **static node** ∗[], §14.

26. **Administration.** The program is now complete except for a few missing organizational details. I will add these after lunch.

27. OK, I'm back; what needs to be done? The main thing is to create the graph itself.

⟨ Choose the vertices and put them into an empty graph 27 ⟩ ≡

  if $(n > characters)$ $n = characters$;
  if $(x > n)$ $x = n$;
  if $(last\_chapter > chapters)$ $last\_chapter = chapters$;
  if $(first\_chapter > last\_chapter)$ $first\_chapter = last\_chapter + 1$;
  $new\_graph = gb\_new\_graph(n - x + (bipartite\ ?\ last\_chapter - first\_chapter + 1 : 0))$;
  if $(new\_graph \equiv \Lambda)$ $panic(no\_room)$;   /* out of memory already */
  $strcpy(new\_graph{\rightarrow}util\_types,$ "IZZIISIZZZZZZZ"$)$;   /* declare the types of utility fields */
  $sprintf(new\_graph{\rightarrow}id,$ "%sbook(\"%s\",%lu,%lu,%lu,%lu,%ld,%ld)"$, bipartite\ ?\ $"bi_"$:\ $""$,$
      $title, n, x, first\_chapter, last\_chapter, in\_weight, out\_weight, seed)$;
  if $(bipartite)$ {
    $mark\_bipartite(new\_graph, n - x)$;
    $chap\_base = new\_graph{\rightarrow}vertices + (new\_graph{\rightarrow}n\_1 - first\_chapter)$;
  }
⟨ Compute the weights and assign vertices to chosen nodes 28 ⟩;

This code is used in section 8.

28. ⟨ Compute the weights and assign vertices to chosen nodes 28 ⟩ ≡

  for $(p = node\_block;\ p < node\_block + characters;\ p{+}{+})$
    $p{\rightarrow}key = in\_weight * (p{\rightarrow}in) + out\_weight * (p{\rightarrow}out) + {}^{\#}40000000$;
  $gb\_linksort(node\_block + characters - 1)$;
  $k = n$;   /* we will look at this many nodes */
  { **register Vertex** $*v = new\_graph{\rightarrow}vertices$;   /* the next vertex to define */
    for $(j = 127;\ j \geq 0;\ j{-}{-})$
      for $(p = ($**node** $*)\ gb\_sorted[j];\ p;\ p = p{\rightarrow}link)$ {
        if $(x > 0)$ $x{-}{-}$;   /* ignore this node */
        else $p{\rightarrow}vert = v{+}{+}$;   /* choose this node */
        if $({-}{-}k \equiv 0)$ **goto** *done*;
      }
  }
done: ;

This code is used in section 27.

29. Once the graph is there, we're ready to fill it in.

⟨ Read the data file more carefully and fill the graph as instructed 29 ⟩ ≡

  if $(gb\_open(file\_name) \neq 0)$ $panic(impossible + 1)$;
     /* this can't happen, because we were successful before */
  ⟨ Read the data about characters again, noting vertex names and the associated descriptions 17 ⟩;
  if $(bipartite)$
    ⟨ Read the chapter information a second time and create the appropriate bipartite edges 20 ⟩
  else
    ⟨ Read the chapter information a second time and create the appropriate edges for encounters 22 ⟩;
  if $(gb\_close(\ ) \neq 0)$ $panic(impossible + 2)$;
     /* again, can hardly happen the second time around */

This code is used in section 8.

GB_DIJK

Important: Before reading GB_DIJK, please read or at least skim the program for GB_GRAPH.

1. Introduction. The GraphBase demonstration routine $dijkstra(uu, vv, gg, hh)$ finds a shortest path from vertex uu to vertex vv in graph gg, with the aid of an optional heuristic function hh. This function implements a version of Dijkstra's algorithm, a general procedure for determining shortest paths in a directed graph that has nonnegative arc lengths [E. W. Dijkstra, "A note on two problems in connexion with graphs," *Numerische Mathematik* **1** (1959), 269–271].

If hh is null, the length of every arc in gg must be nonnegative. If hh is non-null, hh should be a function defined on the vertices of the graph such that the length d of an arc from u to v always satisfies the condition

$$d \geq hh(u) - hh(v).$$

In such a case, we can effectively replace each arc length d by $d - hh(u) + hh(v)$, obtaining a graph with nonnegative arc lengths. The shortest paths between vertices in this modified graph are the same as they were in the original graph.

The basic idea of Dijkstra's algorithm is to explore the vertices of the graph in order of their distance from the starting vertex uu, proceeding until vv is encountered. If the distances have been modified by a heuristic function hh such that $hh(u)$ happens to equal the true distance from u to vv, for all u, then all of the modified distances on shortest paths to vv will be zero. This means that the algorithm will explore all of the most useful arcs first, without wandering off in unfruitful directions. In practice we usually don't know the exact distances to vv in advance, but we can often compute an approximate value $hh(u)$ that will help focus the search.

If the external variable *verbose* is nonzero, *dijkstra* will record its activities on the standard output file by printing the distances from uu to all vertices it visits.

After *dijkstra* has found a shortest path, it returns the length of that path. If no path from uu to vv exists (in particular, if vv is Λ), it returns -1; in such a case, the shortest distances from uu to all vertices reachable from uu will have been computed and stored in the graph. An auxiliary function, *print_dijkstra_result*(vv), can be used to display the actual path found, if one exists.

Examples of the use of *dijkstra* appear in the LADDERS demonstration module.

2. This C module is meant to be loaded as part of another program. It has the following simple structure:

```
#include "gb_graph.h"      /* define the standard GraphBase data structures */
  ⟨ Preprocessor definitions ⟩
  ⟨ Priority queue procedures 16 ⟩
  ⟨ Global declarations 8 ⟩
  ⟨ The dijkstra procedure 9 ⟩
  ⟨ The print_dijkstra_result procedure 14 ⟩
```

3. Users of GB_DIJK should include the header file `gb_dijk.h`:

⟨ `gb_dijk.h` 3 ⟩ ≡
```
  extern long dijkstra( );     /* procedure to calculate shortest paths */
#define print_dijkstra_result  p_dijkstra_result     /* shorthand for linker */
  extern void print_dijkstra_result( );     /* procedure to display the answer */
```
See also sections 5, 6, 7, and 25.

4. The main algorithm.   As Dijkstra's algorithm proceeds, it "knows" shortest paths from
uu to more and more vertices; we will call these vertices "known." Initially only *uu* itself is known.
The procedure terminates when *vv* becomes known, or when all vertices reachable from *uu* are
known.

Dijkstra's algorithm looks at all vertices adjacent to known vertices. A vertex is said to have
been "seen" if it is either known or adjacent to a vertex that's known.

The algorithm proceeds by learning to know all vertices in a greater and greater radius from
the starting point. Thus, if *v* is a known vertex at distance *d* from *uu*, every vertex at distance
less than *d* from *uu* will also be known. (Throughout this discussion the word "distance" actually
means "distance modified by the heuristic function"; we omit mentioning the heuristic because
we can assume that the algorithm is operating on a graph with modified distances.)

The algorithm maintains an auxiliary list of all vertices that have been seen but aren't yet
known. For every such vertex *v*, it remembers the shortest distance *d* from *uu* to *v* by a path
that passes entirely through known vertices except for the very last arc.

This auxiliary list is actually a priority queue, ordered by the *d* values. If *v* is a vertex of the
priority queue having the smallest *d*, we can remove *v* from the queue and consider it known,
because there cannot be a path of length less than *d* from *uu* to *v*. (This is where the assumption
of nonnegative arc length is crucial to the algorithm's validity.)

5.   To implement the ideas just sketched, we use several of the utility fields in vertex records.
Each vertex *v* has a *dist* field *v→dist*, which represents its true distance from *uu* if *v* is known;
otherwise *v→dist* represents the shortest distance from *uu* discovered so far.

Each vertex *v* also has a *backlink* field *v→backlink*, which is non-Λ if and only if *v* has been
seen. In that case *v→backlink* is a vertex one step "closer" to *uu*, on a path from *uu* to *v* that
achieves the current distance *v→dist*. (Exception: Vertex *uu* has a backlink pointing to itself.)
The backlink fields thereby allow us to construct shortest paths from *uu* to all the known vertices,
if desired.

#define *dist* *z.I*   /* distance from *uu*, modified by *hh*, appears in vertex utility field *z* */
#define *backlink* *y.V*   /* pointer to previous vertex appears in utility field *y* */

⟨ `gb_dijk.h` 3 ⟩ +≡
#define *dist* *z.I*
#define *backlink* *y.V*

---

dijkstra: **long** ( ), §9.
gg: **Graph** *, §9.
hh: **long** (*)( ), §9.
I: **long**, GB_GRAPH §8.

p_dijkstra_result = short name, §14.
print_dijkstra_result: **void** ( ), §14.
uu: **Vertex** *, §9.
V: **Vertex** *, GB_GRAPH §8.

verbose: **long**, GB_GRAPH §5.
vv: **Vertex** *, §9.
y: **util**, GB_GRAPH §9.
z: **util**, GB_GRAPH §9.

6. The priority queue is implemented by four procedures:

init_queue(*d*) makes the queue empty and prepares for subsequent keys $\geq d$.

enqueue(*v*, *d*) puts vertex *v* in the queue and assigns it the key value $v\text{-}dist = d$.

requeue(*v*, *d*) takes vertex *v* out of the queue and enters it again with the smaller key value $v\text{-}dist = d$.

del_min( ) removes a vertex with minimum key from the queue and returns a pointer to that vertex. If the queue is empty, Λ is returned.

These procedures are accessed via external pointers, so that the user of GB_DIJK can supply alternate queueing methods if desired.

⟨ gb_dijk.h 3 ⟩ +≡
```
extern void (*init_queue)( );      /* create an empty priority queue for dijkstra */
extern void (*enqueue)( );         /* insert a new element in the priority queue */
extern void (*requeue)( );         /* decrease the key of an element in the queue */
extern Vertex *(*del_min)( );       /* remove an element with smallest key */
```

7. The heuristic function might take awhile to compute, so we avoid recomputation by storing $hh(v)$ in another utility field $v\text{-}hh\_val$ once we've evaluated it.

#define *hh_val* *x.I*   /* computed value of $hh(v)$ */

⟨ gb_dijk.h 3 ⟩ +≡
#define *hh_val* *x.I*

8. If no heuristic function is supplied by the user, we replace it by a dummy function that simply returns 0 in all cases.

⟨ Global declarations 8 ⟩ ≡
```
long dummy(v)
    Vertex *v;
{ return 0; }
```
See also section 15.

This code is used in section 2.

9. Here now is *dijkstra*:

⟨ The *dijkstra* procedure 9 ⟩ ≡
```
long dijkstra(uu, vv, gg, hh)
        Vertex *uu;     /* the starting point */
        Vertex *vv;     /* the ending point */
        Graph *gg;      /* the graph they belong to */
        long (*hh)( );   /* heuristic function */
{ register Vertex *t;    /* current vertex of interest */
    if (hh ≡ Λ) hh = dummy;    /* change to default heuristic */
    ⟨ Make uu the only vertex seen; also make it known 10 ⟩;
    t = uu;
    if (verbose) ⟨ Print initial message 12 ⟩;
    while (t ≠ vv) {
        ⟨ Put all unseen vertices adjacent to t into the queue, and update the distances of other vertices
            adjacent to t 11 ⟩;
```

```
        t = (*del_min)( );
        if (t ≡ Λ) return −1;      /* if the queue becomes empty, there's no way to get to vv */
        if (verbose) ⟨Print the distance to t 13⟩;
    }
    return vv→dist − vv→hh_val + uu→hh_val;      /* true distance from uu to vv */
}
```

This code is used in section 2.

10. As stated above, a vertex is considered seen only when its backlink isn't null, and known only when it is seen but not in the queue.

⟨Make *uu* the only vertex seen; also make it known 10⟩ ≡

```
    for (t = gg→vertices + gg→n − 1; t ≥ gg→vertices; t−−) t→backlink = Λ;
    uu→backlink = uu;
    uu→dist = 0;
    uu→hh_val = (*hh)(uu);
    (*init_queue)(0_L);      /* make the priority queue empty */
```

This code is used in section 9.

11. Here we help the C compiler in case it hasn't got a great optimizer.

⟨Put all unseen vertices adjacent to *t* into the queue, and update the distances of other vertices adjacent to *t* 11⟩ ≡

```
    { register Arc *a;      /* an arc leading from t */
      register long d = t→dist − t→hh_val;

      for (a = t→arcs; a; a = a→next) {
        register Vertex *v = a→tip;      /* a vertex adjacent to t */

        if (v→backlink) {      /* v has already been seen */
          register long dd = d + a→len + v→hh_val;

          if (dd < v→dist) {
            v→backlink = t;
            (*requeue)(v, dd);      /* we found a better way to get there */
          }
        } else {      /* v hasn't been seen before */
          v→hh_val = (*hh)(v);
          v→backlink = t;
          (*enqueue)(v, d + a→len + v→hh_val);
        }
      }
    }
```

This code is used in section 9.

Arc = **struct**, GB_GRAPH §10.
arcs: **Arc** *, GB_GRAPH §9.
backlink = *y.V*, §5.
del_min: **Vertex** *(*)( ), §15.
dist = *z.I*, §5.
enqueue: **void** (*)( ), §15.

Graph = **struct**, GB_GRAPH §20.
I: **long**, GB_GRAPH §8.
init_queue: **void** (*)( ), §15.
len: **long**, GB_GRAPH §10.
n: **long**, GB_GRAPH §20.
next: **Arc** *, GB_GRAPH §10.

requeue: **void** (*)( ), §15.
tip: **Vertex** *, GB_GRAPH §10.
verbose: **long**, GB_GRAPH §5.
Vertex = **struct**, GB_GRAPH §9.
vertices: **Vertex** *, GB_GRAPH §20.
x: **util**, GB_GRAPH §9.

12. The *dist* fields don't contain true distances in the graph; they represent distances modified by the heuristic function. The true distance from *uu* to vertex *v* is $v\text{-}dist - v\text{-}hh\_val + uu\text{-}hh\_val$.

When printing the results, we show true distances. Also, if a nontrivial heuristic is being used, we give the *hh* value in brackets; the user can then observe that vertices are becoming known in order of true distance plus *hh* value.

⟨ Print initial message 12 ⟩ ≡
```
{ printf ("Distances␣from␣%s", uu→name);
  if (hh ≠ dummy) printf ("␣[%ld]", uu→hh_val);
  printf (":\n");
}
```
This code is used in section 9.

13. ⟨ Print the distance to *t* 13 ⟩ ≡
```
{ printf ("␣%ld␣to␣%s", t→dist − t→hh_val + uu→hh_val, t→name);
  if (hh ≠ dummy) printf ("␣[%ld]", t→hh_val);
  printf ("␣via␣%s\n", t→backlink→name);
}
```
This code is used in section 9.

14. After *dijkstra* has found a shortest path, the backlinks from *vv* specify the steps of that path. We want to print the path in the forward direction, so we reverse the links.

We also unreverse them again, just in case the user didn't want the backlinks to be trashed. Indeed, this procedure can be used for any vertex *vv* whose backlink is non-null, not only the *vv* that was a parameter to *dijkstra*.

List reversal is conveniently regarded as a process of popping off one stack and pushing onto another.

#define *print_dijkstra_result* *p_dijkstra_result*   /* shorthand for linker */

⟨ The *print_dijkstra_result* procedure 14 ⟩ ≡
```
  void print_dijkstra_result (vv)
      Vertex *vv;    /* ending vertex */
  { register Vertex *t, *p, *q;    /* registers for reversing links */
    t = Λ, p = vv;
    if (¬p→backlink) {
      printf ("Sorry,␣%s␣is␣unreachable.\n", p→name);
      return;
    }
    do {    /* pop an item from p to t */
      q = p→backlink;
      p→backlink = t;
      t = p;
      p = q;
    } while (t ≠ p);    /* the loop stops with t ≡ p ≡ uu */
    do {
      printf ("%10ld␣%s\n", t→dist − t→hh_val + p→hh_val, t→name);
      t = t→backlink;
    } while (t);
```

```
    t = p;
    do  {       /* pop an item from t to p */
        q = t→backlink;
        t→backlink = p;
        p = t;
        t = q;
    }   while (p ≠ vv);
}
```

This code is used in section 2.

backlink $= y.V$, §5.

dijkstra: **long** ( ), §9.

dist $= z.I$, §5.

dummy: **long** ( ), §8.

hh: **long** (∗)( ), §9.

hh_val $= x.I$, §7.

name: **char** ∗, GB_GRAPH §9.

printf: **int** ( ), <stdio.h>.

t: **register Vertex** ∗, §9.

uu: **Vertex** ∗, §9.

Vertex $=$ **struct**, GB_GRAPH §9.

vv: **Vertex** ∗, §9.

15. Priority queues. Here we provide a simple doubly linked list for queueing; this is a convenient default, good enough for applications that aren't too large. (See MILES_SPAN for implementations of other schemes that are more efficient when the queue gets large.)

The two queue links occupy two of a vertex's remaining utility fields.

#define *llink* *v.V*   /* *llink* is stored in utility field *v* of a vertex */
#define *rlink* *w.V*   /* *rlink* is stored in utility field *w* of a vertex */

⟨ Global declarations 8 ⟩ +≡
  void (*init_queue)( ) = *init_dlist*;   /* create an empty dlist */
  void (*enqueue)( ) = *enlist*;   /* insert a new element in dlist */
  void (*requeue)( ) = *reenlist*;   /* decrease the key of an element in dlist */
  Vertex *(*del_min)( ) = *del_first*;   /* remove element with smallest key */

16. There's a special list head, from which we get to everything else in the queue in decreasing order of keys by following *llink* fields.

The following declaration actually provides for 128 list heads. Only the first of these is used here, but we'll find something to do with the other 127 later.

⟨ Priority queue procedures 16 ⟩ ≡
  Vertex *head*[128];   /* list-head elements that are always present */

  void *init_dlist*(*d*)
    long *d*;
  {
    head→*llink* = *head*→*rlink* = *head*;
    head→*dist* = *d* − 1;   /* a value guaranteed to be smaller than any actual key */
  }

See also sections 17, 18, 19, 21, 22, 23, and 24.

This code is used in section 2.

17. It seems reasonable to assume that an element entering the queue for the first time will tend to have a larger key than the other elements.

Indeed, in the special case that all arcs in the graph have the same length, this strategy turns out to be quite fast. For in that case, every vertex is added to the end of the queue and deleted from the front, without any requeueing; the algorithm produces a strict first-in-first-out queueing discipline and performs a breadth-first search.

⟨ Priority queue procedures 16 ⟩ +≡
  void *enlist*(*v*, *d*)
    Vertex *v;
    long *d*;
  { **register Vertex** *t = *head*→*llink*;

    v→*dist* = *d*;
    while (*d* < *t*→*dist*) *t* = *t*→*llink*;
    v→*llink* = *t*;
    (*v*→*rlink* = *t*→*rlink*)→*llink* = *v*;
    t→*rlink* = *v*;
  }

18. ⟨ Priority queue procedures 16 ⟩ +≡
   void *reenlist*(*v*, *d*)
      Vertex *∗v*;
      long *d*;
  { **register Vertex** *∗t* = *v→llink*;
    (*t→rlink* = *v→rlink*)*→llink* = *v→llink*;   /∗ remove *v* ∗/
    v→dist = *d*;   /∗ we assume that the new *dist* is smaller than it was before ∗/
    while (*d* < *t→dist*) *t* = *t→llink*;
    v→llink = *t*;
    (*v→rlink* = *t→rlink*)*→llink* = *v*;
    t→rlink = *v*;
  }

19. ⟨ Priority queue procedures 16 ⟩ +≡
  Vertex *∗del_first*( )
  { **Vertex** *∗t*;
    t = *head→rlink*;
    if (*t* ≡ *head*) **return** Λ;
    (*head→rlink* = *t→rlink*)*→llink* = *head*;
    return *t*;
  }

---

dist = *z.I*, §5.   *V*: **Vertex** *∗*, GB_GRAPH §8.   *w*: **util**, GB_GRAPH §9.
v: **util**, GB_GRAPH §9.   **Vertex** = **struct**, GB_GRAPH §9.

20. A special case. When the arc lengths in the graph are all fairly small, we can substitute another queueing discipline that does each operation quickly. Suppose the only lengths are 0, 1, ..., $k - 1$; then we can prove easily that the priority queue will never contain more than k different values at once. Moreover, we can implement it by maintaining k doubly linked lists, one for each key value mod k.

For example, let $k = 128$. Here is an alternate set of queue commands, to be used when the arc lengths are known to be less than 128.

21. $\langle$ Priority queue procedures 16 $\rangle$ $+\equiv$
  long *master_key*;   /* smallest key that may be present in the priority queue */

  void *init_128* (*d*)
      long *d*;
  { **register Vertex** *u*;

    master_key = *d*;
    for (*u* = *head*; *u* < *head* + 128; *u*++) *u*→*llink* = *u*→*rlink* = *u*;
  }

22. If the number of lists were not a power of 2, we would calculate a remainder by division instead of by logical-anding.

$\langle$ Priority queue procedures 16 $\rangle$ $+\equiv$
  Vertex *del_128* ( )
  { **long** *d*;
    register Vertex *u*, *t*;

    for (*d* = *master_key*; *d* < *master_key* + 128; *d*++) {
      u = *head* + (*d* & #7f);   /* that's *d* % 128 */
      t = *u*→*rlink*;
      if (*t* ≠ *u*) {   /* we found a nonempty list with minimum key */
        master_key = *d*;
        (*u*→*rlink* = *t*→*rlink*)→*llink* = *u*;
        return *t*;   /* incidentally, *t*→*dist* = *d* */
      }
    }
    return Λ;   /* all 128 lists are empty */
  }

23. $\langle$ Priority queue procedures 16 $\rangle$ $+\equiv$
  void *enq_128* (*v*, *d*)
      Vertex *v*;   /* new vertex for the queue */
      long *d*;   /* its *dist* */
  { **register Vertex** *u* = *head* + (*d* & #7f);

    v→*dist* = *d*;
    (*v*→*llink* = *u*→*llink*)→*rlink* = *v*;
    v→*rlink* = *u*;
    u→*llink* = *v*;
  }

24.   All of these operations have been so simple, one wonders why the lists should be doubly linked. Single linking would indeed be plenty—if we didn't have to support the *requeue* operation.

But requeueing involves deleting an arbitrary element from the middle of its list. And we do seem to need two links for that.

In the application to Dijkstra's algorithm, the new d will always be *master_key* or more. But we want to implement requeueing in general, so that this procedure can be used also for other algorithms such as the calculation of minimum spanning trees (see MILES_SPAN).

⟨ Priority queue procedures 16 ⟩ +≡

  void *req_128* (v, d)
        Vertex $*v$;   /∗ vertex to be moved to another list ∗/
        long d;   /∗ its new *dist* ∗/
  { **register Vertex** $*u = head + (d\ \&\ ^{\#}\mathtt{7f})$;
    $(v{\rightarrow}llink{\rightarrow}rlink = v{\rightarrow}rlink){\rightarrow}llink = v{\rightarrow}llink$;   /∗ remove v ∗/
    $v{\rightarrow}dist = d$;   /∗ the new *dist* is smaller than it was before ∗/
    $(v{\rightarrow}llink = u{\rightarrow}llink){\rightarrow}rlink = v$;
    $v{\rightarrow}rlink = u$;
    $u{\rightarrow}llink = v$;
    if $(d < master\_key)\ master\_key = d$;   /∗ not needed for Dijkstra's algorithm ∗/
  }

25.   The user of GB_DIJK needs to know the names of these queueing procedures if changes to the defaults are made, so we'd better put the necessary info into the header file.

⟨ `gb_dijk.h` 3 ⟩ +≡

  extern void *init_dlist* ( );
  extern void *enlist* ( );
  extern void *reenlist* ( );
  extern Vertex *∗del_first* ( );
  extern void *init_128* ( );
  extern Vertex *∗del_128* ( );
  extern void *enq_128* ( );
  extern void *req_128* ( );

---

del_first: **Vertex** ∗( ), §19.
dist = *z.I*, §5.
enlist: **void** ( ), §17.
head: **Vertex** [ ], §16.

init_dlist: **void** ( ), §16.
llink = *v.V*, §15.
reenlist: **void** ( ), §18.

requeue: **void** (∗)( ), §15.
rlink = *w.V*, §15.
Vertex = **struct**, GB_GRAPH §9.

GB_ECON

Important: Before reading GB_ECON, please read or at least skim the programs for GB_GRAPH and GB_IO.

1. Introduction. This GraphBase module contains the *econ* subroutine, which creates a family of directed graphs related to the flow of money between industries. An example of the use of this procedure can be found in the demo program ECON_ORDER.

⟨ gb_econ.h 1 ⟩ ≡
    extern Graph *∗econ*( );

See also section 5.

2. The subroutine call *econ*(n, *omit*, *threshold*, *seed*) constructs a directed graph based on the information in `econ.dat`. Each vertex of the graph corresponds to one of 81 sectors of the U.S. economy. The data values come from the year 1985; they were derived from tables published in *Survey of Current Business* **70** (1990), 41–56.

If *omit* = *threshold* = 0, the directed graph is a "circulation"; that is, each arc has an associated *flow* value, and the sum of arc flows leaving each vertex is equal to the sum of arc flows entering. This sum is called the "total commodity output" for the sector in question. The flow in an arc from sector j to sector k is the amount of the commodity made by sector j that was used by sector k, rounded to millions of dollars at producers' prices. For example, the total commodity output of the sector called `Apparel` is 54031, meaning that the total cost of making all kinds of apparel in 1985 was about 54 billion dollars. There is an arc from `Apparel` to itself with a flow of 9259, meaning that 9.259 billion dollars' worth of apparel went from one group within the apparel industry to another. There also is an arc of flow 44 from `Apparel` to `Household furniture`, indicating that some 44 million dollars' worth of apparel went into the making of household furniture. By looking at all arcs that leave the `Apparel` vertex, you can see where all that new apparel went; by looking at all arcs that enter `Apparel`, you can see what ingredients the apparel industry needed to make it.

One vertex, called `Users`, represents people like you and me, the non-industrial end users of everything. The arc from `Apparel` to `Users` has flow 42172; this is the "total final demand" for apparel, the amount that didn't flow into other sectors of the economy before it reached people like us. The arc from `Users` to `Apparel` has flow 19409, which is called the "value added" by users; it represents wages and salaries paid to support the manufacturing process. The sum of total final demand over all sectors, which also equals the sum of value added over all sectors, is conventionally called the Gross National Product (GNP). In 1985 the GNP was 3999362, nearly 4 trillion dollars, according to `econ.dat`. (The sum of all arc flows coming out of all vertices was 7198680; this sum overestimates the total economic activity, because it counts some items more than once—statistics are recorded whenever an item passes a statistics gatherer. Economists try to adjust the data so that they avoid double-counting as much as possible.)

Speaking of economists, there is another special vertex called `Adjustments`, included by economists so that GNP is measured more accurately. This vertex takes account of such things as changes in the value of inventories, and imported materials that cannot be obtained within the U.S., as well as work done for the government and for foreign concerns. In 1985, these adjustments accounted for about 11% of the GNP.

Incidentally, some of the "total final demand" arcs are negative. For example, the arc from `Petroleum and natural gas production` to `Users` has flow -27032. This might seem strange at first, but it makes sense when imports are considered, because crude oil and natural gas go more to other industries than to end users. Total final demand does not mean total user demand.

#define *flow*   `a.I`   /* utility field a specifies the flow in an arc */

3.   If *omit* = 1, the `Users` vertex is omitted from the digraph; in particular, this will eliminate all arcs of negative flow. If *omit* = 2, the `Adjustments` vertex is also omitted, thereby leaving 79 sectors with arcs showing inter-industry flow. (The graph is no longer a "circulation," of course, when *omit* > 0.) If `Users` and `Adjustments` are not omitted, `Users` is the last vertex of the graph, and `Adjustments` is next-to-last.

If *threshold* = 0, the digraph has an arc for every nonzero *flow*. But if *threshold* > 0, the digraph becomes more sparse; there is then an arc from j to k if and only if the amount of commodity j used by sector k exceeds *threshold*/65536 times the total input of sector k. (The total input figure always includes value added, even if *omit* > 0.) Thus the arcs go to each sector from that sector's main suppliers. When $n = 79$, *omit* = 2, and *threshold* = 0, the digraph has 4602 arcs out of a possible $79 \times 79 = 6241$. Raising *threshold* to 1 decreases the number of arcs to 4473; raising it to 6000 leaves only 72 arcs. The *len* field in each arc is 1.

The constructed graph will have $\min(n, 81 - omit)$ vertices. If n is less than $81 - omit$, the n vertices will be selected by repeatedly combining related sectors. For example, two of the 81 original sectors are called 'Paper products, except containers' and 'Paperboard containers and boxes'; these might be combined into a sector called 'Paper products'. There is a binary tree with 79 leaves, which describes a fixed hierarchical breakdown of the 79 non-special sectors. This tree is pruned, if necessary, by replacing pairs of leaves by their parent node, which becomes a new leaf; pruning continues until just n leaves remain. Although pruning is a bottom-up process, its effect can also be obtained from the top down if we imagine "growing" the tree, starting out with a whole economy as a single sector and repeatedly subdividing a sector into two parts. For example, if *omit* = 2 and $n = 2$, the two sectors will be called `Goods` and `Services`. If $n = 3$, `Goods` might be subdivided into `Natural Resources` and `Manufacturing`; or `Services` might be subdivided into `Indirect Services` and `Direct Services`.

If *seed* = 0, the binary tree is pruned in such a way that the n resulting sectors are as equal as possible with respect to total input and output, while respecting the tree structure. If *seed* > 0, the pruning is carried out at random, in such a way that all n-leaf subtrees of the original tree are obtained with approximately equal probability (depending on *seed* in a machine-independent fashion). Any *seed* value from 1 to $2^{31} - 1 = 2147483647$ is permissible.

As usual in GraphBase routines, you can set $n = 0$ to get the default situation where n has its maximum value. For example, either *econ*(0, 0, 0, 0) or *econ*(81, 0, 0, 0) produces the full graph; *econ*(0, 2, 0, 0) or *econ*(79, 2, 0, 0) produces the full graph except for the two special vertices.

#define MAX_N 81   /* maximum number of vertices in constructed graph */
#define NORM_N MAX_N $-$ 2   /* the number of normal SIC sectors */
#define ADJ_SEC MAX_N $-$ 1   /* code number for the `Adjustments` sector */

---

a: util, GB_GRAPH §10. I: **long**, GB_GRAPH §8.   *omit*: **unsigned long**, §7.
econ: **Graph** *(), §7.   *len*: **long**, GB_GRAPH §10.   *seed*: **long**, §7.
Graph = **struct**, GB_GRAPH §20. n: **unsigned long**, §7.   *threshold*: **unsigned long**, §7.

4. The U.S. Bureau of Economic Analysis and the U.S. Bureau of the Census have assigned code numbers 1–79 to the individual sectors for which statistics are given in `econ.dat`. These sector numbers are traditionally called Standard Industrial Classification (SIC) codes. If for some reason you want to know the SIC codes for all sectors represented by vertex v of a graph generated by *econ*, you can access them via a list of **Arc** nodes starting at the utility field $v\text{-}SIC\_codes$. This list is linked by *next* fields in the usual way, and each SIC code appears in the *len* field; the *tip* field is unused.

The special vertex `Adjustments` is given code number 80; it is actually a composite of six different SIC categories, numbered 80–86 in their published tables.

For example, if $n = 80$ and *omit* $= 1$, each list will have length 1. Hence $v\text{-}SIC\_codes\text{-}next$ will equal Λ for each v, and $v\text{-}SIC\_codes\text{-}len$ will be v's SIC code, a number between 1 and 80.

The special vertex `Users` has no SIC code; it is the only vertex whose SIC_codes field will be null in the graph returned by *econ*.

#define SIC_codes $z.A$ /∗ utility field z leads to the SIC codes for a vertex ∗/

5. The total output of each sector, which also equals the total input of that sector, is placed in utility field *sector_total* of the corresponding vertex.

#define *sector_total* $y.I$ /∗ utility field y holds the total flow in and out ∗/

⟨ `gb_econ.h` 1 ⟩ +≡
#define *flow* $a.I$ /∗ definitions of utility fields in the header file ∗/
#define SIC_codes $z.A$
#define *sector_total* $y.I$

6. If the *econ* routine encounters a problem, it returns Λ (`NULL`), after putting a nonzero number into the external variable *panic_code*. This code number identifies the type of failure. Otherwise *econ* returns a pointer to the newly created graph, which will be represented with the data structures explained in GB_GRAPH. (The external variable *panic_code* is itself defined in GB_GRAPH.)

#define *panic*(*c*)   { *panic_code* $= c$; *gb_trouble_code* $= 0$; **return** Λ; }

7. The C file `gb_econ.c` has the following overall shape:

```
#include "gb_io.h"      /∗ we will use the GB_IO routines for input ∗/
#include "gb_flip.h"      /∗ we will use the GB_FLIP routines for random numbers ∗/
#include "gb_graph.h"      /∗ and of course we'll use the GB_GRAPH data structures ∗/
  ⟨ Preprocessor definitions ⟩
  ⟨ Type declarations 11 ⟩
  ⟨ Private variables 12 ⟩
  Graph ∗econ(n, omit, threshold, seed)
      unsigned long n;      /∗ number of vertices desired ∗/
      unsigned long omit;      /∗ number of special vertices to omit ∗/
      unsigned long threshold;      /∗ minimum per-64K-age in arcs leading in ∗/
      long seed;      /∗ random number seed ∗/
  { ⟨ Local variables 8 ⟩
    gb_init_rand(seed);
    init_area(working_storage);
    ⟨ Check the parameters and adjust them for defaults 9 ⟩;
```

⟨ Set up a graph with n vertices 10 ⟩;
⟨ Read `econ.dat` and note the binary tree structure 14 ⟩;
⟨ Determine the n sectors to use in the graph 17 ⟩;
⟨ Put the appropriate arcs into the graph 25 ⟩;
if $(gb\_close( ) \neq 0)$ $panic(late\_data\_fault)$;
    /∗ something's wrong with `"econ.dat"`; see io_errors ∗/
$gb\_free(working\_storage)$;
if $(gb\_trouble\_code)$ {
  $gb\_recycle(new\_graph)$;
  $panic(alloc\_fault)$;   /∗ oops, we ran out of memory somewhere back there ∗/
}
return new_graph;
}

8. ⟨ Local variables 8 ⟩ ≡
  Graph ∗new_graph;   /∗ the graph constructed by $econ$ ∗/
  register long j, k;   /∗ all-purpose indices ∗/
  Area $working\_storage$;   /∗ tables needed while $econ$ does its thinking ∗/
See also section 13.
This code is used in section 7.

9. ⟨ Check the parameters and adjust them for defaults 9 ⟩ ≡
  if $(omit > 2)$ $omit = 2$;
  if $(n \equiv 0 \lor n > \texttt{MAX\_N} - omit)$ $n = \texttt{MAX\_N} - omit$;
  else if $(n + omit < 3)$ $omit = 3 - n$;   /∗ we need at least one normal sector ∗/
  if $(threshold > 65536)$ $threshold = 65536$;
This code is used in section 7.

10. ⟨ Set up a graph with n vertices 10 ⟩ ≡
  $new\_graph = gb\_new\_graph(n)$;
  if $(new\_graph \equiv \Lambda)$ $panic(no\_room)$;   /∗ out of memory before we're even started ∗/
  $sprintf(new\_graph{\rightarrow}id, \texttt{"econ(\%lu,\%lu,\%lu,\%ld)"}, n, omit, threshold, seed)$;
  $strcpy(new\_graph{\rightarrow}util\_types, \texttt{"ZZZZIAIZZZZZZZ"})$;
This code is used in section 7.

A: **Arc** ∗, GB_GRAPH §8.
a: **util**, GB_GRAPH §10.
$alloc\_fault = -1$, GB_GRAPH §7.
Arc = **struct**, GB_GRAPH §10.
Area, GB_GRAPH §12.
$flow = a.I$, §2.
gb_close: **long** ( ), GB_IO §39.
gb_free: **void** ( ), GB_GRAPH §16.
gb_init_rand: **void** ( ), GB_FLIP §8.
gb_new_graph: **Graph** ∗( ),
  GB_GRAPH §23.

$gb\_recycle$: **void** ( ), GB_GRAPH §40.
$gb\_trouble\_code$: **long**,
  GB_GRAPH §7.
Graph = **struct**, GB_GRAPH §20.
I: **long**, GB_GRAPH §8.
id: **char** [ ], GB_GRAPH §20.
$init\_area =$ macro ( ), GB_GRAPH §12.
io_errors: **long**, GB_IO §5.
$late\_data\_fault = 11$, GB_GRAPH §7.
len: **long**, GB_GRAPH §10.

$\texttt{MAX\_N} = 81$, §3.
$next$: **Arc** ∗, GB_GRAPH §10.
$no\_room = 1$, GB_GRAPH §7.
$panic\_code$: **long**, GB_GRAPH §5.
$sprintf$: **int** ( ), `<stdio.h>`.
$strcpy$: **char** ∗( ), `<string.h>`.
tip: **Vertex** ∗, GB_GRAPH §10.
$util\_types$: **char** [ ], GB_GRAPH §20.
y: **util**, GB_GRAPH §9.
z: **util**, GB_GRAPH §9.

11. The economic tree. As we read in the data, we construct a sequential list of nodes, each of which represents either a micro-sector of the economy (one of the basic SIC sectors) or a macro-sector (which is the union of two subnodes). In more technical terms, the nodes form an extended binary tree, whose external nodes correspond to micro-sectors and whose internal nodes correspond to macro-sectors. The nodes of the tree appear in preorder. Subsequently we will do a variety of operations on this binary tree, proceeding either top-down (from the beginning of the list to the end) or bottom-up (from the end to the beginning).

Each node is a rather large record, because we will store a complete vector of sector output data in each node.

⟨ Type declarations 11 ⟩ ≡
  typedef struct node_struct {   /∗ records for micro- and macro-sectors ∗/
    struct node_struct ∗*rchild*;   /∗ pointer to right child of macro-sector ∗/
    char *title*[44];   /∗ "Sector␣name" ∗/
    long *table*[MAX_N + 1];   /∗ outputs from this sector ∗/
    unsigned long *total*;   /∗ total input to this sector (= total output) ∗/
    long *thresh*;   /∗ *flow* must exceed *thresh* in arcs to this sector ∗/
    long SIC;   /∗ SIC code number; initially zero in macro-sectors ∗/
    long *tag*;   /∗ 1 if this node will be a vertex in the graph ∗/
    struct node_struct ∗*link*;   /∗ next smallest unexplored sector ∗/
    Arc ∗*SIC_list*;   /∗ first item on list of SIC codes ∗/
  } **node**;
This code is used in section 7.

12. When we read the given data in preorder, we'll need a stack to remember what nodes still need to have their *rchild* pointer filled in. (There is a no need for an *lchild* pointer, because the left child always follows its parent immediately in preorder.)

⟨ Private variables 12 ⟩ ≡
  static node ∗*stack*[NORM_N + NORM_N];
  static node ∗∗*stack_ptr*;   /∗ current position in *stack* ∗/
  static node ∗*node_block*;   /∗ array of nodes, specifies the tree in preorder ∗/
  static node ∗*node_index*[MAX_N + 1];   /∗ which node has a given SIC code ∗/
See also section 26.

This code is used in section 7.

13. ⟨ Local variables 8 ⟩ +≡
  register node ∗*p*, ∗*pl*, ∗*pr*;   /∗ current node and its children ∗/
  register node ∗*q*;   /∗ register for list manipulation ∗/

14. ⟨ Read econ.dat and note the binary tree structure 14 ⟩ ≡
  node_block = *gb_typed_alloc*(2 ∗ MAX_N − 3, **node**, *working_storage*);
  if (*gb_trouble_code*) *panic*(*no_room* + 1);   /∗ no room to copy the data ∗/
  if (*gb_open*("econ.dat") ≠ 0) *panic*(*early_data_fault*);
      /∗ couldn't open "econ.dat" using GraphBase conventions ∗/
  ⟨ Read and store the sector names and SIC numbers 15 ⟩;
  for (*k* = 1; *k* ≤ MAX_N; *k*++) ⟨ Read and store the output coefficients for sector *k* 16 ⟩;
This code is used in section 7.

15.   The first part of `econ.dat` specifies the nodes of the binary tree in preorder. Each line contains a node name followed by a colon, and the colon is followed by the SIC number if that node is a leaf.

The tree is uniquely specified in this way, because of the nature of preorder. (Think of Polish prefix notation, in which a formula like '$+x+xx$' means '$+(x, +(x, x))$'; the parentheses in Polish notation are redundant.)

The two special sector names don't appear in the file; we manufacture them ourselves.

The program here is careful not to clobber itself in the presence of arbitrarily garbled data.

⟨ Read and store the sector names and SIC numbers 15 ⟩ ≡

```
stack_ptr = stack;
for (p = node_block; p < node_block + NORM_N + NORM_N − 1; p++) { register long c;
    gb_string(p→title, ':');
    if (strlen(p→title) > 43) panic(syntax_error);        /* sector name too long */
    if (gb_char() ≠ ':') panic(syntax_error + 1);         /* missing colon */
    p→SIC = c = gb_number(10);
    if (c ≡ 0)        /* macro-sector */
        *stack_ptr++ = p;       /* left child is p + 1, we'll know rchild later */
    else {        /* micro-sector; p + 1 will be somebody's right child */
        node_index[c] = p;
        if (stack_ptr > stack) (*−−stack_ptr)→rchild = p + 1;
    }
    if (gb_char() ≠ '\n') panic(syntax_error + 2);        /* garbage on the line */
    gb_newline();
}
if (stack_ptr ≠ stack) panic(syntax_error + 3);        /* tree malformed */
for (k = NORM_N; k; k−−)
    if (node_index[k] ≡ 0) panic(syntax_error + 4);        /* SIC code not mentioned in the tree */
strcpy(p→title, "Adjustments"); p→SIC = ADJ_SEC; node_index[ADJ_SEC] = p;
strcpy((p + 1)→title, "Users"); node_index[MAX_N] = p + 1;
```

This code is used in section 14.

ADJ_SEC = 80, §3.
Arc = **struct**, GB_GRAPH §10.
early_data_fault = 10, GB_GRAPH §7.
flow = a.I, §2.
gb_char: **char** ( ), GB_IO §22.
gb_newline: **void** ( ), GB_IO §18.
gb_number: **unsigned long** ( ),
  GB_IO §24.

gb_open: **long** ( ), GB_IO §32.
gb_string: **char** *( ), GB_IO §26.
gb_trouble_code: **long**,
  GB_GRAPH §14.
gb_typed_alloc = macro ( ),
  GB_GRAPH §11.
k: **register long**, §8.
MAX_N = 81, §3.

no_room = 1, GB_GRAPH §7.
NORM_N = 79, §3.
panic = macro ( ), §6.
strcpy: **char** *( ), <string.h>.
strlen: **int** ( ), <string.h>.
syntax_error = 20, GB_GRAPH §7.
working_storage: **Area**, §8.

16. The remaining part of `econ.dat` is an 81×80 matrix in which the kth row contains the outputs of sector k to all sectors except `Users`. Each row consists of a blank line followed by 8 data lines; each data line contains 10 numbers separated by commas. Zeroes are represented by "" instead of by "0". For example, the data line

$$8490,2182,42,467,,,,,,$$

follows the initial blank line; it means that sector 1 output 8490 million dollars to itself, \$2182M to sector 2, ..., \$0M to sector 10.

⟨ Read and store the output coefficients for sector k 16 ⟩ ≡

```
{ register long s = 0;      /* row sum */
  register long x;          /* entry read from econ.dat */
  if (gb_char( ) ≠ '\n') panic(syntax_error + 5);     /* blank line missing between rows */
  gb_newline( );
  p = node_index[k];
  for (j = 1; j < MAX_N; j++) {
    p⁃table[j] = x = gb_number(10);  s += x;
    node_index[j]⁃total += x;
    if ((j % 10) ≡ 0) {
      if (gb_char( ) ≠ '\n') panic(syntax_error + 6);     /* out of synch in input file */
      gb_newline( );
    } else if (gb_char( ) ≠ ',') panic(syntax_error + 7);     /* missing comma after entry */
  }
  p⁃table[MAX_N] = s;     /* sum of table[1] through table[80] */
}
```

This code is used in section 14.

17. Growing a subtree.   Once all the data appears in *node_block*, we want to extract from it and combine it as specified by parameters *n*, *omit*, and *seed*. This amalgamation process effectively prunes the tree; it can also be regarded as a procedure that grows a subtree of the full economic tree.

⟨ Determine the *n* sectors to use in the graph 17 ⟩ ≡
   { **long** *l* = *n* + *omit* − 2;   /* the number of leaves in the desired subtree */

     if (*l* ≡ NORM_N) ⟨ Choose all sectors 18 ⟩
     else if (*seed*) ⟨ Grow a random subtree with *l* leaves 21 ⟩
     else ⟨ Grow a subtree with *l* leaves by subdividing largest sectors first 19 ⟩;
   }

This code is used in section 7.

18.   The chosen leaves of our subtree are identified by having their *tag* field set to 1.

⟨ Choose all sectors 18 ⟩ ≡
   for (*k* = NORM_N; *k*; *k*−−) *node_index*[*k*]→*tag* = 1;

This code is used in section 17.

19.   To grow the *l*-leaf subtree when *seed* = 0, we first pass over the tree bottom-up to compute the total input (and output) of each macro-sector; then we proceed from the top down to subdivide sectors in decreasing order of their total input. This process provides a good introduction to the bottom-up and top-down tree methods we will be using in several other parts of the program.

The *special* node is used here for two purposes: It is the head of a linked list of unexplored nodes, sorted by decreasing order of their *total* fields; and it appears at the end of that list, because *special*→*total* = 0.

⟨ Grow a subtree with *l* leaves by subdividing largest sectors first 19 ⟩ ≡
   { **register node** **special* = *node_index*[MAX_N];   /* the **Users** node at the end of *node_block* */

     for (*p* = *node_index*[ADJ_SEC] − 1; *p* ≥ *node_block*; *p*−−)   /* bottom up */
       if (*p*→*rchild*) *p*→*total* = (*p* + 1)→*total* + *p*→*rchild*→*total*;
     special→*link* = *node_block*; *node_block*→*link* = *special*;   /* start at the root */
     k = 1;   /* *k* is the number of nodes we have tagged or put onto the list */
     while (*k* < *l*) ⟨ If the first node on the list is a leaf, delete it and tag it; otherwise replace it by its
         two children 20 ⟩;
     for (*p* = *special*→*link*; *p* ≠ *special*; *p* = *p*→*link*) *p*→*tag* = 1;   /* tag everything on the list */
   }

This code is used in section 17.

ADJ_SEC = 80, §3.
gb_char: **char** ( ), GB_IO §22.
gb_newline: **void** ( ), GB_IO §18.
gb_number: **unsigned long** ( ),
   GB_IO §24.
j: **register long**, §8.
k: **register long**, §8.
link: **struct node_struct** *, §11.

MAX_N = 81, §3.
n: **unsigned long**, §7.
node, §11.
node_block: **static node** *, §12.
node_index: **static node** *[], §12.
NORM_N = 79, §3.
omit: **unsigned long**, §7.
p: **register node** *, §13.

panic = macro ( ), §6.
rchild: **struct node_struct** *, §11.
seed: **long**, §7.
syntax_error = 20, GB_GRAPH §7.
table: **long** [], §11.
tag: **long**, §11.
total: **unsigned long**, §11.

20. ⟨ If the first node on the list is a leaf, delete it and tag it; otherwise replace it by its two
    children 20 ⟩ ≡
```
{
    p = special→link;        /∗ remove p, the node with greatest total ∗/
    special→link = p→link;
    if (p→rchild ≡ 0) p→tag = 1;        /∗ p is a leaf ∗/
    else {
        pl = p + 1;  pr = p→rchild;
        for (q = special;  q→link→total > pl→total;  q = q→link) ;
        pl→link = q→link;  q→link = pl;        /∗ insert left child in its proper place ∗/
        for (q = special;  q→link→total > pr→total;  q = q→link) ;
        pr→link = q→link;  q→link = pr;        /∗ insert right child in its proper place ∗/
        k++;
    }
}
```
This code is used in section 19.

21.   We can obtain a uniformly distributed l-leaf subtree of a given tree by choosing the root
when $l = 1$ or by using the following idea when $l > 1$: Suppose the given tree T has subtrees T_0 and
T_1. Then it has $T(l)$ subtrees with l leaves, where $T(l) = \sum_k T_0(k)T_1(l-k)$. We choose a random
number r between 0 and $T(l) - 1$, and we find the smallest m such that $\sum_{k \leq m} T_0(k)T_1(l-k) > r$.
Then we proceed recursively to compute a random m-leaf subtree of T_0 and a random $(l-m)$-leaf
subtree of T_1.

A difficulty arises when $T(l)$ is 2^{31} or more. But then we can replace $T_0(k)$ and $T_1(l-k)$ in
the formulas above by $\lceil T_0(k)/d_0 \rceil$ and $\lceil T_1(k)/d_1 \rceil$, respectively, where d_0 and d_1 are arbitrary
constants; this yields smaller values $T(l)$ that define approximately the same distribution of k.

The program here computes the $T(l)$ values bottom-up, then grows a random tree top-down.
If node p is not a leaf, its *table*[0] field will be set to the number of leaves below it; and its *table*[l]
field will be set to $T(l)$, for $1 \leq l \leq$ *table*[0].

The data in `econ.dat` is sufficiently simple that most of the $T(l)$ values are less than 2^{31}. We
need to scale them down to avoid overflow only at the root node of the tree; this case is handled
separately.

We set the *tag* field of a node equal to the number of leaves to be grown in the subtree rooted
at that node. This convention is consistent with our previous stipulation that *tag* = 1 should
characterize the nodes that are chosen to be vertices.

⟨ Grow a random subtree with l leaves 21 ⟩ ≡
```
{
    node_block→tag = l;
    for (p = node_index[ADJ_SEC] − 1; p > node_block; p−−)        /∗ bottom up, except root ∗/
        if (p→rchild) ⟨ Compute the T(l) values for subtree p 22 ⟩;
    for (p = node_block; p < node_index[ADJ_SEC]; p++)        /∗ top down, from root ∗/
        if (p→tag > 1) {
            l = p→tag;
            pl = p + 1;  pr = p→rchild;
            if (pl→rchild ≡ Λ) {
                pl→tag = 1;  pr→tag = l − 1;
```

```
    } else if (pr→rchild ≡ Λ) {
        pl→tag = l − 1;  pr→tag = 1;
    } else ⟨Stochastically determine the number of leaves to grow in each of p's children 24⟩;
    }
}
```

This code is used in section 17.

22. Here we are essentially multiplying two generating functions. Suppose $f(z) = \sum_l T(l)z^l$; then we are computing $f_p(z) = z + f_{pl}(z)f_{pr}(z)$.

⟨Compute the $T(l)$ values for subtree p 22⟩ ≡

```
{
    pl = p + 1;  pr = p→rchild;
    p→table[1] = p→table[2] = 1;       /* T(1) and T(2) are always 1 */
    if (pl→rchild ≡ 0) {       /* left child is a leaf */
        if (pr→rchild ≡ 0) p→table[0] = 2;       /* and so is the right child */
        else {       /* no, it isn't */
            for (k = 2;  k ≤ pr→table[0];  k++)  p→table[1 + k] = pr→table[k];
            p→table[0] = pr→table[0] + 1;
        }
    } else if (pr→rchild ≡ 0) {       /* right child is a leaf */
        for (k = 2;  k ≤ pl→table[0];  k++)  p→table[1 + k] = pl→table[k];
        p→table[0] = pl→table[0] + 1;
    } else {       /* neither child is a leaf */
        ⟨Set p→table[2], p→table[3], ... to convolution of pl and pr table entries 23⟩;
        p→table[0] = pl→table[0] + pr→table[0];
    }
}
```

This code is used in section 21.

23. ⟨Set p→table[2], p→table[3], ... to convolution of pl and pr table entries 23⟩ ≡

```
    p→table[2] = 0;
    for (j = pl→table[0];  j;  j−−) {  register long t = pl→table[j];
        for (k = pr→table[0];  k;  k−−)  p→table[j + k] += t * pr→table[k];
    }
```

This code is used in section 22.

ADJ_SEC = 80, §3.
j: **register long**, §8.
k: **register long**, §8.
l: **long**, §17.
link: **struct node_struct** *, §11.
node_block: **static node** *, §12.

node_index: **static node** *[ ], §12.
p: **register node** *, §13.
pl: **register node** *, §13.
pr: **register node** *, §13.
q: **register node** *, §13.

rchild: **struct node_struct** *, §11.
special: **register node** *, §19.
table: **long** [ ], §11.
tag: **long**, §11.
total: **unsigned long**, §11.

24. ⟨ Stochastically determine the number of leaves to grow in each of p's children 24 ⟩ ≡

  { **register long** ss, rr;

    $j = 0$;   /* we will set $j = 1$ if scaling is necessary at the root */

    if $(p \equiv node\_block)$ {

      $ss = 0$;

      if $(l > 29 \wedge l < 67)$ {

        $j = 1$;   /* more than 2^{31} possibilities exist */

        for $(k = (l > pr{\rightarrow}table[0] \text{ ? } l - pr{\rightarrow}table[0] : 1); \; k \le pl{\rightarrow}table[0] \wedge k < l; \; k{+}{+})$

          $ss \mathrel{+{=}} ((pl{\rightarrow}table[k] + {}^{\#}\mathtt{3ff}) \gg 10) * pr{\rightarrow}table[l - k]$;   /* scale with $d_0 = 1024, d_1 = 1$ */

      } **else**

        for $(k = (l > pr{\rightarrow}table[0] \text{ ? } l - pr{\rightarrow}table[0] : 1); \; k \le pl{\rightarrow}table[0] \wedge k < l; \; k{+}{+})$

          $ss \mathrel{+{=}} pl{\rightarrow}table[k] * pr{\rightarrow}table[l - k]$;

    } **else** $ss = p{\rightarrow}table[l]$;

    $rr = gb\_unif\_rand(ss)$;

    if (j)

      for $(ss = 0, k = (l > pr{\rightarrow}table[0] \text{ ? } l - pr{\rightarrow}table[0] : 1); \; ss \le rr; \; k{+}{+})$

        $ss \mathrel{+{=}} ((pl{\rightarrow}table[k] + {}^{\#}\mathtt{3ff}) \gg 10) * pr{\rightarrow}table[l - k]$;

    else

      for $(ss = 0, k = (l > pr{\rightarrow}table[0] \text{ ? } l - pr{\rightarrow}table[0] : 1); \; ss \le rr; \; k{+}{+})$

        $ss \mathrel{+{=}} pl{\rightarrow}table[k] * pr{\rightarrow}table[l - k]$;

    $pl{\rightarrow}tag = k - 1$; $pr{\rightarrow}tag = l - k + 1$;

  }

This code is used in section 21.

25. Arcs. In the general case, we have to combine some of the basic micro-sectors into macro-sectors by adding together the appropriate input/output coefficients. This is a bottom-up pruning process.

Suppose p is being formed as the union of pl and pr. Then the arcs leading out of p are obtaining by summing the numbers on arcs leading out of pl and pr; the arcs leading into p are obtained by summing the numbers on arcs leading into pl and pr; the arcs from p to itself are obtained by summing the four numbers on arcs leading from pl or pr to pl or pr.

We maintain the $node\_index$ table so that its non-Λ entries contain all the currently active nodes. When pl and pr are being pruned in favor of p, node p inherits pl's place in $node\_index$; pr's former place becomes Λ.

$\langle$ Put the appropriate arcs into the graph 25 $\rangle \equiv$
   $\langle$ Prune the sectors that are used in macro-sectors, and form the lists of SIC sector codes 28 $\rangle$;
   $\langle$ Make the special nodes invisible if they are omitted, visible otherwise 30 $\rangle$;
   $\langle$ Compute individual thresholds for each chosen sector 27 $\rangle$;
   { **register Vertex** $*v = new\_graph \rightarrow vertices + n$;

     for $(k = \texttt{MAX\_N};\ k;\ k--)$
      if $((p = node\_index[k]) \neq \Lambda)$ {
       $vert\_index[k] = --v$;
       $v \rightarrow name = gb\_save\_string(p \rightarrow title)$;
       $v \rightarrow SIC\_codes = p \rightarrow SIC\_list$;
       $v \rightarrow sector\_total = p \rightarrow total$;
      } **else** $vert\_index[k] = \Lambda$;
     if $(v \neq new\_graph \rightarrow vertices)\ panic(impossible)$;   /* bug in algorithm; this can't happen */
     for $(j = \texttt{MAX\_N};\ j;\ j--)$
      if $((p = node\_index[j]) \neq \Lambda)$ { **register Vertex** $*u = vert\_index[j]$;

        for $(k = \texttt{MAX\_N};\ k;\ k--)$
         if $((v = vert\_index[k]) \neq \Lambda)$
          if $(p \rightarrow table[k] \neq 0 \wedge p \rightarrow table[k] > node\_index[k] \rightarrow thresh)$ {
           $gb\_new\_arc(u, v, 1_\text{L})$;
           $u \rightarrow arcs \rightarrow flow = p \rightarrow table[k]$;
          }
      }
   }

This code is used in section 7.

---

arcs: **Arc** $*$, GB_GRAPH §9.
$flow = a.I$, §2.
gb_new_arc: **void** ( ),
  GB_GRAPH §30.
gb_save_string: **char** $*($),
  GB_GRAPH §35.
gb_unif_rand: **long** ( ), GB_FLIP §12.
$impossible = 90$, GB_GRAPH §7.
j: **register long**, §8.
k: **register long**, §8.
l: **long**, §17.

MAX_N $= 81$, §3.
n: **unsigned long**, §7.
$name$: **char** $*$, GB_GRAPH §9.
new_graph: **Graph** $*$, §8.
$node\_block$: **static node** $*$, §12.
$node\_index$: **static node** $*[\,]$, §12.
p: **register node** $*$, §13.
$panic =$ macro ( ), §6.
pl: **register node** $*$, §13.
pr: **register node** $*$, §13.
$sector\_total = y.I$, §5.

$SIC\_codes = z.A$, §4.
SIC_list: **Arc** $*$, §11.
$table$: **long** $[\,]$, §11.
tag: **long**, §11.
$thresh$: **long**, §11.
$title$: **char** $[\,]$, §11.
$total$: **unsigned long**, §11.
$vert\_index$: **static Vertex** $*[\,]$, §26.
Vertex $=$ **struct**, GB_GRAPH §9.
$vertices$: **Vertex** $*$, GB_GRAPH §20.

26. ⟨ Private variables 12 ⟩ +≡
   static Vertex *∗vert_index*[MAX_N + 1];   /∗ the vertex assigned to an SIC code ∗/

27. The theory underlying this step is the following, for integers a, b, c, d with $b, d > 0$:

$$\frac{a}{b} > \frac{c}{d} \qquad \Longleftrightarrow \qquad a > \left\lfloor \frac{b}{d} \right\rfloor c + \left\lfloor \frac{(b \bmod d)c}{d} \right\rfloor.$$

In our case, $b = p\text{-}total$ and $c = threshold \le d = 65536 = 2^{16}$, hence the multiplications cannot overflow. (But they can come awfully darn close.)
⟨ Compute individual thresholds for each chosen sector 27 ⟩ ≡
  for $(k = $ MAX_N$;\ k;\ k\text{--})$
    if $((p = node\_index[k]) \ne \Lambda)$ {
      if $(threshold \equiv 0)$ p-*thresh* = −99999999;
      else p-*thresh* = $((p$-*total* ≫ 16$) ∗ threshold) + (((p$-*total* & $^{\#}$**ffff**$) ∗ threshold) ≫ 16$);
    }
This code is used in section 25.

28. ⟨ Prune the sectors that are used in macro-sectors, and form the lists of SIC sector codes 28 ⟩ ≡
  for $(p = node\_index[$ADJ_SEC$];\ p \ge node\_block;\ p\text{--})$ {   /∗ bottom up ∗/
    if $(p$-SIC$)$ {   /∗ original leaf ∗/
      p-*SIC_list* = *gb_virgin_arc*( );
      p-*SIC_list*-*len* = p-SIC;
    } **else** {
      $pl = p + 1$; $pr = p$-*rchild*;
      if $(p$-*tag* $\equiv 0)$ p-*tag* = pl-*tag* + pr-*tag*;
      if $(p$-*tag* $\le 1)$ ⟨ Replace pl and pr by their union, p 29 ⟩;
    }
  }
This code is used in section 25.

29. ⟨ Replace pl and pr by their union, p 29 ⟩ ≡
  { **register Arc** *∗a* = pl-*SIC_list*;
    register long *jj* = pl-SIC, *kk* = pr-SIC;
    p-*SIC_list* = a;
    while $(a$-*next*$)$ $a = a$-*next*;
    a-*next* = pr-*SIC_list*;
    for $(k = $ MAX_N$;\ k;\ k\text{--})$
      if $((q = node\_index[k]) \ne \Lambda)$ {
        if $(q \ne pl \wedge q \ne pr)$ q-*table*[*jj*] += q-*table*[*kk*];
        p-*table*[k] = pl-*table*[k] + pr-*table*[k];
      }
    p-*total* = pl-*total* + pr-*total*;
    p-SIC = *jj*;
    p-*table*[*jj*] += p-*table*[*kk*];
    node_index[*jj*] = p;
    node_index[*kk*] = Λ;
  }
This code is used in section 28.

30. If the `Users` vertex is not omitted, we need to compute each sector's total final demand, which is calculated so that the row sums and column sums of the input/output coefficients come out equal. We've already computed the column sum, $p\text{-}total$; we've also computed $p\text{-}table[1] + \cdots + p\text{-}table[\texttt{ADJ\_SEC}]$, and put it into $p\text{-}table[\texttt{MAX\_N}]$. So now we want to replace $p\text{-}table[\texttt{MAX\_N}]$ by $p\text{-}total - p\text{-}table[\texttt{MAX\_N}]$. As remarked earlier, this quantity might be negative.

In the special node p for the `Users` vertex, the preliminary processing has made $p\text{-}total = 0$; moreover, $p\text{-}table[\texttt{MAX\_N}]$ is the sum of value added, or GNP. We want to switch those fields.

We don't have to set the *tag* fields to 1 in the special nodes, because the remaining parts of the arc-generation algorithm don't look at those fields.

⟨ Make the special nodes invisible if they are omitted, visible otherwise 30 ⟩ ≡
```
if (omit ≡ 2)  node_index[ADJ_SEC] = node_index[MAX_N] = Λ;
else if (omit ≡ 1)  node_index[MAX_N] = Λ;
else {
    for (k = ADJ_SEC; k; k−−)
        if ((p = node_index[k]) ≠ Λ)  p→table[MAX_N] = p→total − p→table[MAX_N];
    p = node_index[MAX_N];      /* the special node */
    p→total = p→table[MAX_N];
    p→table[MAX_N] = 0;
}
```
This code is used in section 25.

ADJ_SEC = 80, §3.
Arc = **struct**, GB_GRAPH §10.
gb_virgin_arc: **Arc** *( ),
   GB_GRAPH §29.
k: **register long**, §8.
len: **long**, GB_GRAPH §10.
MAX_N = 81, §3.
next: **Arc** *, GB_GRAPH §10.

node_block: **static node** *, §12.
node_index: **static node** *[], §12.
omit: **unsigned long**, §7.
p: **register node** *, §13.
pl: **register node** *, §13.
pr: **register node** *, §13.
q: **register node** *, §13.
rchild: **struct node_struct** *, §11.

SIC: **long**, §11.
SIC_list: **Arc** *, §11.
table: **long** [], §11.
tag: **long**, §11.
thresh: **long**, §11.
threshold: **unsigned long**, §7.
total: **unsigned long**, §11.
Vertex = **struct**, GB_GRAPH §9.

GB_FLIP

1. Introduction. This is GB_FLIP, the module used by GraphBase programs to generate random numbers.

To use the routines in this file, first call the function $gb\_init\_rand(seed)$. Subsequent uses of the macro $gb\_next\_rand()$ will then return pseudo-random integers between 0 and $2^{31} - 1$, inclusive.

GraphBase programs are designed to produce identical results on almost all existing computers and operating systems. An improved version of the portable subtractive method recommended in *Seminumerical Algorithms*, Section 3.6, is used to generate random numbers in the routines below. The period length of the generated numbers is at least $2^{55} - 1$, and it is in fact plausibly conjectured to be $2^{85} - 2^{30}$ for all but at most one choice of the *seed* value. The low-order bits of the generated numbers are just as random as the high-order bits.

2. Changes might be needed when these routines are ported to different systems, because the programs have been written to be most efficient on binary computers that use two's complement notation. Almost all modern computers are based on two's complement arithmetic, but if you have a nonconformist machine you might have to revise the code in sections that are listed under 'system dependencies' in the index.

A validation program is provided so that installers can tell if GB_FLIP is working properly. To make the test, simply run `test_flip`.

```
⟨test_flip.c  2⟩ ≡
#include <stdio.h>
#include "gb_flip.h"      /* all users of GB_FLIP should do this */
  int main()
  { long j;
    gb_init_rand(−314159 L);
    if (gb_next_rand() ≠ 119318998) {
      fprintf(stderr, "Failure␣on␣the␣first␣try!\n");
      return −1;
    }
    for (j = 1; j ≤ 133; j++)  gb_next_rand();
    if (gb_unif_rand(#55555555 L) ≠ 748103812) {
      fprintf(stderr, "Failure␣on␣the␣second␣try!\n");
      return −2;
    }
    fprintf(stderr, "OK,␣the␣gb_flip␣routines␣seem␣to␣work!\n");
    return 0;
  }
```

3. The C code for GB_FLIP doesn't have a main routine; it's just a bunch of subroutines to be incorporated into programs at a higher level via the system loading routine. Here is the general outline of `gb_flip.c`:

⟨Private declarations 4⟩
⟨External declarations 5⟩
⟨External functions 7⟩

4. The subtractive method.   If m is any even number and if the numbers a_0, a_1, ..., a_{54} are not all even, then the numbers generated by the recurrence

$$a_n = (a_{n-24} - a_{n-55}) \bmod m$$

have a period length of at least $2^{55} - 1$, because the residues $a_n \bmod 2$ have a period of this length. Furthermore, the numbers 24 and 55 in this recurrence are sufficiently large that deficiencies in randomness due to the simplicity of the recurrence are negligible in most applications.

Here we take $m = 2^{31}$ so that we get the full set of nonnegative numbers on a 32-bit computer. The recurrence is computed by maintaining an array of 55 values, $A[1] \ldots A[55]$. We also set $A[0] = -1$ to act as a sentinel.

⟨ Private declarations 4 ⟩ ≡
    static long $A[56] = \{-1\}$;   /* pseudo-random values */

This code is used in section 3.

5.   Every external variable should be declared twice in this CWEB file: once for GB_FLIP itself (the "real" declaration for storage allocation purposes), and once in `gb_flip.h` (for cross-references by GB_FLIP users).

The pointer variable *gb_fptr* should not be mentioned explicitly by user routines. It is made public only for efficiency, so that the *gb_next_rand* macro can access the private A table.

⟨ External declarations 5 ⟩ ≡
    long *gb_fptr = A;   /* the next A value to be exported */

This code is used in section 3.

6.   Incidentally, we hope that optimizing compilers are smart enough to do the right thing with *gb_next_rand*.

#define *gb_next_rand*()   (*gb_fptr $\geq$ 0 ? *gb_fptr -- : *gb_flip_cycle*())

⟨ `gb_flip.h` 6 ⟩ ≡
  #define *gb_next_rand*()   (*gb_fptr $\geq$ 0 ? *gb_fptr -- : *gb_flip_cycle*())
  extern long *gb_fptr;   /* the next A value to be used */
  extern long *gb_flip_cycle*();   /* compute 55 more pseudo-random numbers */

See also sections 11 and 13.

---

fprintf: **int** (), <stdio.h>.   *gb_init_rand*: **void** (), §8.   *seed*: **long**, §8.
gb_flip_cycle: **long** (), §7.   *gb_unif_rand*: **long** (), §12.   *stderr*: **FILE** *, <stdio.h>.

7. The user is not supposed to call *gb_flip_cycle* directly either. It is a routine invoked by the macro *gb_next_rand*( ) when *gb_fptr* points to the negative value in $A[0]$.

The purpose of *gb_flip_cycle* is to do 55 more steps of the basic recurrence, at high speed, and to reset *gb_fptr*.

The nonnegative remainder of $(x - y)$ mod 2^{31} is computed here by doing a logical-and with the constant #7fffffff. This technique doesn't work on computers that do not perform two's complement arithmetic. An alternative for such machines is to add the value 2^{30} twice to $(x - y)$, when $(x - y)$ turns out to be negative. Careful calculations are essential because the GraphBase results must be identical on all computer systems.

The sequence of random numbers returned by successive calls of *gb_next_rand*( ) isn't really a_n, $a_{n+1}, \ldots$, as defined by the basic recurrence above. Blocks of 55 consecutive values are essentially being "flipped" or "reflected"—output in reverse order—because *gb_next_rand*( ) makes the value of *gb_fptr* decrease instead of increase. But such flips don't make the results any less random.

#define *mod_diff*(x, y) $(((x) - (y))$ & #7fffffff)   /* difference modulo 2^{31} */

⟨ External functions 7 ⟩ ≡

```
long gb_flip_cycle( )
{ register long *ii, *jj;
    for (ii = &A[1], jj = &A[32]; jj ≤ &A[55]; ii++, jj++) *ii = mod_diff(*ii, *jj);
    for (jj = &A[1]; ii ≤ &A[55]; ii++, jj++) *ii = mod_diff(*ii, *jj);
    gb_fptr = &A[54];
    return A[55];
}
```

See also sections 8 and 12.

This code is used in section 3.

8.   Initialization.   To get everything going, we use a scheme like that recommended in *Seminumerical Algorithms*, but revised so that the least significant bits of the starting values depend on the entire seed, not just on the seed's least significant bits.

Notice that we jump around in the array by increments of 21, a number that is relatively prime to 55. Repeated skipping by steps of 21 mod 55 keeps the values we're computing spread out as far from each other as possible in the array, since 21, 34, and 55 are consecutive Fibonacci numbers (see the discussion of Fibonacci hashing in Section 6.4 of *Sorting and Searching*). Our initialization mechanism would be rather poor if we didn't do something like that to disperse the values (see *Seminumerical Algorithms*, exercise 3.2.2–2).

⟨ External functions 7 ⟩ +≡
  void *gb_init_rand* (*seed*)
     long *seed*;
  { **register long** *i*;
    register long *prev* = *seed*, *next* = 1;
    seed = *prev* = *mod_diff* (*prev*, 0);   /* strip off the sign */
    $A[55]$ = *prev*;
    for ($i = 21$; i; $i = (i + 21)$ % 55) {
      $A[i]$ = *next*;
      ⟨ Compute a new *next* value, based on *next*, *prev*, and *seed* 9 ⟩;
      prev = $A[i]$;
    }
    ⟨ Get the array values "warmed up" 10 ⟩;
  }

9.   Incidentally, if `test_flip` fails, the person debugging these routines will want to know some of the intermediate numbers computed during initialization. The first nontrivial values calculated by *gb_init_rand* are $A[42] = 2147326568$, $A[8] = 1073977445$, and $A[29] = 536517481$. Once you get those right, the rest should be easy.

An early version of this routine simply said '*seed* ≫ 1' instead of making *seed* shift cyclically. This method had an interesting flaw: When the original *seed* was a number of the form $4s + 1$, the first 54 elements $A[1]$, ..., $A[54]$ were set to exactly the same values as when *seed* was $4s + 2$. Therefore one out of every four seed values was effectively being wasted.

⟨ Compute a new *next* value, based on *next*, *prev*, and *seed* 9 ⟩ ≡
  next = *mod_diff* (*prev*, *next*);
  if (*seed* & 1) *seed* = #40000000 + (*seed* ≫ 1);
  else *seed* ≫= 1;   /* cyclic shift right 1 */
  next = *mod_diff* (*next*, *seed*);
This code is used in section 8.

A: **static long** [ ], §4.   *gb_fptr*: **long** *, §5.   *gb_next_rand* = macro ( ), §6.

10.   After the first 55 values have been computed as a function of *seed*, they aren't random enough for us to start using them right away. For example, we have set $A[21] = 1$ in order to ensure that at least one starting value is an odd number. But once the sequence a_n gets going far enough from its roots, the initial transients become imperceptible. Therefore we call *gb_flip_cycle* five times, effectively skipping past the first 275 elements of the sequence; this has the desired effect. It also initializes *gb_fptr*.

Note: It is possible to express the least significant bit of the generated numbers as a linear combination mod 2 of the 31 bits of *seed* and of the constant 1. For example, the first generated number turns out to be odd if and only if

$$s_{24} + s_{23} + s_{22} + s_{21} + s_{19} + s_{18} + s_{15} + s_{14} + s_{13} + s_{11} + s_{10} + s_8 + s_7 + s_6 + s_2 + s_1 + s_0$$

is odd, when $seed = (s_{31} \ldots s_1 s_0)_2$. We can represent this linear combination conveniently by the hexadecimal number $^\#$01ecedc7; the 1 stands for s_{24} and the final 7 stands for $s_2 + s_1 + s_0$. The first ten least-significant bits turn out to be respectively $^\#$01ecedc7, $^\#$dbbdc362, $^\#$400e0b06, $^\#$0eb73780, $^\#$da0d66ae, $^\#$002b63bc, $^\#$adb801ed, $^\#$8077bbbc, $^\#$803d9db5, and $^\#$401a0eda in this notation (using the sign bit to indicate cases when 1 must be added to the sum).

We must admit that these ten 32-bit patterns do not look at all random; the number of b's, d's, and 0's is unusually high. (Before the "warmup cycles," the patterns are even more regular.) This phenomenon eventually disappears, however, as the sequence proceeds; and it does not seem to imply any serious deficiency in practice, even at the beginning of the sequence, once we've done the warmup exercises.

⟨ Get the array values "warmed up" 10 ⟩ ≡

   (**void**) *gb_flip_cycle*( );
   (**void**) *gb_flip_cycle*( );
   (**void**) *gb_flip_cycle*( );
   (**void**) *gb_flip_cycle*( );
   (**void**) *gb_flip_cycle*( );

This code is used in section 8.

11.   ⟨ gb_flip.h   6 ⟩ +≡
   extern void *gb_init_rand*( );

12.   Uniform integers.   Here is a simple routine that produces a uniform integer between 0 and $m-1$, inclusive, when m is any positive integer less than 2^{31}. It avoids the bias toward small values that would occur if we simply calculated $gb\_next\_rand(\,)\,\%\,m$. (The bias is insignificant when m is small, but it can be serious when m is large. For example, if $m \approx 2^{32}/3$, the simple remainder algorithm would give an answer less than $m/2$ about 2/3 of the time.)

This routine consumes fewer than two random numbers, on the average, for any fixed m.

In the `test_flip` program (*main*), this routine should compute $t = m$, then it should reject the values $r = 2081307921, 1621414801$, and 1469108743 before returning the answer 748103812.

#define *two_to_the_31*   ((**unsigned long**) $^\#$80000000)

⟨External functions 7⟩ +≡
 long *gb_unif_rand*(*m*)
   long *m*;
 { **register unsigned long** $t = two\_to\_the\_31 - (two\_to\_the\_31\,\%\,m)$;
  register long *r*;
  do {
   $r = gb\_next\_rand(\,)$;
  } **while** ($t \leq$ (**unsigned long**) *r*);
  return $r\,\%\,m$;
 }

13.   ⟨`gb_flip.h` 6⟩ +≡
 extern long *gb_unif_rand*(\,);

A: **static long** [ ], §4.
gb_flip_cycle: **long** (\,), §7.
gb_fptr: **long** ∗, §5.

gb_init_rand: **void** (\,), §8.
gb_next_rand = macro (\,), §6.

main: **int** (\,), §2.
seed: **long**, §8.

GB_GAMES

Important: Before reading GB_GAMES, please read or at least skim the programs for GB_GRAPH and GB_IO.

1. Introduction. This GraphBase module contains the *games* subroutine, which creates a family of undirected graphs based on college football scores. An example of the use of this procedure can be found in the demo program FOOTBALL.

⟨ gb_games.h 1 ⟩ ≡
  extern Graph *games*( );

See also section 5.

2. The subroutine call *games*(n, *ap0_weight*, *upi0_weight*, *ap1_weight*, *upi1_weight*, *first_day*, *last_day*, *seed*) constructs a graph based on the information in `games.dat`. Each vertex of the graph corresponds to one of 120 football teams at American colleges and universities (more precisely, to the 106 college football teams of division I-A together with the 14 division I-AA teams of the Ivy League and the Patriot League). Each edge of the graph corresponds to one of the 638 games played between those teams during the 1990 season.

An arc from vertex u to vertex v is assigned a length representing the number of points scored by u when playing v. Thus the graph isn't really "undirected," although it is true that its arcs are paired (i.e., that u played v if and only if v played u). A truly undirected graph with the same vertices and edges can be obtained by applying the *copy* routine of GB_BASIC.

The constructed graph will have $\min(n, 120)$ vertices. If n is less than 120, the n teams will be selected by assigning a weight to each team and choosing the n with largest weight, using random numbers to break ties in case of equal weights. Weights are computed by the formula

$$ap0\_weight \cdot ap0 + upi0\_weight \cdot upi0 + ap1\_weight \cdot ap1 + upi1\_weight \cdot upi1 ,$$

where *ap0* and *upi0* are the point scores given to a team in the Associated Press and United Press International polls at the beginning of the season, and *ap1* and *upi1* are the similar scores given at the end of the season. (The *ap* scores were obtained by asking 60 sportswriters to choose and rank the top 25 teams, assigning 25 points to a team ranked 1st and 1 point to a team ranked 25th; thus the total of each of the *ap* scores, summed over all teams, is 19500. The *upi* scores were obtained by asking football coaches to choose and rank the top 15 teams, assigning 15 points to a team ranked 1st and 1 point to a team ranked 15th. In the case of *upi0*, there were 48 coaches voting, making 5760 points altogether; but in the case of *upi1*, 59 coaches were polled, yielding a total of 7080 points. The coaches agreed not to vote for any team that was on probation for violating NCAA rules, but the sportswriters had no such policy.)

Parameters *first_day* and *last_day* can be used to vary the number of edges; only games played between *first_day* and *last_day*, inclusive, will be included in the constructed graph. Day 0 was August 26, 1990, when Colorado and Tennessee competed in the Disneyland Pigskin Classic. Day 128 was January 1, 1991, when the final end-of-season bowl games were played. About half of each team's games were played between day 0 and day 50. If *last_day* = 0, the value of *last_day* is automatically increased to 128.

As usual in GraphBase routines, you can set $n = 0$ to get the default situation where n has its maximum value. For example, either $games(0, 0, 0, 0, 0, 0, 0, 0)$ or $games(120, 0, 0, 0, 0, 0, 0, 0)$ produces the full graph; $games(0, 0, 0, 0, 0, 50, 0, 0)$ or $games(120, 0, 0, 0, 0, 50, 0, 0)$ or $games(120, 0, 0, 0, 0, 50, 128, 0)$ produces the graph for the last half of the season. One way to select a subgraph containing the 30 "best" teams is to ask for $games(30, 0, 0, 1, 2, 0, 0, 0)$, which adds the votes of the sportswriters to the votes of the coaches (considering that a coach's first choice is worth 30 points while a sportswriter's first choice is worth only 25). It turns out that 67 of the teams did not receive votes in any of the four polls; the subroutine call $games(53, 1, 1, 1, 1, 0, 0, 0)$ will pick out the 53 teams that were selected at least once by some sportswriter or coach, and $games(67, -1, -1, -1, -1, 0, 0, 0)$ will pick out the 67 that were not. A random selection of 60 teams can be obtained by calling $games(60, 0, 0, 0, 0, 0, 0, s)$. Different choices of the seed number s will produce different selections in a system-independent manner; any value of s between 0 and $2^{31} - 1$ is permissible. If you ask for $games(120, 0, 0, 0, 0, 0, 0, s)$ with different choices of s, you always get the full graph, but the vertices will appear in different (random) orderings depending on s.

Parameters *ap0_weight*, *upi0_weight*, *ap1_weight*, and *upi1_weight* must be at most $2^{17} = 131072$ in absolute value.

```
#define  MAX_N  120
#define  MAX_DAY  128
#define  MAX_WEIGHT  131072
#define  ap  u.I      /* Associated Press scores: (ap0 ≪ 16) + ap1 */
#define  upi  v.I     /* United Press International scores (upi0 ≪ 16) + upi1 */
```

3. Most of the teams belong to a "conference," and they play against almost every other team that belongs to the same conference. For example, Stanford and nine other teams belong to the Pacific Ten conference. Eight of Stanford's eleven games were against other teams of the Pacific Ten; the other three were played against Colorado (from the Big Eight), San José State (from the Big West) and Notre Dame (which is independent). The graphs produced by *games* therefore illustrate "cliquey" patterns of social interaction.

Eleven different conferences are included in `games.dat`. Utility field $z.S$ of a vertex is set to the name of a team's conference, or to Λ if that team is independent. (Exactly 24 of the I-A football teams were independent in 1990.) Two teams u and v belong to the same conference if and only if $u\text{-}conference \equiv v\text{-}conference$ and $u\text{-}conference \neq \Lambda$.

```
#define  conference  z.S
```

---

ap0_weight: **long**, §7.	*I*: **long**, GB_GRAPH §8.	*u*: **util**, GB_GRAPH §9.
ap1_weight: **long**, §7.	*last_day*: **long**, §7.	*upi0_weight*: **long**, §7.
copy: parameter, GB_BASIC §74.	*n*: **unsigned long**, §7.	*upi1_weight*: **long**, §7.
first_day: **long**, §7.	*S*: **char** *, GB_GRAPH §8.	*v*: **util**, GB_GRAPH §9.
games: **Graph** *(), §7.	*seed*: **long**, §7.	*z*: **util**, GB_GRAPH §9.
Graph = struct, GB_GRAPH §20.		

4. Each team has a nickname, which is recorded in utility field $y.S$. For example, Georgia Tech's team is called the Yellow Jackets. Six teams (Auburn, Clemson, Memphis State, Missouri, Pacific, and Princeton) are called the Tigers, and five teams (Fresno State, Georgia, Louisiana Tech, Mississippi State, Yale) are called the Bulldogs. But most of the teams have a unique nickname, and 94 distinct nicknames exist.

A shorthand code for team names is also provided, in the *abbr* field.

```
#define  nickname  y.S
#define  abbr  x.S
```

5. If a points to an arc from u to v, utility field $a\text{-}a.I$ contains the value 3 if u was the home team, 1 if v was the home team, and 2 if both teams played on neutral territory. The date of that game, represented as a integer number of days after August 26, 1990, appears in utility field $a\text{-}b.I$. The arcs in each vertex list $v\text{-}arcs$ appear in reverse order of their dates: last game first and first game last.

```
#define  HOME  1
#define  NEUTRAL  2      /* this value is halfway between HOME and AWAY */
#define  AWAY  3
#define  venue  a.I
#define  date  b.I
```

⟨ gb_games.h 1 ⟩ +≡
```
#define  ap  u.I      /* repeat the definitions in the header file */
#define  upi  v.I
#define  abbr  x.S
#define  nickname  y.S
#define  conference  z.S
#define HOME  1
#define NEUTRAL  2
#define AWAY  3
#define  venue  a.I
#define  date  b.I
```

6. If the *games* routine encounters a problem, it returns Λ (NULL), after putting a code number into the external variable *panic_code*. This code number identifies the type of failure. Otherwise *games* returns a pointer to the newly created graph, which will be represented with the data structures explained in GB_GRAPH. (The external variable *panic_code* is itself defined in GB_GRAPH.)

```
#define  panic(c)  { panic_code = c; gb_trouble_code = 0; return Λ; }
```

7. The C file `gb_games.c` has the following overall shape:

```
#include "gb_io.h"      /* we will use the GB_IO routines for input */
#include "gb_flip.h"      /* we will use the GB_FLIP routines for random numbers */
#include "gb_graph.h"      /* we will use the GB_GRAPH data structures */
#include "gb_sort.h"      /* and gb_linksort for sorting */
  ⟨ Preprocessor definitions ⟩
  ⟨ Type declarations 11 ⟩
  ⟨ Private variables 13 ⟩
```

⟨ Private functions 23 ⟩

Graph *$games$(n, $ap0\_weight$, $upi0\_weight$, $ap1\_weight$, $upi1\_weight$, $first\_day$, $last\_day$, $seed$)
   unsigned long n; /* number of vertices desired */
   long $ap0\_weight$; /* coefficient of $ap0$ in the weight function */
   long $ap1\_weight$; /* coefficient of $ap1$ in the weight function */
   long $upi0\_weight$; /* coefficient of $upi0$ in the weight function */
   long $upi1\_weight$; /* coefficient of $upi1$ in the weight function */
   long $first\_day$; /* lower cutoff for games to be considered */
   long $last\_day$; /* upper cutoff for games to be considered */
   long $seed$; /* random number seed */
{ ⟨ Local variables 8 ⟩
 gb_init_rand($seed$);
 ⟨ Check that the parameters are valid 9 ⟩;
 ⟨ Set up a graph with n vertices 10 ⟩;
 ⟨ Read the first part of `games.dat` and compute team weights 14 ⟩;
 ⟨ Determine the n teams to use in the graph 19 ⟩;
 ⟨ Put the appropriate edges into the graph 21 ⟩;
 if (gb_close( ) $\neq 0$) $panic$($late\_data\_fault$);
   /* something's wrong with `"games.dat"`; see io_errors */
 gb_free($working\_storage$);
 if ($gb\_trouble\_code$) {
  $gb\_recycle$(new_graph);
  $panic$($alloc\_fault$); /* oops, we ran out of memory somewhere back there */
 }
 return new_graph;
}

8. ⟨ Local variables 8 ⟩ $\equiv$
 Graph *new_graph; /* the graph constructed by $games$ */
 register long j, k; /* all-purpose indices */
This code is used in section 7.

---

a: **util**, GB_GRAPH §10.
$alloc\_fault = -1$, GB_GRAPH §7.
$ap = u.I$, §2.
$ap0$: part of ap, §2.
$ap1$: part of ap, §2.
$arcs$: **Arc** *, GB_GRAPH §9.
b: **util**, GB_GRAPH §10.
$conference = z.S$, §3.
gb_close: **long** ( ), GB_IO §39.
gb_free: **void** ( ), GB_GRAPH §16.

gb_init_rand: **void** ( ), GB_FLIP §8.
$gb\_linksort$: **void** ( ), GB_SORT §5.
$gb\_recycle$: **void** ( ), GB_GRAPH §40.
$gb\_trouble\_code$: **long**,
 GB_GRAPH §14.
Graph = **struct**, GB_GRAPH §20.
I: **long**, GB_GRAPH §8.
io_errors: **long**, GB_IO §5.
$late\_data\_fault = 11$, GB_GRAPH §7.
$panic\_code$: **long**, GB_GRAPH §5.

S: **char** *, GB_GRAPH §8.
u: **util**, GB_GRAPH §9.
$upi = v.I$, §2.
$upi0$: part of upi, §2.
$upi1$: part of upi, §2.
v: **util**, GB_GRAPH §9.
$working\_storage$: **static Area**, §13.
x: **util**, GB_GRAPH §9.
y: **util**, GB_GRAPH §9.
z: **util**, GB_GRAPH §9.

9. ⟨ Check that the parameters are valid 9 ⟩ ≡

   if $(n \equiv 0 \lor n >$ MAX_N$)$ $n =$ MAX_N;

   if $(ap0\_weight >$ MAX_WEIGHT $\lor ap0\_weight < -$MAX_WEIGHT $\lor upi0\_weight >$ MAX_WEIGHT $\lor upi0\_weight <$

        $-$MAX_WEIGHT $\lor ap1\_weight >$ MAX_WEIGHT $\lor ap1\_weight < -$MAX_WEIGHT $\lor upi1\_weight >$

        MAX_WEIGHT $\lor upi1\_weight < -$MAX_WEIGHT$)$ *panic*(*bad_specs*);

       /* the magnitude of at least one weight is too big */

   if $(first\_day < 0)$ $first\_day = 0$;

   if $(last\_day \equiv 0 \lor last\_day >$ MAX_DAY$)$ $last\_day =$ MAX_DAY;

This code is used in section 7.

10. ⟨ Set up a graph with n vertices 10 ⟩ ≡

   $new\_graph = gb\_new\_graph(n)$;

   if $(new\_graph \equiv \Lambda)$ *panic*(*no_room*);   /* out of memory before we're even started */

   sprintf$(new\_graph\text{-}id,$ `"games(%lu,%ld,%ld,%ld,%ld,%ld,%ld,%ld)"`$, n, ap0\_weight, upi0\_weight,$

      $ap1\_weight, upi1\_weight, first\_day, last\_day, seed)$;

   strcpy$(new\_graph\text{-}util\_types,$ `"IIZSSSIIZZZZZ"`$)$;

This code is used in section 7.

11. Vertices. As we read in the data, we construct a list of nodes, each of which contains a team's name, nickname, conference, and weight. After this list has been sorted by weight, the top n entries will be the vertices of the new graph.

⟨ Type declarations 11 ⟩ ≡

```
typedef struct node_struct {      /* records to be sorted by gb_linksort */
    long key;      /* the nonnegative sort key (weight plus 2³⁰) */
    struct node_struct *link;      /* pointer to next record */
    char name[24];      /* "College␣Name" */
    char nick[22];      /* "Team␣Nickname" */
    char abb[6];      /* "ABBR" */
    long a0, u0, a1, u1;      /* team scores in press polls */
    char *conf;      /* pointer to conference name */
    struct node_struct *hash_link;      /* pointer to next ABBR in hash list */
    Vertex *vert;      /* vertex corresponding to this team */
} node;
```

This code is used in section 7.

12. The data in `games.dat` appears in two parts. The first 120 lines have the form

`ABBR College Name(Team Nickname)Conference;a0,u0;a1,u1`

and they give basic information about the teams. An internal abbreviation code `ABBR` is used to identify each team in the second part of the data.

The second part presents scores of the games, and it contains two kinds of lines. If the first character of a line is '`>`', it means "change the current date," and the remaining characters specify a date as a one-letter month code followed by the day of the month. Otherwise the line gives scores of a game, using the `ABBR` codes for two teams. The scores are separated by '`@`' if the second team was the home team and by '`,`' if both teams were on neutral territory.

For example, two games were played on December 8, namely the annual Army-Navy game and the California Raisin Bowl game. These are recorded in three lines of `games.dat` as follows:

`>D8`
`NAVY20@ARMY30`
`SJSU48,CMICH24`

We deduce that Navy played at Army's home stadium, losing 20 to 30; moreover, San José State played Central Michigan on neutral territory and won, 48 to 24. (The California Raisin Bowl is traditionally a playoff between the champions of the Big West and Mid-American conferences.)

---

ap0_weight: **long**, §7.
ap1_weight: **long**, §7.
bad_specs = 30, GB_GRAPH §7.
first_day: **long**, §7.
gb_linksort: **void** ( ), GB_SORT §5.
gb_new_graph: **Graph** *( ),
  GB_GRAPH §23.
id: **char** [ ], GB_GRAPH §20.

last_day: **long**, §7.
MAX_DAY = 128, §2.
MAX_N = 120, §2.
MAX_WEIGHT = 131072, §2.
n: **unsigned long**, §7.
new_graph: **Graph** *, §8.
no_room = 1, GB_GRAPH §7.
panic = macro ( ), §6.

seed: **long**, §7.
sprintf: **int** ( ), <stdio.h>.
strcpy: **char** *( ), <string.h>.
upi0_weight: **long**, §7.
upi1_weight: **long**, §7.
util_types: **char** [ ], GB_GRAPH §20.
Vertex = **struct**, GB_GRAPH §9.

13. In order to map `ABBR` codes to team names, we use a simple hash coding scheme. Two abbreviations with the same hash address are linked together via the *hash_link* address in their node.

The constants defined here are taken from the specific data in `games.dat`, because this routine is not intended to be perfectly general.

#**define** `HASH_PRIME` 1009

⟨Private variables 13⟩ ≡

  static long *ma0* = 1451, *mu0* = 666, *ma1* = 1475, *mu1* = 847;
    /∗ maximum poll values in the data ∗/
  static node ∗*node_block*;   /∗ array of nodes holding team info ∗/
  static node ∗∗*hash_block*;   /∗ array of heads of hash code lists ∗/
  static Area *working_storage*;   /∗ memory needed only while *games* is working ∗/
  static char ∗∗*conf_block*;   /∗ array of conference names ∗/
  static long *m*;   /∗ the number of conference names known so far ∗/

This code is used in section 7.

14. ⟨Read the first part of `games.dat` and compute team weights 14⟩ ≡

  node_block = *gb_typed_alloc*(`MAX_N` + 2, **node**, *working_storage*);
    /∗ leave room for string overflow ∗/
  hash_block = *gb_typed_alloc*(`HASH_PRIME`, **node** ∗, *working_storage*);
  conf_block = *gb_typed_alloc*(`MAX_N`, **char** ∗, *working_storage*);
  m = 0;
  if (*gb_trouble_code*) {
    gb_free(*working_storage*);
    panic(*no_room* + 1);   /∗ nowhere to copy the data ∗/
  }
  if (*gb_open*("games.dat") ≠ 0) *panic*(*early_data_fault*);
      /∗ couldn't open "games.dat" using GraphBase conventions; *io_errors* tells why ∗/
  for (*k* = 0; *k* < `MAX_N`; *k*++) ⟨Read and store data for team *k* 15⟩;

This code is used in section 7.

15. ⟨Read and store data for team *k* 15⟩ ≡

  { **register node** ∗*p*;
    register char ∗*q*;

    p = *node_block* + *k*;
    if (*k*) *p*⇀*link* = *p* − 1;
    q = *gb_string*(*p*⇀*abb*, '␣');
    if (*q* > &*p*⇀*abb*[6] ∨ *gb_char*( ) ≠ '␣') *panic*(*syntax_error*);   /∗ out of sync in `games.dat` ∗/
    ⟨Enter *p*⇀*abb* in the hash table 16⟩;
    q = *gb_string*(*p*⇀*name*, '(');
    if (*q* > &*p*⇀*name*[24] ∨ *gb_char*( ) ≠ '(') *panic*(*syntax_error* + 1);   /∗ team name too long ∗/
    q = *gb_string*(*p*⇀*nick*, ')');
    if (*q* > &*p*⇀*nick*[22] ∨ *gb_char*( ) ≠ ')') *panic*(*syntax_error* + 2);   /∗ team nickname too long ∗/
    ⟨Read the conference name for *p* 17⟩;
    ⟨Read the press poll scores for *p* and compute *p*⇀*key* 18⟩;
    gb_newline( );
  }

This code is used in section 14.

16. ⟨ Enter $p\text{-}abb$ in the hash table 16 ⟩ ≡
  { **long** $h = 0$;   /∗ the hash code ∗/
    for $(q = p\text{-}abb; \ ∗q; \ q\text{++}) \ h = (h + h + ∗q) \ \% \ \texttt{HASH\_PRIME}$;
    $p\text{-}hash\_link = hash\_block[h]$;
    $hash\_block[h] = p$;
  }

This code is used in section 15.

17. ⟨ Read the conference name for p 17 ⟩ ≡
  {
    $gb\_string(str\_buf, ';')$;
    if $(gb\_char() \neq ';')$ $panic(syntax\_error + 3)$;   /∗ conference name clobbered ∗/
    if $(strcmp(str\_buf, \texttt{"Independent"}) \neq 0)$ {
      for $(j = 0; \ j < m; \ j\text{++})$
        if $(strcmp(str\_buf, conf\_block[j]) \equiv 0)$ **goto** *found*;
      $conf\_block[m\text{++}] = gb\_save\_string(str\_buf)$;
    found: $p\text{-}conf = conf\_block[j]$;
    }
  }

This code is used in section 15.

18. The key value computed here will be between 0 and 2^{31}, because of the bound we've imposed on the weight parameters.

⟨ Read the press poll scores for p and compute $p\text{-}key$ 18 ⟩ ≡
  $p\text{-}a0 = gb\_number(10)$;
  if $(p\text{-}a0 > ma0 \lor gb\_char() \neq ',')$ $panic(syntax\_error + 4)$;   /∗ first AP score clobbered ∗/
  $p\text{-}u0 = gb\_number(10)$;
  if $(p\text{-}u0 > mu0 \lor gb\_char() \neq ';')$ $panic(syntax\_error + 5)$;   /∗ first UPI score clobbered ∗/
  $p\text{-}a1 = gb\_number(10)$;
  if $(p\text{-}a1 > ma1 \lor gb\_char() \neq ',')$ $panic(syntax\_error + 6)$;   /∗ second AP score clobbered ∗/
  $p\text{-}u1 = gb\_number(10)$;
  if $(p\text{-}u1 > mu1 \lor gb\_char() \neq \texttt{'\textbackslash n'})$ $panic(syntax\_error + 7)$;   /∗ second UPI score clobbered ∗/
  $p\text{-}key = ap0\_weight∗(p\text{-}a0) + upi0\_weight∗(p\text{-}u0) + ap1\_weight∗(p\text{-}a1) + upi1\_weight∗(p\text{-}u1) + {}^{\#}40000000$;

This code is used in section 15.

$a0$: **long**, §11.
$a1$: **long**, §11.
abb: **char** [], §11.
$ap0\_weight$: **long**, §7.
$ap1\_weight$: **long**, §7.
Area, GB_GRAPH §12.
$conf$: **char** ∗, §11.
$early\_data\_fault = 10$, GB_GRAPH §7.
$games$: **Graph** ∗(), §7.
gb_char: **char** (), GB_IO §22.
gb_free: **void** (), GB_GRAPH §16.
$gb\_newline$: **void** (), GB_IO §18.
gb_number: **unsigned long** (),
  GB_IO §24.

gb_open: **long** (), GB_IO §32.
gb_save_string: **char** ∗(),
  GB_GRAPH §35.
gb_string: **char** ∗(), GB_IO §26.
$gb\_trouble\_code$: **long**,
  GB_GRAPH §14.
$gb_typed_alloc = $ **macro** (),
  GB_GRAPH §11.
$hash\_link$: **struct node_struct** ∗,
  §11.
io_errors: **long**, GB_IO §5.
j: **register long**, §8.
k: **register long**, §8.
key: **long**, §11.

$link$: **struct node_struct** ∗, §11.
$\texttt{MAX\_N} = 120$, §2.
$name$: **char** [], §11.
$nick$: **char** [], §11.
$no\_room = 1$, GB_GRAPH §7.
node, §11.
$panic = $ **macro** (), §6.
str_buf: **char** [], GB_IO §26.
$strcmp$: **int** (), <string.h>.
$syntax\_error = 20$, GB_GRAPH §7.
$u0$: **long**, §11.
$u1$: **long**, §11.
$upi0\_weight$: **long**, §7.
$upi1\_weight$: **long**, §7.

19. Once all the nodes have been set up, we can use the *gb_linksort* routine to sort them into the desired order. It builds 128 lists from which the desired nodes are readily accessed in decreasing order of weight, using random numbers to break ties.

We set the abbreviation code to zero in every team that isn't chosen. Then games involving that team will be excluded when edges are generated below.

⟨ Determine the n teams to use in the graph 19 ⟩ ≡

```
{ register node *p;      /* the current node being considered */
  register Vertex *v = new_graph→vertices;      /* the next vertex to use */

  gb_linksort(node_block + MAX_N − 1);
  for (j = 127; j ≥ 0; j−−)
    for (p = (node *) gb_sorted[j]; p; p = p→link) {
      if (v < new_graph→vertices + n) ⟨ Add team p to the graph 20 ⟩
      else p→abb[0] = '\0';      /* this team is not being used */
    }
}
```

This code is used in section 7.

20. ⟨ Add team p to the graph 20 ⟩ ≡

```
{
  v→ap = ((long) (p→a0) ≪ 16) + p→a1;
  v→upi = ((long) (p→u0) ≪ 16) + p→u1;
  v→abbr = gb_save_string(p→abb);
  v→nickname = gb_save_string(p→nick);
  v→conference = p→conf;
  v→name = gb_save_string(p→name);
  p→vert = v++;
}
```

This code is used in section 19.

21. Arcs. Finally, we read through the rest of **games.dat**, adding a pair of arcs for each game that belongs to the selected time interval and was played by two of the selected teams.

⟨ Put the appropriate edges into the graph 21 ⟩ ≡

```
{ register Vertex *u, *v;
  register long today = 0;        /* current day of play */
  long su, sv;        /* points scored by each team */
  long ven;        /* HOME if v is home team, NEUTRAL if on neutral ground */
  while (¬gb_eof ( )) {
    if (gb_char ( ) ≡ '>') ⟨Change the current date 22⟩
    else gb_backup ( );
    u = team_lookup ( );
    su = gb_number (10);
    ven = gb_char ( );
    if (ven ≡ '@')  ven = HOME;
    else if (ven ≡ ',')  ven = NEUTRAL;
    else panic (syntax_error + 8);        /* bad syntax in game score line */
    v = team_lookup ( );
    sv = gb_number (10);
    if (gb_char ( ) ≠ '\n') panic (syntax_error + 9);        /* bad syntax in game score line */
    if (u ≠ Λ ∧ v ≠ Λ ∧ today ≥ first_day ∧ today ≤ last_day) ⟨Enter a new edge 24⟩;
    gb_newline ( );
  }
}
```

This code is used in section 7.

---

22. ⟨ Change the current date 22 ⟩ ≡
  { **register char** $c = gb\_char(\,)$;   /∗ month code ∗/
    register long d;   /∗ day of football season ∗/
    switch (c) {
    case 'A': $d = -26$; **break**;   /∗ August ∗/
    case 'S': $d = 5$; **break**;   /∗ thirty days hath September ∗/
    case 'O': $d = 35$; **break**;   /∗ October ∗/
    case 'N': $d = 66$; **break**;   /∗ November ∗/
    case 'D': $d = 96$; **break**;   /∗ December ∗/
    case 'J': $d = 127$; **break**;   /∗ January ∗/
    default: $d = 1000$;
    }
    $d \mathrel{+}= gb\_number(10)$;
    if $(d < 0 \vee d > \texttt{MAX\_DAY})$ $panic(syntax\_error - 1)$;   /∗ date was clobbered ∗/
    $today = d$;
    $gb\_newline(\,)$;   /∗ now ready to read a non-date line ∗/
  }

This code is used in section 21.

23. ⟨ Private functions 23 ⟩ ≡
  static Vertex ∗$team\_lookup(\,)$ /∗ read and decode an abbreviation ∗/
  { **register char** ∗$q = str\_buf$;   /∗ position in str_buf ∗/
    register long $h = 0$;   /∗ hash code ∗/
    register node ∗p;   /∗ position in hash list ∗/
    while $(gb\_digit(10) < 0)$ {
      ∗$q = gb\_char(\,)$;
      $h = (h + h + {∗}q) \mathbin{\%} \texttt{HASH\_PRIME}$;
      $q{+}{+}$;
    }
    $gb\_backup(\,)$;   /∗ prepare to re-scan the digit following the abbreviation ∗/
    ∗$q =$ '\0';   /∗ null-terminate the abbreviation just scanned ∗/
    for $(p = hash\_block[h];\; p;\; p = p{\to}hash\_link)$
      if $(strcmp(p{\to}abb, str\_buf) \equiv 0)$ **return** $p{\to}vert$;
    return Λ;   /∗ not found ∗/
  }

This code is used in section 7.

24. We retain the convention of GB_GRAPH that the arc from v to u appears immediately after a matching arc from u to v when $u < v$.

⟨ Enter a new edge 24 ⟩ ≡
  { **register Arc** ∗a;
    if $(u > v)$ { **register Vertex** ∗w;
      register long sw;
      $w = u;\; u = v;\; v = w$;
      $sw = su;\; su = sv;\; sv = sw$;
      $ven = \texttt{HOME} + \texttt{AWAY} - ven$;
    }

```
    gb_new_arc(u, v, su);
    gb_new_arc(v, u, sv);
    a = u→arcs;      /* a pointer to the new arc */
    if (v→arcs ≠ a + 1) panic(impossible + 9);      /* can't happen */
    a→venue = ven;  (a + 1)→venue = HOME + AWAY − ven;
    a→date = (a + 1)→date = today;
}
```

This code is used in section 21.

GB_GATES _____

Important: Before reading GB_GATES, please read or at least skim the program for GB_GRAPH.

1. Introduction. This GraphBase module provides six external subroutines:

> *risc*, a routine that creates a directed acyclic graph based on the logic of a simple RISC computer;
>
> *prod*, a routine that creates a directed acyclic graph based on the logic of parallel multiplication circuits;
>
> *print_gates*, a routine that outputs a symbolic representation of such directed acyclic graphs;
>
> *gate_eval*, a routine that evaluates such directed acyclic graphs by assigning boolean values to each gate;
>
> *partial_gates*, a routine that extracts a subgraph by assigning random values to some of the input gates;
>
> *run_risc*, a routine that can be used to play with the output of *risc*.

Examples of the use of these routines can be found in the demo programs TAKE_RISC and MULTIPLY.

⟨gb_gates.h 1⟩ ≡
```
#define print_gates  p_gates      /* abbreviation for Procrustean linkers */
  extern Graph *risc( );       /* make a network for a microprocessor */
  extern Graph *prod( );        /* make a network for high-speed multiplication */
  extern void print_gates( );      /* write a network to standard output file */
  extern long gate_eval( );      /* evaluate a network */
  extern Graph *partial_gates( );       /* reduce network size */
  extern long run_risc( );      /* simulate the microprocessor */
  extern unsigned long risc_state[ ];       /* the output of run_risc */
```
See also sections 2 and 50.

2. The directed acyclic graphs produced by GB_GATES are GraphBase graphs with special conventions related to logical networks. Each vertex represents a gate of a network, and utility field *val* is a boolean value associated with that gate. Utility field *typ* is an ASCII code that tells what kind of gate is present:

'I' denotes an input gate, whose value is specified externally.

'&' denotes an AND gate, whose value is the logical AND of two or more previous gates (namely, 1 if all those gates are 1, otherwise 0).

'|' denotes an OR gate, whose value is the logical OR of two or more previous gates (namely, 0 if all those gates are 0, otherwise 1).

'^' denotes an XOR gate, whose value is the logical EXCLUSIVE-OR of two or more previous gates (namely, their sum modulo 2).

'~' denotes an inverter, whose value is the logical complement of the value of a single previous gate.

'L' denotes a latch, whose value depends on past history; it is the value that was assigned to
a subsequent gate when the network was most recently evaluated. Utility field *alt* points
to that subsequent gate.

Latches can be used to include "state" information in a circuit; for example, they correspond to
registers of the RISC machine constructed by *risc*. The *prod* procedure does not use latches.

The vertices of the directed acyclic graph appear in a special "topological" order convenient
for evaluation: All the input gates come first, followed by all the latches; then come the other
types of gates, whose values are computed from their predecessors. The arcs of the graph run
from each gate to its arguments, and all arguments to a gate precede that gate.

If g points to such a graph of gates, the utility field $g\text{-}outs$ points to a list of **Arc** records,
denoting "outputs" that might be used in certain applications. For example, the outputs of the
graphs created by *prod* correspond to the bits of the product of the numbers represented in the
input gates.

A special convention is used so that the routines will support partial evaluation: The *tip* fields
in the output list either point to a vertex or hold one of the constant values 0 or 1 when regarded
as an unsigned long integer.

```
#define val    x.I      /* the field containing a boolean value */
#define typ    y.I      /* the field containing the gate type */
#define alt    z.V      /* the field pointing to another related gate */
#define outs   zz.A     /* the field pointing to the list of output gates */
#define is_boolean(v)  ((unsigned long) (v) ≤ 1)      /* is a tip field constant? */
#define the_boolean(v)  ((long) (v))      /* if so, this is its value */
#define tip_value(v)   (is_boolean(v) ? the_boolean(v) : (v)→val)
#define AND    '&'
#define OR     '|'
#define NOT    '~'
#define XOR    '^'
```

⟨ gb_gates.h 1 ⟩ +≡
```
#define val    x.I        /* the definitions are repeated in the header file */
#define typ    y.I
#define alt    z.V
#define outs   zz.A
#define is_boolean(v)      ((unsigned long) (v) ≤ 1)
#define the_boolean(v)     ((long) (v))
#define tip_value(v)    (is_boolean(v) ? the_boolean(v) : (v)→val)
#define AND       '&'
#define OR        '|'
#define NOT       '~'
#define XOR       '^'
```

A: **Arc** *, GB_GRAPH §8.
Arc = **struct**, GB_GRAPH §10.
gate_eval: **long** ( ), §3.
Graph = **struct**, GB_GRAPH §20.
I: **long**, GB_GRAPH §8.
p_gates = short name, §49.

partial_gates: **Graph** *( ), §84.
print_gates: **void** ( ), §49.
prod: **Graph** *( ), §66.
risc: **Graph** *( ), §8.
risc_state: **unsigned long** [ ], §48.
run_risc: **long** ( ), §43.

tip: **Vertex** *, GB_GRAPH §10.
V: **Vertex** *, GB_GRAPH §8.
x: util, GB_GRAPH §9.
y: util, GB_GRAPH §9.
z: util, GB_GRAPH §9.
zz: util, GB_GRAPH §20.

3. Let's begin with the *gate_eval* procedure, because it is quite simple and because it illustrates the conventions just explained. Given a gate graph g and optional pointers *in_vec* and *out_vec*, the procedure *gate_eval* will assign values to each gate of g. If *in_vec* is non-null, it should point to a string of characters, each '0' or '1', that will be assigned to the first gates of the network, in order; otherwise *gate_eval* assumes that all input gates have already received appropriate values and it will not change them. New values are computed for each gate after the bits of *in_vec* have been consumed.

If *out_vec* is non-null, it should point to a memory area capable of receiving $m + 1$ characters, where m is the number of outputs of g; a string containing the respective output values will be deposited there.

If *gate_eval* encounters an unknown gate type, it terminates execution prematurely and returns the value -1. Otherwise it returns 0.

⟨ The *gate_eval* routine 3 ⟩ ≡
```
long gate_eval(g, in_vec, out_vec)
       Graph *g;        /* graph with gates as vertices */
       char *in_vec;      /* string for input values, or Λ */
       char *out_vec;     /* string for output values, or Λ */
{ register Vertex *v;      /* the current vertex of interest */
  register Arc *a;        /* the current arc of interest */
  register char t;        /* boolean value being computed */
  if (¬g) return −2;      /* no graph supplied! */
  v = g→vertices;
  if (in_vec) ⟨Read a sequence of input values from in_vec 4⟩;
  for ( ; v < g→vertices + g→n; v++) {
    switch (v→typ) {       /* branch on type of gate */
    case 'I': continue;      /* this input gate's value should be externally set */
    case 'L': t = v→alt→val; break;
  ⟨Compute the value t of a classical logic gate 6⟩;
    default: return −1;      /* unknown gate type! */
    }
    v→val = t;        /* assign the computed value */
  }
  if (out_vec) ⟨Store the sequence of output values in out_vec 5⟩;
  return 0;
}
```
This code is used in section 7.

4. ⟨ Read a sequence of input values from *in_vec* 4 ⟩ ≡
```
  while (*in_vec ∧ v < g→vertices + g→n) (v++)→val = *in_vec++ − '0';
```
This code is used in section 3.

5. ⟨ Store the sequence of output values in *out_vec* 5 ⟩ ≡
```
  {
    for (a = g→outs; a; a = a→next) *out_vec++ = '0' + tip_value(a→tip);
    *out_vec = 0;      /* terminate the string */
  }
```
This code is used in section 3.

6. ⟨ Compute the value t of a classical logic gate 6 ⟩ ≡

case AND: $t = 1$;
   for $(a = v{\to}arcs;\ a;\ a = a{\to}next)\ t\ \&= a{\to}tip{\to}val$;
   break;
case OR: $t = 0$;
   for $(a = v{\to}arcs;\ a;\ a = a{\to}next)\ t\ |= a{\to}tip{\to}val$;
   break;
case XOR: $t = 0$;
   for $(a = v{\to}arcs;\ a;\ a = a{\to}next)\ t\ \oplus= a{\to}tip{\to}val$;
   break;
case NOT: $t = 1 - v{\to}arcs{\to}tip{\to}val$;
   break;

This code is used in section 3.

7. Here now is an outline of the entire GB_GATES module, as seen by the C compiler:

#include "gb_flip.h"   /∗ we will use the GB_FLIP routines for random numbers ∗/
#include "gb_graph.h"   /∗ and we will use the GB_GRAPH data structures ∗/
   ⟨ Preprocessor definitions ⟩

   ⟨ Private variables 12 ⟩
   ⟨ Global variables 48 ⟩
   ⟨ Internal subroutines 11 ⟩
   ⟨ The *gate_eval* routine 3 ⟩
   ⟨ The *print_gates* routine 49 ⟩;
   ⟨ The *risc* routine 8 ⟩
   ⟨ The *run_risc* routine 43 ⟩
   ⟨ The *prod* routine 66 ⟩
   ⟨ The *partial_gates* routine 84 ⟩

$alt = z.V$, §2.
AND $= $ '**&**', §2.
Arc $=$ **struct**, GB_GRAPH §10.
arcs: **Arc** ∗, GB_GRAPH §9.
Graph $=$ **struct**, GB_GRAPH §20.
n: **long**, GB_GRAPH §20.
next: **Arc** ∗, GB_GRAPH §10.
NOT $= $ '**~**', §2.

OR $= $ '**|**', §2.
outs $= zz.A$, §2.
partial_gates: **Graph** ∗( ), §84.
print_gates: **void** ( ), §49.
prod: **Graph** ∗( ), §66.
risc: **Graph** ∗( ), §8.
run_risc: **long** ( ), §43.

tip: **Vertex** ∗, GB_GRAPH §10.
tip_value $= $ macro ( ), §2.
$typ = y.I$, §2.
$val = x.I$, §2.
Vertex $=$ **struct**, GB_GRAPH §9.
vertices: **Vertex** ∗, GB_GRAPH §20.
XOR $= $ '**~**', §2.

8. The RISC netlist. The subroutine call *risc*(*regs*) creates a gate graph having *regs* registers; the value of *regs* must be between 2 and 16, inclusive, otherwise *regs* is set to 16. This gate graph describes the circuitry for a small RISC computer, defined below. The total number of gates turns out to be $1400 + 115 * regs$; thus it lies between 1630 (when *regs* = 2) and 3240 (when *regs* = 16). EXCLUSIVE-OR gates are not used; the effect of xoring is obtained where needed by means of ANDs, ORs, and inverters.

If *risc* cannot do its thing, it returns Λ (NULL) and sets *panic_code* to indicate the problem. Otherwise *risc* returns a pointer to the graph.

#define *panic*(*c*) { *panic_code* = *c*; *gb_trouble_code* = 0; **return** Λ; }

⟨ The *risc* routine 8 ⟩ ≡
  Graph **risc*(*regs*)
    unsigned long *regs*;   /* number of registers supported */
  { ⟨ Local variables for *risc* 9 ⟩

    ⟨ Initialize *new_graph* to an empty graph of the appropriate size 16 ⟩;
    ⟨ Add the RISC data to *new_graph* 17 ⟩;
    if (*gb_trouble_code*) {
      gb_recycle(*new_graph*);
      panic(*alloc_fault*);   /* oops, we ran out of memory somewhere back there */
    }
    return *new_graph*;
  }

This code is used in section 7.

9. ⟨ Local variables for *risc* 9 ⟩ ≡
  Graph **new_graph*;   /* the graph constructed by *risc* */
  register long *k*, *r*;   /* all-purpose indices */

See also sections 18, 20, 25, 28, 33, 37, and 40.

This code is used in section 8.

10. This RISC machine works with 16-bit registers and 16-bit data words. It cannot write into memory, but it assumes the existence of an external read-only memory. The circuit has 16 outputs, representing the 16 bits of a memory address register. It also has 17 inputs, the last 16 of which are supposed to be set to the contents of the memory address computed on the previous cycle. Thus we can run the machine by accessing memory between calls of *gate_eval*. The first input bit, called RUN, is normally set to 1; if it is 0, the other inputs are effectively ignored and all registers and outputs will be cleared to 0. Input bits for the memory appear in "little-endian order," that is, least significant bit first; but the output bits for the memory address register appear in "big-endian order," most significant bit first.

Words read from memory are interpreted as instructions having the following format:

DST	MOD	OP	A	SRC
15 14 13 12	11 10 9 8	7 6	5 4	3 2 1 0

The SRC and A fields specify a "source" value. If A = 0, the source is SRC, treated as a 16-bit signed number between −8 and +7 inclusive. If A = 1, the source is the contents of register DST plus the (signed) value of SRC. If $A = 2$, the source is the contents of register SRC. And if $A = 3$, the source is the contents of the memory location whose address is the contents of register SRC. Thus, for example, if $DST = 3$ and $SRC = 10$, and if r3 contains 17 while r10 contains 1009, the source value will be -6 if $A = 0$, or $17 - 6 = 11$ if $A = 1$, or 1009 if $A = 2$, or the contents of memory location 1009 if $A = 3$.

The DST field specifies the number of the destination register. This register receives a new value based on its previous value and the source value, as prescribed by the operation defined in the OP and MOD fields. For example, when $OP = 0$, a general logical operation is performed, as follows: Suppose the bits of MOD are called $\mu_{11}\mu_{10}\mu_{01}\mu_{00}$ from left to right. Then if the kth bit of the destination register currently is equal to i and the kth bit of the source value is equal to j, the general logical operator changes the kth bit of the destination register to μ_{ij}. If the MOD bits are, for example, 1010, the source value is simply copied to the destination register; if $MOD = 0110$, an exclusive-or is done; if $MOD = 0011$, the destination register is complemented and the source value is effectively ignored.

The machine contains four status bits called S (sign), N (nonzero), K (carry), and V (overflow). Every general logical operation sets S equal to the sign of the new result transferred to the destination register; this is bit 15, the most significant bit. A general logical operation also sets N to 1 if any of the other 15 bits are 1, to 0 if all of the other bits are 0. Thus S and N both become zero if and only if the new result is entirely zero. Logical operations do not change the values of K and V; the latter are affected only by the arithmetic operations described below.

The status of the S and N bits can be tested by using the conditional load operator, $OP = 2$: This operation loads the source value into the destination register if and only if MOD bit $\mu_{ij} = 1$, where i and j are the current values of S and N, respectively. For example, if $MOD = 0011$, the source value is loaded if and only if $S = 0$, which means that the last value affecting S and N was greater than or equal to zero. If $MOD = 1111$, loading is always done; this option provides a way to move source to destination without affecting S or N.

A second conditional load operator, $OP = 3$, is similar, but it is used for testing the status of K and V instead of S and N. For example, a command having $MOD = 1010$, $OP = 3$, $A = 1$, and $SRC = 1$ adds the current overflow bit to the destination register. (Please take a moment to understand why this is true.)

---

alloc_fault $= -1$, GB_GRAPH §7. GB_GRAPH §14. *new_graph*: **Graph** $*$, §51.
gb_recycle: **void** ( ), GB_GRAPH §40. **Graph** $=$ **struct**, GB_GRAPH §20. *panic_code*: **long**, GB_GRAPH §5.
gb_trouble_code: **long**,

10. (continued)   We have now described all the operations except those that are performed when OP = 1. As you might expect, our machine is able to do rudimentary arithmetic. The general addition and subtraction operators belong to this final case, together with various shift operators, depending on the value of MOD.

Eight of the OP = 1 operations set the destination register to a shifted version of the source value: MOD = 0 means "shift left 1," which is equivalent to multiplying the source by 2; MOD = 1 means "cyclic shift left 1," which is the same except that it also adds the previous sign bit to the result; MOD = 2 means "shift left 4," which is equivalent to multiplying by 16; MOD = 3 means "cyclic shift left 4"; MOD = 4 means "shift right 1," which is equivalent to dividing the source by 2 and rounding down to the next lower integer if there was a remainder; MOD = 5 means "unsigned shift right 1," which is the same except that the most significant bit is always set to zero instead of retaining the previous sign; MOD = 6 means "shift right 4," which is equivalent to dividing the source by 16 and rounding down; MOD = 7 means "unsigned shift right 4." Each of these shift operations affects S and N, as in the case of logical operations. They also affect K and V, as follows: Shifting left sets K to 1 if and only if at least one of the bits shifted off the left was nonzero, and sets V to 1 if and only if the corresponding multiplication would cause overflow. Shifting right 1 sets K to the value of the bit shifted out, and sets V to 0; shifting right 4 sets K to the value of the last bit shifted out, and sets V to the logical OR of the other three lost bits. The same values of K and V arise from cyclic or unsigned shifts as from ordinary shifts.

When OP = 1 and MOD = 8, the source value is added to the destination register. This sets S, N, and V as you would expect; and it sets K to the carry you would get if you were treating the operands as 16-bit unsigned integers. Another addition operation, having MOD = 9, is similar, but the current value of K is also added to the result; in this case, the new value of N will be zero if and only if the 15 non-sign bits of the result are zero and the previous values of S and N were also zero. This means that you can use the first addition operation on the lower halves of a 32-bit number and the second operation on the upper halves, thereby obtaining a correct 32-bit result, with appropriate sign, nonzero, carry, and overflow bits set. Higher precision (48 bits, 64 bits, etc.) can be obtained in a similar way.

When OP = 1 and MOD = 10, the source value is subtracted from the destination register. Again, S, N, K, and V are set; the K value in this case represents the "borrow" bit. An auxiliary subtraction operation, having MOD = 11, subtracts also the current value of K, thereby allowing for correct 32-bit subtraction.

The operations for OP = 1 and MOD = 12, 13, and 14 are "reserved for future expansion." Actually they will never change, however, since this RISC chip is purely academic. If you check out the logic below, you will find that they simply set the destination register and the four status bits all to zero.

A final operation, called JUMP, will be explained momentarily. It has OP = 1 and MOD = 15. It does not affect S, N, K, or V.

If the RISC is made with fewer than 16 registers, the higher-numbered ones will effectively contain zero whenever their values are fetched. But if you use them as destination registers, you will set S, N, K, and V as if actual numbers were being stored.

Register 0 is different from the other 15 registers: It is the location of the current instruction. Therefore if you change the contents of register 0, you are changing the control flow of the program. If you do not change register 0, it automatically increases by 1.

Special treatment occurs when A = 3 and SRC = 0. In such a case, the normal rules given above say that the source value should be the contents of the memory location specified by register 0. But that memory location holds the current instruction; so the machine uses the *following* location instead, as a 16-bit source operand. If the contents of register 0 are not changed by such a two-word instruction, register 0 will increase by 2 instead of 1.

We have now discussed everything about the machine except the operation of the JUMP command. This command moves the source value to register 0, thereby changing the flow of control. Furthermore, if DST $\neq$ 0, it also sets register DST to the location of the instruction following the JUMP. Assembly language programmers will recognize this as a convenient way to jump to a subroutine.

Example programs can be found in the TAKE_RISC module, which includes a simple subroutine for multiplication and division.

11. A few auxiliary functions will ameliorate the task of constructing the RISC logic. First comes a routine that "christens" a new gate, assigning it a name and a type. The name is constructed from a prefix and a serial number, where the prefix indicates the current portion of logic being created.

⟨ Internal subroutines 11 ⟩ ≡
```
static Vertex *new_vert(t)
      char t;        /* the type of the new gate */
{ register Vertex *v;

   v = next_vert ++;
   if (count < 0) v→name = gb_save_string(prefix);
   else {
      sprintf(name_buf, "%s%ld", prefix, count);
      v→name = gb_save_string(name_buf);
      count ++;
   }
   v→typ = t;
   return v;
}
```
See also sections 13, 14, 15, 38, and 51.

This code is used in section 7.

12. **#define** *start_prefix*(s)   strcpy(prefix, s); count = 0
#define *numeric_prefix*(a, b)   sprintf(prefix, "%c%ld:", a, b); count = 0;

⟨ Private variables 12 ⟩ ≡
```
   static Vertex *next_vert;      /* the first vertex not yet assigned a name */
   static char prefix[5];        /* prefix string for vertex names */
   static long count;           /* serial number for vertex names */
   static char name_buf[100];     /* place to form vertex names */
```
This code is used in section 7.

---

gate_eval: **long** ( ), §3.
gb_save_string: **char** *( ),
   GB_GRAPH §35.

name: **char** *, GB_GRAPH §9.
sprintf: **int** ( ), <stdio.h>.
strcpy: **char** *( ), <string.h>.

typ = y.I, §2.
Vertex = **struct**, GB_GRAPH §9.

13. Here are some trivial routines to create gates with 2, 3, or more arguments. The arcs from such a gate to its inputs are assigned length 100. Other routines, defined below, assign length 1 to the arc between an inverter and its unique input. This convention makes the lengths of shortest paths in the resulting network a bit more interesting than they would otherwise be.

#define DELAY 100$_L$

⟨ Internal subroutines 11 ⟩ +≡

```
static Vertex *make2(t, v1, v2)
      char t;      /* the type of the new gate */
      Vertex *v1, *v2;
{ register Vertex *v = new_vert(t);

  gb_new_arc(v, v1, DELAY);
  gb_new_arc(v, v2, DELAY);
  return v;
}

static Vertex *make3(t, v1, v2, v3)
      char t;      /* the type of the new gate */
      Vertex *v1, *v2, *v3;
{ register Vertex *v = new_vert(t);

  gb_new_arc(v, v1, DELAY);
  gb_new_arc(v, v2, DELAY);
  gb_new_arc(v, v3, DELAY);
  return v;
}

static Vertex *make4(t, v1, v2, v3, v4)
      char t;      /* the type of the new gate */
      Vertex *v1, *v2, *v3, *v4;
{ register Vertex *v = new_vert(t);

  gb_new_arc(v, v1, DELAY);
  gb_new_arc(v, v2, DELAY);
  gb_new_arc(v, v3, DELAY);
  gb_new_arc(v, v4, DELAY);
  return v;
}

static Vertex *make5(t, v1, v2, v3, v4, v5)
      char t;      /* the type of the new gate */
      Vertex *v1, *v2, *v3, *v4, *v5;
{ register Vertex *v = new_vert(t);

  gb_new_arc(v, v1, DELAY);
  gb_new_arc(v, v2, DELAY);
  gb_new_arc(v, v3, DELAY);
  gb_new_arc(v, v4, DELAY);
  gb_new_arc(v, v5, DELAY);
  return v;
}
```

14. We use utility field $w.V$ to store a pointer to the complement of a gate, if that complement has been formed. This trick prevents the creation of excessive gates that are equivalent to each other. The following subroutine returns a pointer to the complement of a given gate.

#define *bar* $w.V$ /* field pointing to complement, if known to exist */
#define *even_comp*(s, v) $((s) \& 1 ? v : comp(v))$
⟨ Internal subroutines 11 ⟩ +≡
    static Vertex *∗comp*(v)
        Vertex *∗v*;
  { **register Vertex** *∗u*;

    if $(v\text{-}bar)$ **return** $v\text{-}bar$;
    $u = next\_vert\text{++}$;
    $u\text{-}bar = v$; $v\text{-}bar = u$;
    sprintf$(name\_buf, \texttt{"\%s\textasciitilde"}, v\text{-}name)$;
    $u\text{-}name = gb\_save\_string(name\_buf)$;
    $u\text{-}typ = \texttt{NOT}$;
    gb_new_arc$(u, v, 1_\text{L})$;
    return u;
  }

15. To create a gate for the EXCLUSIVE-OR of two arguments, we can either construct the OR of two ANDs or the AND of two ORs. We choose the former alternative:

⟨ Internal subroutines 11 ⟩ +≡
    static Vertex *∗make_xor*(u, v)
        Vertex *∗u, ∗v*;
  { **register Vertex** *∗t1, ∗t2*;

    $t1 = make2(\texttt{AND}, u, comp(v))$;
    $t2 = make2(\texttt{AND}, comp(u), v)$;
    return $make2(\texttt{OR}, t1, t2)$;
  }

16. OK, let's get going.

⟨ Initialize *new_graph* to an empty graph of the appropriate size 16 ⟩ ≡
    if $(regs < 2 \lor regs > 16)$ $regs = 16$;
    $new\_graph = gb\_new\_graph(1400 + 115 * regs)$;
    if $(new\_graph \equiv \Lambda)$ *panic*(no_room);   /* out of memory before we're even started */
    sprintf$(new\_graph\text{-}id, \texttt{"risc(\%lu)"}, regs)$;
    strcpy$(new\_graph\text{-}util\_types, \texttt{"ZZZIIVZZZZZZZA"})$;
    $next\_vert = new\_graph\text{-}vertices$;
This code is used in section 8.

AND = '&', §2.
gb_new_arc: **void** ( ),
  GB_GRAPH §30.
gb_new_graph: **Graph** ∗( ),
  GB_GRAPH §23.
gb_save_string: **char** ∗( ),
  GB_GRAPH §35.
id: **char** [ ], GB_GRAPH §20.
name: **char** ∗, GB_GRAPH §9.

name_buf: **static char** [ ], §12.
new_graph: **Graph** ∗, §9.
new_vert: **static Vertex** ∗( ), §11.
next_vert: **static Vertex** ∗, §12.
no_room = 1, GB_GRAPH §7.
NOT = '~', §2.
OR = '|', §2.
panic = macro ( ), §8.
regs: **unsigned long**, §8.

sprintf: **int** ( ), <stdio.h>.
strcpy: **char** ∗( ), <string.h>.
typ = $y.I$, §2.
util_types: **char** [ ], GB_GRAPH §20.
V: **Vertex** ∗, GB_GRAPH §8.
Vertex = **struct**, GB_GRAPH §9.
vertices: **Vertex** ∗, GB_GRAPH §20.
w: **util**, GB_GRAPH §9.

17. ⟨Add the RISC data to *new_graph* 17⟩ ≡
⟨Create the inputs and latches 19⟩;
⟨Create gates for instruction decoding 21⟩;
⟨Create gates for fetching the source value 22⟩;
⟨Create gates for the general logic operation 26⟩;
⟨Create gates for the conditional load operations 27⟩;
⟨Create gates for the arithmetic operations 41⟩;
⟨Create gates that bring everything together properly 29⟩;
if (*next_vert* ≠ *new_graph*⇁*vertices* + *new_graph*⇁*n*) *panic*(*impossible*);
        /∗ oops, we miscounted; this should be impossible ∗/
This code is used in section 8.

18. Internal names will make it convenient to refer to the most important gates. Here are the names of inputs and latches.

⟨Local variables for *risc* 9⟩ +≡
  Vertex ∗*run_bit*;   /∗ the RUN input ∗/
  Vertex ∗*mem*[16];   /∗ 16 bits of input from read-only memory ∗/
  Vertex ∗*prog*;   /∗ first of 10 bits in the program register ∗/
  Vertex ∗*sign*;   /∗ the latched value of S ∗/
  Vertex ∗*nonzero*;   /∗ the latched value of N ∗/
  Vertex ∗*carry*;   /∗ the latched value of K ∗/
  Vertex ∗*overflow*;   /∗ the latched value of V ∗/
  Vertex ∗*extra*;   /∗ latched status bit: are we doing an extra memory cycle? ∗/
  Vertex ∗*reg*[16];   /∗ the least-significant bit of a given register ∗/

19. **#define** *first_of*(*n*, *t*) *new_vert*(*t*); **for** (*k* = 1; *k* < *n*; *k*++) *new_vert*(*t*);
⟨Create the inputs and latches 19⟩ ≡
  strcpy(*prefix*, "RUN"); *count* = −1; *run_bit* = *new_vert*('I');
  start_prefix("M"); **for** (*k* = 0; *k* < 16; *k*++) *mem*[*k*] = *new_vert*('I');
  start_prefix("P"); *prog* = *first_of*(10, 'L');
  strcpy(*prefix*, "S"); *count* = −1; *sign* = *new_vert*('L');
  strcpy(*prefix*, "N"); *nonzero* = *new_vert*('L');
  strcpy(*prefix*, "K"); *carry* = *new_vert*('L');
  strcpy(*prefix*, "V"); *overflow* = *new_vert*('L');
  strcpy(*prefix*, "X"); *extra* = *new_vert*('L');
  for (*r* = 0; *r* < *regs*; *r*++) {
    numeric_prefix('R', *r*); *reg*[*r*] = *first_of*(16, 'L');
  }
This code is used in section 17.

20. The order of evaluation of function arguments is not defined in C, so we introduce a few macros that force left-to-right order.

#define *do2*(*result*, *t*, *v1*, *v2*)
        { *t1* = *v1*; *t2* = *v2*;
            result = *make2*(*t*, *t1*, *t2*); }
#define *do3*(*result*, *t*, *v1*, *v2*, *v3*)
          { *t1* = *v1*; *t2* = *v2*; *t3* = *v3*;
              result = *make3*(*t*, *t1*, *t2*, *t3*); }

```
#define  do4(result, t, v1, v2, v3, v4)
            { t1 = v1; t2 = v2; t3 = v3; t4 = v4;
              result = make4(t, t1, t2, t3, t4); }
#define  do5(result, t, v1, v2, v3, v4, v5)
            { t1 = v1; t2 = v2; t3 = v3; t4 = v4; t5 = v5;
              result = make5(t, t1, t2, t3, t4, t5); }
```

⟨ Local variables for *risc* 9 ⟩ +≡
  Vertex *t1, *t2, *t3, *t4, *t5;   /* temporary holds to force evaluation order */
  Vertex *tmp[16];   /* additional holding places for partial results */
  Vertex *imm;   /* is the source value immediate (a given constant)? */
  Vertex *rel;   /* is the source value relative to the current destination register? */
  Vertex *dir;   /* should the source value be fetched directly from a source register? */
  Vertex *ind;   /* should the source value be fetched indirectly from memory? */
  Vertex *op;   /* least significant bit of OP */
  Vertex *cond;   /* most significant bit of OP */
  Vertex *mod[4];   /* the MOD bits */
  Vertex *dest[4];   /* the DEST bits */

21.   The sixth line of the program here can be translated into the logic equation

$$op = (extra \land prog) \lor (\overline{extra} \land mem[6]).$$

Once you see why, you'll be able to read the rest of this curious code.

⟨ Create gates for instruction decoding 21 ⟩ ≡
  start_prefix("D");
  do3(*imm*, AND, *comp*(*extra*), *comp*(*mem*[4]), *comp*(*mem*[5]));   /* A = 0 */
  do3(*rel*, AND, *comp*(*extra*), *mem*[4], *comp*(*mem*[5]));   /* A = 1 */
  do3(*dir*, AND, *comp*(*extra*), *comp*(*mem*[4]), *mem*[5]);   /* A = 2 */
  do3(*ind*, AND, *comp*(*extra*), *mem*[4], *mem*[5]);   /* A = 3 */
  do2(*op*, OR, *make2*(AND, *extra*, *prog*), *make2*(AND, *comp*(*extra*), *mem*[6]));
  do2(*cond*, OR, *make2*(AND, *extra*, *prog* + 1), *make2*(AND, *comp*(*extra*), *mem*[7]));
  for (*k* = 0; *k* < 4; *k*++) {
    do2(*mod*[*k*], OR, *make2*(AND, *extra*, *prog* + 2 + *k*), *make2*(AND, *comp*(*extra*), *mem*[8 + *k*]));
    do2(*dest*[*k*], OR, *make2*(AND, *extra*, *prog* + 6 + *k*), *make2*(AND, *comp*(*extra*), *mem*[12 + *k*]));
  }
This code is used in section 17.

AND = '&', §2.
comp: **static Vertex** *(), §14.
count: **static long**, §12.
impossible = 90, GB_GRAPH §7.
k: **register long**, §9.
make2: **static Vertex** *(), §13.
make3: **static Vertex** *(), §13.
make4: **static Vertex** *(), §13.
make5: **static Vertex** *(), §13.

n: **long**, GB_GRAPH §20.
new_graph: **Graph** *, §9.
new_vert: **static Vertex** *(), §11.
next_vert: **static Vertex** *, §12.
numeric_prefix = macro (), §12.
OR = '|', §2.
panic = macro (), §8.
prefix: **static char** [], §12.

r: **register long**, §9.
regs: **unsigned long**, §8.
result: **Vertex** *[], §25.
risc: **Graph** *(), §8.
start_prefix = macro (), §8.
strcpy: **char** *(), <string.h>.
Vertex = **struct**, GB_GRAPH §9.
vertices: **Vertex** *, GB_GRAPH §20.

22. ⟨ Create gates for fetching the source value 22 ⟩ ≡
   start_prefix (`"F"`);
   ⟨ Set *old_dest* to the present value of the destination register 23 ⟩;
   ⟨ Set *old_src* to the present value of the source register 24 ⟩;
   ⟨ Set *inc_dest* to *old_dest* plus SRC 39 ⟩;
   for ($k = 0$; $k < 16$; $k{+}{+}$)
      do4 (*source*[k], OR, *make2* (AND, *imm*, *mem*[$k < 4$? k : 3]), *make2* (AND, *rel*, *inc_dest*[k]),
         make2 (AND, *dir*, *old_src*[k]), *make2* (AND, *extra*, *mem*[k])));

This code is used in section 17.

23. Here and in the immediately following section we create OR gates *old_dest*[k] and *old_src*[k]
that might have as many as 16 inputs. (The actual number of inputs is *regs*.) All other gates in
the network will have at most five inputs.

⟨ Set *old_dest* to the present value of the destination register 23 ⟩ ≡
   for ($r = 0$; $r < regs$; $r{+}{+}$)
      do4 (*dest_match*[r], AND, *even_comp* (r, *dest*[0]), *even_comp* ($r \gg 1$, *dest*[1]),
         even_comp ($r \gg 2$, *dest*[2]), *even_comp* ($r \gg 3$, *dest*[3]));
   for ($k = 0$; $k < 16$; $k{+}{+}$) {
      for ($r = 0$; $r < regs$; $r{+}{+}$)
         tmp[r] = *make2* (AND, *dest_match*[r], *reg*[r] + k);
      old_dest[k] = *new_vert* (OR);
      for ($r = 0$; $r < regs$; $r{+}{+}$) *gb_new_arc* (*old_dest*[k], *tmp*[r], DELAY);
   }

This code is used in section 22.

24. ⟨ Set *old_src* to the present value of the source register 24 ⟩ ≡
   for ($k = 0$; $k < 16$; $k{+}{+}$) {
      for ($r = 0$; $r < regs$; $r{+}{+}$)
         do5 (*tmp*[r], AND, *reg*[r] + k, *even_comp* (r, *mem*[0]), *even_comp* ($r \gg 1$, *mem*[1]), *even_comp* ($r \gg 2$,
            mem[2]), *even_comp* ($r \gg 3$, *mem*[3]));
      old_src[k] = *new_vert* (OR);
      for ($r = 0$; $r < regs$; $r{+}{+}$) *gb_new_arc* (*old_src*[k], *tmp*[r], DELAY);
   }

This code is used in section 22.

25. ⟨ Local variables for *risc* 9 ⟩ +≡
   Vertex *∗dest_match*[16];   /∗ *dest_match*[r] ≡ 1 iff DST = r ∗/
   Vertex *∗old_dest*[16];   /∗ contents of destination register before operation ∗/
   Vertex *∗old_src*[16];   /∗ contents of source register before operation ∗/
   Vertex *∗inc_dest*[16];   /∗ *old_dest* plus the SRC field ∗/
   Vertex *∗source*[16];   /∗ source value for the operation ∗/
   Vertex *∗log*[16];   /∗ result of general logic operation ∗/
   Vertex *∗shift*[18];   /∗ result of shift operation, with carry and overflow ∗/
   Vertex *∗sum*[18];   /∗ *old_dest* plus *source* plus optional carry ∗/
   Vertex *∗diff*[18];   /∗ *old_dest* minus *source* minus optional borrow ∗/
   Vertex *∗next_loc*[16];   /∗ contents of register 0, plus 1 ∗/
   Vertex *∗next_next_loc*[16];   /∗ contents of register 0, plus 2 ∗/
   Vertex *∗result*[18];   /∗ result of operating on *old_dest* and *source* ∗/

26.   ⟨ Create gates for the general logic operation 26 ⟩ ≡
  start_prefix ("L");
  for ($k = 0$; $k < 16$; $k{+}{+}$)
    do4 (*log* [k], OR,
        make3 (AND, *mod* [0], *comp* (*old_dest* [k]), *comp* (*source* [k])),
        make3 (AND, *mod* [1], *comp* (*old_dest* [k]), *source* [k]),
        make3 (AND, *mod* [2], *old_dest* [k], *comp* (*source* [k])),
        make3 (AND, *mod* [3], *old_dest* [k], *source* [k]));
This code is used in section 17.

27.   ⟨ Create gates for the conditional load operations 27 ⟩ ≡
  start_prefix ("C");
  do4 (*tmp* [0], OR,
      make3 (AND, *mod* [0], *comp* (*sign*), *comp* (*nonzero*)),
      make3 (AND, *mod* [1], *comp* (*sign*), *nonzero*),
      make3 (AND, *mod* [2], *sign*, *comp* (*nonzero*)),
      make3 (AND, *mod* [3], *sign*, *nonzero*));
  do4 (*tmp* [1], OR,
      make3 (AND, *mod* [0], *comp* (*carry*), *comp* (*overflow*)),
      make3 (AND, *mod* [1], *comp* (*carry*), *overflow*),
      make3 (AND, *mod* [2], *carry*, *comp* (*overflow*)),
      make3 (AND, *mod* [3], *carry*, *overflow*));
  do3 (*change*, OR, *comp* (*cond*), *make2* (AND, *tmp* [0], *comp* (*op*)), *make2* (AND, *tmp* [1], *op*));
This code is used in section 17.

28.   ⟨ Local variables for *risc* 9 ⟩ +≡
  Vertex **change*;   /* is the destination register supposed to change? */

AND = '&', §2.
carry: **Vertex** *, §18.
comp: **static Vertex** *( ), §14.
cond: **Vertex** *, §20.
DELAY = 100 L, §13.
dest: **Vertex** *[ ], §20.
dir: **Vertex** *, §20.
do3 = macro ( ), §20.
do4 = macro ( ), §20.
do5 = macro ( ), §20.
even_comp = macro ( ), §14.
extra: **Vertex** *, §18.

gb_new_arc: **void** ( ),
  GB_GRAPH §30.
imm: **Vertex** *, §20.
k: **register long**, §9.
make2: **static Vertex** *( ), §13.
make3: **static Vertex** *( ), §13.
mem: **Vertex** *[ ], §18.
mod: **Vertex** *[ ], §20.
new_vert: **static Vertex** *( ), §11.
nonzero: **Vertex** *, §18.
op: **Vertex** *, §20.

OR = '|', §2.
overflow: **Vertex** *, §18.
r: **register long**, §9.
reg: **Vertex** *[ ], §18.
regs: **unsigned long**, §8.
rel: **Vertex** *, §20.
risc: **Graph** *( ), §8.
sign: **Vertex** *, §18.
start_prefix = macro ( ), §12.
tmp: **Vertex** *[ ], §20.
Vertex = struct, GB_GRAPH §9.

29. Hardware is like software except that it performs all the operations all the time and then selects only the results it needs. (If you think about it, this is a profound observation about economics, society, and nature. Gosh.)

⟨ Create gates that bring everything together properly 29 ⟩ ≡
    start_prefix ("Z");
    ⟨ Create gates for the *next_loc* and *next_next_loc* bits 30 ⟩;
    ⟨ Create gates for the *result* bits 31 ⟩;
    ⟨ Create gates for the new values of registers 1 to *regs* 34 ⟩;
    ⟨ Create gates for the new values of S, N, K, and V 35 ⟩;
    ⟨ Create gates for the new values of the program register and *extra* 32 ⟩;
    ⟨ Create gates for the new values of register 0 and the memory address register 36 ⟩;
This code is used in section 17.

30. ⟨ Create gates for the *next_loc* and *next_next_loc* bits 30 ⟩ ≡
    $next\_loc[0] = comp(reg[0]);$ $next\_next\_loc[0] = reg[0];$
    $next\_loc[1] = make\_xor(reg[0] + 1, reg[0]);$ $next\_next\_loc[1] = comp(reg[0] + 1);$
    for $(t5 = reg[0] + 1, k = 2;\ k < 16;\ t5 = make2(\text{AND}, t5, reg[0] + k\text{++}))$ {
        $next\_loc[k] = make\_xor(reg[0] + k, make2(\text{AND}, reg[0], t5));$
        $next\_next\_loc[k] = make\_xor(reg[0] + k, t5);$
    }
This code is used in section 29.

31. ⟨ Create gates for the *result* bits 31 ⟩ ≡
    $jump = make5(\text{AND}, op, mod[0], mod[1], mod[2], mod[3]);$ /* assume *cond* = 0 */
    for $(k = 0;\ k < 16;\ k\text{++})$ {
        $do5(result[k], \text{OR},$
            $make2(\text{AND}, comp(op), log[k]),$
            $make2(\text{AND}, jump, next\_loc[k]),$
            $make3(\text{AND}, op, comp(mod[3]), shift[k]),$
            $make5(\text{AND}, op, mod[3], comp(mod[2]), comp(mod[1]), sum[k]),$
            $make5(\text{AND}, op, mod[3], comp(mod[2]), mod[1], diff[k]));$
        $do2(result[k], \text{OR},$
            $make3(\text{AND}, cond, change, source[k]),$
            $make2(\text{AND}, comp(cond), result[k]));$
    }
    for $(k = 16;\ k < 18;\ k\text{++})$ /* carry and overflow bits of the result */
        $do3(result[k], \text{OR},$
            $make3(\text{AND}, op, comp(mod[3]), shift[k]),$
            $make5(\text{AND}, op, mod[3], comp(mod[2]), comp(mod[1]), sum[k]),$
            $make5(\text{AND}, op, mod[3], comp(mod[2]), mod[1], diff[k]));$
This code is used in section 29.

32. The program register *prog* and the *extra* bit are needed for the case when we must spend an extra cycle to fetch a word from memory. On the first cycle, *ind* is true, so a "result" is calculated but not actually used. On the second cycle, *extra* is true.

A slight optimization has been introduced in order to make the circuit a bit more interesting: If a conditional load instruction occurs with indirect addressing and a false condition, the extra cycle is not taken. (The *next_next_loc* values were computed for this reason.)

#define *latchit*(*u, latch*) (*latch*)→*alt* = *make2*(AND, *u, run_bit*)
    /* *u* & *run_bit* is new value for *latch* */
⟨ Create gates for the new values of the program register and *extra* 32 ⟩ ≡
   for (*k* = 0; *k* < 10; *k*++) *latchit*(*mem*[*k* + 6], *prog* + *k*);
   do2(*nextra*, OR, *make2*(AND, *ind*, *comp*(*cond*)), *make2*(AND, *ind*, *change*));
   latchit(*nextra*, *extra*);
   nzs = *make4*(OR, *mem*[0], *mem*[1], *mem*[2], *mem*[3]);
   nzd = *make4*(OR, *dest*[0], *dest*[1], *dest*[2], *dest*[3]);
This code is used in section 29.

33. ⟨ Local variables for *risc* 9 ⟩ +≡
   Vertex *jump; /* is this command a JUMP, assuming *cond* is false? */
   Vertex *nextra; /* must we take an extra cycle? */
   Vertex *nzs; /* is the SRC field nonzero? */
   Vertex *nzd; /* is the DST field nonzero? */

34. ⟨ Create gates for the new values of registers 1 to *regs* 34 ⟩ ≡
   t5 = *make2*(AND, *change*, *comp*(*ind*)); /* should destination register change? */
   for (*r* = 1; *r* < *regs*; *r*++) {
    t4 = *make2*(AND, *t5*, *dest_match*[*r*]); /* should register *r* change? */
    for (*k* = 0; *k* < 16; *k*++) {
     do2(*t3*, OR, *make2*(AND, *t4*, *result*[*k*]), *make2*(AND, *comp*(*t4*), *reg*[*r*] + *k*));
     latchit(*t3*, *reg*[*r*] + *k*);
    }
   }
This code is used in section 29.

---

alt = *z.V*, §2.
AND = '&', §2.
change: **Vertex** *, §28.
comp: **static Vertex** *( ), §14.
cond: **Vertex** *, §20.
dest: **Vertex** *[ ], §20.
dest_match: **Vertex** *[ ], §25.
diff: **Vertex** *[ ], §25.
do2 = macro ( ), §20.
do3 = macro ( ), §20.
do5 = macro ( ), §20.
extra: **Vertex** *, §18.
ind: **Vertex** *, §20.
k: **register long**, §9.

log: **Vertex** *[ ], §25.
make_xor: **static Vertex** *( ), §15.
make2: **static Vertex** *( ), §13.
make3: **static Vertex** *( ), §13.
make4: **static Vertex** *( ), §13.
make5: **static Vertex** *( ), §13.
mem: **Vertex** *[ ], §18.
mod: **Vertex** *[ ], §20.
next_loc: **Vertex** *[ ], §25.
next_next_loc: **Vertex** *[ ], §25.
op: **Vertex** *, §20.
OR = '|', §2.
prog: **Vertex** *, §18.
r: **register long**, §9.

reg: **Vertex** *[ ], §18.
regs: **unsigned long**, §8.
result: **Vertex** *[ ], §25.
risc: **Graph** *( ), §8.
run_bit: **Vertex** *, §18.
shift: **Vertex** *[ ], §25.
source: **Vertex** *[ ], §25.
start_prefix = macro ( ), §12.
sum: **Vertex** *[ ], §25.
t3: **Vertex** *, §20.
t4: **Vertex** *, §20.
t5: **Vertex** *, §20.
Vertex = **struct**, GB_GRAPH §9.

35. ⟨ Create gates for the new values of S, N, K, and V 35 ⟩ ≡
  do4 (*t5*, OR,
      make2 (AND, *sign*, *cond*),
      make2 (AND, *sign*, *jump*),
      make2 (AND, *sign*, *ind*),
      make4 (AND, *result*[15], *comp*(*cond*), *comp*(*jump*), *comp*(*ind*)));
  latchit (*t5*, *sign*);
  do4 (*t5*, OR,
      make4 (OR, *result*[0], *result*[1], *result*[2], *result*[3]),
      make4 (OR, *result*[4], *result*[5], *result*[6], *result*[7]),
      make4 (OR, *result*[8], *result*[9], *result*[10], *result*[11]),
      make4 (OR, *result*[12], *result*[13], *result*[14],
              make5 (AND, *make2* (OR, *nonzero*, *sign*), *op*, *mod*[0], *comp*(*mod*[2]), *mod*[3])));
  do4 (*t5*, OR,
      make2 (AND, *nonzero*, *cond*),
      make2 (AND, *nonzero*, *jump*),
      make2 (AND, *nonzero*, *ind*),
      make4 (AND, *t5*, *comp*(*cond*), *comp*(*jump*), *comp*(*ind*)));
  latchit (*t5*, *nonzero*);
  do5 (*t5*, OR,
      make2 (AND, *overflow*, *cond*),
      make2 (AND, *overflow*, *jump*),
      make2 (AND, *overflow*, *comp*(*op*)),
      make2 (AND, *overflow*, *ind*),
      make5 (AND, *result*[17], *comp*(*cond*), *comp*(*jump*), *comp*(*ind*), *op*));
  latchit (*t5*, *overflow*);
  do5 (*t5*, OR,
      make2 (AND, *carry*, *cond*),
      make2 (AND, *carry*, *jump*),
      make2 (AND, *carry*, *comp*(*op*)),
      make2 (AND, *carry*, *ind*),
      make5 (AND, *result*[16], *comp*(*cond*), *comp*(*jump*), *comp*(*ind*), *op*));
  latchit (*t5*, *carry*);
This code is used in section 29.

36. As usual, we have left the hardest case for last, hoping that we will have learned enough tricks to handle it when the time of reckoning finally arrives.

The most subtle part of the logic here is perhaps the case of a JUMP command with $A = 3$. We want to increase register 0 by 1 during the first cycle of such a command, if $SRC = 0$, so that the *result* will be correct on the next cycle.

⟨ Create gates for the new values of register 0 and the memory address register 36 ⟩ ≡
  skip = *make2* (AND, *cond*, *comp*(*change*));   /* false conditional? */
  hop = *make2* (AND, *comp*(*cond*), *jump*);   /* JUMP command? */
  do4 (*normal*, OR,
      make2 (AND, *skip*, *comp*(*ind*)),
      make2 (AND, *skip*, *nzs*),
      make3 (AND, *comp*(*skip*), *ind*, *comp*(*nzs*)),
      make3 (AND, *comp*(*skip*), *comp*(*hop*), *nzd*));

```
special = make3 (AND, comp(skip), ind, nzs);
for (k = 0; k < 16; k++) {
    do4 (t5 , OR,
            make2 (AND, normal, next_loc[k]),
            make4 (AND, skip, ind, comp(nzs), next_next_loc[k]),
            make3 (AND, hop, comp(ind), source[k]),
            make5 (AND, comp(skip), comp(hop), comp(ind), comp(nzd), result[k]));
    do2 (t4 , OR,
            make2 (AND, special, reg[0] + k),
            make2 (AND, comp(special), t5 ));
    latchit (t4 , reg[0] + k);
    do2 (t4 , OR,
            make2 (AND, special, old_src[k]),
            make2 (AND, comp(special), t5 ));
    { register Arc *a = gb_virgin_arc( );
        a→tip = make2 (AND, t4 , run_bit);
        a→next = new_graph→outs;
        new_graph→outs = a;      /* pointer to memory address bit */
    }
}            /* arcs for output bits will appear in big-endian order */
```
This code is used in section 29.

37. ⟨ Local variables for *risc* 9 ⟩ +≡
 Vertex **skip*; /* are we skipping a conditional load operation? */
 Vertex **hop*; /* are we doing a JUMP? */
 Vertex **normal*; /* is this a case where register 0 is simply incremented? */
 Vertex **special*;
  /* is this a case where register 0 and the memory address register will not coincide? */

AND = '&', §2.
Arc = **struct**, GB_GRAPH §10.
carry: **Vertex** *, §18.
change: **Vertex** *, §28.
comp: **static Vertex** *( ), §14.
cond: **Vertex** *, §20.
do2 = macro ( ), §20.
do4 = macro ( ), §20.
do5 = macro ( ), §20.
gb_virgin_arc: **Arc** *( ),
 GB_GRAPH §29.
ind: **Vertex** *, §20.
jump: **Vertex** *, §33.
k: **register long**, §9.

latchit = macro ( ), §32.
make2: **static Vertex** *( ), §13.
make3: **static Vertex** *( ), §13.
make4: **static Vertex** *( ), §13.
make5: **static Vertex** *( ), §13.
mod: **Vertex** *[], §20.
new_graph: **Graph** *, §9.
next: **Arc** *, GB_GRAPH §10.
next_loc: **Vertex** *[], §25.
next_next_loc: **Vertex** *[], §25.
nonzero: **Vertex** *, §18.
nzd: **Vertex** *, §33.
nzs: **Vertex** *, §33.
old_src: **Vertex** *[], §25.

op: **Vertex** *, §20.
OR = '|', §2.
outs = zz.A, §2.
overflow: **Vertex** *, §18.
reg: **Vertex** *[], §18.
result: **Vertex** *[], §25.
risc: **Graph** *( ), §8.
run_bit: **Vertex** *, §18.
sign: **Vertex** *, §18.
source: **Vertex** *[], §25.
t4 : **Vertex** *, §20.
t5 : **Vertex** *, §20.
tip: **Vertex** *, GB_GRAPH §10.
Vertex = **struct**, GB_GRAPH §9.

38. Serial addition. We haven't yet specified the parts of *risc* that deal with addition and subtraction; somehow, those parts wanted to be separate from the rest. To complete our mission, we will use subroutine calls of the form '*make_adder*$(n, x, y, z, carry, add)$', where x and y are n-bit arrays of input gates and z is an $(n+1)$-bit array of output gates. If $add \neq 0$, the subroutine computes $x + y$, otherwise it computes $x - y$. If $carry \neq 0$, the *carry* gate is effectively added to y before the operation.

A simple n-stage serial scheme, which reduces the problem of n-bit addition to $(n - 1)$-bit addition, is adequate for our purposes here. (A parallel adder, which gains efficiency by reducing the problem size from n to n/ϕ, can be found in the *prod* routine below.)

The handy identity $x - y = \overline{\overline{x} + y}$ is used to reduce subtraction to addition.

⟨ Internal subroutines 11 ⟩ +≡
```
    static void make_adder(n, x, y, z, carry, add)
        unsigned long n;       /* number of bits */
        Vertex *x[], *y[];       /* input gates */
        Vertex *z[];       /* output gates */
        Vertex *carry;       /* add this to y, unless it's null */
        char add;       /* should we add or subtract? */
    { register long k;
      Vertex *t1, *t2, *t3, *t4;       /* temporary storage used by do4 */
      if (¬carry) {
        z[0] = make_xor(x[0], y[0]);
        carry = make2(AND, even_comp(add, x[0]), y[0]);
        k = 1;
      } else  k = 0;
      for ( ; k < n; k++) {
        comp(x[k]);  comp(y[k]);  comp(carry);       /* generate inverse gates */
        do4(z[k], OR,
              make3(AND, x[k], comp(y[k]), comp(carry)),
              make3(AND, comp(x[k]), y[k], comp(carry)),
              make3(AND, comp(x[k]), comp(y[k]), carry),
              make3(AND, x[k], y[k], carry));
        do3(carry, OR,
              make2(AND, even_comp(add, x[k]), y[k]),
              make2(AND, even_comp(add, x[k]), carry),
              make2(AND, y[k], carry));
      }
      z[n] = carry;
    }
```

39. OK, now we can add. What good does that do us? In the first place, we need a 4-bit adder to compute the least significant bits of *old_dest* + SRC. The other 12 bits of that sum are simpler.

⟨ Set *inc_dest* to *old_dest* plus SRC 39 ⟩ ≡
```
    make_adder(4 L, old_dest, mem, inc_dest, Λ, 1);
    up = make2(AND, inc_dest[4], comp(mem[3]));       /* remaining bits must increase */
    down = make2(AND, comp(inc_dest[4]), mem[3]);       /* remaining bits must decrease */
    for (k = 4; ; k++) {
      comp(up);  comp(down);
```

```
do3 (inc_dest[k], OR,
    make2 (AND, comp(old_dest[k]), up),
    make2 (AND, comp(old_dest[k]), down),
    make3 (AND, old_dest[k], comp(up), comp(down)));
if (k < 15) {
    up = make2 (AND, up, old_dest[k]);
    down = make2 (AND, down, comp(old_dest[k]));
} else break;
}
```

This code is used in section 22.

40. ⟨ Local variables for *risc* 9 ⟩ +≡
Vertex *up, *down;   /∗ gates used when computing *inc_dest* ∗/

41. In the second place, we need a 16-bit adder and a 16-bit subtracter for the four addition/subtraction commands.
⟨ Create gates for the arithmetic operations 41 ⟩ ≡
```
start_prefix ("A");
⟨ Create gates for the shift operations 42 ⟩;
make_adder (16 L, old_dest, source, sum, make2 (AND, carry, mod[0]), 1);      /∗ adder ∗/
make_adder (16 L, old_dest, source, diff, make2 (AND, carry, mod[0]), 0);     /∗ subtracter ∗/
do2 (sum[17], OR,
    make3 (AND, old_dest[15], source[15], comp(sum[15])),
    make3 (AND, comp(old_dest[15]), comp(source[15]), sum[15]));      /∗ overflow ∗/
do2 (diff[17], OR,
    make3 (AND, old_dest[15], comp(source[15]), comp(diff[15])),
    make3 (AND, comp(old_dest[15]), source[15], diff[15]));      /∗ overflow ∗/
```
This code is used in section 17.

AND = '&', §2.
carry: **Vertex** ∗, §18.
comp: **static Vertex** ∗( ), §14.
diff: **Vertex** ∗[], §25.
do2 = macro ( ), §20.
do3 = macro ( ), §20.
do4 = macro ( ), §20.
even_comp = macro ( ), §14.

inc_dest: **Vertex** ∗[], §25.
k: **register long**, §9.
make_xor: **static Vertex** ∗( ), §15.
make2: **static Vertex** ∗( ), §13.
make3: **static Vertex** ∗( ), §13.
mem: **Vertex** ∗[], §18.
mod: **Vertex** ∗[], §20.
old_dest: **Vertex** ∗[], §25.

OR = '|', §2.
prod: **Graph** ∗( ), §66.
risc: **Graph** ∗( ), §8.
source: **Vertex** ∗[], §25.
start_prefix = macro ( ), §12.
sum: **Vertex** ∗[], §25.
Vertex = **struct**, GB_GRAPH §9.

42. ⟨ Create gates for the shift operations 42 ⟩ ≡

  for (*k* = 0; *k* < 16; *k*++)

    do4 (*shift* [*k*], OR,

        (*k* ≡ 0 ? *make4* (AND, *source* [15], *mod* [0], *comp* (*mod* [1]), *comp* (*mod* [2])) :

                make3 (AND, *source* [*k* − 1], *comp* (*mod* [1]), *comp* (*mod* [2])))),

        (*k* < 4 ? *make4* (AND, *source* [*k* + 12], *mod* [0], *mod* [1], *comp* (*mod* [2])) :

                make3 (AND, *source* [*k* − 4], *mod* [1], *comp* (*mod* [2])))),

        (*k* ≡ 15 ? *make4* (AND, *source* [15], *comp* (*mod* [0]), *comp* (*mod* [1]), *mod* [2]) :

                make3 (AND, *source* [*k* + 1], *comp* (*mod* [1]), *mod* [2])),

        (*k* > 11 ? *make4* (AND, *source* [15], *comp* (*mod* [0]), *mod* [1], *mod* [2]) :

                make3 (AND, *source* [*k* + 4], *mod* [1], *mod* [2]))));

  do4 (*shift* [16], OR,

    make2 (AND, *comp* (*mod* [2]), *source* [15]),

    make3 (AND, *comp* (*mod* [2]), *mod* [1], *make3* (OR, *source* [14], *source* [13], *source* [12])),

    make3 (AND, *mod* [2], *comp* (*mod* [1]), *source* [0]),

    make3 (AND, *mod* [2], *mod* [1], *source* [3]));   /* "carry" */

  do3 (*shift* [17], OR,

    make3 (AND, *comp* (*mod* [2]), *comp* (*mod* [1]), *make_xor* (*source* [15], *source* [14])),

    make4 (AND, *comp* (*mod* [2]), *mod* [1],

          make5 (OR, *source* [15], *source* [14], *source* [13], *source* [12], *source* [11]),

          make5 (OR, *comp* (*source* [15]), *comp* (*source* [14]), *comp* (*source* [13]),

                comp (*source* [12]), *comp* (*source* [11])))),

    make3 (AND, *mod* [2], *mod* [1], *make3* (OR, *source* [0], *source* [1], *source* [2])));   /* "overflow" */

This code is used in section 41.

43. RISC management. The *run_risc* procedure takes a gate graph output by *risc* and simulates its behavior, given the contents of its read-only memory. (See the demonstration program TAKE_RISC, which appears in a module by itself, for a typical illustration of how *run_risc* might be used.)

This procedure clears the simulated machine and begins executing the program that starts at address 0. It stops when it gets to an address greater than the size of read-only memory supplied. One way to stop it is therefore to execute a command such as #0f00, which will transfer control to location #ffff; even better is #0f8f, which transfers to location #ffff without changing the status of S and N. However, if the given read-only memory contains a full set of 2^{16} words, *run_risc* will never stop.

When *run_risc* does stop, it returns 0 and puts the final contents of the simulated registers into the global array *risc_state*. Or, if g was not a decent graph, *run_risc* returns a negative value and leaves *risc_state* untouched.

⟨ The *run_risc* routine 43 ⟩ ≡
```
long run_risc(g, rom, size, trace_regs)
    Graph *g;        /* graph output by risc */
    unsigned long rom[];       /* contents of read-only memory */
    unsigned long size;      /* length of rom vector */
    unsigned long trace_regs;      /* if nonzero, this many registers will be traced */
{ register unsigned long l;      /* memory address */
  register unsigned long m;       /* memory or register contents */
  register Vertex *v;       /* the current gate of interest */
  register Arc *a;       /* the current output list element of interest */
  register long k, r;       /* general-purpose indices */
  long x, s, n, c, o;       /* status bits */

  if (trace_regs) ⟨Print a headline 44⟩;
  r = gate_eval(g, "0", Λ);       /* reset the RISC by turning off the RUN bit */
  if (r < 0) return r;       /* not a valid gate graph! */
  g→vertices→val = 1;       /* turn the RUN bit on */
  while (1) {
    for (a = g→outs, l = 0; a; a = a→next) l = 2 * l + a→tip→val;       /* set l = memory address */
    if (trace_regs) ⟨Print register contents 46⟩;
    if (l ≥ size) break;       /* stop if memory check occurs */
    for (v = g→vertices + 1, m = rom[l]; v ≤ g→vertices + 16; v++, m ≫= 1) v→val = m & 1;
        /* store bits of memory word in the input gates */
    gate_eval(g, Λ, Λ);       /* do another RISC cycle */
  }
  if (trace_regs) ⟨Print a footline 45⟩;
  ⟨Dump the register contents into risc_state 47⟩;
  return 0;
}
```
This code is used in section 7.

44. If tracing is requested, we write on the standard output file.

⟨ Print a headline 44 ⟩ ≡
```
{
    for (r = 0; r < trace_regs; r++) printf ("␣r%-2ld␣",r);       /* register names */
    printf ("␣P␣XSNKV␣MEM\n");       /* prog, extra, status bits, memory */
}
```
This code is used in section 43.

45. ⟨ Print a footline 45 ⟩ ≡
```
printf ("Execution␣terminated␣with␣memory␣address␣%04lx.\n", l);
```
This code is used in section 43.

46. Here we peek inside the circuit to see what values are about to be latched.

⟨ Print register contents 46 ⟩ ≡
```
{
    for (r = 0; r < trace_regs; r++) {
        v = g⃗vertices + (16 * r + 47);       /* most significant bit of register r */
        m = 0;
        if (v⃗typ ≡ 'L')
            for (k = 0, m = 0; k < 16; k++, v--)  m = 2 * m + v⃗alt⃗val;
        printf ("%04lx␣", m);
    }
    for (k = 0, m = 0, v = g⃗vertices + 26; k < 10; k++, v--)  m = 2 * m + v⃗alt⃗val;       /* prog */
    x = (g⃗vertices + 31)⃗alt⃗val;       /* extra */
    s = (g⃗vertices + 27)⃗alt⃗val;       /* sign */
    n = (g⃗vertices + 28)⃗alt⃗val;       /* nonzero */
    c = (g⃗vertices + 29)⃗alt⃗val;       /* carry */
    o = (g⃗vertices + 30)⃗alt⃗val;       /* overflow */
    printf ("%03lx%c%c%c%c%c␣", m ≪ 2, x ? 'X' : '.', s ? 'S' : '.', n ? 'N' : '.', c ? 'K' : '.',
            o ? 'V' : '.');
    if (l ≥ size)  printf ("????\n");
    else  printf ("%04lx\n", rom[l]);
}
```
This code is used in section 43.

---

alt = *z.V*, §2.
Arc = **struct**, GB_GRAPH §10.
carry: **Vertex** *, §18.
extra: **Vertex** *, §18.
gate_eval: **long** ( ), §3.
Graph = **struct**, GB_GRAPH §20.
next: **Arc** *, GB_GRAPH §10.

nonzero: **Vertex** *, §18.
outs = *zz.A*, §2.
overflow: **Vertex** *, §18.
printf: **int** ( ), <stdio.h>.
prog: **Vertex** *, §18.
risc: **Graph** *( ), §8.
risc_state: **unsigned long** [ ], §48.

sign: **Vertex** *, §18.
tip: **Vertex** *, GB_GRAPH §10.
typ = *y.I*, §2.
val = *x.I*, §2.
Vertex = **struct**, GB_GRAPH §9.
vertices: **Vertex** *, GB_GRAPH §20.

47. ⟨ Dump the register contents into *risc_state* 47 ⟩ ≡

   for $(r = 0; \ r < 16; \ r{+}{+})$ {

      $v = g{\rightarrow}vertices + (16 * r + 47);$ /* most significant bit of register r */

      $m = 0;$

      if $(v{\rightarrow}typ \equiv \text{'L'})$

         for $(k = 0, m = 0; \ k < 16; \ k{+}{+}, v{-}{-}) \ m = 2 * m + v{\rightarrow}alt{\rightarrow}val;$

      $risc\_state[r] = m;$

   }

   for $(k = 0, m = 0, v = g{\rightarrow}vertices + 26; \ k < 10; \ k{+}{+}, v{-}{-}) \ m = 2 * m + v{\rightarrow}alt{\rightarrow}val;$ /* *prog* */

   $m = 4 * m + (g{\rightarrow}vertices + 31){\rightarrow}alt{\rightarrow}val;$ /* *extra* */

   $m = 2 * m + (g{\rightarrow}vertices + 27){\rightarrow}alt{\rightarrow}val;$ /* *sign* */

   $m = 2 * m + (g{\rightarrow}vertices + 28){\rightarrow}alt{\rightarrow}val;$ /* *nonzero* */

   $m = 2 * m + (g{\rightarrow}vertices + 29){\rightarrow}alt{\rightarrow}val;$ /* *carry* */

   $m = 2 * m + (g{\rightarrow}vertices + 30){\rightarrow}alt{\rightarrow}val;$ /* *overflow* */

   $risc\_state[16] = m;$ /* program register and status bits go here */

   $risc\_state[17] = l;$ /* this is the out-of-range address that caused termination */

This code is used in section 43.

48. ⟨ Global variables 48 ⟩ ≡

   unsigned long *risc_state*[18];

This code is used in section 7.

49. Generalized gate graphs. For intermediate computations, it is convenient to allow two additional types of gates:

'C' denotes a constant gate of value $z.I$.

'=' denotes a copy of a previous gate; utility field *alt* points to that previous gate.

Such gates might appear anywhere in the graph, possibly interspersed with the inputs and latches.

Here is a simple subroutine that prints a symbolic representation of a generalized gate graph on the standard output file:

#define *bit* $z.I$ /* field containing the constant value of a 'C' gate */
#define *print_gates* *p_gates*   /* abbreviation makes chopped-off name unique */

⟨ The *print_gates* routine 49 ⟩ ≡
  static void *pr_gate*(v)
      Vertex *v;
  { **register Arc** *a;

    printf("%s␣=␣", v→*name*);
    switch (v→*typ*) {
    case 'I': *printf*("input"); **break**;
    case 'L': *printf*("latch");
      if (v→*alt*) *printf*("ed␣%s", v→*alt*→*name*);
      break;
    case '~': *printf*("~␣"); **break**;
    case 'C': *printf*("constant␣%ld", v→*bit*);
      break;
    case '=': *printf*("copy␣of␣%s", v→*alt*→*name*);
    }
    for (a = v→*arcs*; a; a = a→*next*) {
      if (a ≠ v→*arcs*) *printf*("␣%c␣", (**char**) v→*typ*);
      printf(a→*tip*→*name*);
    }
    printf("\n");
  }

  void *print_gates*(g)
      Graph *g;
  { **register Vertex** *v;
    register Arc *a;

    for (v = g→*vertices*; v < g→*vertices* + g→*n*; v++) *pr_gate*(v);
    for (a = g→*outs*; a; a = a→*next*)
      if (*is_boolean*(a→*tip*)) *printf*("Output␣%ld\n", *the_boolean*(a→*tip*));
      else *printf*("Output␣%s\n", a→*tip*→*name*);
  }

This code is used in section 7.

50. ⟨ gb_gates.h 1 ⟩ +≡
#define *bit* $z.I$

51. The *reduce* routine takes a generalized graph g and uses the identities $\bar{\bar{x}} = x$ and

$$
\begin{aligned}
x \wedge 0 = 0, &\quad x \wedge 1 = x, &\quad x \wedge x = x, &\quad x \wedge \bar{x} = 0, \\
x \vee 0 = x, &\quad x \vee 1 = 1, &\quad x \vee x = x, &\quad x \vee \bar{x} = 1, \\
x \oplus 0 = x, &\quad x \oplus 1 = \bar{x}, &\quad x \oplus x = 0, &\quad x \oplus \bar{x} = 1,
\end{aligned}
$$

to create an equivalent graph having no 'C' or '=' or obviously redundant gates. The reduced graph also excludes any gates that are not used directly or indirectly in the computation of the output values.

⟨ Internal subroutines 11 ⟩ +≡

```
  static Graph *reduce(g)
      Graph *g;
  { register Vertex *u, *v;      /* the current vertices of interest */
    register Arc *a, *b;      /* the current arcs of interest */
    Arc *aa, *bb;      /* their predecessors */
    Vertex *latch_ptr;      /* top of the latch list */
    long n = 0;      /* the number of marked gates */
    Graph *new_graph;      /* the reduced gate graph */
    Vertex *next_vert = Λ, *max_next_vert = Λ;      /* allocation of new vertices */
    Arc *avail_arc = Λ;      /* list of recycled arcs */
    Vertex *sentinel;      /* end of the vertices */
    if (g ≡ Λ) panic(missing_operand);      /* where is g? */
    sentinel = g↠vertices + g↠n;
    while (1) {
      latch_ptr = Λ;
      for (v = g↠vertices; v < sentinel; v++)
        ⟨ Reduce gate v, if possible, or put it on the latch list 53 ⟩;
      ⟨ Check to see if any latch has become constant; if not, break 52 ⟩;
    }
    ⟨ Mark all gates that are used in some output 60 ⟩;
    ⟨ Copy all marked gates to a new graph 62 ⟩;
    gb_recycle(g);
    return new_graph;
  }
```

52. We will link latches together via their $v.V$ fields.

⟨ Check to see if any latch has become constant; if not, **break** 52 ⟩ ≡
```
{ char no_constants_yet = 1;
    for (v = latch_ptr; v; v = v⃗v.V) {
        u = v⃗alt;      /* the gate whose value will be latched */
        if (u⃗typ ≡ '=') v⃗alt = u⃗alt;
        else if (u⃗typ ≡ 'C') {
            v⃗typ = 'C'; v⃗bit = u⃗bit; no_constants_yet = 0;
        }
    }
    if (no_constants_yet) break;
}
```
This code is used in section 51.

53. **#define** *foo* $x.V$ /* link field used to find all the gates later */

⟨ Reduce gate v, if possible, or put it on the latch list 53 ⟩ ≡
```
{
    switch (v⃗typ) {
    case 'L': v⃗v.V = latch_ptr; latch_ptr = v; break;
    case 'I': case 'C': break;
    case '=': u = v⃗alt;
        if (u⃗typ ≡ '=') v⃗alt = u⃗alt;
        else if (u⃗typ ≡ 'C') {
            v⃗bit = u⃗bit; goto make_v_constant;
        }
        break;
    case NOT: ⟨ Try to reduce an inverter, then goto done 54 ⟩;
    case AND: ⟨ Try to reduce an AND gate 55 ⟩; goto test_single_arg;
    case OR: ⟨ Try to reduce an OR gate 56 ⟩; goto test_single_arg;
    case XOR: ⟨ Try to reduce an EXCLUSIVE-OR gate 57 ⟩;
    test_single_arg:
        if (v⃗arcs⃗next) break;
        v⃗alt = v⃗arcs⃗tip;
    make_v_eq: v⃗typ = '='; goto make_v_arcless;
    make_v_1: v⃗bit = 1; goto make_v_constant;
    make_v_0: v⃗bit = 0;
    make_v_constant: v⃗typ = 'C';
    make_v_arcless: v⃗arcs = Λ;
    }
    v⃗bar = Λ;      /* this field will point to the complement, if computed later */
  done: v⃗foo = v + 1;      /* this field will link all the vertices together */
}
```
This code is used in section 51.

54. ⟨ Try to reduce an inverter, then **goto** *done* 54 ⟩ ≡
```
u = v⃗arcs⃗tip;
if (u⃗typ ≡ '=') u = v⃗arcs⃗tip = u⃗alt;
if (u⃗typ ≡ 'C') {
```

$v{\rightarrow}bit = 1 - u{\rightarrow}bit$; **goto** *make_v_constant*;
  } **else if** $(u{\rightarrow}bar)$ {   /* this inverse already computed */
  $v{\rightarrow}alt = u{\rightarrow}bar$; **goto** *make_v_eq*;
  } **else** {
  $u{\rightarrow}bar = v$; $v{\rightarrow}bar = u$; **goto** *done*;
  }

This code is used in section 53.

55. ⟨ Try to reduce an AND gate 55 ⟩ ≡
  for $(a = v{\rightarrow}arcs, aa = \Lambda;\ a;\ a = a{\rightarrow}next)$ {
    $u = a{\rightarrow}tip$;
    if $(u{\rightarrow}typ \equiv {'='})\ u = a{\rightarrow}tip = u{\rightarrow}alt$;
    if $(u{\rightarrow}typ \equiv {'C'})$ { **if** $(u{\rightarrow}bit \equiv 0)$ **goto** *make_v_0*;
      goto *bypass_and*;
    } **else for** $(b = v{\rightarrow}arcs;\ b \neq a;\ b = b{\rightarrow}next)$ {
      if $(b{\rightarrow}tip \equiv u)$ **goto** *bypass_and*;
      if $(b{\rightarrow}tip \equiv u{\rightarrow}bar)$ **goto** *make_v_0*;
    }
    $aa = a$; **continue**;
  bypass_and: **if** $(aa)\ aa{\rightarrow}next = a{\rightarrow}next$; **else** $v{\rightarrow}arcs = a{\rightarrow}next$;
  }
  if $(v{\rightarrow}arcs \equiv \Lambda)$ **goto** *make_v_1*;

This code is used in section 53.

56. ⟨ Try to reduce an OR gate 56 ⟩ ≡
  for $(a = v{\rightarrow}arcs, aa = \Lambda;\ a;\ a = a{\rightarrow}next)$ {
    $u = a{\rightarrow}tip$;
    if $(u{\rightarrow}typ \equiv {'='})\ u = a{\rightarrow}tip = u{\rightarrow}alt$;
    if $(u{\rightarrow}typ \equiv {'C'})$ { **if** $(u{\rightarrow}bit)$ **goto** *make_v_1*;
      goto *bypass_or*;
    } **else for** $(b = v{\rightarrow}arcs;\ b \neq a;\ b = b{\rightarrow}next)$ {
      if $(b{\rightarrow}tip \equiv u)$ **goto** *bypass_or*;
      if $(b{\rightarrow}tip \equiv u{\rightarrow}bar)$ **goto** *make_v_1*;
    }
    $aa = a$; **continue**;
  bypass_or: **if** $(aa)\ aa{\rightarrow}next = a{\rightarrow}next$; **else** $v{\rightarrow}arcs = a{\rightarrow}next$;
  }
  if $(v{\rightarrow}arcs \equiv \Lambda)$ **goto** *make_v_0*;

This code is used in section 53.

---

a: **register Arc** *, §51.
aa: **Arc** *, §51.
alt $= z.V$, §2.
AND $= {'\&'}$, §2.
arcs: **Arc** *, GB_GRAPH §9.
b: **register Arc** *, §51.
bar $= w.V$, §14.

bit $= z.I$, §49.
latch_ptr: **Vertex** *, §51.
next: **Arc** *, GB_GRAPH §10.
NOT $= {'\sim'}$, §2.
OR $= {'|'}$, §2.
tip: **Vertex** *, GB_GRAPH §10.
typ $= y.I$, §2.

u: **register Vertex** *, §51.
v: **register Vertex** *, §51.
v: **util**, GB_GRAPH §9.
V: **Vertex** *, GB_GRAPH §8.
x: **util**, GB_GRAPH §9.
XOR $= {'\hat{}'}$, §2.

57. ⟨ Try to reduce an EXCLUSIVE-OR gate 57 ⟩ ≡
  { **long** $cmp = 0$;
    for $(a = v{\to}arcs, aa = \Lambda;\ a;\ a = a{\to}next)$ {
      $u = a{\to}tip$;
      if $(u{\to}typ \equiv {}'{=}')\ u = a{\to}tip = u{\to}alt$;
      if $(u{\to}typ \equiv {}'\mathtt{C}')$ {
        if $(u{\to}bit)\ cmp = 1 - cmp$;
        goto *bypass_xor*;
      } **else for** $(bb = \Lambda, b = v{\to}arcs;\ b \neq a;\ b = b{\to}next)$ {
          if $(b{\to}tip \equiv u)$ **goto** *double_bypass*;
          if $(b{\to}tip \equiv u{\to}bar)$ {
            $cmp = 1 - cmp$;
            goto *double_bypass*;
          }
          $bb = b$; **continue**;
        double_bypass:
          if $(bb)\ bb{\to}next = b{\to}next$;
          else $v{\to}arcs = b{\to}next$;
          goto *bypass_xor*;
        }
      $aa = a$; **continue**;
    bypass_xor:
      if $(aa)\ aa{\to}next = a{\to}next$;
      else $v{\to}arcs = a{\to}next$;
      $a{\to}a.A = avail\_arc$;
      $avail\_arc = a$;
    }
    if $(v{\to}arcs \equiv \Lambda)$ {
      $v{\to}bit = cmp$;
      goto *make_v_constant*;
    }
    if (cmp) ⟨ Complement one argument of v 58 ⟩;
  }
This code is used in section 53.

58. ⟨ Complement one argument of v 58 ⟩ ≡
  {
    for $(a = v{\to}arcs;\ ;\ a = a{\to}next)$ {
      $u = a{\to}tip$;
      if $(u{\to}bar)$ **break**;   /* good, the complement is already known */
      if $(a{\to}next \equiv \Lambda)$ {   /* oops, this is our last chance */
        ⟨ Create a new vertex for complement of u 59 ⟩;
        break;
      }
    }
    $a{\to}tip = u{\to}bar$;
  }
This code is used in section 57.

59. Here we've come to a subtle point: If a lot of XOR gates involve an input that is set to the constant value 1, the "reduced" graph might actually be larger than the original, in the sense of having more vertices (although fewer arcs). Therefore we must have the ability to allocate new vertices during the reduction phase of *reduce*. At least one arc has been added to the *avail_arc* list whenever we reach this portion of the program.

⟨ Create a new vertex for complement of *u* 59 ⟩ ≡
```
    if (next_vert ≡ max_next_vert) {
        next_vert = gb_typed_alloc(7, Vertex, g⃗aux_data);
        if (next_vert ≡ Λ) {
            gb_recycle(g);
            panic(no_room + 1);        /* can't get auxiliary storage! */
        }
        max_next_vert = next_vert + 7;
    }
    next_vert⃗typ = NOT;
    sprintf(name_buf, "%s~", u⃗name);
    next_vert⃗name = gb_save_string(name_buf);
    next_vert⃗arcs = avail_arc;        /* this is known to be non-Λ */
    avail_arc⃗tip = u;
    avail_arc = avail_arc⃗a.A;
    next_vert⃗arcs⃗next = Λ;
    next_vert⃗bar = u;
    next_vert⃗foo = u⃗foo;
    u⃗foo = u⃗bar = next_vert++;
```
This code is used in section 58.

60. During the marking phase, we will use the $w.V$ field to link the list of nodes-to-be-marked. That field will turn out to be non-Λ only in the marked nodes. (We no longer use its former meaning related to complementation, so we call it *lnk* instead of *bar*.)

#define *lnk* $w.V$ /* stack link for marking */

$\langle$ Mark all gates that are used in some output 60 $\rangle \equiv$

```
{
    for (v = g⃗vertices; v ≠ sentinel; v = v⃗foo) v⃗lnk = Λ;
    for (a = g⃗outs; a; a = a⃗next) {
        v = a⃗tip;
        if (is_boolean(v)) continue;
        if (v⃗typ ≡ '=') v = a⃗tip = v⃗alt;
        if (v⃗typ ≡ 'C') {      /* this output is constant, so make it boolean */
            a⃗tip = (Vertex *) v⃗bit;
            continue;
        }
        ⟨ Mark all gates that are used to compute v 61 ⟩;
    }
}
```

This code is used in section 51.

61. $\langle$ Mark all gates that are used to compute v 61 $\rangle \equiv$

```
if (v⃗lnk ≡ Λ) {
    v⃗lnk = sentinel;      /* v now represents the top of the stack of nodes to be marked */
    do {
        n++;
        b = v⃗arcs;
        if (v⃗typ ≡ 'L') {
            u = v⃗alt;      /* latch vertices have a "hidden" dependency */
            if (u < v) n++;      /* latched input value will get a special gate */
            if (u⃗lnk ≡ Λ) {
                u⃗lnk = v⃗lnk;
                v = u;
            } else v = v⃗lnk;
        } else v = v⃗lnk;
        for ( ; b; b = b⃗next) {
            u = b⃗tip;
            if (u⃗lnk ≡ Λ) {
                u⃗lnk = v;
                v = u;
            }
        }
    } while (v ≠ sentinel);
}
```

This code is used in section 60.

62. It is easier to copy a directed acyclic graph than to copy a general graph, but we do have to contend with the feedback in latches.

```
#define reverse_arc_list(alist)
        { for (aa = alist, b = Λ; aa; b = aa, aa = a) {
            a = aa→next;
            aa→next = b;
        }
        alist = b; }
```

⟨ Copy all marked gates to a new graph 62 ⟩ ≡
   new_graph = *gb_new_graph*(*n*);
   if (*new_graph* ≡ Λ) {
     gb_recycle(*g*);
     panic(*no_room* + 2);   /∗ out of memory ∗/
   }
   strcpy(*new_graph*→*id*, *g*→*id*);
   strcpy(*new_graph*→*util_types*, "ZZZIIVZZZZZZZA");
   next_vert = *new_graph*→*vertices*;
   for (*v* = *g*→*vertices*, *latch_ptr* = Λ; *v* ≠ *sentinel*; *v* = *v*→*foo*) {
     if (*v*→*lnk*) {   /∗ yes, *v* is marked ∗/
       u = *v*→*lnk* = *next_vert* ++;   /∗ make note of where we've copied it ∗/
       ⟨ Make *u* a copy of *v*; put it on the latch list if it's a latch 63 ⟩;
     }
   }
   ⟨ Fix up the *alt* fields of the newly copied latches 64 ⟩;
   reverse_arc_list(*g*→*outs*);
   for (*a* = *g*→*outs*; *a*; *a* = *a*→*next*) {
     b = *gb_virgin_arc*();
     b→*tip* = *is_boolean*(*a*→*tip*) ? *a*→*tip* : *a*→*tip*→*lnk*;
     b→*next* = *new_graph*→*outs*;
     new_graph→*outs* = *b*;
   }

This code is used in section 51.

---

a: **register Arc** ∗, §51.
aa: **Arc** ∗, §51.
alt = *z.V*, §2.
arcs: **Arc** ∗, GB_GRAPH §9.
b: **register Arc** ∗, §51.
bar = *w.V*, §14.
bit = *z.I*, §49.
foo = *x.V*, §53.
g: **Graph** ∗, §51.
gb_new_graph: **Graph** ∗(),
   GB_GRAPH §23.
gb_recycle: **void** (), GB_GRAPH §40.

gb_virgin_arc: **Arc** ∗(),
   GB_GRAPH §29.
id: **char** [], GB_GRAPH §20.
is_boolean = macro (), §2.
latch_ptr: **Vertex** ∗, §51.
n: **long**, §51.
new_graph: **Graph** ∗, §51.
next: **Arc** ∗, GB_GRAPH §10.
next_vert: **Vertex** ∗, §51.
no_room = 1, GB_GRAPH §7.
outs = *zz.A*, §2.
panic = macro (), §8.

sentinel: **Vertex** ∗, §51.
strcpy: **char** ∗(), <string.h>.
tip: **Vertex** ∗, GB_GRAPH §10.
typ = *y.I*, §2.
u: **register Vertex** ∗, §51.
util_types: **char** [], GB_GRAPH §20.
v: **register Vertex** ∗, §51.
V: **Vertex** ∗, GB_GRAPH §8.
Vertex = struct, GB_GRAPH §9.
vertices: **Vertex** ∗, GB_GRAPH §20.
w: **util**, GB_GRAPH §9.

63. ⟨Make u a copy of v; put it on the latch list if it's a latch 63⟩ ≡

  $u{\to}name = gb\_save\_string(v{\to}name);$
  $u{\to}typ = v{\to}typ;$
  if $(v{\to}typ \equiv \text{'L'})$ {
    $u{\to}alt = latch\_ptr;\ latch\_ptr = v;$
  }
  $reverse\_arc\_list(v{\to}arcs);$
  for $(a = v{\to}arcs;\ a;\ a = a{\to}next)\ gb\_new\_arc(u, a{\to}tip{\to}lnk, a{\to}len);$

This code is used in section 62.

64. ⟨Fix up the *alt* fields of the newly copied latches 64⟩ ≡

  while $(latch\_ptr)$ {
    $u = latch\_ptr{\to}lnk;$ /* the copy of a latch */
    $v = u{\to}alt;$
    $u{\to}alt = latch\_ptr{\to}alt{\to}lnk;$
    $latch\_ptr = v;$
    if $(u{\to}alt < u)$ ⟨Replace $u{\to}alt$ by a new gate that copies an input 65⟩;
  }

This code is used in section 62.

65. Suppose we had a latch whose value was originally the AND of two inputs, where one of those inputs has now been set to 1. Then the latch should still refer to a subsequent gate, equal to the value of the other input on the previous cycle. We create such a gate here, making it an OR of two identical inputs. We do this because we're not supposed to leave any '=' in the result of *reduce*, and because every OR is supposed to have at least two inputs.

⟨ Replace $u{\rightarrow}alt$ by a new gate that copies an input 65 ⟩ ≡
```
  {
    v = u→alt;      /* the input gate that should be copied for latching */
    u→alt = next_vert ++;
    sprintf (name_buf, "%s>%s", v→name, u→name);
    u = u→alt;
    u→name = gb_save_string (name_buf);
    u→typ = OR;
    gb_new_arc(u, v, DELAY);  gb_new_arc(u, v, DELAY);
  }
```
This code is used in section 64.

---

a: **register Arc** *∗, §51.*
alt = *z.V*, §2.
arcs: **Arc** *∗*, GB_GRAPH §9.
DELAY = 100ᴸ, §13.
gb_new_arc: **void** ( ),
    GB_GRAPH §30.
gb_save_string: **char** *∗*( ),
    GB_GRAPH §35.

latch_ptr: **Vertex** *∗, §51.*
len: **long**, GB_GRAPH §10.
lnk = *w.V*, §60.
name: **char** *∗*, GB_GRAPH §9.
name_buf: **static char** [ ], §12.
next: **Arc** *∗*, GB_GRAPH §10.
next_vert: **Vertex** *∗, §51.*
OR = '|', §2.

reduce: **static Graph** *∗*( ), §51.
reverse_arc_list = macro ( ), §62.
sprintf: **int** ( ), <stdio.h>.
tip: **Vertex** *∗*, GB_GRAPH §10.
typ = *y.I*, §2.
u: **register Vertex** *∗, §51.*
v: **register Vertex** *∗, §51.*

66. Parallel multiplication. Now comes the *prod* routine, which constructs a rather different network of gates, based this time on a divide-and-conquer paradigm. Let's take a breather before we tackle it.

(Deep breath.)

The subroutine call $prod(m, n)$ creates a network for the binary multiplication of unsigned m-bit numbers by n-bit numbers, assuming that $m \geq 2$ and $n \geq 2$. There is no upper limit on the sizes of m and n, except of course the limits imposed by the size of memory in which this routine is run.

The overall strategy used by *prod* is to start with a generalized gate graph for multiplication in which many of the gates are identically zero or copies of other gates. Then the *reduce* routine will perform local optimizations leading to the desired result. Since there are no latches, some of the complexities of the general *reduce* routine are avoided.

All of the AND, OR, and XOR gates of the network returned by *prod* have exactly two inputs. The depth of the circuit (i.e., the length of its longest path) is $3 \log m / \log 1.5 + \log(m+n) / \log \phi + O(1)$, where $\phi = (1 + \sqrt{5})/2$ is the golden ratio. The grand total number of gates is $6mn + 5m^2 + O((m+n)\log(m+n))$.

There is a demonstration program called MULTIPLY that uses *prod* to compute products of large integers.

⟨ The *prod* routine 66 ⟩ ≡
  Graph *prod*(m, n)
      unsigned long m, n;   /* lengths of the binary numbers to be multiplied */
  { ⟨ Local variables for *prod* 68 ⟩
    if (m < 2) m = 2;
    if (n < 2) n = 2;
    ⟨ Allocate space for a temporary graph g and for auxiliary tables 67 ⟩;
    ⟨ Fill g with generalized gates that do parallel multiplication 70 ⟩;
    if (*gb_trouble_code*) {
      gb_recycle(g); *panic*(*alloc_fault*);   /* too big */
    }
    g = *reduce*(g);
    return g;   /* if g ≡ Λ, the *panic_code* was set by *reduce* */
  }

This code is used in section 7.

67. The divide-and-conquer recurrences used in this network lead to interesting patterns. First we use a method for parallel column addition that reduces the sum of three numbers to the sum of two numbers. Repeated use of this reduction makes it possible to reduce the sum of m numbers to a sum of just two numbers, with a total circuit depth that satisfies the recurrence $T(3N) = T(2N) + O(1)$. Then when the result has been reduced to a sum of two numbers, we use a parallel addition scheme based on recursively "golden sectioning the data"; in other words, the recursion partitions the data into two parts such that the ratio of the larger part to the smaller part is approximately ϕ. This technique proves to be slightly better than a binary partition would be, both asymptotically and for small values of $m + n$.

We define flog N, the Fibonacci logarithm of N, to be the smallest nonnegative integer k such that $N \leq F_{k+1}$. Let $N = m + n$. Our parallel adder for two numbers of N bits will turn out to have depth at most $2 + \mathrm{flog}\, N$. The unreduced graph g in our circuit for multiplication will have fewer than $(6m + 3\, \mathrm{flog}\, N)N$ gates.

⟨ Allocate space for a temporary graph g and for auxiliary tables 67 ⟩ ≡
  $m\_plus\_n = m + n$; ⟨ Compute $f = \mathrm{flog}(m + n)$ 69 ⟩;
  $g = gb\_new\_graph((6 * m - 7 + 3 * f) * m\_plus\_n)$;
  if $(g \equiv \Lambda)$ $panic(no\_room)$;   /∗ out of memory before we're even started ∗/
  $sprintf(g{\rightarrow}id, \texttt{"prod(\%lu,\%lu)"}, m, n)$;
  $strcpy(g{\rightarrow}util\_types, \texttt{"ZZZIIVZZZZZZA"})$;
  $long\_tables = gb\_typed\_alloc(2 * m\_plus\_n + f, \textbf{long}, g{\rightarrow}aux\_data)$;
  $vert\_tables = gb\_typed\_alloc(f * m\_plus\_n, \textbf{Vertex} *, g{\rightarrow}aux\_data)$;
  if $(gb\_trouble\_code)$ {
    $gb\_recycle(g)$;
    $panic(no\_room + 1)$;   /∗ out of memory trying to create auxiliary tables ∗/
  }
This code is used in section 66.

68. ⟨ Local variables for *prod* 68 ⟩ ≡
  unsigned long m_plus_n;   /∗ guess what this variable holds ∗/
  long f;   /∗ initially $\mathrm{flog}(m + n)$, later flog of other things ∗/
  Graph ∗g;   /∗ graph of generalized gates, to be reduced eventually ∗/
  long ∗$long\_tables$;   /∗ beginning of auxiliary array of **long** numbers ∗/
  Vertex ∗∗$vert\_tables$;   /∗ beginning of auxiliary array of gate pointers ∗/
See also sections 71 and 77.

This code is used in section 66.

69. ⟨ Compute $f = \mathrm{flog}(m + n)$ 69 ⟩ ≡
  $f = 4$; $j = 3$; $k = 5$;   /∗ $j = F_f$, $k = F_{f+1}$ ∗/
  while $(k < m\_plus\_n)$ {
    $k = k + j$;
    $j = k - j$;
    $f{+}{+}$;
  }
This code is used in section 67.

$alloc\_fault = -1$, GB_GRAPH §7.
AND = '&', §2.
aux_data: **Area**, GB_GRAPH §20.
gb_new_graph: **Graph** ∗(),
  GB_GRAPH §23.
$gb\_recycle$: **void** (), GB_GRAPH §40.
$gb\_trouble\_code$: **long**,
  GB_GRAPH §14.

gb_typed_alloc = macro (),
  GB_GRAPH §11.
Graph = **struct**, GB_GRAPH §20.
id: **char** [], GB_GRAPH §20.
j: **register long**, §71.
k: **register long**, §71.
$no\_room = 1$, GB_GRAPH §7.
OR = '|', §2.

$panic$ = macro (), §8.
$panic\_code$: **long**, GB_GRAPH §5.
$reduce$: **static Graph** ∗(), §51.
$sprintf$: **int** (), <stdio.h>.
$strcpy$: **char** ∗(), <string.h>.
$util\_types$: **char** [], GB_GRAPH §20.
Vertex = **struct**, GB_GRAPH §9.
XOR = '^', §2.

70. The well-known formulas for a "full adder,"

$$x + y + z = s + 2c, \qquad \text{where } s = x \oplus y \oplus z \text{ and } c = xy \vee yz \vee zx,$$

can be applied to each bit of an N-bit number, thereby providing us with a way to reduce the sum of three numbers to the sum of two.

The input gates of our network will be called $x_0, x_1, \ldots, x_{m-1}, y_0, y_1, \ldots, y_{n-1}$, and the outputs will be called $z_0, z_1, \ldots, z_{m+n-1}$. The logic of the *prod* network will compute

$$(z_{m+n-1} \ldots z_1 z_0)_2 = (x_{m-1} \ldots x_1 x_0)_2 \cdot (y_{n-1} \ldots y_1 y_0)_2 \,,$$

by first considering the product to be the m-fold sum $A_0 + A_1 + \cdots + A_{m-1}$, where

$$A_j = 2^j x_j \cdot (y_{n-1} \ldots y_1 y_0)_2 \,, \qquad 0 \le j < m.$$

Then the three-to-two rule for addition is used to define further numbers $A_m, A_{m+1}, \ldots, A_{3m-5}$ by the scheme

$$A_{m+2j} + A_{m+2j+1} = A_{3j} + A_{3j+1} + A_{3j+2} \,, \qquad 0 \le j \le m - 3.$$

[A similar but slightly less efficient scheme was used by Pratt and Stockmeyer in *Journal of Computer and System Sciences* **12** (1976), Proposition 5.3. The recurrence used here is related to the Josephus problem with step-size 3; see *Concrete Mathematics*, §3.3.] For this purpose, we compute intermediate results P_j, Q_j, and R_j by the rules

$$P_j = A_{3j} \oplus A_{3j+1} \,;$$
$$Q_j = A_{3j} \wedge A_{3j+1} \,;$$
$$A_{m+2j} = P_j \oplus A_{3j+2} \,;$$
$$R_j = P_j \wedge A_{3j+2} \,;$$
$$A_{m+2j+1} = 2(Q_j \vee R_j) \,.$$

Finally we let

$$U = A_{3m-6} \oplus A_{3m-5} \,,$$
$$V = A_{3m-6} \wedge A_{3m-5} \,;$$

these are the values that would be P_{m-2} and Q_{m-2} if the previous formulas were allowed to run past $j = m - 3$. The final result $Z = (z_{m+n-1} \ldots z_1 z_0)_2$ can now be expressed as

$$Z = U + 2V \,.$$

The gates of the first part of the network are conveniently obtained in groups of $N = m + n$, representing the bits of the quantities A_j, P_j, Q_j, R_j, U, and V. We will put the least significant bit of A_j in gate position $g\text{-}vertices + a(j) * N$, where $a(j) = j + 1$ for $0 \le j < m$ and $a(m + 2j + t) = m + 5j + 3 + 2t$ for $0 \le j \le m - 3$, $0 \le t \le 1$.

⟨ Fill g with generalized gates that do parallel multiplication 70 ⟩ ≡
  next_vert = $g\text{-}vertices$;
  start_prefix ("X"); $x = $ *first_of* $(m, \text{'I'})$;
  start_prefix ("Y"); $y = $ *first_of* $(n, \text{'I'})$;
  ⟨ Define A_j for $0 \le j < m$ 72 ⟩;
  ⟨ Define P_j, Q_j, A_{m+2j}, R_j, and A_{m+2j+1} for $0 \le j \le m - 3$ 73 ⟩;
  ⟨ Define U and V 74 ⟩;
  ⟨ Compute the final result Z by parallel addition 75 ⟩;
This code is used in section 66.

71. ⟨ Local variables for *prod* 68 ⟩ +≡
   register long i, j, k, l;   /∗ all-purpose indices ∗/
   register Vertex ∗v;   /∗ current vertex of interest ∗/
   Vertex ∗x, ∗y;   /∗ least-significant bits of the input gates ∗/
   Vertex ∗*alpha*, ∗*beta*;   /∗ least-significant bits of arguments ∗/

72. ⟨ Define A_j for $0 \le j < m$ 72 ⟩ ≡
   for ($j = 0$; $j < m$; $j{+}{+}$) {
     numeric_prefix(`'A'`, j);
     for ($k = 0$; $k < j$; $k{+}{+}$) {
      $v = new\_vert($`'C'`$)$; $v{\rightarrow}bit = 0$;   /∗ this gate is the constant 0 ∗/
     }
     for ($k = 0$; $k < n$; $k{+}{+}$) *make2*(AND, $x + j$, $y + k$);
     for ($k = j + n$; $k < m\_plus\_n$; $k{+}{+}$) {
      $v = new\_vert($`'C'`$)$; $v{\rightarrow}bit = 0$;   /∗ this gate is the constant 0 ∗/
     }
   }

This code is used in section 70.

AND = `'&'`, §2.
bit = *z.I*, §49.
first_of = macro ( ), §19.
g: **Graph** ∗, §68.
m: **unsigned long**, §66.

m_plus_n: **unsigned long**, §68.
make2: **static Vertex** ∗( ), §13.
n: **unsigned long**, §66.
new_vert: **static Vertex** ∗( ), §11.
next_vert: **static Vertex** ∗, §12.

numeric_prefix = macro ( ), §12.
prod: **Graph** ∗( ), §66.
start_prefix = macro ( ), §12.
Vertex = **struct**, GB_GRAPH §9.
vertices: **Vertex** ∗, GB_GRAPH §20.

73. Since m is **unsigned**, it is necessary to say '$j < m - 2$' here instead of '$j \leq m - 3$'.

#define $a\_pos(j)$ $(j < m\ ?\ j + 1 : m + 5 * ((j - m) \gg 1) + 3 + (((j - m)\ \&\ 1) \ll 1))$

$\langle$ Define P_j, Q_j, A_{m+2j}, R_j, and A_{m+2j+1} for $0 \leq j \leq m - 3$ 73 $\rangle \equiv$
  for $(j = 0;\ j < m - 2;\ j{+}{+})$ {
    $alpha = g{\to}vertices + (a\_pos(3 * j) * m\_plus\_n);$
    $beta = g{\to}vertices + (a\_pos(3 * j + 1) * m\_plus\_n);$
    $numeric\_prefix(\texttt{'P'}, j);$
    for $(k = 0;\ k < m\_plus\_n;\ k{+}{+})\ make2(\texttt{XOR}, alpha + k, beta + k);$
    $numeric\_prefix(\texttt{'Q'}, j);$
    for $(k = 0;\ k < m\_plus\_n;\ k{+}{+})\ make2(\texttt{AND}, alpha + k, beta + k);$
    $alpha = next\_vert - 2 * m\_plus\_n;$
    $beta = g{\to}vertices + (a\_pos(3 * j + 2) * m\_plus\_n);$
    $numeric\_prefix(\texttt{'A'}, (\textbf{long})\ m + 2 * j);$
    for $(k = 0;\ k < m\_plus\_n;\ k{+}{+})\ make2(\texttt{XOR}, alpha + k, beta + k);$
    $numeric\_prefix(\texttt{'R'}, j);$
    for $(k = 0;\ k < m\_plus\_n;\ k{+}{+})\ make2(\texttt{AND}, alpha + k, beta + k);$
    $alpha = next\_vert - 3 * m\_plus\_n;$
    $beta = next\_vert - m\_plus\_n;$
    $numeric\_prefix(\texttt{'A'}, (\textbf{long})\ m + 2 * j + 1);$
    $v = new\_vert(\texttt{'C'});\ v{\to}bit = 0;$ /* another 0, it multiplies $Q \vee R$ by 2 */
    for $(k = 0;\ k < m\_plus\_n - 1;\ k{+}{+})\ make2(\texttt{OR}, alpha + k, beta + k);$
  }

This code is used in section 70.

74. Actually v_{m+n-1} will never be used (it has to be zero); but we compute it anyway. We don't have to worry about such nitty gritty details because *reduce* will get rid of all the obvious redundancy.

$\langle$ Define U and V 74 $\rangle \equiv$
  $alpha = g{\to}vertices + (a\_pos(3 * m - 6) * m\_plus\_n);$
  $beta = g{\to}vertices + (a\_pos(3 * m - 5) * m\_plus\_n);$
  $start\_prefix(\texttt{"U"});$
  for $(k = 0;\ k < m\_plus\_n;\ k{+}{+})\ make2(\texttt{XOR}, alpha + k, beta + k);$
  $start\_prefix(\texttt{"V"});$
  for $(k = 0;\ k < m\_plus\_n;\ k{+}{+})\ make2(\texttt{AND}, alpha + k, beta + k);$

This code is used in section 70.

75. Parallel addition. It's time now to take another deep breath. We have finished the parallel multiplier except for one last step, the design of a parallel adder.

The adder is based on the following theory: We want to perform the binary addition

$$u_{N-1} \ldots u_2 \; u_1 \; u_0$$
$$v_{N-2} \ldots v_1 \; v_0$$
$$\overline{z_{N-1} \ldots z_2 \; z_1 \; z_0}$$

where we know that $u_k + v_k \leq 1$ for all k. It follows that $z_k = u_k \oplus w_k$, where $w_0 = 0$ and

$$w_k = v_{k-1} \lor u_{k-1}v_{k-2} \lor u_{k-1}u_{k-2}v_{k-3} \lor \cdots \lor u_{k-1}\ldots u_1 v_0$$

for $k > 0$. The problem has therefore been reduced to the evaluation of $w_1, w_2, \ldots, w_{N-1}$.

Let c_k^j denote the OR of the first j terms in the formula that defines w_k, and let d_k^j denote the j-fold product $u_{k-1}u_{k-2}\ldots u_{k-j}$. Then $w_k = c_k^k$, and we can use a recursive scheme of the form

$$c_k^j = c_k^i \lor d_k^i c_{k-i}^{j-i}, \qquad d_k^j = d_k^i d_{k-i}^{j-i}, \qquad j \geq 2,$$

to do the evaluation.

It turns out that this recursion behaves very nicely if we choose $i = \text{down}[j]$, where $\text{down}[j]$ is defined for $j > 1$ by the formula

$$\text{down}[j] = j - F_{(\text{flog } j)-1}.$$

For example, flog $18 = 7$ because $F_7 = 13 < 18 \leq 21 = F_8$, hence $\text{down}[18] = 18 - F_6 = 10$.

Let us write $j \to \text{down}[j]$, and consider the oriented tree on the set of all positive integers that is defined by this relation. One of the paths in this tree, for example, is $18 \to 10 \to 5 \to 3 \to 2 \to 1$. Our recurrence for $w_{18} = c_{18}^{18}$ involves c_{18}^{10}, which involves c_{18}^5, which involves c_{18}^3, and so on. In general, we will compute c_k^j for all j with $k \to^* j$, and we will compute d_k^j for all j with $k \to^+ j$. It is not difficult to prove that

$$k \to^* j \to i \qquad \text{implies} \qquad k - i \to^* j - i;$$

therefore the auxiliary factors c_{k-i}^{j-i} and d_{k-i}^{j-i} needed in the recurrence scheme will already have been evaluated. (Indeed, one can prove more: Let $l = \text{flog } k$. If the complete path from k to 1 in the tree is $k = k_0 \to k_1 \to \cdots \to k_t = 1$, then the differences $k_0 - k_1, k_1 - k_2, \ldots, k_{t-2} - k_{t-1}$ will consist of precisely the Fibonacci numbers $F_{l-1}, F_{l-2}, \ldots, F_2$, except for the numbers that appear when $F_{l+1} - k$ is written as a sum of non-consecutive Fibonacci numbers.)

It can also be shown that, when $k > 1$, we have

$$\text{flog } k = \min_{0 < j < n} \max\bigl(1 + \text{flog } j, \; 2 + \text{flog}(k - j)\bigr),$$

and that $\text{down}[k]$ is the smallest j such that the minimum is achieved in this equation. Therefore the depth of the circuit for computing w_k from the u's and v's is exactly flog k.

In particular, we can be sure that at most $3 \text{ flog } N$ gates will be created when computing z_k, and that there will be at most $3N \text{ flog } N$ gates in the parallel addition portion of the circuit.

⟨ Compute the final result Z by parallel addition 75 ⟩ ≡
  ⟨ Set up auxiliary tables to handle Fibonacci-based recurrences 76 ⟩;
  ⟨ Create the gates for W, remembering intermediate results that might be reused later 78 ⟩;
  ⟨ Compute the last gates $Z = U \oplus W$, and record their locations as outputs of the network 83 ⟩;
  g-n = next_vert − g-vertices; /* reduce to the actual number of gates used */
This code is used in section 70.

76.   After we have created a gate for w_k, we will store its address as the value of $w[k]$ in an
auxiliary table. After we've created a gate for c_k^i where $i < k$ is a Fibonacci number F_{l+1} and
$l = \text{flog}\, i \geq 2$, we will store its address as the value of $c[k + (l-2)N]$; the gate d_k^i will immediately
follow this one. Tables of $\text{flog}\, j$ and $\text{down}[j]$ will facilitate all these manipulations.

⟨ Set up auxiliary tables to handle Fibonacci-based recurrences 76 ⟩ ≡
```
    w = vert_tables;
    c = w + m_plus_n;
    flog = long_tables;
    down = flog + m_plus_n + 1;
    anc = down + m_plus_n;
    flog[1] = 0;  flog[2] = 2;
    down[1] = 0;  down[2] = 1;
    for (i = 3, j = 2, k = 3, l = 3;  l ≤ m_plus_n;  l++) {
       if (l > k) {
          k = k + j;
          j = k - j;
          i++;      /* Fᵢ = j < l ≤ k = Fᵢ₊₁ */
       }
       flog[l] = i;
       down[l] = l - k + j;
    }
```
This code is used in section 75.

77.   ⟨ Local variables for *prod* 68 ⟩ +≡
Vertex $*uu$, $*vv$;   /* pointer to u_0 and v_0 */
Vertex $**w$;   /* table of pointers to w_k */
Vertex $**c$;   /* table of pointers to potentially important intermediate values c_k^i */
Vertex $*cc$, $*dd$;   /* pointers to c_k^i and d_k^i */
long $*flog$;   /* table of flog values */
long $*down$;   /* table of down values */
long $*anc$;   /* table of ancestors of the current k */

---

g: **Graph** $*$, §68.
i: **register long**, §71.
j: **register long**, §71.
k: **register long**, §71.
l: **register long**, §71.

long_tables: **long** $*$, §68.
m_plus_n: **unsigned long**, §68.
n: **long**, GB_GRAPH §20.
next_vert: **static Vertex** $*$, §12.

prod: **Graph** $*()$, §66.
vert_tables: **Vertex** $**$, §68.
Vertex = **struct**, GB_GRAPH §9.
vertices: **Vertex** $*$, GB_GRAPH §20.

78. ⟨Create the gates for W, remembering intermediate results that might be reused later 78⟩ ≡

$vv = next\_vert - m\_plus\_n$; $uu = vv - m\_plus\_n$;
$start\_prefix(\texttt{"W"})$;
$v = new\_vert(\texttt{'C'})$; $v{\to}bit = 0$; $w[0] = v$;   /* $w_0 = 0$ */
$v = new\_vert(\texttt{'='})$; $v{\to}alt = vv$; $w[1] = v$;   /* $w_1 = v_0$ */
for $(k = 2;\ k < m\_plus\_n;\ k{+}{+})$ {
   ⟨Set the anc table to a list of the ancestors of k in decreasing order, stopping with $anc[l] = 2$ 79⟩;
   $i = 1$; $cc = vv + k - 1$; $dd = uu + k - 1$;
   while (1) {
      $j = anc[l]$;   /* now $i = down[j]$ */
      ⟨Compute the gate $b_k^j = d_k^i \wedge c_{k-i}^{j-i}$ 80⟩;
      ⟨Compute the gate $c_k^j = c_k^i \vee b_k^j$ 81⟩;
      if $(flog[j] < flog[j + 1])$ /* j is a Fibonacci number */
         $c[k + (flog[j] - 2) * m\_plus\_n] = v$;
      if $(l \equiv 0)$ **break**;
      $cc = v$;
      ⟨Compute the gate $d_k^j = d_k^i \wedge d_{k-i}^{j-i}$ 82⟩;
      $dd = v$; $i = j$; $l{-}{-}$;
   }
   $w[k] = v$;
}

This code is used in section 75.

79. If $k \to j$, we call j an "ancestor" of k because we are thinking of the tree defined by '→'; this tree is rooted at $2 \to 1$.

⟨Set the anc table to a list of the ancestors of k in decreasing order, stopping with $anc[l] = 2$ 79⟩ ≡
   for $(l = 0, j = k;\ ;\ l{+}{+}, j = down[j])$ {
      $anc[l] = j$;
      if $(j \equiv 2)$ **break**;
   }

This code is used in section 78.

80.   **#define** $spec\_gate(v, a, k, j, t)$ $v = next\_vert{+}{+}$;
       $sprintf(name\_buf, \texttt{"\%c\%ld:\%ld"}, a, k, j)$;
       $v{\to}name = gb\_save\_string(name\_buf)$;
       $v{\to}typ = t$;

⟨Compute the gate $b_k^j = d_k^i \wedge c_{k-i}^{j-i}$ 80⟩ ≡
   $spec\_gate(v, \texttt{'B'}, k, j, \texttt{AND})$;
   $gb\_new\_arc(v, dd, \texttt{DELAY})$;   /* first argument is d_k^i */
   $f = flog[j - i]$;   /* get ready to compute the second argument, c_{k-i}^{j-i} */
   $gb\_new\_arc(v, f > 0\ ?\ c[k - i + (f - 2) * m\_plus\_n] : vv + k - i - 1, \texttt{DELAY})$;

This code is used in section 78.

81. ⟨Compute the gate $c_k^j = c_k^i \vee b_k^j$ 81⟩ ≡
   if (l) {
      $spec\_gate(v, \texttt{'C'}, k, j, \texttt{OR})$;
   } **else** $v = new\_vert(\texttt{OR})$;   /* if l is zero, this gate is $c_k^k = w_k$ */

$gb\_new\_arc(v, cc, \text{DELAY})$;   /* first argument is c_k^i */
$gb\_new\_arc(v, next\_vert - 2, \text{DELAY})$;   /* second argument is b_k^j */

This code is used in section 78.

82.   Here we reuse the value $f = \text{flog}(j - i)$ computed a minute ago.

⟨ Compute the gate $d_k^j = d_k^i \wedge d_{k-i}^{j-i}$ 82 ⟩ ≡
  $spec\_gate(v, \text{'D'}, k, j, \text{AND})$;
  $gb\_new\_arc(v, dd, \text{DELAY})$;   /* first argument is d_k^i */
  $gb\_new\_arc(v, f > 0 \,?\, c[k - i + (f - 2) * m\_plus\_n] + 1 : uu + k - i - 1, \text{DELAY})$;   /* d_{k-i}^{j-i} */

This code is used in section 78.

83.   The output list will contain the gates in "big-endian order" z_{m+n-1}, ..., z_1, z_0, because we insert them into the *outs* list in little-endian order.

⟨ Compute the last gates $Z = U \oplus W$, and record their locations as outputs of the network 83 ⟩ ≡
  $start\_prefix(\text{"Z"})$;
  for $(k = 0;\ k < m\_plus\_n;\ k\text{++})$ { **register Arc** $*a = gb\_virgin\_arc(\,)$;

  $a\text{→}tip = make2(\text{XOR}, uu + k, w[k])$;
  $a\text{→}next = g\text{→}outs$;
  $g\text{→}outs = a$;
  }

This code is used in section 75.

$alt = z.V$, §2.
anc: **long** $*$, §77.
$\text{AND} = \text{'\&'}$, §2.
Arc = **struct**, GB_GRAPH §10.
$bit = z.I$, §49.
c: **Vertex** $**$, §77.
cc: **Vertex** $*$, §77.
dd: **Vertex** $*$, §77.
$\text{DELAY} = 100_{\text{L}}$, §13.
$down$: **long** $*$, §77.
f: **long**, §68.
$flog$: **long** $*$, §77.
g: **Graph** $*$, §68.
gb_new_arc: **void** ( ),

GB_GRAPH §30.
gb_save_string: **char** $*()$,
  GB_GRAPH §35.
gb_virgin_arc: **Arc** $*()$,
  GB_GRAPH §29.
i: **register long**, §71.
j: **register long**, §71.
k: **register long**, §71.
l: **register long**, §71.
m_plus_n: **unsigned long**, §68.
$make2$: **static Vertex** $*()$, §13.
$name$: **char** $*$, GB_GRAPH §9.
$name\_buf$: **static char** [ ], §12.
new_vert: **static Vertex** $*()$, §11.

$next$: **Arc** $*$, GB_GRAPH §10.
$next\_vert$: **static Vertex** $*$, §12.
$\text{OR} = \text{'|'}$, §2.
$outs = zz.A$, §2.
$sprintf$: **int** ( ), <stdio.h>.
$start\_prefix$ = macro ( ), §12.
tip: **Vertex** $*$, GB_GRAPH §10.
$typ = y.I$, §2.
uu: **Vertex** $*$, §77.
v: **register Vertex** $*$, §71.
vv: **Vertex** $*$, §77.
w: **Vertex** $**$, §77.
$\text{XOR} = \text{'\^'}$, §2.

84. Partial evaluation. The subroutine call *partial_gates*(*g*, *r*, *prob*, *seed*, *buf*) creates a new gate graph from a given gate graph *g* by "partial evaluation," i.e., by setting some of the inputs to constant values and simplifying the result. The new graph is usually smaller than *g*; it might, in fact, be a great deal smaller. Graph *g* is destroyed in the process.

The first *r* inputs of *g* are retained unconditionally. Each remaining input is retained with probability *prob*/65536, and if not retained it is assigned a random constant value. For example, about half of the inputs will become constant if *prob* = 32768. The *seed* parameter defines a machine-independent source of random numbers, and it may be given any value between 0 and $2^{31} - 1$.

If the *buf* parameter is non-null, it should be the address of a string. In such a case, *partial_gates* will put a record of its partial evaluation into that string; *buf* will contain one character for each input gate after the first *r*, namely '*' if the input was retained, '0' if it was set to 0, or '1' if it was set to 1.

The new graph will contain only gates that contribute to the computation of at least one output value. Therefore some input gates might disappear even though they were supposedly "retained," i.e., even though their value has not been set constant. The *name* field of a vertex can be used to determine exactly which input gates have survived.

If graph *g* was created by *risc*, users will probably want to make $r \geq 1$, since the whole RISC circuit collapses to zero whenever its first input 'RUN' is set to 0.

An interesting class of graphs is produced by the function call *partial_gates*(*prod*(*m*, *n*), *m*, 0, *seed*, Λ), which creates a graph corresponding to a circuit that multiplies a given *m*-bit number by a fixed (but randomly selected) *n*-bit constant. If the constant is not zero, all *m* of the "retained" input gates necessarily survive. The demo program called MULTIPLY illustrates such circuits.

The graph *g* might be a generalized network; that is, it might involve the 'C' or '=' gates described earlier. Notice that if *r* is sufficiently large, *partial_gates* becomes equivalent to the *reduce* routine. Therefore we need not make that private routine public.

As usual, the result will be Λ, and *panic_code* will be set, if *partial_gates* is unable to complete its task.

⟨ The *partial_gates* routine 84 ⟩ ≡

  Graph *partial_gates($g, r, prob, seed, buf$)

      Graph *g;   /* generalized gate graph */

      unsigned long *r*;   /* the number of initial gates to leave untouched */

      unsigned long *prob*;   /* scaled probability of not touching subsequent input gates */

      long *seed*;   /* seed value for random number generation */

      char *buf;   /* optional parameter for information about partial assignment */

  { **register Vertex** *v;   /* the current gate of interest */

      if ($g \equiv \Lambda$) *panic*(*missing_operand*);   /* where is g? */

      gb_init_rand(*seed*);   /* get them random numbers rolling */

      for ($v = g{\rightarrow}vertices + r$; $v < g{\rightarrow}vertices + g{\rightarrow}n$; $v{+}{+}$)

        switch ($v{\rightarrow}typ$) {

        case 'C': **case** '=': **continue**;   /* input gates might still follow */

        case 'I':

          if (($gb\_next\_rand(\,) \gg 15) \geq prob$) {

            $v{\rightarrow}typ$ = 'C'; $v{\rightarrow}bit = gb\_next\_rand(\,) \gg 30$;

            if (*buf*) *buf$+{+}$ = $v{\rightarrow}bit$ + '0';

          } **else if** (*buf*) *buf$+{+}$ = '*';

          break;

        default: **goto** *done*;   /* no more input gates can follow */

        }

    done:

      if (*buf*) *buf = 0;   /* terminate the string */

      $g = reduce(g)$;

      ⟨ Give the reduced graph a suitable *id* 85 ⟩;

      return *g*;   /* if ($g \equiv \Lambda$), a *panic_code* has been set by *reduce* */

  }

This code is used in section 7.

85. The *buf* parameter is not recorded in the graph's *id* field, since it has no effect on the graph itself.

⟨ Give the reduced graph a suitable *id* 85 ⟩ ≡

  if (*g*) {

    strcpy(*name_buf*, $g{\rightarrow}id$);

    if (*strlen*(*name_buf*) > 54) *strcpy*(*name_buf* + 51, "...");

    sprintf($g{\rightarrow}id$, "partial_gates(%s,%lu,%lu,%ld)", *name_buf*, *r*, *prob*, *seed*);

  }

This code is used in section 84.

---

bit = *z.I*, §49.

gb_init_rand: **void** ( ), GB_FLIP §8.

gb_next_rand = macro ( ),
  GB_FLIP §6.

Graph = **struct**, GB_GRAPH §20.

id: **char** [ ], GB_GRAPH §20.

missing_operand = 50,
  GB_GRAPH §7.

n: **long**, GB_GRAPH §20.

name: **char** *, GB_GRAPH §9.

name_buf: **static char** [ ], §12.

panic = macro ( ), §8.

panic_code: **long**, GB_GRAPH §5.

prod: **Graph** *( ), §66.

reduce: **static Graph** *( ), §51.

risc: **Graph** *( ), §8.

sprintf: **int** ( ), <stdio.h>.

strcpy: **char** *( ), <string.h>.

strlen: **int** ( ), <string.h>.

typ = *y.I*, §2.

Vertex = **struct**, GB_GRAPH §9.

vertices: **Vertex** *, GB_GRAPH §20.

GB_GRAPH

1. Introduction. This is GB_GRAPH, the data-structure module used by all GraphBase routines to allocate memory. The basic data types for graph representation are also defined here.

Many examples of how to use these conventions appear in other GraphBase modules. The best introduction to such examples can probably be found in GB_BASIC, which contains subroutines for generating and transforming various classical graphs.

2. The code below is believed to be system-independent; it should produce equivalent results on all systems, assuming that the standard *calloc* and *free* functions of C are available.

However, a test program helps build confidence that everything does in fact work as it should. To make such a test, simply compile and run `test_graph`. This particular test is fairly rudimentary, but it should be passed before more elaborate routines are tested.

⟨ `test_graph.c` 2 ⟩ ≡
#include `"gb_graph.h"`   /∗ all users of GB_GRAPH should do this ∗/
  ⟨ Declarations of test variables 19 ⟩
  int *main*( )
  {
    ⟨ Create a small graph 36 ⟩;
    ⟨ Test some intentional errors 18 ⟩;
    ⟨ Check that the small graph is still there 38 ⟩;
    printf(`"OK,␣the␣gb_graph␣routines␣seem␣to␣work!\n"`); **return** 0;
  }

3. The C code for GB_GRAPH doesn't have a main routine; it's just a bunch of subroutines waiting to be incorporated into programs at a higher level via the system loading routine. Here is the general outline of `gb_graph.c`:

#ifdef SYSV
#include `<string.h>`
#else
#include `<strings.h>`
#endif
#include `<stdio.h>`
  ⟨ Preprocessor definitions ⟩
  ⟨ Type declarations 8 ⟩
  ⟨ Private declarations 28 ⟩
  ⟨ External declarations 5 ⟩
  ⟨ External functions 13 ⟩

4. The type declarations of GB_GRAPH appear also in the header file `gb_graph.h`. For convenience, that header file also incorporates the standard system headers for input/output and string manipulation.

Some system header files define an unsafe macro called *min*, which will interfere with Graph-Base use of a useful identifier. We scotch that.

⟨ `gb_graph.h` 4 ⟩ ≡
#include `<stdio.h>`

```
#ifdef SYSV
#include <string.h>
#else
#include <strings.h>
#endif
#undef min
```
⟨ Type declarations 8 ⟩

See also sections 6, 7, 15, 17, 22, 25, 33, 41, and 42.

5. GraphBase programs often have a "verbose" option, which needs to be enabled by the setting of an external variable. They also tend to have a variable called *panic_code*, which helps identify unusual errors. We might as well declare those variables here.

⟨ External declarations 5 ⟩ ≡
    long *verbose* = 0;   /∗ nonzero if "verbose" output is desired ∗/
    long *panic_code* = 0;   /∗ set nonzero if graph generator returns null pointer ∗/

See also sections 14, 24, and 32.

This code is used in section 3.

6. Every external variable should be declared twice in this CWEB file: once for GB_GRAPH itself (the "real" declaration for storage allocation purposes) and once in `gb_graph.h` (for cross-references by GB_GRAPH users).

⟨ `gb_graph.h` 4 ⟩ +≡
    extern long *verbose*;   /∗ nonzero if "verbose" output is desired ∗/
    extern long *panic_code*;   /∗ set nonzero if graph generator panics ∗/

7. When *panic_code* is assigned a nonzero value, one of the symbolic names defined here is used to help pinpoint the problem. Small values indicate memory limitations; values in the 10s and 20s indicate input/output anomalies; values in the 30s and 40s indicate errors in the parameters to a subroutine. Some panic codes stand for cases the author doesn't think will ever arise, although the program checks for them just to be extra safe. Multiple instances of the same type of error within a single subroutine are distinguished by adding an integer; for example, '*syntax_error* + 1' and '*syntax_error* + 2' identify two different kinds of syntax error, as an aid in trouble-shooting. The *early_data_fault* and *late_data_fault* codes are explained further by the value of *io_errors*.

⟨ `gb_graph.h` 4 ⟩ +≡
```
#define alloc_fault   (−1)     /∗ a previous memory request failed ∗/
#define no_room   1     /∗ the current memory request failed ∗/
#define early_data_fault   10     /∗ error detected at beginning of .dat file ∗/
#define late_data_fault   11     /∗ error detected at end of .dat file ∗/
#define syntax_error   20     /∗ error detected while reading .dat file ∗/
#define bad_specs   30     /∗ parameter out of range or otherwise disallowed ∗/
#define very_bad_specs   40     /∗ parameter far out of range or otherwise stupid ∗/
#define missing_operand   50     /∗ graph parameter is Λ ∗/
#define invalid_operand   60     /∗ graph parameter doesn't obey assumptions ∗/
#define impossible   90     /∗ "this can't happen" ∗/
```

---

calloc: **void** ∗( ), `<stdlib.h>`.   *io_errors*: **long**, GB_IO §5.   SYSV, UNIX System 5.
free: **void** ( ), `<stdlib.h>`.   *printf*: **int** ( ), `<stdio.h>`.

8. Representation of graphs. The GraphBase programs employ a simple and flexible set of data structures to represent and manipulate graphs in computer memory. Vertices appear in a sequential array of **Vertex** records, and the arcs emanating from each vertex appear in a linked list of **Arc** records. There is also a **Graph** record, to provide information about the graph as a whole.

The structure layouts for **Vertex**, **Arc**, and **Graph** records include a number of utility fields that can be used for any purpose by algorithms that manipulate the graphs. Each utility field is a union type that can be either a pointer of various kinds or a (long) integer.

Let's begin the formal definition of these data structures by declaring the union type **util**. The suffixes *.V*, *.A*, *.G*, and *.S* on the name of a utility variable mean that the variable is a pointer to a vertex, arc, graph, or string, respectively; the suffix *.I* means that the variable is an integer. (We use one-character names because such names are easy to type when debugging.)

⟨ Type declarations 8 ⟩ ≡
    typedef union {
        struct vertex_struct *$*V$;   /* pointer to **Vertex** */
        struct arc_struct *$*A$;   /* pointer to **Arc** */
        struct graph_struct *$*G$;   /* pointer to **Graph** */
        char *$*S$;   /* pointer to string */
        long I;   /* integer */
    } **util**;
See also sections 9, 10, 12, 20, and 34.
This code is used in sections 3 and 4.

9. Each **Vertex** has two standard fields and six utility fields; hence it occupies 32 bytes on most systems, not counting the memory needed for supplementary string data. The standard fields are

                arcs,   a pointer to an **Arc**;
                name,   a pointer to a string of characters.

If v points to a **Vertex** and $v{\rightarrow}arcs$ is Λ, there are no arcs emanating from v. But if $v{\rightarrow}arcs$ is non-Λ, it points to an **Arc** record representing an arc from v, and that record has a *next* field that points in the same way to the representations of all other arcs from v.

The utility fields are called u, v, w, x, y, z. Macros can be used to give them syntactic sugar in particular applications. They are typically used to record such things as the in-degree or out-degree, or whether a vertex is 'marked'. Utility fields might also link the vertex to other vertices or arcs in one or more lists.

⟨ Type declarations 8 ⟩ +≡
    typedef struct vertex_struct {
        struct arc_struct *$*arcs$;   /* linked list of arcs coming out of this vertex */
        char *$*name$;   /* string identifying this vertex symbolically */
        util u, v, w, x, y, z;   /* multipurpose fields */
    } **Vertex**;

10. Each **Arc** has three standard fields and two utility fields. Thus it occupies 20 bytes on most computer systems. The standard fields are

                tip,   a pointer to a **Vertex**;
                next, a pointer to an **Arc**;
                len,   a (long) integer.

If a points to an **Arc** in the list of arcs from vertex v, it represents an arc of length $a\text{-}len$ from v to $a\text{-}tip$, and the next arc from v in the list is represented by $a\text{-}next$.

The utility fields are called a and b.

⟨ Type declarations 8 ⟩ +≡

```
typedef struct arc_struct {
    struct vertex_struct *tip;      /* the arc points to this vertex */
    struct arc_struct *next;        /* another arc pointing from the same vertex */
    long len;       /* length of this arc */
    util a, b;      /* multipurpose fields */
} Arc;
```

---

graph_struct: struct, §20.

11. Storage allocation. Memory space must be set aside dynamically for vertices, arcs, and their attributes. The GraphBase routines provided by GB_GRAPH accomplish this task with reasonable ease and efficiency by using the concept of memory "areas." The user should first declare an **Area** variable by saying, for example,

$$\textbf{Area } s;$$

and if this variable isn't static or otherwise known to be zero, it must be cleared initially by saying '*init_area*(*s*)'. Then any number of subroutine calls of the form '*gb_alloc*(*n, s*)' can be given; *gb_alloc* will return a pointer to a block of *n* consecutive bytes, all cleared to zero. Finally, the user can issue the command

$$gb\_free(s);$$

this statement will return all memory blocks currently allocated to area *s*, making them available for future allocation.

The number of bytes *n* specified to *gb_alloc* must be positive, and it should usually be 1000 or more, since this will reduce the number of system calls. Other routines are provided below to allocate smaller amounts of memory, such as the space needed for a single new **Arc**.

If no memory of the requested size is presently available, *gb_alloc* returns the null pointer Λ. In such cases *gb_alloc* also sets the external variable *gb_trouble_code* to a nonzero value. The user can therefore discover whether any one of an arbitrarily long series of allocation requests has failed by making a single test, '**if** (*gb_trouble_code*)'. The value of *gb_trouble_code* should be cleared to zero by every graph generation subroutine; therefore it need not be initialized to zero.

A special macro *gb_typed_alloc*(*n, t, s*) makes it convenient to allocate the space for *n* items of type *t* in area *s*.

#define *gb_typed_alloc*(*n, t, s*) (*t**)*gb_alloc*((**long**) ((*n*) * **sizeof**(*t*)), *s*)

12. The implementation of this scheme is almost ridiculously easy. The value of *n* is increased by twice the number of bytes in a pointer, and the resulting number is rounded upwards if necessary so that it's a multiple of 256. Then memory is allocated using *calloc*. The extra bytes will contain two pointers, one to the beginning of the block and one to the next block associated with the same area variable.

The **Area** type is defined to be an array of length 1. This makes it possible for users to say just '*s*' instead of '&*s*' when using an area variable as a parameter.

⟨ Type declarations 8 ⟩ +≡
#define *init_area*(*s*)   *s = Λ
    struct area_pointers {
        char *first;   /* address of the beginning of this block */
        struct area_pointers *next;   /* address of area pointers in the previously allocated block */
    };
    typedef struct area_pointers ***Area**[1];

13. First we round *n* up, if necessary, so that it's a multiple of the size of a pointer variable. Then we know we can put **area_pointers** into memory at a position *n* after any address returned by *calloc*. (This logic should work whenever the number of bytes in a pointer variable is a divisor of 256.)

The upper limit on n here is governed by old C conventions in which the first parameter to *calloc* must be less than 2^{16}. Users who need graphs with more than half a million vertices might want to raise this limit on their systems, but they would probably be better off representing large graphs in a more compact way.

⟨ External functions 13 ⟩ ≡

```
char *gb_alloc(n, s)
    long n;       /* number of consecutive bytes desired */
    Area s;       /* storage area that will contain the new block */
{ long m = sizeof(char *);      /* m is the size of a pointer variable */
  Area t;       /* a temporary pointer */
  char *loc;      /* the block address */
  if (n ≤ 0 ∨ n > #ffff00 − 2 ∗ m) {
    gb_trouble_code |= 2;      /* illegal request */
    return Λ;
  }
  n = ((n + m − 1)/m) ∗ m;      /* round up to multiple of m */
  loc = (char *) calloc((unsigned) ((n + 2 ∗ m + 255)/256), 256);
  if (loc) {
    *t = (struct area_pointers *) (loc + n);
    (*t)⁻first = loc;
    (*t)⁻next = *s;
    *s = *t;
  } else gb_trouble_code |= 1;
  return loc;
}
```

See also sections 16, 23, 26, 27, 29, 30, 31, 35, 39, 40, 44, 46, 47, and 48.

This code is used in section 3.

14. ⟨ External declarations 5 ⟩ +≡

```
long gb_trouble_code = 0;      /* did gb_alloc return Λ? */
```

15. ⟨ gb_graph.h 4 ⟩ +≡

```
extern long gb_trouble_code;      /* anomalies noted by gb_alloc */
```

---

calloc: **void** *( ), <stdlib.h>.   *gb_free*: **void** ( ), §16.

16. Notice that *gb_free*(*s*) can be called twice in a row, because the list of blocks is cleared out of the area variable *s*.

⟨ External functions 13 ⟩ +≡
  void *gb_free*(*s*)
      Area *s*;
  { **Area** *t*;
    while (∗*s*) {
      ∗*t* = (∗*s*)⃗*next*;
      free((∗*s*)⃗*first*);
      ∗*s* = ∗*t*;
    }
  }

17. The two external procedures we've defined above should be mentioned in the header file, so let's do that before we forget.

⟨ gb_graph.h 4 ⟩ +≡
  extern char ∗*gb_alloc*( );   /∗ allocate another block for an area ∗/
  #**define** *gb_typed_alloc*(*n*, *t*, *s*)   (*t*∗)*gb_alloc*((**long**) ((*n*) ∗ **sizeof**(*t*)), *s*)
  extern void *gb_free*( );   /∗ deallocate all blocks for an area ∗/

18. Here we try to allocate 10 million bytes of memory. If we succeed, fine; if not, we verify that the error was properly reported.

(An early draft of this program attempted to allocate memory until all space was exhausted. That tactic provided a more thorough test, but it was a bad idea because it brought certain large systems to their knees; it was terribly unfriendly to other users who were innocently trying to do their own work on the same machine.)

⟨ Test some intentional errors 18 ⟩ ≡
  if (*gb_alloc*(0$_L$, *s*) ≠ Λ ∨ *gb_trouble_code* ≠ 2) {
    fprintf(*stderr*, "Allocation␣error␣2␣wasn't␣reported␣properly!\n"); **return** −2;
  }
  for (; *g*⃗*vv*.*I* < 100; *g*⃗*vv*.*I*++) **if** (*gb_alloc*(100000$_L$, *s*)) *g*⃗*uu*.*I*++;
  if (*g*⃗*uu*.*I* < 100 ∧ *gb_trouble_code* ≠ 3) {
    fprintf(*stderr*, "Allocation␣error␣1␣wasn't␣reported␣properly!\n"); **return** −1;
  }
  if (*g*⃗*uu*.*I* ≡ 0) {
    fprintf(*stderr*, "I␣couldn't␣allocate␣any␣memory!\n"); **return** −3;
  }
  gb_free(*s*);   /∗ we've exhausted memory, let's put some back ∗/
  printf("Hey,␣I␣allocated␣%ld00000␣bytes␣successfully.␣Terrific...\n", *g*⃗*uu*.*I*);
  gb_trouble_code = 0;

This code is used in section 2.

19. ⟨ Declarations of test variables 19 ⟩ ≡
  Area *s*;   /∗ temporary allocations in the test routine ∗/

See also section 37.

This code is used in section 2.

20. Growing a graph.   Now we're ready to look at the **Graph** type. This is a data structure that can be passed to an algorithm that operates on graphs—to find minimum spanning trees, or strong components, or whatever.

A **Graph** record has seven standard fields and six utility fields. The standard fields are

> *vertices*,   a pointer to an array of **Vertex** records;
>
> *n*,   the total number of vertices;
>
> *m*,   the total number of arcs;
>
> *id*,   a symbolic identification giving parameters of the GraphBase procedure that generated this graph;
>
> *util_types*, a symbolic representation of the data types in utility fields;
>
> *data*,   an **Area** used for **Arc** storage and string storage;
>
> *aux_data*, an **Area** used for auxiliary information that some users might want to discard.

The utility fields are called *uu*, *vv*, *ww*, *xx*, *yy*, and *zz*.

As a consequence of these conventions, we can visit all arcs of a graph *g* by using the following program:

$$\textbf{Vertex } *v;$$
$$\textbf{Arc } *a;$$
$$\textbf{for } (v = g\text{-}vertices; \ v < g\text{-}vertices + g\text{-}n; \ v\text{++})$$
$$\quad\textbf{for } (a = v\text{-}arcs; \ a; \ a = a\text{-}next)$$
$$\qquad visit(v, a);$$

⟨ Type declarations 8 ⟩ +≡

```
#define ID_FIELD_SIZE  161
  typedef struct graph_struct {
    Vertex *vertices;      /* beginning of the vertex array */
    long n;     /* total number of vertices */
    long m;     /* total number of arcs */
    char id[ID_FIELD_SIZE];     /* GraphBase identification */
    char util_types[15];     /* usage of utility fields */
    Area data;     /* the main data blocks */
    Area aux_data;     /* subsidiary data blocks */
    util uu, vv, ww, xx, yy, zz;     /* multipurpose fields */
  } Graph;
```

Arc = struct **arc_struct**, §10.
arcs: struct **arc_struct** *, §9.
Area = struct **area_pointers** *[ ],
   §12.
first: **char** *, §12.
fprintf: **int** ( ), <stdio.h>.
free: **void** ( ), <stdlib.h>.

g: **Graph** *, §37.
gb_alloc: **char** *( ), §13.
gb_trouble_code: **long**, §14.
gb_typed_alloc = macro ( ), §11.
I: **long**, §8.
next: struct **area_pointers** *, §12.

next: struct **arc_struct** *, §10.
printf: **int** ( ), <stdio.h>.
stderr: **FILE** *, <stdio.h>.
util = **union**, §8.
Vertex = struct **vertex_struct**,
   §9.

21. The *util_types* field should always hold a string of length 14, followed as usual by a null character to terminate that string. The first six characters of *util_types* specify the usage of utility fields u, v, w, x, y, and z in **Vertex** records; the next two characters give the format of the utility fields in **Arc** records; the last six give the format of the utility fields in **Graph** records. Each character should be either I (denoting a **long** integer), S (denoting a pointer to a string), V (denoting a pointer to a **Vertex**), A (denoting a pointer to an **Arc**), G (denoting a pointer to a **Graph**), or Z (denoting an unused field that remains zero). The default for *util_types* is "ZZZZZZZZZZZZZZ", when none of the utility fields is being used.

For example, suppose that a bipartite graph g is using field $g\text{-}uu.I$ to specify the size of its first part. Suppose further that g has a string in utility field a of each **Arc** and uses utility field w of **Vertex** records to point to an **Arc**. If g leaves all other utility fields untouched, its *util_types* should be "ZZAZZZSZIZZZZZ".

The *util_types* string is presently examined only by the *save_graph* and *restore_graph* routines, which convert GraphBase graphs from internal data structures to symbolic external files and vice versa. Therefore users need not update the *util_types* when they write algorithms to manipulate graphs, unless they are going to use *save_graph* to output a graph in symbolic form, or unless they are using some other GraphBase-related software that might rely on the conventions of *util_types*. (Such software is not part of the "official" Stanford GraphBase, but it might conceivably exist some day.)

22. Some applications of bipartite graphs require all vertices of the first part to appear at the beginning of the *vertices* array. In such cases, utility field $uu.I$ is traditionally given the symbolic name n_1, and it is set equal to the size of that first part. The size of the other part is then $g\text{-}n - g\text{-}n\_1$.

#**define** n_1 $uu.I$ /∗ utility field uu may denote size of bipartite first part ∗/

⟨gb_graph.h 4⟩ +≡
#**define** n_1 $uu.I$
#**define** $mark\_bipartite(g, n1)$ $g\text{-}n\_1 = n1, g\text{-}util\_types[8] = \text{'I'}$

23. A new graph is created by calling $gb\_new\_graph(n)$, which returns a pointer to a **Graph** record for a graph with n vertices and no arcs. This function also initializes several private variables that are used by the gb_new_arc, gb_new_edge, gb_virgin_arc, and gb_save_string procedures below.

We actually reserve space for $n + extra\_n$ vertices, although claiming only n, because several graph manipulation algorithms like to add a special vertex or two to the graphs they deal with.

⟨External functions 13⟩ +≡
```
  Graph *gb_new_graph(n)
      long n;     /* desired number of vertices */
  {
    cur_graph = (Graph *) calloc(1, sizeof(Graph));
    if (cur_graph) {
      cur_graph→vertices = gb_typed_alloc(n + extra_n, Vertex, cur_graph→data);
      if (cur_graph→vertices) {
        Vertex *p;

        cur_graph→n = n;
```

$$\textbf{for } (p = \mathit{cur\_graph} \rightarrow \mathit{vertices} + n + \mathit{extra\_n} - 1; \ p \geq \mathit{cur\_graph} \rightarrow \mathit{vertices}; \ p\mathord{-}\mathord{-})$$
$$\qquad p \rightarrow \mathit{name} = \mathit{null\_string};$$
$$\mathit{sprintf}(\mathit{cur\_graph} \rightarrow \mathit{id}, \texttt{"gb\_new\_graph(\%ld)"}, n);$$
$$\mathit{strcpy}(\mathit{cur\_graph} \rightarrow \mathit{util\_types}, \texttt{"ZZZZZZZZZZZZZZ"});$$
$$\} \ \textbf{else } \{$$
$$\quad \mathit{free}((\textbf{char } *) \ \mathit{cur\_graph});$$
$$\quad \mathit{cur\_graph} = \Lambda;$$
$$\}$$
$$\}$$
$$\mathit{next\_arc} = \mathit{bad\_arc} = \Lambda;$$
$$\mathit{next\_string} = \mathit{bad\_string} = \Lambda;$$
$$\mathit{gb\_trouble\_code} = 0;$$
$$\textbf{return } \mathit{cur\_graph};$$
$$\}$$

24. The value of *extra_n* is ordinarily 4, and it should probably always be at least 4.

⟨ External declarations 5 ⟩ +≡

long *extra_n* = 4;   /∗ the number of shadow vertices allocated by *gb_new_graph* ∗/
char *null_string*[1];   /∗ a null string constant ∗/

25. ⟨ gb_graph.h 4 ⟩ +≡

extern long *extra_n*;   /∗ the number of shadow vertices allocated by *gb_new_graph* ∗/
extern char *null_string*[ ];   /∗ a null string constant ∗/
extern void *make_compound_id*( );   /∗ routine to set one *id* field from another ∗/
extern void *make_double_compound_id*( );   /∗ ditto, but from two others ∗/

---

a: util, §10.
Arc = struct **arc_struct**, §10.
bad_arc: static **Arc** ∗, §28.
bad_string: static **char** ∗, §28.
calloc: **void** ∗( ), <stdlib.h>.
cur_graph: static **Graph** ∗, §28.
data: **Area**, §20.
free: **void** ( ), <stdlib.h>.
gb_new_arc: **void** ( ), §30.
gb_new_edge: **void** ( ), §31.
gb_save_string: **char** ∗( ), §35.
gb_trouble_code: **long**, §14.
gb_typed_alloc = macro ( ), §11.
gb_virgin_arc: **Arc** ∗( ), §29.

Graph = struct **graph_struct**,
    §20.
I: **long**, §8.
id: **char** [ ], §20.
make_compound_id: **void** ( ), §26.
make_double_compound_id: **void**
    ( ), §27.
n: **long**, §20.
name: **char** ∗, §9.
next_arc: static **Arc** ∗, §28.
next_string: static **char** ∗, §28.
restore_graph: **Graph** ∗( ),
    GB_SAVE §4.
save_graph: **long** ( ), GB_SAVE §20.

sprintf: **int** ( ), <stdio.h>.
strcpy: **char** ∗( ), <string.h>.
u: util, §9.
util_types: **char** [ ], §20.
uu: util, §9.
v: util, §9.
Vertex = struct **vertex_struct**,
    §9.
vertices: **Vertex** ∗, §20.
w: util, §9.
x: util, §9.
y: util, §9.
z: util, §9.

26. The *id* field of a graph is sometimes manufactured from the *id* field of another graph. The following routines do this without allowing the string to get too long after repeated copying.

⟨ External functions 13 ⟩ +≡

  void *make_compound_id*(*g*, *s1*, *gg*, *s2*)   /∗ *sprintf*(*g*→*id*, "%s%s%s", *s1*, *gg*→*id*, *s2*) ∗/

    Graph ∗*g*;   /∗ graph whose *id* is to be set ∗/

    char ∗*s1*;   /∗ string for the beginning of the new *id* ∗/

    Graph ∗*gg*;   /∗ graph whose *id* is to be copied ∗/

    char ∗*s2*;   /∗ string for the end of the new *id* ∗/

  { **int** *avail* = ID_FIELD_SIZE − *strlen*(*s1*) − *strlen*(*s2*);

    char *tmp*[ID_FIELD_SIZE];

    strcpy(*tmp*, *gg*→*id*);

    if (*strlen*(*tmp*) < *avail*) *sprintf*(*g*→*id*, "%s%s%s", *s1*, *tmp*, *s2*);

    else *sprintf*(*g*→*id*, "%s%.∗s...)%s", *s1*, *avail* − 5, *tmp*, *s2*);

  }

27. ⟨ External functions 13 ⟩ +≡

  void *make_double_compound_id*(*g*, *s1*, *gg*, *s2*, *ggg*, *s3*)

      /∗ *sprintf*(*g*→*id*, "%s%s%s%s%s", *s1*, *gg*→*id*, *s2*, *ggg*→*id*, *s3*) ∗/

    Graph ∗*g*;   /∗ graph whose *id* is to be set ∗/

    char ∗*s1*;   /∗ string for the beginning of the new *id* ∗/

    Graph ∗*gg*;   /∗ first graph whose *id* is to be copied ∗/

    char ∗*s2*;   /∗ string for the middle of the new *id* ∗/

    Graph ∗*ggg*;   /∗ second graph whose *id* is to be copied ∗/

    char ∗*s3*;   /∗ string for the end of the new *id* ∗/

  { **int** *avail* = ID_FIELD_SIZE − *strlen*(*s1*) − *strlen*(*s2*) − *strlen*(*s3*);

    if (*strlen*(*gg*→*id*) + *strlen*(*ggg*→*id*) < *avail*) *sprintf*(*g*→*id*, "%s%s%s%s%s", *s1*, *gg*→*id*, *s2*, *ggg*→*id*, *s3*);

    else *sprintf*(*g*→*id*, "%s%.∗s...)%s%.∗s...)%s", *s1*, *avail*/2 − 5, *gg*→*id*, *s2*, (*avail* − 9)/2, *ggg*→*id*, *s3*);

  }

28. But how do the arcs get there? That's where the private variables in *gb_new_graph* come in. If *next_arc* is unequal to *bad_arc*, it points to an unused **Arc** record in a previously allocated block of **Arc** records. Similarly, *next_string* and *bad_string* are addresses used to place strings into a block of memory allocated for that purpose.

⟨ Private declarations 28 ⟩ ≡

  static Arc ∗*next_arc*;   /∗ the next **Arc** available for allocation ∗/

  static Arc ∗*bad_arc*;   /∗ but if *next_arc* = *bad_arc*, that **Arc** isn't there ∗/

  static char ∗*next_string*;   /∗ the next byte available for storing a string ∗/

  static char ∗*bad_string*;   /∗ but if *next_string* = *bad_string*, don't byte ∗/

  static Arc *dummy_arc*[2];   /∗ an **Arc** record to point to in an emergency ∗/

  static Graph *dummy_graph*;   /∗ a **Graph** record that's normally unused ∗/

  static Graph ∗*cur_graph* = &*dummy_graph*;   /∗ the **Graph** most recently created ∗/

This code is used in section 3.

29. All new **Arc** records that are created by the automatic *next_arc*/*bad_arc* scheme originate in a procedure called *gb_virgin_arc*, which returns the address of a new record having type **Arc**.

When a new block of **Arc** records is needed, we create 102 of them at once. This strategy causes exactly 2048 bytes to be allocated on most computer systems—a nice round number. The routine will still work, however, if 102 is replaced by any positive even number. The new block goes into the *data* area of *cur_graph*.

Graph-building programs do not usually call *gb_virgin_arc* directly; they generally invoke one of the higher-level routines *gb_new_arc* or *gb_new_edge* described below.

If memory space has been exhausted, *gb_virgin_arc* will return a pointer to *dummy_arc*, so that the calling procedure can safely refer to fields of the result even though *gb_trouble_code* is nonzero.

#define *arcs_per_block*   102

⟨ External functions 13 ⟩ +≡
  Arc *∗gb_virgin_arc*( )
  { **register Arc** *∗cur_arc* = *next_arc*;
    if (*cur_arc* ≡ *bad_arc*) {
      cur_arc = *gb_typed_alloc*(*arcs_per_block*, **Arc**, *cur_graph→data*);
      if (*cur_arc* ≡ Λ) *cur_arc* = *dummy_arc*;
      else {
        next_arc = *cur_arc* + 1;
        bad_arc = *cur_arc* + *arcs_per_block*;
      }
    }
    else *next_arc*++;
    return *cur_arc*;
  }

Arc = **struct arc_struct**, §10.
data: **Area**, §20.
gb_new_arc: **void** ( ), §30.
gb_new_edge: **void** ( ), §31.
gb_new_graph: **Graph** ∗( ), §23.

gb_trouble_code: **long**, §14.
gb_typed_alloc = macro ( ), §11.
Graph = **struct graph_struct**, §20.
id: **char** [], §20.

ID_FIELD_SIZE = 161, §20.
sprintf: **int** ( ), <stdio.h>.
strcpy: **char** ∗( ), <string.h>.
strlen: **int** ( ), <string.h>.

30. The routine $gb\_new\_arc(u, v, len)$ creates a new arc of length *len* from vertex u to vertex v. The arc becomes part of the graph that was most recently created by gb_new_graph—the graph pointed to by the private variable *cur_graph*. This routine assumes that u and v are both vertices in *cur_graph*.

The new arc will be pointed to by $u\text{-}arcs$, immediately after $gb\_new\_arc(u, v, len)$ has acted. If there is no room for the new arc, $gb\_trouble\_code$ is set nonzero, but $u\text{-}arcs$ will point to the non-Λ record *dummy_arc* so that additional information can safely be stored in its utility fields without risking system crashes before $gb\_trouble\_code$ is tested. However, the linking structure of arcs is apt to be fouled up in such cases; programs should make sure that $gb\_trouble\_code \equiv 0$ before doing any extensive computation on a graph.

⟨ External functions 13 ⟩ +≡

```
void gb_new_arc(u, v, len)
    Vertex *u, *v;      /* a newly created arc will go from u to v */
    long len;           /* its length */
{ register Arc *cur_arc = gb_virgin_arc( );
    cur_arc→tip = v;  cur_arc→next = u→arcs;  cur_arc→len = len;
    u→arcs = cur_arc;
    cur_graph→m++;
}
```

31. An undirected graph has "edges" instead of arcs. We represent an edge by two arcs, one going each way.

The fact that *arcs_per_block* is an even number means that the gb_new_edge routine needs to call gb_virgin_arc only once instead of twice.

Caveats: This routine, like gb_new_arc, should be used only after gb_new_graph has caused the private variable *cur_graph* to point to the graph containing the new edge. The routine gb_new_edge must not be used together with gb_new_arc or gb_virgin_arc when building a graph, unless gb_new_arc and gb_virgin_arc have been called an even number of times before gb_new_edge is invoked.

The new edge will be pointed to by $u\text{-}arcs$ and by $v\text{-}arcs$ immediately after gb_new_edge has created it, assuming that $u \neq v$. The two arcs appear next to each other in memory; indeed, gb_new_edge rigs things so that $v\text{-}arcs$ is $u\text{-}arcs + 1$ when $u < v$.

On many computers it turns out that the first **Arc** record of every such pair of arcs will have an address that is a multiple of 8, and the second **Arc** record will have an address that is not a multiple of 8 (because the first **Arc** will be 20 bytes long, and because *calloc* always returns a multiple of 8). However, it is not safe to assume this when writing portable code. Algorithms for undirected graphs can still make good use of the fact that arcs for edges are paired, without needing any mod 8 assumptions, if all edges have been created and linked into the graph by gb_new_edge: The inverse of an arc a from u to v will be arc $a + 1$ if and only if $u < v$ or $a\text{-}next = a + 1$; it will be arc $a - 1$ if and only if $u \geq v$ and $a\text{-}next \neq a + 1$. The condition $a\text{-}next = a + 1$ can hold only if $u = v$.

```
#define gb_new_graph  gb_nugraph     /* abbreviations for Procrustean linkers */
#define gb_new_arc    gb_nuarc
#define gb_new_edge   gb_nuedge
```

⟨ External functions 13 ⟩ +≡

```
void gb_new_edge(u, v, len)
    Vertex *u, *v;      /* new arcs will go from u to v and from v to u */
    long len;      /* their length */
{ register Arc *cur_arc = gb_virgin_arc( );
  if (cur_arc ≠ dummy_arc) next_arc++;
  if (u < v) {
    cur_arc⁺tip = v;  cur_arc⁺next = u⁺arcs;
    (cur_arc + 1)⁺tip = u;  (cur_arc + 1)⁺next = v⁺arcs;
    u⁺arcs = cur_arc;
    v⁺arcs = cur_arc + 1;
  } else {
    (cur_arc + 1)⁺tip = v;  (cur_arc + 1)⁺next = u⁺arcs;
    u⁺arcs = cur_arc + 1;      /* do this now in case u ≡ v */
    cur_arc⁺tip = u;  cur_arc⁺next = v⁺arcs;
    v⁺arcs = cur_arc;
  }
  cur_arc⁺len = (cur_arc + 1)⁺len = len;
  cur_graph⁺m += 2;
}
```

32. Sometimes (let us hope rarely) we might need to use a dirty trick hinted at in the previous discussion. On most computers, the mate to arc a will be $a - 1$ if and only if *edge_trick* & (**siz_t**) a is nonzero.

⟨ External declarations 5 ⟩ +≡
  siz_t *edge_trick* = **sizeof**(**Arc**) − (**sizeof**(**Arc**) & (**sizeof**(**Arc**) − 1));

33. ⟨ gb_graph.h 4 ⟩ +≡
  extern siz_t *edge_trick*;   /* least significant 1 bit in **sizeof**(**Arc**) */

34. The type **siz_t** just mentioned should be the type returned by C's **sizeof** operation; it's the basic unsigned type for machine addresses in pointers. ANSI standard C calls this type **size_t**, but we cannot safely use **size_t** in all the GraphBase programs because some older C systems mistakenly define **size_t** to be a signed type.

⟨ Type declarations 8 ⟩ +≡
  typedef unsigned long siz_t;   /* basic machine address, as signless integer */

Arc = struct arc_struct, §10.
arcs: struct arc_struct *, §9.
arcs_per_block = 102, §29.
calloc: void *( ), <stdlib.h>.
cur_graph: static Graph *, §28.

dummy_arc: static Arc [], §28.
gb_new_graph: Graph *( ), §23.
gb_trouble_code: long, §14.
gb_virgin_arc: Arc *( ), §29.
m: long, §20.

next: struct arc_struct *, §10.
next_arc: static Arc *, §28.
tip: struct vertex_struct *, §10.
Vertex = struct vertex_struct,
    §9.

35. Vertices generally have a symbolic name, and we need a place to put such names. The *gb_save_string* function is a convenient utility for this purpose: Given a null-terminated string of any length, *gb_save_string* stashes it away in a safe place and returns a pointer to that place. Memory is conserved by combining strings from the current graph into largish blocks of a convenient size.

Note that *gb_save_string* should be used only after *gb_new_graph* has provided suitable initialization, because the private variable *cur_graph* must point to the graph for which storage is currently being allocated, and because the private variables *next_string* and *bad_string* must also have suitable values.

#define *string_block_size* 1016 /* 1024 − 8 is usually efficient */
⟨ External functions 13 ⟩ +≡
  char **gb_save_string*(*s*)
      register char **s*; /* the string to be copied */
  { **register char** **p = s*;
    register long *len*; /* length of the string and the following null character */
    while (**p++*) ; /* advance to the end of the string */
    len = p − s;
    p = next_string;
    if (*p + len > bad_string*) { /* not enough room in the current block */
      long *size = string_block_size*;
      if (*len > size*) *size = len*;
      p = gb_alloc(*size*, *cur_graph→data*);
      if (*p ≡ Λ*) **return** *null_string*; /* return a pointer to "" if memory ran out */
      bad_string = p + size;
    }
    while (**s*) **p++ = *s++*; /* copy the non-null bytes of the string */
    p++ = '\0'; / and append a null character */
    next_string = p;
    return *p − len*;
  }

36. The test routine illustrates some of these basic maneuvers.

⟨ Create a small graph 36 ⟩ ≡
  g = gb_new_graph(2 ʟ);
  if (*g ≡ Λ*) {
    fprintf(*stderr*, "Oops,␣I␣couldn't␣even␣create␣a␣trivial␣graph!\n");
    return −4;
  }
  u = g→vertices; *v = u + 1*;
  u→name = gb_save_string("vertex␣0");
  v→name = gb_save_string("vertex␣1");
This code is used in section 2.

37. ⟨ Declarations of test variables 19 ⟩ +≡
  Graph **g*;
  Vertex **u*, **v*;

38. If the "edge trick" fails, the standard GraphBase routines are unaffected except for the demonstration program MILES_SPAN. (And that program uses *edge_trick* only when printing verbose comments.)

⟨ Check that the small graph is still there 38 ⟩ ≡

    if (*strncmp*(*u*→*name*, *v*→*name*, 7)) {

        fprintf(*stderr*, "Something␣is␣fouled␣up␣in␣the␣string␣storage␣machinery!\n");

        return −5;

    }

    gb_new_edge(*v*, *u*, −1ʟ);

    gb_new_edge(*u*, *u*, 1ʟ);

    gb_new_arc(*v*, *u*, −1ʟ);

    if ((*edge_trick* & (**siz_t**) (*u*→*arcs*)) ∨ (*edge_trick* & (**siz_t**) (*u*→*arcs*→*next*→*next*)) ∨ ¬(*edge_trick* & (**siz_t**)

        (*v*→*arcs*→*next*))) *printf*("Warning:␣The␣\"edge␣trick\"␣failed!\n");

    if (*v*→*name*[7] + *g*→*n* ≠ *v*→*arcs*→*next*→*tip*→*name*[7] + *g*→*m* − 2) {   /∗ '1' + 2 ≠ '0' + 5 − 2 ∗/

        fprintf(*stderr*, "Sorry,␣the␣graph␣data␣structures␣aren't␣working␣yet.\n");

        return −6;

    }

This code is used in section 2.

---

arcs: **struct arc_struct** ∗, §9.
bad_string: **static char** ∗, §28.
cur_graph: **static Graph** ∗, §28.
data: **Area**, §20.
edge_trick: **siz_t**, §32.
fprintf: **int** ( ), <stdio.h>.
gb_alloc: **char** ∗( ), §13.
gb_new_arc: **void** ( ), §30.
gb_new_edge: **void** ( ), §31.

gb_new_graph: **Graph** ∗( ), §23.
Graph = **struct graph_struct**, §20.
m: **long**, §20.
n: **long**, §20.
name: **char** ∗, §9.
next: **struct arc_struct** ∗, §10.
next_string: **static char** ∗, §28.
null_string: **char** [ ], §24.

printf: **int** ( ), <stdio.h>.
siz_t = **unsigned long**, §34.
stderr: **FILE** ∗, <stdio.h>.
strncmp: **int** ( ), <string.h>.
tip: **struct vertex_struct** ∗, §10.
Vertex = **struct vertex_struct**, §9.
vertices: **Vertex** ∗, §20.

39. Some applications might need to add arcs to several graphs at a time, violating the assumptions stated above about *cur_graph* and the other private variables. The *switch_to_graph* function gets around that restriction, by using the utility slots *ww*, *xx*, *yy*, and *zz* of **Graph** records to save and restore the private variables.

Just say *switch_to_graph*(g) in order to make *cur_graph* be *g* and to restore the other private variables that are needed by *gb_new_arc*, *gb_virgin_arc*, *gb_new_edge*, and *gb_save_string*. Restriction: The graph *g* being switched to must have previously been switched from; that is, it must have been *cur_graph* when *switch_to_graph* was called previously. Otherwise its private allocation variables will not have been saved. To meet this restriction, you should say *switch_to_graph*(Λ) just before calling *gb_new_graph*, if you intend to switch back to the current graph later.

(The swap-in-swap-out nature of these conventions may seem inelegant, but convenience and efficiency are more important than elegance when most applications do not need the ability to switch between graphs.)

⟨ External functions 13 ⟩ +≡
```
  void switch_to_graph(g)
        Graph *g;
  {
      cur_graph→ww.A = next_arc;  cur_graph→xx.A = bad_arc;
      cur_graph→yy.S = next_string;  cur_graph→zz.S = bad_string;
      cur_graph = (g ? g : &dummy_graph);
      next_arc = cur_graph→ww.A;  bad_arc = cur_graph→xx.A;
      next_string = cur_graph→yy.S;  bad_string = cur_graph→zz.S;
      cur_graph→ww.A = Λ;  cur_graph→xx.A = Λ;  cur_graph→yy.S = Λ;  cur_graph→zz.S = Λ;
  }
```

40. Finally, here's a routine that obliterates an entire graph when it is no longer needed:

⟨ External functions 13 ⟩ +≡
```
  void gb_recycle(g)
        Graph *g;
  {
    if (g) {
        gb_free(g→data);  gb_free(g→aux_data);
        free((char *) g);       /* the user must not refer to g again */
    }
  }
```

41. ⟨ gb_graph.h 4 ⟩ +≡
```
#define gb_new_graph  gb_nugraph      /* abbreviations for external linkage */
#define gb_new_arc  gb_nuarc
#define gb_new_edge  gb_nuedge
  extern Graph *gb_new_graph( );      /* create a new graph structure */
  extern void gb_new_arc( );      /* append an arc to the current graph */
  extern Arc *gb_virgin_arc( );      /* allocate a new Arc record */
  extern void gb_new_edge( );      /* append an edge (two arcs) to the current graph */
  extern char *gb_save_string( );      /* store a string in the current graph */
  extern void switch_to_graph( );      /* save allocation variables, swap in others */
  extern void gb_recycle( );      /* delete a graph structure */
```

42. Searching for vertices. We sometimes want to be able to find a vertex, given its name, and it is nice to do this in a standard way. The following simple subroutines can be used:

> *hash_in*(*v*) puts the name of vertex *v* into the hash table;
>
> *hash_out*(*s*) finds a vertex named *s*, if present in the hash table;
>
> *hash_setup*(*g*) prepares a hash table for all vertices of graph *g*;
>
> *hash_lookup*(*s, g*) looks up the name *s* in the hash table of *g*.

Routines *hash_in* and *hash_out* apply to the current graph being created, while *hash_setup* and *hash_lookup* apply to arbitrary graphs.

Important: Utility fields *u* and *v* of each vertex are reserved for use by the search routine when hashing is active. You can crash the system if you try to fool around with these values yourself, or if you use any subroutines that change those fields. The first two characters in the current graph's *util_types* field should be VV if the hash table information is to be saved by GB_SAVE.

Warning: Users of this hash scheme must preserve the number of vertices *g→n* in the current graph *g*. If *g→n* is changed, the hash table will be worthless, unless *hash_setup* is used to rehash everything.

⟨ gb_graph.h 4 ⟩ +≡
    extern void *hash_in*( );   /* input a name to the hash table of current graph */
    extern Vertex **hash_out*( );   /* find a name in hash table of current graph */
    extern void *hash_setup*( );   /* create a hash table for a given graph */
    extern Vertex **hash_lookup*( );   /* find a name in a given graph */

43. The lookup scheme is quite simple. We compute a more-or-less random value *h* based on the vertex name, where $0 \le h < n$, assuming that the graph has *n* vertices. There is a list of all vertices whose hash address is *h*, starting at (*g→vertices* + *h*)→*hash_head* and linked together in the *hash_link* fields, where *hash_head* and *hash_link* are utility fields *u.V* and *v.V*.

#define *hash_link*   *u.V*
#define *hash_head* *v.V*

A: **struct arc_struct** ***, §8.
Arc = **struct arc_struct**, §10.
aux_data: **Area**, §20.
bad_arc: **static Arc** ***, §28.
bad_string: **static char** ***, §28.
cur_graph: **static Graph** ***, §28.
data: **Area**, §20.
dummy_graph: **static Graph**, §28.
free: **void** ( ), <stdlib.h>.
gb_free: **void** ( ), §16.
gb_new_arc: **void** ( ), §30.
gb_new_edge: **void** ( ), §31.
gb_new_graph: **Graph** ***( ), §23.
gb_nuarc = short name, §31.

gb_nuedge = short name, §31.
gb_nugraph = short name, §31.
gb_save_string: **char** ***( ), §35.
gb_virgin_arc: **Arc** ***( ), §29.
Graph = **struct graph_struct**, §20.
h: **register long**, §45.
hash_in: **void** ( ), §44.
hash_lookup: **Vertex** ***( ), §48.
hash_out: **Vertex** ***( ), §46.
hash_setup: **void** ( ), §47.
n: **long**, §20.
next_arc: **static Arc** ***, §28.

next_string: **static char** ***, §28.
S: **char** ***, §8.
u: **util**, §9.
util_types: **char** [ ], §20.
v: **util**, §9.
V: **struct vertex_struct** ***, §8.
Vertex = **struct vertex_struct**, §9.
vertices: **Vertex** ***, §20.
ww: **util**, §20.
xx: **util**, §20.
yy: **util**, §20.
zz: **util**, §20.

44. ⟨ External functions 13 ⟩ +≡

void $hash\_in(v)$
  Vertex $*v$;
{ **register char** $*t = v \rightarrow name$;
 register Vertex $*u$;
 ⟨ Find vertex u, whose location is the hash code for string t 45 ⟩;
 $v \rightarrow hash\_link = u \rightarrow hash\_head$;
 $u \rightarrow hash\_head = v$;
}

45. The hash code for a string $c_1 c_2 \ldots c_l$ of length l is a nonlinear function of the characters; this function appears to produce reasonably random results between 0 and the number of vertices in the current graph. Simpler approaches were noticeably poorer in the author's tests.

 Caution: This hash coding scheme is system-dependent, because it uses the system's character codes. If you create a graph on a machine with ASCII code and save it with GB_SAVE, and if you subsequently ship the resulting text file to some friend whose machine does not use ASCII code, your friend will have to rebuild the hash structure with *hash_setup* before being able to use *hash_lookup* successfully.

#define HASH_MULT 314159   /* random multiplier */
#define HASH_PRIME 516595003   /* the 27182818th prime; it's less than 2^{29} */

⟨ Find vertex u, whose location is the hash code for string t 45 ⟩ ≡
 { **register long** h;
  for $(h = 0; \ *t; \ t{+}{+})$ {
   $h \mathrel{+}= (h \oplus (h \gg 1)) + \text{HASH\_MULT} * (\textbf{unsigned char}) \ *t$;
   while $(h \geq \text{HASH\_PRIME}) \ h \mathrel{-}= \text{HASH\_PRIME}$;
  }
  $u = cur\_graph \rightarrow vertices + (h \ \% \ cur\_graph \rightarrow n)$;
 }

This code is used in sections 44 and 46.

46. If the hash function were truly random, the average number of string comparisons made would be less than $(e^2 + 7)/8 \approx 1.80$ on a successful search, and less than $(e^2 + 1)/4 \approx 2.10$ on an unsuccessful search [*Sorting and Searching*, Section 6.4, Eqs. (15) and (16)].

⟨ External functions 13 ⟩ +≡
 Vertex $*hash\_out(s)$
   char $*s$;
 { **register char** $*t = s$;
  register Vertex $*u$;
  ⟨ Find vertex u, whose location is the hash code for string t 45 ⟩;
  for $(u = u \rightarrow hash\_head; \ u; \ u = u \rightarrow hash\_link)$
   if $(strcmp(s, u \rightarrow name) \equiv 0)$ **return** u;
  return Λ;   /* not found */
 }

47. ⟨External functions 13⟩ +≡
  void *hash_setup*(*g*)
      Graph *∗g*;
  { **Graph** *∗save_cur_graph*;
    if (*g* ∧ *g⁀n* > 0) { **register Vertex** *∗v*;
      save_cur_graph = *cur_graph*;
      cur_graph = *g*;
      for (*v* = *g⁀vertices*; *v* < *g⁀vertices* + *g⁀n*; *v*++) *v⁀hash_head* = Λ;
      for (*v* = *g⁀vertices*; *v* < *g⁀vertices* + *g⁀n*; *v*++) *hash_in*(*v*);
      g⁀util_types[0] = *g⁀util_types*[1] = 'V';   /∗ indicate usage of *hash_head* and *hash_link* ∗/
      cur_graph = *save_cur_graph*;
    }
  }

48. ⟨External functions 13⟩ +≡
  Vertex *∗hash_lookup*(*s*, *g*)
      char *∗s*;
      Graph *∗g*;
  { **Graph** *∗save_cur_graph*;
    if (*g* ∧ *g⁀n* > 0) { **register Vertex** *∗v*;
      save_cur_graph = *cur_graph*;
      cur_graph = *g*;
      v = *hash_out*(*s*);
      cur_graph = *save_cur_graph*;
      return *v*;
    }
    else return Λ;
  }

---

cur_graph: **static Graph** ∗, §28.
Graph = **struct graph_struct**,
  §20.
hash_head = *v*.*V*, §43.
hash_link = *u*.*V*, §43.

hash_lookup: **Vertex** ∗( ), §48.
n: **long**, §20.
name: **char** ∗, §9.
strcmp: **int** ( ), **<string.h>**.

util_types: **char** [ ], §20.
Vertex = **struct vertex_struct**,
  §9.
vertices: **Vertex** ∗, §20.

GB_IO

1. Introduction. This is GB_IO, the input/output module used by all GraphBase routines to access data files. It doesn't actually do any output; but somehow 'input/output' sounds like a more useful title than just 'input'.

All files of GraphBase data are designed to produce identical results on almost all existing computers and operating systems. Each line of each file contains at most 79 characters. Each character is either a blank or a digit or an uppercase letter or a lowercase letter or a standard punctuation mark. Blank characters at the end of each line are "invisible"; that is, they have no perceivable effect. Hence identical results will be obtained on record-oriented systems that pad every line with blanks.

The data is carefully sum-checked so that defective input files have little chance of being accepted.

2. Changes might be needed when these routines are ported to different systems. Sections of the program that are most likely to require such changes are listed under 'system dependencies' in the index.

A validation program is provided so that installers can tell if GB_IO is working properly. To make the test, simply run `test_io`.

⟨ `test_io.c` 2 ⟩ ≡
#include `"gb_io.h"`   /* all users of GB_IO should include this header file */
#define *exit_test*(*m*)   /* we invoke this macro if something goes wrong */
  { *fprintf*(*stderr*, `"%s!\n(Error␣code␣=␣%ld)\n"`, *m*, *io_errors*); **return** −1; }

int *main*( )
{
  ⟨ Test the *gb_open* routine; exit if there's trouble 28 ⟩;
  ⟨ Test the sample data lines; exit if there's trouble 27 ⟩;
  ⟨ Test the *gb_close* routine; exit if there's trouble 38 ⟩;
  printf(`"OK,␣the␣gb_io␣routines␣seem␣to␣work!\n"`);
  return 0;
}

3. The external variable *io_errors* mentioned in the previous section will be set nonzero if any anomalies are detected. Errors won't occur in normal use of GraphBase programs, so no attempt has been made to provide a user-friendly way to decode the nonzero values that *io_errors* might assume. Information is simply gathered in binary form; system wizards who might need to do a bit of troubleshooting should be able to decode *io_errors* without great pain.

#define *cant_open_file*   `#1`   /* bit set in *io_errors* if *fopen* fails */
#define *cant_close_file*   `#2`   /* bit set if *fclose* fails */
#define *bad_first_line*   `#4`   /* bit set if the data file's first line isn't legit */
#define *bad_second_line*   `#8`   /* bit set if the second line doesn't pass muster */
#define *bad_third_line*   `#10`   /* bit set if the third line is awry */
#define *bad_fourth_line*   `#20`   /* guess when this bit is set */
#define *file_ended_prematurely* `#40` /* bit set if *fgets* fails */
#define *missing_newline*   `#80`   /* bit set if line is too long or '\n' is missing */

#define *wrong_number_of_lines* #100   /∗ bit set if the line count is wrong ∗/
#define *wrong_checksum* #200   /∗ bit set if the check sum is wrong ∗/
#define *no_file_open* #400   /∗ bit set if user tries to close an unopened file ∗/
#define *bad_last_line* #800   /∗ bit set if final line has incorrect form ∗/

4. The C code for GB_IO doesn't have a main routine; it's just a bunch of subroutines to be incorporated into programs at a higher level via the system loading routine. Here is the general outline of **gb_io.c**:

⟨ Header files to include 7 ⟩
⟨ Preprocessor definitions ⟩
⟨ External declarations 5 ⟩
⟨ Private declarations 8 ⟩
⟨ Internal functions 9 ⟩
⟨ External functions 12 ⟩

5. Every external variable is declared twice in this **CWEB** file: once for GB_IO itself (the "real" declaration for storage allocation purposes) and once in **gb_io.h** (for cross-references by GB_IO users).

⟨ External declarations 5 ⟩ ≡
    long *io_errors*;   /∗ record of anomalies noted by GB_IO routines ∗/
This code is used in section 4.

6. ⟨ **gb_io.h** 6 ⟩ ≡
    ⟨ Header files to include 7 ⟩
    extern long *io_errors*;   /∗ record of anomalies noted by GB_IO routines ∗/
See also sections 13, 16, 19, 21, 23, 25, 29, and 41.

7. We will stick to standard C-type input conventions. We'll also have occasion to use some of the standard string operations.

⟨ Header files to include 7 ⟩ ≡
#include <stdio.h>
#ifdef SYSV
#include <string.h>
#else
#include <strings.h>
#endif
This code is used in sections 4 and 6.

---

fclose: **int** ( ), <stdio.h>.
fgets: **char** ∗( ), <stdio.h>.
fopen: **FILE** ∗( ), <stdio.h>.
fprintf: **int** ( ), <stdio.h>.
gb_close: **long** ( ), §39.
gb_open: **long** ( ), §32.
printf: **int** ( ), <stdio.h>.
stderr: **FILE** ∗, <stdio.h>.
SYSV, UNIX System 5.

8. Inputting a line. The GB_IO routines get their input from an array called *buffer*. This array is internal to GB_IO—its contents are hidden from user programs. We make it 81 characters long, since the data is supposed to have at most 79 characters per line, followed by newline and null.

⟨ Private declarations 8 ⟩ ≡
    static char *buffer*[81];   /* the current line of input */
    static char *∗cur_pos* = *buffer*;   /* the current character of interest */
    static FILE *∗cur_file*;   /* current file, or Λ is none is open */

See also sections 10, 11, and 33.

This code is used in section 4.

9. Here's a basic subroutine to fill the *buffer*. The main feature of interest is the removal of trailing blanks. We assume that *cur_file* is open.

    Notice that a line of 79 characters (followed by '\n') will just fit into the buffer, and will cause no errors. A line of 80 characters will be split into two lines and the *missing_newline* message will occur, because of the way *fgets* is defined. A *missing_newline* error will also occur if the file ends in the middle of a line, or if a null character ('\0') occurs within a line.

⟨ Internal functions 9 ⟩ ≡
    static void *fill_buf* ( )
    { **register char** *∗p*;
      if (¬*fgets*(*buffer*, 81, *cur_file*)) {
        io_errors |= *file_ended_prematurely*;
        buffer[0] = *more_data* = 0;
      }
      for (*p* = *buffer*; *∗p*; *p*++) ;   /* advance to first null character */
      if (*p*-- ≡ *buffer* ∨ *∗p* ≠ '\n') {
        io_errors |= *missing_newline*;
        p++;
      }
      while (--*p* ≥ *buffer* ∧ *∗p* ≡ '␣') ;   /* move back over trailing blanks */
      ∗++p = '\n';
      ∗++p = 0;   /* newline and null are always present at end of line */
      cur_pos = *buffer*;   /* get ready to read *buffer*[0] */
    }

See also section 15.

This code is used in section 4.

10. Checksums. Each data file has a "magic number," which is defined to be

$$\left(\sum_l 2^l c_l\right) \bmod p.$$

Here p is a large prime number, and c_l denotes the internal code corresponding to the lth-from-last data character read (including newlines but not nulls).

The "internal codes" c_l are computed in a system-independent way: Each character c in the actual encoding scheme being used has a corresponding *icode*, which is the same on all systems. For example, the *icode* of '0' is zero, regardless of whether '0' is actually represented in ASCII or EBCDIC or some other scheme. (We assume that every modern computer system is capable of printing at least 95 different characters, including a blank space.)

We will accept a data file as error-free if it has the correct number of lines and ends with the proper magic number.

⟨ Private declarations 8 ⟩ +≡
    static char *icode*[256];   /* mapping of characters to internal codes */
    static long *checksum_prime* = $(1_L \ll 30) - 83$;
      /* large prime such that $2p + 100$ won't overflow */
    static long *magic*;   /* current checksum value */
    static long *line_no*;   /* current line number in file */
    static long *final_magic*;   /* desired final magic number */
    static long *tot_lines*;   /* total number of data lines */
    static char *more_data*;   /* is there data still waiting to be read? */

11. The *icode* mapping is defined by a single string, *imap*, such that character *imap*[k] has *icode* value k. There are 96 characters in *imap*, namely the 94 standard visible ASCII codes plus space and newline. If EBCDIC code is used instead of ASCII, the cents sign ¢ should take the place of single-left-quote ', and ¬ should take the place of ~.

All characters that don't appear in *imap* are given the same *icode* value, called *unexpected_char*. Such characters should be avoided in GraphBase files whenever possible. (If they do appear, they can still get into a user's data, but we don't distinguish them from each other for checksumming purposes.)

The *icode* table actually plays a dual role, because we've rigged it so that codes 0–15 come from the characters "0123456789ABCDEF". This facilitates conversion of decimal and hexadecimal data. We can also use it for radices higher than 16.

#define *unexpected_char* 127   /* default *icode* value */

⟨ Private declarations 8 ⟩ +≡
    static char *∗imap* = "0123456789ABCDEFGHIJKLMNOPQRSTUVWXYZabcdefghijklmnopqrstuvw\
      xyz_^~&@,;.:?!%#$+-*/|\\<=>()[]{}‘'\"␣\n";

---

fgets: **char** ∗(), <stdio.h>.   *file_ended_prematurely* = #40, §3.   *k*: **register long**, §15.
FILE, <stdio.h>.   *io_errors*: **long**, §5.   *missing_newline* = #80, §3.

12. Users of GB_IO can look at the *imap*, but they can't change it.

⟨ External functions 12 ⟩ ≡
> **char** *imap_chr*(d)
>> **long** d;
>
> {
>> **return** d < 0 ∨ d > strlen(imap) ? '\0' : imap[d];
>
> }
> **long** *imap_ord*(c)
>> **char** c;
>
> {
>> ⟨ Make sure that *icode* has been initialized 14 ⟩;
>> **return** (c < 0 ∨ c > 255) ? *unexpected_char* : *icode*[c];
>
> }

See also sections 17, 18, 20, 22, 24, 26, 30, 32, 39, and 42.

This code is used in section 4.

13. ⟨ gb_io.h 6 ⟩ +≡
> #**define** *unexpected_char*   127
> **extern char** *imap_chr*();   /* the character that maps to a given character */
> **extern long** *imap_ord*();   /* the ordinal number of a given character */

14. ⟨ Make sure that *icode* has been initialized 14 ⟩ ≡
> **if** (¬*icode*['1']) *icode_setup*();

This code is used in sections 12 and 30.

15. ⟨ Internal functions 9 ⟩ +≡
> **static void** *icode_setup*()
> { **register long** k;
>> **register char** *p;
>> **for** (k = 0; k < 256; k++) *icode*[k] = *unexpected_char*;
>> **for** (p = imap, k = 0; *p; p++, k++) *icode*[*p] = k;
>
> }

16. Now we're ready to specify some external subroutines that do input. Calling *gb_newline*( ) will read the next line of data into *buffer* and update the magic number accordingly.

⟨ gb_io.h 6 ⟩ +≡
> **extern void** *gb_newline*();   /* advance to next line of the data file */
> **extern long** *new_checksum*();   /* compute change in magic number */

17. Users can compute checksums as *gb_newline* does, but they can't change the (private) value of *magic*.

⟨ External functions 12 ⟩ +≡
> **long** *new_checksum*(s, old_checksum)
>> **char** *s;   /* a string */
>> **long** old_checksum;
>
> { **register long** a = old_checksum;
>> **register char** *p;
>> **for** (p = s; *p; p++) a = (a + a + icode[*p]) % *checksum_prime*;

```
      return a;
   }
```

18.   The magic checksum is not affected by lines that begin with *.

⟨ External functions 12 ⟩ +≡

```
   void gb_newline( )
   {
      if (++line_no > tot_lines) more_data = 0;
      if (more_data) {
         fill_buf( );
         if (buffer[0] ≠ '*') magic = new_checksum(buffer, magic);
      }
   }
```

19.   Another simple routine allows a user to read (but not write) the variable *more_data*.

⟨ gb_io.h 6 ⟩ +≡

```
   extern long gb_eof( );      /* has the data all been read? */
```

20.   ⟨ External functions 12 ⟩ +≡

```
   long gb_eof( )
   {
      return ¬more_data;
   }
```

---

buffer: **static char** [ ], §8.
checksum_prime: **static long**, §10.
fill_buf: **static void** ( ), §9.
icode: **static char** [ ], §10.

imap: **static char** *, §11.
line_no: **static long**, §10.
magic: **static long**, §10.
more_data: **static char**, §10.

strlen: **int** ( ), <string.h>.
tot_lines: **static long**, §10.
unexpected_char = 127, §11.

21. Parsing a line. The user can input characters from the buffer in several ways. First, there's a basic *gb_char*( ) routine, which returns a single character. The character is '\n' if the last character on the line has already been read (and it continues to be '\n' until the user calls *gb_newline*).

The current position in the line, *cur_pos*, always advances when *gb_char* is called, unless *cur_pos* was already at the end of the line. There's also a *gb_backup*( ) routine, which moves *cur_pos* one place to the left unless it was already at the beginning.

⟨ `gb_io.h` 6 ⟩ +≡
> **extern char** *gb_char*( ); /* get next character of current line, or '\n' */
> **extern void** *gb_backup*( ); /* move back ready to scan a character again */

22. ⟨ External functions 12 ⟩ +≡
> **char** *gb_char*( )
> {
>   **if** (**cur_pos*) **return** (**cur_pos* ++);
>   **return** '\n';
> }
> **void** *gb_backup*( )
> {
>   **if** (*cur_pos* > *buffer*) *cur_pos* −−;
> }

23. There are two ways to read numerical data. The first, *gb_digit*(*d*), expects to read a single character in radix *d*, using *icode* values to specify digits greater than 9. (Thus, for example, 'A' represents the hexadecimal digit for decimal 10.) If the next character is a valid *d*-git, *cur_pos* moves to the next character and the numerical value is returned. Otherwise *cur_pos* stays in the same place and −1 is returned.

The second routine, *gb_number*(*d*), reads characters and forms an unsigned radix-*d* number until the first non-digit is encountered. The resulting number is returned; it is zero if no digits were found. No errors are possible with this routine, because it uses **unsigned long** arithmetic.

⟨ `gb_io.h` 6 ⟩ +≡
> **extern long** *gb_digit*( ); /* *gb_digit*(*d*) reads a digit between 0 and *d* − 1 */
> **extern unsigned long** *gb_number*( ); /* *gb_number*(*d*) reads a radix-*d* number */

24. The value of *d* should be at most 127, if users want their programs to be portable, because C does not treat larger **char** values in a well-defined manner. In most applications, *d* is of course either 10 or 16.

⟨ External functions 12 ⟩ +≡
> **long** *gb_digit*(*d*)
>   **char** *d*;
> {
>   **if** (*icode*[**cur_pos*] < *d*) **return** *icode*[**cur_pos* ++];
>   **return** −1;
> }

```
unsigned long gb_number(d)
      char d;
   { register unsigned long a = 0;
      icode[0] = d;      /* make sure '\0' is a nondigit */
      while (icode[*cur_pos] < d)  a = a * d + icode[*cur_pos ++];
      return a;
   }
```

25. The final subroutine for fetching data is $gb\_string(p, c)$, which stores a null-terminated string into locations starting at p. The string starts at *cur_pos* and ends just before the first appearance of character c. If $c \equiv$ '\n', the string will stop at the end of the line. If c doesn't appear in the buffer at or after *cur_pos*, the last character of the string will be the '\n' that is always inserted at the end of a line, unless the entire line has already been read. (If the entire line has previously been read, the empty string is always returned.) After the string has been copied, *cur_pos* advances past it.

In order to use this routine safely, the user should first check that there is room to store up to 81 characters beginning at location p. A suitable place to put the result, called *str_buf*, is provided for the user's convenience.

The location following the stored string is returned. Thus, if the stored string has length l (not counting the null character that is stored at the end), the value returned will be $p + l + 1$.

⟨ gb_io.h 6 ⟩ +≡
```
#define STR_BUF_LENGTH  160
   extern char str_buf[];      /* safe place to receive output of gb_string */
   extern char *gb_string();
   /* gb_string(p, c) reads a string delimited by c into bytes starting at p */
```

26. #**define** STR_BUF_LENGTH 160
⟨ External functions 12 ⟩ +≡
```
   char str_buf[STR_BUF_LENGTH];      /* users can put strings here if they wish */
   char *gb_string(p, c)
        char *p;      /* where to put the result */
        char c;       /* character following the string */
   {
      while (*cur_pos ∧ *cur_pos ≠ c)  *p++ = *cur_pos ++;
      *p++ = 0;
      return p;
   }
```

---

buffer: **static char** [], §8.   *gb_newline*: **void** ( ), §18.   *icode*: **static char** [], §10.
cur_pos: **static char** *, §8.

27. Here's how we test those routines in `test_io`: The first line of test data consists of 79 characters, beginning with 64 zeroes and ending with '`123456789ABCDEF`'. The second line is completely blank. The third and final line says '`Oops:(intentional mistake)`'.

⟨ Test the sample data lines; exit if there's trouble 27 ⟩ ≡

    if $(gb\_number(10) \neq 123456789)$ io_errors $|= 1_L \ll 20$;   /∗ decimal number not working ∗/

    if $(gb\_digit(16) \neq 10)$ io_errors $|= 1_L \ll 21$;   /∗ we missed the **A** following the decimal number ∗/

    $gb\_backup(\,)$; $gb\_backup(\,)$;   /∗ get set to read '9A' again ∗/

    if $(gb\_number(16) \neq {}^\#9ABCDEF)$ io_errors $|= 1_L \ll 22$;   /∗ hexadecimal number not working ∗/

    $gb\_newline(\,)$;   /∗ now we should be scanning a blank line ∗/

    if $(gb\_char(\,) \neq$ '\n'$)$ io_errors $|= 1_L \ll 23$;   /∗ newline not inserted at end ∗/

    if $(gb\_char(\,) \neq$ '\n'$)$ io_errors $|= 1_L \ll 24$;   /∗ newline not implied after end ∗/

    if $(gb\_number(60) \neq 0)$ io_errors $|= 1_L \ll 25$;   /∗ number should stop at null character ∗/

    { **char** $temp[100]$;

        if $(gb\_string(temp,$ '\n'$) \neq temp + 1)$ io_errors $|= 1_L \ll 26$;

        /∗ string should be null after end of line ∗/

    $gb\_newline(\,)$;

        if $(gb\_string(temp,$ ':'$) \neq temp + 5 \lor strcmp(temp, $"Oops"$))$ io_errors $|= 1_L \ll 27$;

        /∗ string not read properly ∗/

    }

    if (io_errors) $exit\_test($"Sorry,␣it␣failed.␣Look␣at␣the␣error␣code␣for␣clues"$)$;

    if $(gb\_digit(10) \neq -1)$ $exit\_test($"Digit␣error␣not␣detected"$)$;

    if $(gb\_char(\,) \neq$ ':'$)$ io_errors $|= 1_L \ll 28$;   /∗ lost synch after gb_string and gb_digit ∗/

    if $(gb\_eof(\,))$ io_errors $|= 1_L \ll 29$;   /∗ premature end-of-file indication ∗/

    $gb\_newline(\,)$;

    if $(\neg gb\_eof(\,))$ io_errors $|= 1_L \ll 30$;   /∗ postmature end-of-file indication ∗/

This code is used in section 2.

28. Opening a file. The call *gb_raw_open*("foo") will open file "foo" and initialize the checksumming process. If the file cannot be opened, *io_errors* will be set to *cant_open_file*, otherwise *io_errors* will be initialized to zero.

The call *gb_open*("foo") is a stronger version of *gb_raw_open*, which is used for standard GraphBase data files like "words.dat" to make doubly sure that they have not been corrupted. It returns the current value of *io_errors*, which will be nonzero if any problems were detected at the beginning of the file.

⟨ Test the *gb_open* routine; exit if there's trouble 28 ⟩ ≡
    if (*gb_open*("test.dat") ≠ 0) *exit_test*("Can't␣open␣test.dat");

This code is used in section 2.

29. #define *gb_raw_open* *gb_r_open*   /∗ abbreviation for Procrustean external linkage ∗/
⟨ gb_io.h 6 ⟩ +≡
#define *gb_raw_open* *gb_r_open*
    extern void *gb_raw_open*( );   /∗ open a file for GraphBase input ∗/
    extern long *gb_open*( );   /∗ open a GraphBase data file; return 0 if OK ∗/

30. ⟨ External functions 12 ⟩ +≡
    void *gb_raw_open*(*f*)
        char ∗*f*;
    {
      ⟨ Make sure that *icode* has been initialized 14 ⟩;
      ⟨ Try to open *f* 31 ⟩;
      if (*cur_file*) {
        io_errors = 0;
        more_data = 1;
        line_no = *magic* = 0;
        tot_lines = #7fffffff;   /∗ allow "infinitely many" lines ∗/
        fill_buf( );
      } **else** *io_errors* = *cant_open_file*;
    }

---

cant_open_file = #1, §3.
cur_file: **static FILE** ∗, §8.
exit_test: **int** ( ), §2.
fill_buf: **static void** ( ), §9.
gb_backup: **void** ( ), §22.
gb_char: **char** ( ), §22.
gb_digit: **long** ( ), §24.

gb_eof: **long** ( ), §20.
gb_newline: **void** ( ), §18.
gb_number: **unsigned long** ( ), §24.
gb_open: **long** ( ), §32.
gb_string: **char** ∗( ), §26.
icode: **static char** [ ], §10.

io_errors: **long**, §5.
line_no: **static long**, §10.
magic: **static long**, §10.
more_data: **static char**, §10.
strcmp: **int** ( ), <string.h>.
tot_lines: **static long**, §10.

31. Here's a possibly system-dependent part of the code: We try first to open the data file by using the file name itself as the path name; failing that, we try to prefix the file name with the name of the standard directory for GraphBase data, if the program has been compiled with `DATA_DIRECTORY` defined.

⟨ Try to open f 31 ⟩ ≡
  $cur\_file = fopen(f, "\mathtt{r}");$
#**ifdef** `DATA_DIRECTORY`
  if $(\neg cur\_file \wedge (strlen(\mathtt{DATA\_DIRECTORY}) + strlen(f) < \mathtt{STR\_BUF\_LENGTH}))$ {
    $sprintf(str\_buf, "\mathtt{\%s\%s}", \mathtt{DATA\_DIRECTORY}, f);$
    $cur\_file = fopen(str\_buf, "\mathtt{r}");$
  }
#**endif** `DATA_DIRECTORY`

This code is used in section 30.

32. ⟨ External functions 12 ⟩ +≡
  long $gb\_open(f)$
      char $*f;$
  {
    $strncpy(file\_name, f, 19);$ /* save the name for use by gb_close */
    $gb\_raw\_open(f);$
    if (cur_file) {
      ⟨ Check the first line; return if unsuccessful 34 ⟩;
      ⟨ Check the second line; return if unsuccessful 35 ⟩;
      ⟨ Check the third line; return if unsuccessful 36 ⟩;
      ⟨ Check the fourth line; return if unsuccessful 37 ⟩;
      $gb\_newline();$ /* the first line of real data is now in the buffer */
    }
    return $io\_errors;$
  }

33. ⟨ Private declarations 8 ⟩ +≡
  static char $file\_name[20];$ /* name of the data file, without a prefix */

34. The first four lines of a typical data file should look something like this:

```
* File "words.dat" from the Stanford GraphBase (C) 1993 Stanford University
* A database of English 5-letter words
* This file may be freely copied but please do not change it in any way!
* (Checksum parameters 5757,526296596)
```

We actually verify only that the first four lines of a data file named `"foo"` begin respectively with the characters

```
* File "foo"
*
*
* (Checksum parameters l,m)
```

where l and m are decimal numbers. The values of l and m are stored away as *tot_lines* and *final_magic*, to be matched at the end of the file.

⟨ Check the first line; return if unsuccessful 34 ⟩ ≡
    sprintf (*str_buf*, "*␣File␣\"%s\""*, f);
    if (*strncmp* (*buffer*, *str_buf*, *strlen* (*str_buf*))) **return** (*io_errors* |= *bad_first_line*);

This code is used in section 32.

35.   ⟨ Check the second line; return if unsuccessful 35 ⟩ ≡
    fill_buf ( );
    if (*∗buffer* ≠ '*') **return** (*io_errors* |= *bad_second_line*);

This code is used in section 32.

36.   ⟨ Check the third line; return if unsuccessful 36 ⟩ ≡
    fill_buf ( );
    if (*∗buffer* ≠ '*') **return** (*io_errors* |= *bad_third_line*);

This code is used in section 32.

37.   ⟨ Check the fourth line; return if unsuccessful 37 ⟩ ≡
    fill_buf ( );
    if (*strncmp* (*buffer*, "*␣(Checksum␣parameters␣"*, 23)) **return** (*io_errors* |= *bad_fourth_line*);
    cur_pos += 23;
    tot_lines = *gb_number* (10);
    if (*gb_char* ( ) ≠ ',') **return** (*io_errors* |= *bad_fourth_line*);
    final_magic = *gb_number* (10);
    if (*gb_char* ( ) ≠ ')') **return** (*io_errors* |= *bad_fourth_line*);

This code is used in section 32.

---

bad_first_line = #4, §3.
bad_fourth_line = #20, §3.
bad_second_line = #8, §3.
bad_third_line = #10, §3.
buffer: **static char** [], §8.
cur_file: **static FILE** ∗, §8.
cur_pos: **static char** ∗, §8.
DATA_DIRECTORY, Makefile.
f: **char** ∗, §30.

fill_buf: **static void** ( ), §9.
final_magic: **static long**, §10.
fopen: **FILE** ∗( ), <stdio.h>.
gb_char: **char** ( ), §22.
gb_close: **long** ( ), §39.
gb_newline: **void** ( ), §18.
gb_number: **unsigned long** ( ), §24.
gb_raw_open: **void** ( ), §30.

io_errors: **long**, §5.
sprintf: **int** ( ), <stdio.h>.
str_buf: **char** [], §26.
STR_BUF_LENGTH = 160, §26.
strlen: **int** ( ), <string.h>.
strncmp: **int** ( ), <string.h>.
strncpy: **char** ∗( ), <string.h>.
tot_lines: **static long**, §10.

38. Closing a file.   After all data has been input, or should have been input, we check that the file was open and that it had the correct number of lines, the correct magic number, and a correct final line. The subroutine *gb_close*, like *gb_open*, returns the value of *io_errors*, which will be nonzero if at least one problem was noticed.

⟨Test the *gb_close* routine; exit if there's trouble 38⟩ ≡

   if (*gb_close*( ) ≠ 0) *exit_test*("Bad␣checksum,␣or␣difficulty␣closing␣the␣file");

This code is used in section 2.

39. ⟨External functions 12⟩ +≡

   long *gb_close*( )
   {
     if (¬*cur_file*) **return** (*io_errors* |= *no_file_open*);
     fill_buf( );
     sprintf(*str_buf*, "*␣End␣of␣file␣\"%s\"", *file_name*);
     if (*strncmp*(*buffer*, *str_buf*, *strlen*(*str_buf*))) *io_errors* |= *bad_last_line*;
     more_data = *buffer*[0] = 0;   /* now the GB_IO routines are effectively shut down */
        /* we have *cur_pos* = *buffer* */
     if (*fclose*(*cur_file*) ≠ 0) **return** (*io_errors* |= *cant_close_file*);
     cur_file = Λ;
     if (*line_no* ≠ *tot_lines* + 1) **return** (*io_errors* |= *wrong_number_of_lines*);
     if (*magic* ≠ *final_magic*) **return** (*io_errors* |= *wrong_checksum*);
     return *io_errors*;
   }

40.   There is also a less paranoid routine, *gb_raw_close*, that closes user-generated files. It simply closes the current file, if any, and returns the value of the *magic* checksum.

   Example: The *restore_graph* subroutine in GB_SAVE uses *gb_raw_open* and *gb_raw_close* to provide system-independent input that is almost as foolproof as the reading of standard GraphBase data.

41.   #define *gb_raw_close   gb_r_close*   /* for Procrustean external linkage */

⟨gb_io.h 6⟩ +≡
#**define** *gb_raw_close   gb_r_close*
   extern long *gb_close*( );   /* close a GraphBase data file; return 0 if OK */
   extern long *gb_raw_close*( );   /* close file and return the checksum */

42.   ⟨External functions 12⟩ +≡

   long *gb_raw_close*( )
   {
     if (*cur_file*) {
       fclose(*cur_file*);
       more_data = *buffer*[0] = 0;
       cur_pos = *buffer*;
       cur_file = Λ;
     }
     return *magic*;
   }

GB_LISA

Important: Before reading GB_LISA, please read or at least skim the programs for GB_GRAPH and GB_IO.

1. Introduction. This GraphBase module contains the *lisa* subroutine, which creates rectangular matrices of data based on Leonardo da Vinci's *Gioconda* (aka Mona Lisa). It also contains the *plane_lisa* subroutine, which constructs undirected planar graphs based on *lisa*, and the *bi_lisa* subroutine, which constructs undirected bipartite graphs. Another example of the use of *lisa* can be found in the demo program ASSIGN_LISA.

#define *plane_lisa* *p_lisa*   /* abbreviation for Procrustean external linkage */

⟨ gb_lisa.h 1 ⟩ ≡
#define *plane_lisa* *p_lisa*
  extern long *∗lisa*( );
  extern Graph *∗plane_lisa*( );
  extern Graph *∗bi_lisa*( );

See also sections 3 and 25.

2. The subroutine call $lisa(m, n, d, m0, m1, n0, n1, d0, d1, area)$ constructs an $m \times n$ matrix of integers in the range $[0 .. d]$, based on the information in `lisa.dat`. Storage space for the matrix is allocated in the memory area called *area*, using the normal GraphBase conventions explained in GB_GRAPH. The entries of the matrix can be regarded as pixel data, with 0 representing black and d representing white, and with intermediate values representing shades of gray.

The data in `lisa.dat` has 360 rows and 250 columns. The rows are numbered 0 to 359 from top to bottom, and the columns are numbered 0 to 249 from left to right. The output of *lisa* is generated from a rectangular section of the picture consisting of $m1 - m0$ rows and $n1 - n0$ columns; more precisely, *lisa* uses the data in positions (k, l) for $m0 \le k < m1$ and $n0 \le l < n1$.

One way to understand the process of mapping $M = m1 - m0$ rows and $N = n1 - n0$ columns of input into m rows and n columns of output is to imagine a giant matrix of mM rows and nN columns in which the original input data has been replicated as an $M \times N$ array of submatrices of size $m \times n$; each of the submatrices contains mn identical pixel values. We can also regard the giant matrix as an $m \times n$ array of submatrices of size $M \times N$. The pixel values to be output are obtained by averaging the MN pixel values in the submatrices of this second interpretation.

More precisely, the output pixel value in a given row and column is obtained in two steps. First we sum the MN entries in the corresponding submatrix of the giant matrix, obtaining a value D between 0 and $255MN$. Then we scale the value D linearly into the desired final range $[0 .. d]$ by setting the result to 0 if $D < d0$, to d if $D \ge d1$, and to $\lfloor d(D - d0)/(d1 - d0) \rfloor$ if $d0 \le D < d1$.

#define MAX_M 360   /* the total number of rows of input data */
#define MAX_N 250   /* the total number of columns of input data */
#define MAX_D 255   /* maximum pixel value in the input data */

3. Default parameter values are automatically substituted when m, n, d, $m1$, $n1$, and/or $d1$ are given as 0: If $m1 = 0$ or $m1 > 360$, $m1$ is changed to 360; if $n1 = 0$ or $n1 > 250$, $n1$ is changed to 250. Then if m is zero, it is changed to $m1 - m0$; if n is zero, it is changed to $n1 - n0$.

If d is zero, it is changed to 255. If $d1$ is zero, it is changed to $255(m1 - m0)(n1 - n0)$. After these substitutions have been made, the parameters must satisfy

$$m0 < m1, \qquad n0 < n1, \qquad \text{and} \qquad d0 < d1.$$

Examples: The call $lisa\_pix = lisa(0, 0, 0, 0, 0, 0, 0, 0, 0, area)$ is equivalent to the call $lisa\_pix = lisa(360, 250, 255, 0, 360, 0, 250, 0, 255 * 360 * 250, area)$; this special case delivers the original `lisa.dat` data as a 360×250 array of integers in the range $[0 \mathinner{.\,.} 255]$. You can access the pixel in row k and column l by writing

$$*(lisa\_pix + n * k + l),$$

where n in this case is 250. A square array extracted from the top part of the picture, leaving out Mona's hands at the bottom, can be obtained via $lisa(250, 250, 255, 0, 250, 0, 250, 0, 0, area)$.

The call $lisa(36, 25, 25500, 0, 0, 0, 0, 0, 0, area)$ gives a 36×25 array of pixel values in the range $[0 \mathinner{.\,.} 25500]$, obtained by summing 10×10 subsquares of the original data.

The call $lisa(100, 100, 100, 0, 0, 0, 0, 0, 0, area)$ gives a 100×100 array of pixel values in the range $[0 \mathinner{.\,.} 100]$; in this case the original data is effectively broken into subpixels and averaged appropriately. Notice that each output pixel in this example comes from 3.6 input rows and 2.5 input columns; therefore the image is being distorted (compressed vertically). However, our GraphBase applications are generally interested more in combinatorial test data, not in images per se. If $(m1 - m0)/m = (n1 - n0)/n$, the output of $lisa$ will represent "square pixels." But if $(m1 - m0)/m < (n1 - n0)/n$, a halftone generated from the output will be compressed in the horizontal dimension; if $(m1 - m0)/m > (n1 - n0)/n$, it will be compressed in the vertical dimension.

If you want to reduce the original image to binary data, with the value 0 wherever the original pixels are less than some threshold value t and the value 1 whenever they are t or more, call $lisa(m, n, 1, m0, m1, n0, n1, 0, t * (m1 - m0) * (n1 - n0), area)$.

The subroutine call $lisa(1000, 1000, 255, 0, 250, 0, 250, 0, 0, area)$ produces a million pixels from the upper part of the original image. This matrix contains more entries than the original data in `lisa.dat`, but of course it is not any more accurate; it has simply been obtained by linear interpolation—in fact, by replicating the original data in 4×4 subarrays.

Mona Lisa's famous smile appears in the 16×32 subarray defined by $m0 = 94$, $m1 = 110$, $n0 = 97$, $n1 = 129$. The *smile* macro makes this easily accessible. (See also *eyes*.)

A string $lisa\_id$ is constructed, showing the actual parameter values used by $lisa$ after defaults have been supplied. The *area* parameter is omitted from this string.

⟨ `gb_lisa.h` 1 ⟩ +≡
#define *smile* $m0 = 94, m1 = 110, n0 = 97, n1 = 129$ /* 16×32 */
#define *eyes* $m0 = 61, m1 = 80, n0 = 91, n1 = 140$ /* 20×50 */
   extern char $lisa\_id[\,]$;

---

area: **Area**, §6.
bi_lisa: **Graph** *(), §33.
d: **unsigned long**, §6.
d0: **unsigned long**, §6.
d1: **unsigned long**, §6.

Graph = **struct**, GB_GRAPH §20.
lisa: **long** *(), §6.
lisa_id: **char** [], §4.
m: **unsigned long**, §6.
m0: **unsigned long**, §6.

m1: **unsigned long**, §6.
n: **unsigned long**, §6.
n0: **unsigned long**, §6.
n1: **unsigned long**, §6.

4. ⟨ Global variables 4 ⟩ ≡
   char *lisa_id*[] = "lisa(360,250,9999999999,359,360,249,250,9999999999,9999999999)";

This code is used in section 6.

5. If the *lisa* routine encounters a problem, it returns Λ (NULL), after putting a nonzero number into the external variable *panic_code*. This code number identifies the type of failure. Otherwise *lisa* returns a pointer to the newly created array. (The external variable *panic_code* is defined in GB_GRAPH.)

#define *panic*(*c*) { *panic_code* = *c*; *gb_trouble_code* = 0; **return** Λ; }

6. The C file `gb_lisa.c` begins as follows. (Other subroutines come later.)

```
#include "gb_io.h"      /* we will use the GB_IO routines for input */
#include "gb_graph.h"   /* we will use the GB_GRAPH data structures */
```
  ⟨ Preprocessor definitions ⟩
  ⟨ Global variables 4 ⟩
  ⟨ Private variables 16 ⟩
  ⟨ Private subroutines 15 ⟩

  long *lisa*(*m*, *n*, *d*, *m0*, *m1*, *n0*, *n1*, *d0*, *d1*, *area*)
      unsigned long *m*, *n*;   /* number of rows and columns desired */
      unsigned long *d*;   /* maximum pixel value desired */
      unsigned long *m0*, *m1*;   /* input will be from rows [*m0* .. *m1*) */
      unsigned long *n0*, *n1*;   /* and from columns [*n0* .. *n1*) */
      unsigned long *d0*, *d1*;   /* lower and upper threshold of raw pixel scores */
      Area *area*;   /* where to allocate the matrix that will be output */
  { ⟨ Local variables for *lisa* 7 ⟩
  ⟨ Check the parameters and adjust them for defaults 8 ⟩;
  ⟨ Allocate the matrix 9 ⟩;
  ⟨ Read `lisa.dat` and map it to the desired output form 10 ⟩;
  return *matx*;
  }

7. ⟨ Local variables for *lisa* 7 ⟩ ≡
  long *matx* = Λ;   /* the matrix constructed by *lisa* */
  register long *k*, *l*;   /* the current row and column of output */
  register long *i*, *j*;   /* all-purpose indices */
  long *cap_M*, *cap_N*;   /* *m1* − *m0* and *n1* − *n0*, dimensions of the input */
  long *cap_D*;   /* *d1* − *d0*, scale factor */

See also sections 11 and 14.

This code is used in section 6.

8. ⟨ Check the parameters and adjust them for defaults 8 ⟩ ≡
  if (*m1* ≡ 0 ∨ *m1* > MAX_M) *m1* = MAX_M;
  if (*m1* ≤ *m0*) *panic*(*bad_specs* + 1);   /* *m0* must be less than *m1* */
  if (*n1* ≡ 0 ∨ *n1* > MAX_N) *n1* = MAX_N;
  if (*n1* ≤ *n0*) *panic*(*bad_specs* + 2);   /* *n0* must be less than *n1* */
  cap_M = *m1* − *m0*; *cap_N* = *n1* − *n0*;
  if (*m* ≡ 0) *m* = *cap_M*;
  if (*n* ≡ 0) *n* = *cap_N*;

if $(d \equiv 0)$ $d = \text{MAX\_D}$;
if $(d1 \equiv 0)$ $d1 = \text{MAX\_D} * cap\_M * cap\_N$;
if $(d1 \leq d0)$ $panic(bad\_specs + 3)$;   /* $d0$ must be less than $d1$ */
if $(d1 \geq {}^{\#}80000000)$ $panic(bad\_specs + 4)$;   /* $d1$ must be less than 2^{31} */
$cap\_D = d1 - d0$;
$sprintf(lisa\_id, \texttt{"lisa(\%lu,\%lu,\%lu,\%lu,\%lu,\%lu,\%lu,\%lu,\%lu)"}, m, n, d, m0, m1, n0, n1, d0, d1)$;

This code is used in section 6.

9. ⟨ Allocate the matrix 9 ⟩ ≡
$matx = gb\_typed\_alloc(m * n, \textbf{long}, area)$;
  if $(gb\_trouble\_code)$ $panic(no\_room + 1)$;   /* no room for the output data */

This code is used in section 6.

10.   ⟨ Read `lisa.dat` and map it to the desired output form 10 ⟩ ≡
  ⟨ Open the data file, skipping unwanted rows at the beginning 19 ⟩;
  ⟨ Generate the m rows of output 13 ⟩;
  ⟨ Close the data file, skipping unwanted rows at the end 20 ⟩;

This code is used in section 6.

Area, GB_GRAPH §12.
bad_specs = 30, GB_GRAPH §7.
gb_trouble_code: **long**,
  GB_GRAPH §14.

gb_typed_alloc = macro ( ),
  GB_GRAPH §11.
MAX_D = 255, §2.
MAX_M = 360, §2.

MAX_N = 250, §2.
no_room = 1, GB_GRAPH §7.
panic_code: **long**, GB_GRAPH §5.
sprintf: **int** ( ), <stdio.h>.

11.   Elementary image processing.   As mentioned in the introduction, we can visualize the input as a giant $mM \times nN$ matrix, into which an $M \times N$ image is placed by replication of pixel values, and from which an $m \times n$ image is derived by summation of pixel values and subsequent scaling. Here $M = m1 - m0$ and $N = n1 - n0$.

Let (κ, λ) be a position in the giant matrix, where $0 \leq \kappa < mM$ and $0 \leq \lambda < nN$. The corresponding indices of the input image are then $\left(m0 + \lfloor \kappa/m \rfloor, n0 + \lfloor \lambda/n \rfloor\right)$, and the corresponding indices of the output image are $\left(\lfloor \kappa/M \rfloor, \lfloor \lambda/N \rfloor\right)$. Our main job is to compute the sum of all pixel values that lie in each given row k and column l of the output image. Many elements are repeated in the sum, so we want to use multiplication instead of simple addition whenever possible.

For example, let's consider the inner loop first, the loop on l and λ. Suppose $n = 3$, and suppose the input pixels in the current row of interest are $\langle a_0, \ldots, a_{N-1} \rangle$. Then if $N = 3$, we want to compute the output pixels $\langle 3a_0, 3a_1, 3a_2 \rangle$; if $N = 4$, we want to compute $\langle 3a_0 + a_1, 2a_1 + 2a_2, a_2 + 3a_3 \rangle$; if $N = 2$, we want to compute $\langle 2a_1, a_0 + a_1, 2a_1 \rangle$. The logic for doing this computation with the proper timing can be expressed conveniently in terms of four local variables:

⟨ Local variables for *lisa* 7 ⟩ +≡
   long **cur_pix*;   /* current position within *in_row* */
   long *lambda*;   /* right boundary in giant for the input pixel in *cur_pix* */
   long *lam*;   /* the first giant column not yet used in the current row */
   long *next_lam*;   /* right boundary in giant for the output pixel in column *l* */

12.   ⟨ Process one row of pixel sums, multiplying them by f 12 ⟩ ≡
   lambda = n;   *cur_pix = in_row + n0*;
   for (*l = lam = 0*; *l < n*; *l++*) { **register long** *sum = 0*;
      next_lam = lam + cap_N;
      do { **register long** *nl*;   /* giant column where something new might happen */
         if (*lam ≥ lambda*) *cur_pix++, lambda += n*;
         if (*lambda < next_lam*) *nl = lambda*;
         else *nl = next_lam*;
         sum += (nl - lam) * (*cur_pix);
         lam = nl;
      } **while** (*lam < next_lam*);
      *(out_row + l) += f * sum;
   }

This code is used in section 13.

13.   The outer loop (on k and κ) is similar but slightly more complicated, because it deals with a vector of sums instead of a single sum and because it must invoke the input routine when we're done with a row of input data.

⟨ Generate the m rows of output 13 ⟩ ≡
   kappa = 0;
   out_row = matx;
   for (*k = kap = 0*; *k < m*; *k++*) {
      for (*l = 0*; *l < n*; *l++*) **(out_row + l) = 0*;   /* clear the vector of sums */
      next_kap = kap + cap_M;
      do { **register long** *nk*;   /* giant row where something new might happen */

```
        if (kap ≥ kappa) {
            ⟨ Read a row of input into in_row 21 ⟩;
            kappa += m;
        }
        if (kappa < next_kap)  nk = kappa;
        else  nk = next_kap;
        f = nk − kap;
        ⟨ Process one row of pixel sums, multiplying them by f 12 ⟩;
        kap = nk;
    }  while (kap < next_kap);
    for (l = 0;  l < n;  l++, out_row ++)      /* note that out_row will advance by n */
        ⟨ Scale the sum found in *out_row 18 ⟩;
}
```

This code is used in section 10.

14. ⟨ Local variables for *lisa* 7 ⟩ +≡

    long *kappa*;   /* bottom boundary in giant for the input pixels in *in_row* */

    long *kap*;   /* the first giant row not yet used */

    long *next_kap*;   /* bottom boundary in giant for the output pixel in row k */

    long *f*;   /* factor by which current input sums should be replicated */

    long **out_row*;   /* current position in *matx* */

---

cap_M: **long**, §7.
cap_N: **long**, §7.
in_row: **static long** [], §22.
k: **register long**, §7.
l: **register long**, §7.

lisa: **long** *( )*, §6.
m: **unsigned long**, §6.
m0: **unsigned long**, §6.
m1: **unsigned long**, §6.

matx: **long** *, §7.
n: **unsigned long**, §6.
n0: **unsigned long**, §6.
n1: **unsigned long**, §6.

15. Integer scaling. Here's a general-purpose routine to compute $\lfloor na/b \rfloor$ exactly without risking integer overflow, given integers $n \geq 0$ and $0 < a \leq b$. The idea is to solve the problem first for $n/2$, if n is too large.

We are careful to precompute values so that integer overflow cannot occur when b is very large.

#define *el_gordo* #7fffffff   /* $2^{31} - 1$, the largest single-precision **long** */

⟨ Private subroutines 15 ⟩ ≡
  static long *na_over_b*(*n*, *a*, *b*)
      long *n*, *a*, *b*;
  { **long** *nmax* = *el_gordo*/*a*;   /* the largest *n* such that *na* doesn't overflow */
    register long *r*, *k*, *q*, *br*;
    long *a_thresh*, *b_thresh*;

    if $(n \leq nmax)$ **return** $(n * a)/b$;
    a_thresh = $b - a$; *b_thresh* = $(b + 1) \gg 1$;   /* $\lceil b/2 \rceil$ */
    $k = 0$;
    do { *bit*[*k*] = *n* & 1;   /* save the least significant bit of *n* */
      $n \gg= 1$;   /* and shift it out */
      k++;
    } **while** $(n > nmax)$;
    $r = n * a$; $q = r/b$; $r = r - q * b$;
    ⟨ Maintain quotient *q* and remainder *r* while increasing *n* back to its original value
        $2^k n + (bit[k-1] \ldots bit[0])_2$ 17 ⟩;
    return *q*;
  }

See also section 32.

This code is used in section 6.

16. ⟨ Private variables 16 ⟩ ≡
  static long *bit*[30];   /* bits shifted out of *n* */

See also section 22.

This code is used in section 6.

17. ⟨ Maintain quotient *q* and remainder *r* while increasing *n* back to its original value
    $2^k n + (bit[k-1] \ldots bit[0])_2$ 17 ⟩ ≡
  do { *k*--; $q \ll= 1$;
    if $(r < b\_thresh)$ $r \ll= 1$; **else** *q*++, *br* = $(b - r) \ll 1$, $r = b - br$;
    if (*bit*[*k*]) {
      if $(r < a\_thresh)$ $r += a$; **else** *q*++, $r -= a\_thresh$;
    }
  } **while** (*k*);

This code is used in section 15.

18. ⟨ Scale the sum found in *out_row 18 ⟩ ≡
  if (**out_row* ≤ *d0*) **out_row* = 0;
  else if (**out_row* ≥ *d1*) **out_row* = *d*;
  else **out_row* = *na_over_b*(*d*, **out_row* − *d0*, *cap_D*);

This code is used in section 13.

19. Input data format. The file `lisa.dat` contains 360 rows of pixel data, and each row appears on five consecutive lines of the file. The first four lines contain the data for 60 pixels; each sequence of four pixels is represented by five radix-85 digits, using the *icode* mapping of GB_IO. The fifth and final line of each row contains $4 + 4 + 2 = 10$ more pixels, represented as $5 + 5 + 3$ radix-85 digits.

⟨ Open the data file, skipping unwanted rows at the beginning 19 ⟩ ≡
    if $(gb\_open(\texttt{"lisa.dat"}) \neq 0)$ *panic*(*early_data_fault*);
        /* couldn't open the file; *io_errors* tells why */
    for $(i = 0;\ i < m0;\ i{+}{+})$
        for $(j = 0;\ j < 5;\ j{+}{+})$ *gb_newline*( );   /* ignore one row of data */

This code is used in section 10.

20. ⟨ Close the data file, skipping unwanted rows at the end 20 ⟩ ≡
    for $(i = m1;\ i < \texttt{MAX\_M};\ i{+}{+})$
        for $(j = 0;\ j < 5;\ j{+}{+})$ *gb_newline*( );   /* ignore one row of data */
    if $(gb\_close(\,) \neq 0)$ *panic*(*late_data_fault*);
        /* check sum or other failure in data file; see *io_errors* */

This code is used in section 10.

21. ⟨ Read a row of input into *in_row* 21 ⟩ ≡
    { **register long** *dd*;
        for $(j = 15, cur\_pix = \&in\_row[0];\ ;\ cur\_pix\ {+}{=}\ 4)$ {
            $dd = gb\_digit(85);\ dd = dd * 85 + gb\_digit(85);\ dd = dd * 85 + gb\_digit(85);$
            if $(cur\_pix \equiv \&in\_row[\texttt{MAX\_N} - 2])$ **break**;
            $dd = dd * 85 + gb\_digit(85);\ dd = dd * 85 + gb\_digit(85);$
            $*(cur\_pix + 3) = dd\ \&\ {}^{\#}\texttt{ff};\ dd = (dd \gg 8)\ \&\ {}^{\#}\texttt{ffffff};$
            $*(cur\_pix + 2) = dd\ \&\ {}^{\#}\texttt{ff};\ dd \gg{=} 8;$
            $*(cur\_pix + 1) = dd\ \&\ {}^{\#}\texttt{ff};\ *cur\_pix = dd \gg 8;$
            if $({-}{-}j \equiv 0)$ *gb_newline*( ), $j = 15;$
        }
        $*(cur\_pix + 1) = dd\ \&\ {}^{\#}\texttt{ff};\ *cur\_pix = dd \gg 8;$ *gb_newline*( );
    }

This code is used in section 13.

22. ⟨ Private variables 16 ⟩ +≡
    static long *in_row*[MAX_N];

---

cap_D: **long**, §7.
cur_pix: **long** *, §11.
d: **unsigned long**, §6.
d0: **unsigned long**, §6.
d1: **unsigned long**, §6.
early_data_fault $= 10$, GB_GRAPH §7.
gb_close: **long** ( ), GB_IO §39.

gb_digit: **long** ( ), GB_IO §24.
gb_newline: **void** ( ), GB_IO §18.
gb_open: **long** ( ), GB_IO §32.
i: **register long**, §7.
icode: **static char** [], GB_IO §10.
io_errors: **long**, GB_IO §5.
j: **register long**, §7.

late_data_fault $= 11$, GB_GRAPH §7.
m0: **unsigned long**, §6.
m1: **unsigned long**, §6.
MAX_M $= 360$, §2.
MAX_N $= 250$, §2.
out_row: **long** *, §14.
panic $=$ macro ( ), §5.

23. Planar graphs.   We can obtain a large family of planar graphs based on digitizations of Mona Lisa by using the following simple scheme: Each matrix of pixels defines a set of connected regions containing pixels of the same value. (Two pixels are considered adjacent if they share an edge.) These connected regions are taken to be vertices of an undirected graph; two vertices are adjacent if the corresponding regions have at least one pixel edge in common.

We can also state the construction another way. If we take any planar graph and collapse two adjacent vertices, we obtain another planar graph. Suppose we start with the planar graph having mn vertices $[k, l]$ for $0 \le k < m$ and $0 \le l < n$, where $[k, l]$ is adjacent to $[k, l-1]$ when $l > 0$ and to $[k-1, l]$ when $k > 0$. Then we can attach pixel values to each vertex, after which we can repeatedly collapse adjacent vertices whose pixel values are equal. The resulting planar graph is the same as the graph of connected regions that was described in the previous paragraph.

The subroutine call $plane\_lisa(m, n, d, m0, m1, n0, n1, d0, d1)$ constructs the planar graph associated with the digitization produced by $lisa$. The description of $lisa$, given earlier, explains the significance of parameters m, n, d, $m0$, $m1$, $n0$, $n1$, $d0$, and $d1$. There will be at most mn vertices, and the graph will be simply an $m \times n$ grid unless d is small enough to permit adjacent pixels to have equal values. The graph will also become rather trivial if d is too small.

Utility fields $first\_pixel$ and $last\_pixel$ give, for each vertex, numbers of the form $k * n + l$, identifying the topmost/leftmost and bottommost/rightmost positions $[k, l]$ in the region corresponding to that vertex. Utility fields $matrix\_rows$ and $matrix\_cols$ in the **Graph** record contain the values of m and n; thus, in particular, the value of n needed to decompose $first\_pixel$ and $last\_pixel$ into individual coordinates can be found in $g\text{-}matrix\_cols$.

The original pixel value of a vertex is placed into its $pixel\_value$ utility field.

```
#define  pixel_value   x.I
#define  first_pixel    y.I
#define  last_pixel     z.I
#define  matrix_rows    uu.I
#define  matrix_cols    vv.I
    Graph *plane_lisa(m, n, d, m0, m1, n0, n1, d0, d1)
        unsigned long m, n;      /* number of rows and columns desired */
        unsigned long d;         /* maximum value desired */
        unsigned long m0, m1;    /* input will be from rows [m0 .. m1) */
        unsigned long n0, n1;    /* and from columns [n0 .. n1) */
        unsigned long d0, d1;    /* lower and upper threshold of raw pixel scores */
    { ⟨ Local variables for plane_lisa 24 ⟩
      init_area(working_storage);
      ⟨ Figure out the number of connected regions, regs 26 ⟩;
      ⟨ Set up a graph with regs vertices 29 ⟩;
      ⟨ Put the appropriate edges into the graph 30 ⟩;
    trouble:  gb_free(working_storage);
      if (gb_trouble_code) {
        gb_recycle(new_graph);
        panic(alloc_fault);      /* oops, we ran out of memory somewhere back there */
      }
      return new_graph;
    }
```

24. ⟨ Local variables for *plane_lisa* 24 ⟩ ≡
    Graph *∗new_graph*;   /∗ the graph constructed by *plane_lisa* ∗/
    register long *j*, *k*, *l*;   /∗ all-purpose indices ∗/
    Area *working_storage*;   /∗ tables needed while *plane_lisa* does its thinking ∗/
    long *∗a*;   /∗ the matrix constructed by *lisa* ∗/
    long *regs* = 0;   /∗ number of vertices generated so far ∗/
See also sections 27 and 31.

This code is used in section 23.

25. ⟨ gb_lisa.h 1 ⟩ +≡
#define *pixel_value* *x.I*   /∗ definitions for the header file ∗/
#define *first_pixel* *y.I*
#define *last_pixel* *z.I*
#define *matrix_rows* *uu.I*
#define *matrix_cols* *vv.I*

---

alloc_fault = −1, GB_GRAPH §7.
Area, GB_GRAPH §12.
gb_free: **void** ( ), GB_GRAPH §16.
gb_recycle: **void** ( ), GB_GRAPH §40.
gb_trouble_code: **long**,
   GB_GRAPH §14.

Graph = **struct**, GB_GRAPH §20.
I: **long**, GB_GRAPH §8.
init_area = macro ( ), GB_GRAPH §12.
lisa: **long** ∗( ), §6.
panic = macro ( ), §5.

uu: **util**, GB_GRAPH §20.
vv: **util**, GB_GRAPH §20.
x: **util**, GB_GRAPH §9.
y: **util**, GB_GRAPH §9.
z: **util**, GB_GRAPH §9.

26. The following algorithm for counting the connected regions considers the array elements $a[k, l]$ to be linearly ordered as they appear in memory. Thus we can speak of the n elements preceding a given element $a[k, l]$, if $k > 0$; these are the elements $a[k, l - 1]$, ..., $a[k, 0]$, $a[k - 1, n - 1]$, ..., $a[k - 1, l]$. These n elements appear in n different columns.

During the algorithm, we move through the array from bottom right to top left, maintaining an auxiliary table $\langle f[0], \ldots, f[n - 1] \rangle$ with the following significance: Whenever two of the n elements preceding our current position $[k, l]$ are connected to each other by a sequence of pixels with equal value, where the connecting links do not involve pixels more than n steps before our current position, those elements will be linked together in the f array. More precisely, we will have $f[c_1] = c_2$, ..., $f[c_{j-1}] = c_j$, and $f[c_j] = c_j$, when there are j equivalent elements in columns c_1, ..., c_j. Here c_1 will be the "last" column and c_j the "first," in wraparound order; each element with $f[c] \neq c$ points to an earlier element.

The main function of the f table is to identify the topmost/leftmost pixel of a region. If we are at position $[k, l]$ and if we find $f[l] = l$ while $a[k - 1, l] \neq a[k, l]$, there is no way to connect $[k, l]$ to earlier positions, so we create a new vertex for it.

We also change the a matrix, to facilitate another algorithm below. If position $[k, l]$ is the topmost/leftmost pixel of a region, we set $a[k, l] = -1 - a[k, l]$; otherwise we set $a[k, l] = f[l]$, the column of a preceding element belonging to the same region.

$\langle$ Figure out the number of connected regions, *regs* 26 $\rangle \equiv$
  $a = lisa(m, n, d, m0, m1, n0, n1, d0, d1, working\_storage)$;
  if $(a \equiv \Lambda)$ **return** Λ;   /* *panic_code* has been set by *lisa* */
  $sscanf(lisa\_id, \texttt{"lisa(\%lu,\%lu,"}, \&m, \&n)$;   /* adjust for defaults */
  $f = gb\_typed\_alloc(n, \textbf{unsigned long}, working\_storage)$;
  if $(f \equiv \Lambda)$ {
    $gb\_free(working\_storage)$;   /* recycle the a matrix */
    $panic(no\_room + 2)$;   /* there's no room for the f vector */
  }
  $\langle$ Pass over the a matrix from bottom right to top left, looking for the beginnings of connected
      regions 28 $\rangle$;

This code is used in section 23.

27. $\langle$ Local variables for *plane_lisa* 24 $\rangle$ +=
  unsigned long $*f$;   /* beginning of array f; $f[j]$ is the column of an equivalent element */
  long $*apos$;   /* the location of $a[k, l]$ */

28. We maintain a pointer *apos* equal to $\&a[k, l]$, so that $*(apos - 1) = a[k, l - 1]$ and $*(apos - n) = a[k - 1, l]$ when $l > 0$ and $k > 0$.

The loop that replaces $f[j]$ by j can cause this algorithm to take time mn^2. We could improve the worst case by using path compression, but the extra complication is rarely worth the trouble.

$\langle$ Pass over the a matrix from bottom right to top left, looking for the beginnings of connected
      regions 28 $\rangle \equiv$
  for $(k = m, apos = a + n * (m + 1) - 1; k \geq 0; k--)$
    for $(l = n - 1; l \geq 0; l--, apos--)$ {
      if $(k < m)$ {
        if $(k > 0 \wedge *(apos - n) \equiv *apos)$ {
          for $(j = l; f[j] \neq j; j = f[j])$;   /* find the first element */

$f[j] = l;$ /* link it to the new first element */
$*apos = l;$
} **else if** $(f[l] \equiv l)$ $*apos = -1 - *apos, regs{+}{+};$ /* new region found */
else $*apos = f[l];$
}
if $(k > 0 \wedge l < n - 1 \wedge *(apos - n) \equiv *(apos - n + 1))$ $f[l+1] = l;$
$f[l] = l;$
}

This code is used in section 26.

29. ⟨ Set up a graph with *regs* vertices 29 ⟩ ≡
$new\_graph = gb\_new\_graph(regs);$
if $(new\_graph \equiv \Lambda)$ $panic(no\_room);$ /* out of memory before we're even started */
$sprintf(new\_graph\text{-}id, \texttt{"plane\_\%s"}, lisa\_id);$
$strcpy(new\_graph\text{-}util\_types, \texttt{"ZZZIIIZZIIZZZZ"});$
$new\_graph\text{-}matrix\_rows = m;$
$new\_graph\text{-}matrix\_cols = n;$

This code is used in section 23.

---

a: **long** *, §24.
d: **unsigned long**, §23.
d0: **unsigned long**, §23.
d1: **unsigned long**, §23.
gb_free: **void** ( ), GB_GRAPH §16.
gb_new_graph: **Graph** *( ),
    GB_GRAPH §23.
gb_typed_alloc = macro ( ),
    GB_GRAPH §11.
id: **char** [ ], GB_GRAPH §20.
j: **register long**, §24.
k: **register long**, §24.

l: **register long**, §24.
lisa: **long** *( ), §6.
lisa_id: **char** [ ], §4.
m: **unsigned long**, §23.
m0: **unsigned long**, §23.
m1: **unsigned long**, §23.
matrix_cols = *vv.I*, §23.
matrix_rows = *uu.I*, §23.
n: **unsigned long**, §23.
n0: **unsigned long**, §23.
n1: **unsigned long**, §23.

new_graph: **Graph** *, §24.
no_room = 1, GB_GRAPH §7.
panic = macro ( ), §5.
panic_code: **long**, GB_GRAPH §5.
plane_lisa: **Graph** *( ), §23.
regs: **long**, §24.
sprintf: **int** ( ), <stdio.h>.
sscanf: **int** ( ), <stdio.h>.
strcpy: **char** *( ), <string.h>.
util_types: **char** [ ], GB_GRAPH §20.
working_storage: **Area**, §24.

30. Now we make another pass over the matrix, this time from top left to bottom right. An auxiliary vector of length n is once again sufficient to tell us when one region is adjacent to a previous one. In this case the vector is called u, and it contains pointers to the vertices in the n positions before our current position. We assume that a pointer to a **Vertex** takes the same amount of memory as an **unsigned long**, hence u can share the space formerly occupied by f; if this is not the case, a system-dependent change should be made here.

The vertex names are simply integers, starting with 0.

⟨ Put the appropriate edges into the graph 30 ⟩ ≡

```
regs = 0;
u = (Vertex **) f;
for (l = 0; l < n; l++) u[l] = Λ;
for (k = 0, apos = a, aloc = 0; k < m; k++)
    for (l = 0; l < n; l++, apos++, aloc++) {
        w = u[l];
        if (*apos < 0) {
            sprintf (str_buf , "%ld", regs );
            v = new_graph→vertices + regs;
            v→name = gb_save_string (str_buf );
            v→pixel_value = −*apos − 1;
            v→first_pixel = aloc;
            regs ++;
        } else  v = u[*apos];
        u[l] = v;
        v→last_pixel = aloc;
        if (gb_trouble_code) goto trouble;
        if (k > 0 ∧ v ≠ w)  adjac(v, w);
        if (l > 0 ∧ v ≠ u[l − 1])  adjac(v, u[l − 1]);
    }
```

This code is used in section 23.

31. ⟨ Local variables for *plane_lisa* 24 ⟩ +≡

```
Vertex **u;      /* table of vertices for previous n pixels */
Vertex *v;       /* vertex corresponding to position [k, l] */
Vertex *w;       /* vertex corresponding to position [k − 1, l] */
long aloc;       /* k * n + l */
```

32. The *adjac* routine makes two vertices adjacent, if they aren't already. A faster way to recognize duplicates would probably speed things up.

⟨ Private subroutines 15 ⟩ +≡

```
static void adjac(u, v)
    Vertex *u, *v;
{ Arc *a;
    for (a = u→arcs; a; a = a→next)
        if (a→tip ≡ v) return;
    gb_new_edge (u, v, 1_L);
}
```

33. Bipartite graphs. An even simpler class of Mona-Lisa-based graphs is obtained by considering the m rows and n columns to be individual vertices, with a row adjacent to a column if the associated pixel value is sufficiently large or sufficiently small. All edges have length 1.

The subroutine call $bi\_lisa(m, n, m0, m1, n0, n1, thresh, c)$ constructs the bipartite graph corresponding to the $m \times n$ digitization produced by *lisa*, using parameters $(m0, m1, n0, n1)$ to define a rectangular subpicture as described earlier. The threshold parameter *thresh* should be between 0 and 65535. If the pixel value in row k and column l is at least $thresh/65535$ of its maximum, vertices k and l will be adjacent. If $c \neq 0$, however, the convention is reversed; vertices are then adjacent when the corresponding pixel value is *smaller* than $thresh/65535$. Thus adjacencies come from "light" areas of da Vinci's painting when $c = 0$ and from "dark" areas when $c \neq 0$. There are $m + n$ vertices and up to $m \times n$ edges.

The actual pixel value is recorded in utility field $b.I$ of each arc, and scaled to be in the range $[0, 65535]$.

```
Graph *bi_lisa(m, n, m0, m1, n0, n1, thresh, c)
     unsigned long m, n;        /* number of rows and columns desired */
     unsigned long m0, m1;      /* input will be from rows [m0 .. m1) */
     unsigned long n0, n1;      /* and from columns [n0 .. n1) */
     unsigned long thresh;      /* threshold defining adjacency */
     long c;      /* should we prefer dark pixels to light pixels? */
{ ⟨Local variables for bi_lisa 34⟩
  init_area(working_storage);
  ⟨Set up a bipartite graph with m + n vertices 35⟩;
  ⟨Put the appropriate edges into the bigraph 36⟩;
  gb_free(working_storage);
  if (gb_trouble_code) {
    gb_recycle(new_graph);
    panic(alloc_fault);      /* oops, we ran out of memory somewhere back there */
  }
  return new_graph;
}
```

---

a: **long** *, §24.
alloc_fault $= -1$, GB_GRAPH §7.
apos: **long** *, §27.
Arc = **struct**, GB_GRAPH §10.
arcs: **Arc** *, GB_GRAPH §9.
b: **util**, GB_GRAPH §10.
f: **unsigned long** *, §27.
first_pixel $= y.I$, §23.
gb_free: **void** ( ), GB_GRAPH §16.
gb_new_edge: **void** ( ),
  GB_GRAPH §31.
gb_recycle: **void** ( ), GB_GRAPH §40.
gb_save_string: **char** *( ),
  GB_GRAPH §35.

gb_trouble_code: **long**,
  GB_GRAPH §14.
Graph = **struct**, GB_GRAPH §20.
I: **long**, GB_GRAPH §8.
init_area = macro ( ), GB_GRAPH §12.
k: **register long**, §24.
l: **register long**, §24.
last_pixel $= z.I$, §23.
lisa: **long** *( ), §6.
m: **unsigned long**, §23.
n: **unsigned long**, §23.
name: **char** *, GB_GRAPH §9.
new_graph: **Graph** *, §24.

new_graph: **Graph** *, §34.
next: **Arc** *, GB_GRAPH §10.
panic = macro ( ), §5.
pixel_value $= x.I$, §23.
plane_lisa: **Graph** *( ), §23.
regs: **long**, §24.
sprintf: **int** ( ), <stdio.h>.
str_buf: **char** [ ], GB_IO §26.
tip: **Vertex** *, GB_GRAPH §10.
trouble: label, §23.
Vertex = **struct**, GB_GRAPH §9.
vertices: **Vertex** *, GB_GRAPH §20.
working_storage: **Area**, §34.

34. ⟨ Local variables for *bi_lisa* 34 ⟩ ≡
  Graph *∗new_graph*;   /∗ the graph constructed by *bi_lisa* ∗/
  register long *k*, *l*;   /∗ all-purpose indices ∗/
  Area *working_storage*;   /∗ tables needed while *bi_lisa* does its thinking ∗/
  long *∗a*;   /∗ the matrix constructed by *lisa* ∗/
  long *∗apos*;   /∗ the location of *a[k, l]* ∗/
  register Vertex *∗u*, *∗v*;   /∗ current vertices of interest ∗/
This code is used in section 33.

35. ⟨ Set up a bipartite graph with $m + n$ vertices 35 ⟩ ≡
$a = lisa(m, n, 65535\,{\rm L}, m0, m1, n0, n1, 0\,{\rm L}, 0\,{\rm L}, working\_storage);$
if $(a \equiv \Lambda)$ **return** Λ;   /∗ *panic_code* has been set by *lisa* ∗/
$sscanf(lisa\_id, \texttt{"lisa(\%lu,\%lu,65535,\%lu,\%lu,\%lu,\%lu"}, \&m, \&n, \&m0, \&m1, \&n0, \&n1);$
$new\_graph = gb\_new\_graph(m + n);$
if $(new\_graph \equiv \Lambda)$ $panic(no\_room);$ /∗ out of memory before we're even started ∗/
$sprintf(new\_graph\text{-}id, \texttt{"bi\_lisa(\%lu,\%lu,\%lu,\%lu,\%lu,\%lu,\%lu,\%c)"}, m, n, m0, m1, n0, n1, thresh,$
    $c \mathbin{?} \texttt{'1'} : \texttt{'0'});$
$new\_graph\text{-}util\_types[6] = \texttt{'I'};$
$mark\_bipartite(new\_graph, m);$
for $(k = 0, v = new\_graph\text{-}vertices; \ k < m; \ k{+}{+}, v{+}{+})$ {
  $sprintf(str\_buf, \texttt{"r\%ld"}, k);$ /∗ row vertices are called \texttt{"r0"}, \texttt{"r1"}, etc. ∗/
  $v\text{-}name = gb\_save\_string(str\_buf);$
}
for $(l = 0; \ l < n; \ l{+}{+}, v{+}{+})$ {
  $sprintf(str\_buf, \texttt{"c\%ld"}, l);$ /∗ column vertices are called \texttt{"c0"}, \texttt{"c1"}, etc. ∗/
  $v\text{-}name = gb\_save\_string(str\_buf);$
}
This code is used in section 33.

36. Since we've called *lisa* with $d = 65535$, the determination of adjacency is simple.
⟨ Put the appropriate edges into the bigraph 36 ⟩ ≡
for $(u = new\_graph\text{-}vertices, apos = a; \ u < new\_graph\text{-}vertices + m; \ u{+}{+})$
  for $(v = new\_graph\text{-}vertices + m; \ v < new\_graph\text{-}vertices + m + n; \ apos{+}{+}, v{+}{+})$ {
    if $(c \mathbin{?} *apos < thresh : *apos \geq thresh)$ {
      $gb\_new\_edge(u, v, 1\,{\rm L});$
      $u\text{-}arcs\text{-}b.I = v\text{-}arcs\text{-}b.I = *apos;$
    }
  }
This code is used in section 33.

GB_MILES

Important: Before reading GB_MILES, please read or at least skim the programs for GB_GRAPH and GB_IO.

1. Introduction. This GraphBase module contains the *miles* subroutine, which creates a family of undirected graphs based on highway mileage data between North American cities. Examples of the use of this procedure can be found in the demo programs MILES_SPAN and GB_PLANE.

⟨ gb_miles.h 1 ⟩ ≡
  extern Graph *∗miles*( );

See also sections 2, 16, and 21.

2. The subroutine call

$$miles(n, north\_weight, west\_weight, pop\_weight, max\_distance, max\_degree, seed)$$

constructs a graph based on the information in `miles.dat`. Each vertex of the graph corresponds to one of the 128 cities whose name is alphabetically greater than or equal to 'Ravenna, Ohio' in the 1949 edition of Rand McNally & Company's *Standard Highway Mileage Guide*. Edges between vertices are assigned lengths representing distances between cities, in miles. In most cases these mileages come from the Rand McNally Guide, but several dozen entries needed to be changed drastically because they were obviously too large or too small; in such cases an educated guess was made. Furthermore, about 5% of the entries were adjusted slightly in order to ensure that all distances satisfy the "triangle inequality": The graph generated by *miles* has the property that the distance from u to v plus the distance from v to w always exceeds or equals the distance from u to w.

The constructed graph will have $\min(n, 128)$ vertices; the default value $n = 128$ is substituted if $n = 0$. If n is less than 128, the n cities will be selected by assigning a weight to each city and choosing the n with largest weight, using random numbers to break ties in case of equal weights. Weights are computed by the formula

$$north\_weight \cdot lat + west\_weight \cdot lon + pop\_weight \cdot pop,$$

where *lat* is latitude north of the equator, *lon* is longitude west of Greenwich, and *pop* is the population in 1980. Both *lat* and *lon* are given in "decidegrees" (hundredths of degrees). For example, San Francisco has $lat = 3778$, $lon = 12242$, and $pop = 678974$; this means that, before the recent earthquake, it was located at 37.78° north latitude and 122.42° west longitude, and that it had 678,974 residents in the 1980 census. The weight parameters must satisfy

$$|north\_weight| \leq 100{,}000, \quad |west\_weight| \leq 100{,}000, \quad |pop\_weight| \leq 100.$$

The constructed graph will be "complete"—that is, it will have edges between every pair of vertices—unless special values are given to the parameters *max_distance* or *max_degree*. If

$max\_distance \neq 0$, edges with more than $max\_distance$ miles will not appear; if $max\_degree \neq 0$, each vertex will be limited to at most max_degree of its shortest edges.

Vertices of the graph will appear in order of decreasing weight. The *seed* parameter defines the pseudo-random numbers used wherever a "random" choice between equal-weight vertices or equal-length edges needs to be made.

#**define** MAX_N 128

⟨ gb_miles.h 1 ⟩ +≡
 #**define** MAX_N 128 /∗ maximum and default number of cities ∗/

3. Examples: The call $miles(100, 0, 0, 1, 0, 0, 0)$ will construct a complete graph on 100 vertices, representing the 100 most populous cities in the database. It turns out that San Diego, with a population of 875,538, is the winning city by this criterion, followed by San Antonio (population 786,023), San Francisco (678,974), and Washington D.C. (638,432).

To get n cities in the western United States and Canada, you can say $miles(n, 0, 1, 0, \ldots)$; to get n cities in the Northeast, use a call like $miles(n, 1, -1, 0, \ldots)$. A parameter setting like $(50, -500, 0, 1, \ldots)$ produces mostly Southern cities, except for a few large metropolises in the north.

If you ask for $miles(n, a, b, c, 0, 1, 0)$, you get an edge between cities if and only if each city is the nearest to the other, among the n cities selected. (The graph is always undirected: There is an arc from u to v if and only if there's an arc of the same length from v to u.)

A random selection of cities can be obtained by calling $miles(n, 0, 0, 0, m, d, s)$. Different choices of the seed number s will produce different selections, in a system-independent manner; identical results will be obtained on all computers when identical parameters have been specified. Equivalent experiments on algorithms for graph manipulation can therefore be performed by researchers in different parts of the world. Any value of s between 0 and $2^{31} - 1$ is permissible.

4. If the *miles* routine encounters a problem, it returns Λ (NULL), after putting a code number into the external variable *panic_code*. This code number identifies the type of failure. Otherwise *miles* returns a pointer to the newly created graph, which will be represented with the data structures explained in GB_GRAPH. (The external variable *panic_code* is itself defined in GB_GRAPH.)

#**define** $panic(c)$ { $panic\_code = c$; $gb\_trouble\_code = 0$; **return** Λ; }

---

gb_trouble_code: **long**, GB_GRAPH §14.
Graph = **struct**, GB_GRAPH §20.
max_degree: **unsigned long**, §5.

max_distance: **unsigned long**, §5.
miles: **Graph** ∗( ), §5.
n: **unsigned long**, §5.
north_weight: **long**, §5.

panic_code: **long**, GB_GRAPH §5.
pop_weight: **long**, §5.
seed: **long**, §5.
west_weight: **long**, §5.

5. The C file `gb_miles.c` has the following overall shape:

```
#include "gb_io.h"     /* we will use the GB_IO routines for input */
#include "gb_flip.h"    /* we will use the GB_FLIP routines for random numbers */
#include "gb_graph.h"    /* we will use the GB_GRAPH data structures */
#include "gb_sort.h"    /* and the linksort routine */
```
⟨ Preprocessor definitions ⟩
⟨ Type declarations 9 ⟩
⟨ Private variables 10 ⟩

Graph $*miles(n, north\_weight, west\_weight, pop\_weight, max\_distance, max\_degree, seed)$
    unsigned long n;   /* number of vertices desired */
    long $north\_weight$;   /* coefficient of latitude in the weight function */
    long $west\_weight$;   /* coefficient of longitude in the weight function */
    long pop_weight;   /* coefficient of population in the weight function */
    unsigned long $max\_distance$;   /* maximum distance in an edge, if nonzero */
    unsigned long max_degree;   /* maximum number of edges per vertex, if nonzero */
    long $seed$;   /* random number seed */
{ ⟨ Local variables 6 ⟩
  $gb\_init\_rand(seed)$;
  ⟨ Check that the parameters are valid 7 ⟩;
  ⟨ Set up a graph with n vertices 8 ⟩;
  ⟨ Read the data file `miles.dat` and compute city weights 11 ⟩;
  ⟨ Determine the n cities to use in the graph 14 ⟩;
  ⟨ Put the appropriate edges into the graph 17 ⟩;
  if $(gb\_trouble\_code)$ { $gb\_recycle(new\_graph)$;
    $panic(alloc\_fault)$;   /* oops, we ran out of memory somewhere back there */
  }
  return new_graph;
}

6. ⟨ Local variables 6 ⟩ ≡
  Graph $*new\_graph$;   /* the graph constructed by $miles$ */
  register long j, k;   /* all-purpose indices */
This code is used in section 5.

7. ⟨ Check that the parameters are valid 7 ⟩ ≡
  if $(n \equiv 0 \lor n > \texttt{MAX\_N})$ $n = \texttt{MAX\_N}$;
  if $(max\_degree \equiv 0 \lor max\_degree \geq n)$ $max\_degree = n - 1$;
  if $(north\_weight > 100000 \lor west\_weight > 100000 \lor pop\_weight > 100$
    $\lor north\_weight < -100000 \lor west\_weight < -100000 \lor pop\_weight < -100)$ $panic(bad\_specs)$;
    /* the magnitude of at least one weight is too big */
This code is used in section 5.

8. ⟨ Set up a graph with n vertices 8 ⟩ ≡
  $new\_graph = gb\_new\_graph(n)$;
  if $(new\_graph \equiv \Lambda)$ $panic(no\_room)$;   /* out of memory before we're even started */
  $sprintf(new\_graph\text{-}id, \texttt{"miles(\%lu,\%ld,\%ld,\%ld,\%lu,\%lu,\%ld)"}, n, north\_weight, west\_weight,$
    $pop\_weight, max\_distance, max\_degree, seed)$;
  $strcpy(new\_graph\text{-}util\_types, \texttt{"ZZIIIIZZZZZZZ"})$;
This code is used in section 5.

9. Vertices. As we read in the data, we construct a list of nodes, each of which contains a city's name, latitude, longitude, population, and weight. These nodes conform to the specifications stipulated in the GB_SORT module. After the list has been sorted by weight, the top n entries will be the vertices of the new graph.

⟨ Type declarations 9 ⟩ ≡

```
typedef struct node_struct {      /* records to be sorted by gb_linksort */
    long key;      /* the nonnegative sort key (weight plus 2^30) */
    struct node_struct *link;      /* pointer to next record */
    long kk;      /* index of city in the original database */
    long lat, lon, pop;      /* latitude, longitude, population */
    char name[30];      /* "City␣Name,␣ST" */
} node;
```

This code is used in section 5.

10. The constants defined here are taken from the specific data in `miles.dat`, because this routine is not intended to be perfectly general.

⟨ Private variables 10 ⟩ ≡

```
long min_lat = 2672, max_lat = 5042, min_lon = 7180, max_lon = 12312, min_pop = 2521,
    max_pop = 875538;      /* tight bounds on data entries */
node *node_block;      /* array of nodes holding city info */
long *distance;      /* array of distances */
```

This code is used in section 5.

---

alloc_fault = −1, GB_GRAPH §7.
bad_specs = 30, GB_GRAPH §7.
gb_init_rand: **void** ( ), GB_FLIP §8.
gb_linksort: **void** ( ), GB_SORT §5.
gb_new_graph: **Graph** *( ),
    GB_GRAPH §23.

gb_recycle: **void** ( ), GB_GRAPH §40.
gb_trouble_code: **long**,
    GB_GRAPH §14.
Graph = **struct**, GB_GRAPH §20.
id: **char** [], GB_GRAPH §20.
MAX_N = 128, §2.

no_room = 1, GB_GRAPH §7.
panic = macro ( ), §4.
sprintf: **int** ( ), <stdio.h>.
strcpy: **char** *( ), <string.h>.
util_types: **char** [], GB_GRAPH §20.

11. The data in `miles.dat` appears in 128 groups of lines, one for each city, in reverse alphabetical order. These groups have the general form

```
City Name, ST[lat,lon]pop
d1 d2 d3 d4 d5 d6 ...  (possibly several lines' worth)
```

where `City Name` is the name of the city (possibly including spaces); `ST` is the two-letter state code; `lat` and `lon` are latitude and longitude in hundredths of degrees; `pop` is the population; and the remaining numbers `d1`, `d2`, ... are distances to the previously named cities in reverse order. Each distance is separated from the previous item by either a blank space or a newline character. For example, the line

```
San Francisco, CA[3778,12242]678974
```

specifies the data about San Francisco that was mentioned earlier. From the first few groups

```
Youngstown, OH[4110,8065]115436
Yankton, SD[4288,9739]12011
966
Yakima, WA[4660,12051]49826
1513 2410
Worcester, MA[4227,7180]161799
2964 1520 604
```

we learn that the distance from Worcester, Massachusetts, to Yakima, Washington, is 2964 miles; from Worcester to Youngstown it is 604 miles.

The following two-letter "state codes" are used for Canadian provinces: `BC` = British Columbia, `MB` = Manitoba, `ON` = Ontario, `SA` = Saskatchewan. (Please don't ask what code would have been used to distinguish New Brunswick from Nebraska if the need had arisen.)

⟨ Read the data file `miles.dat` and compute city weights 11 ⟩ ≡

   $node\_block = gb\_typed\_alloc(\texttt{MAX\_N}, \textbf{node}, new\_graph \rightarrow aux\_data)$;
   $distance = gb\_typed\_alloc(\texttt{MAX\_N} * \texttt{MAX\_N}, \textbf{long}, new\_graph \rightarrow aux\_data)$;
   if $(gb\_trouble\_code)$ {
    $gb\_free(new\_graph \rightarrow aux\_data)$;
    $panic(no\_room + 1)$;   /∗ no room to copy the data ∗/
   }
   if $(gb\_open(\texttt{"miles.dat"}) \neq 0)$ $panic(early\_data\_fault)$;
    /∗ couldn't open `"miles.dat"` using GraphBase conventions; *io_errors* tells why ∗/
   for $(k = \texttt{MAX\_N} - 1;\ k \geq 0;\ k--)$ ⟨ Read and store data for city k 12 ⟩;
   if $(gb\_close() \neq 0)$ $panic(late\_data\_fault)$;
    /∗ something's wrong with `"miles.dat"`; see *io_errors* ∗/

This code is used in section 5.

12. The bounds we've imposed on *north_weight*, *west_weight*, and *pop_weight* guarantee that the key value computed here will be between 0 and 2^{31}.

⟨ Read and store data for city k 12 ⟩ ≡

```
{ register node *p;
  p = node_block + k;
  p→kk = k;
  if (k) p→link = p − 1;
  gb_string(p→name, '[');
  if (gb_char() ≠ '[') panic(syntax_error);        /* out of sync in miles.dat */
  p→lat = gb_number(10);
  if (p→lat < min_lat ∨ p→lat > max_lat ∨ gb_char() ≠ ',') panic(syntax_error + 1);
       /* latitude data was clobbered */
  p→lon = gb_number(10);
  if (p→lon < min_lon ∨ p→lon > max_lon ∨ gb_char() ≠ ']') panic(syntax_error + 2);
       /* longitude data was clobbered */
  p→pop = gb_number(10);
  if (p→pop < min_pop ∨ p→pop > max_pop) panic(syntax_error + 3);
       /* population data was clobbered */
  p→key = north_weight * (p→lat − min_lat) + west_weight * (p→lon − min_lon) + pop_weight * (p→pop −
       min_pop) + #40000000;
  ⟨ Read the mileage data for city k 13 ⟩;
  gb_newline();
}
```

This code is used in section 11.

13.   **#define** $d(j, k)$ $*(distance + (\texttt{MAX\_N} * j + k))$

⟨ Read the mileage data for city k 13 ⟩ ≡

```
{
  for (j = k + 1; j < MAX_N; j++) {
    if (gb_char() ≠ '␣') gb_newline();
    d(j, k) = d(k, j) = gb_number(10);
  }
}
```

This code is used in section 12.

---

14. Once all the nodes have been set up, we can use the *gb_linksort* routine to sort them into the desired order. This routine, which is part of the *gb_graph* module, builds 128 lists from which the desired nodes are readily accessed in decreasing order of weight, using random numbers to break ties.

We set the population to zero in every city that isn't chosen. Then that city will be excluded when edges are examined later.

⟨ Determine the *n* cities to use in the graph 14 ⟩ ≡
```
{ register node *p;        /* the current node being considered */
  register Vertex *v = new_graph⃗vertices;        /* the first unfilled vertex */

  gb_linksort(node_block + MAX_N − 1);
  for (j = 127; j ≥ 0; j−−)
    for (p = (node *) gb_sorted[j]; p; p = p⃗link) {
      if (v < new_graph⃗vertices + n) ⟨ Add city p⃗kk to the graph 15 ⟩
      else  p⃗pop = 0;        /* this city is not being used */
    }
}
```
This code is used in section 5.

15. Utility fields *x* and *y* for each vertex are set to coordinates that can be used in geometric computations; these coordinates are obtained by simple linear transformations of latitude and longitude (not by any kind of sophisticated polyconic projection). We will have

$$0 \le x \le 5132, \qquad 0 \le y \le 3555.$$

Utility field *z* is set to the city's index number (0 to 127) in the original database. Utility field *w* is set to the city's population.

The coordinates computed here are compatible with those in the T_EX file `cities.texmap`. Users might want to incorporate edited copies of that file into documents that display results obtained with *miles* graphs.

```
#define  x_coord   x.I
#define  y_coord   y.I
#define  index_no  z.I
#define  people    w.I
```
⟨ Add city *p⃗kk* to the graph 15 ⟩ ≡
```
{
  v⃗x_coord = max_lon − p⃗lon;        /* x coordinate is complement of longitude */
  v⃗y_coord = p⃗lat − min_lat;
  v⃗y_coord += (v⃗y_coord) ≫ 1;        /* y coordinate·is 1.5 times latitude */
  v⃗index_no = p⃗kk;
  v⃗people = p⃗pop;
  v⃗name = gb_save_string(p⃗name);
  v++;
}
```
This code is used in section 14.

16. ⟨ gb_miles.h 1 ⟩ +≡

#**define** *x_coord*   *x.I*   /∗ utility field definitions for the header file ∗/
#**define** *y_coord*   *y.I*
#**define** *index_no*   *z.I*
#**define** *people*   *w.I*

---

17. Arcs. We make the distance negative in the matrix entry for an arc that is not to be included. Nothing needs to be done in this regard unless the user has specified a maximum degree or a maximum edge length.

⟨ Put the appropriate edges into the graph 17 ⟩ ≡

```
if (max_distance > 0 ∨ max_degree > 0) ⟨Prune unwanted edges by negating their distances 18⟩;
{ register Vertex *u, *v;
  for (u = new_graph→vertices; u < new_graph→vertices + n; u++) {
    j = u→index_no;
    for (v = u + 1; v < new_graph→vertices + n; v++) {
      k = v→index_no;
      if (d(j, k) > 0 ∧ d(k, j) > 0) gb_new_edge(u, v, d(j, k));
    }
  }
}
```

This code is used in section 5.

18. ⟨ Prune unwanted edges by negating their distances 18 ⟩ ≡

```
{ register node *p;
  if (max_degree ≡ 0) max_degree = MAX_N;
  if (max_distance ≡ 0) max_distance = 30000;
  for (p = node_block; p < node_block + MAX_N; p++)
    if (p→pop) {      /* this city not deleted */
      k = p→kk;
      ⟨Blank out all undesired edges from city k 19⟩;
    }
}
```

This code is used in section 17.

19. Here we reuse the key fields of the nodes, storing complementary distances there instead of weights. We also let the sorting routine change the link fields. The other fields, however—especially *pop*—remain unchanged. Yes, the author knows this is a wee bit tricky, but why not?

⟨ Blank out all undesired edges from city k 19 ⟩ ≡

```
{ register node *q;
  register node *s = Λ;      /* list of nodes containing edges from city k */
  for (q = node_block; q < node_block + MAX_N; q++)
    if (q→pop ∧ q ≠ p) {      /* another city not deleted */
      j = d(k, q→kk);      /* distance from p to q */
      if (j > max_distance) d(k, q→kk) = −j;
      else { q→key = max_distance − j; q→link = s; s = q;
      }
    }
  gb_linksort(s);      /* now all the surviving edges from p are in the list gb_sorted[0] */
  j = 0;      /* j counts how many edges have been accepted */
  for (q = (node *) gb_sorted[0]; q; q = q→link)
    if (++j > max_degree) d(k, q→kk) = −d(k, q→kk);
}
```

This code is used in section 18.

20.   Random access to the distance matrix is provided to users via the external function *miles_distance*. Caution: This function can be used only with the graph most recently made by *miles*, and only when the graph's *aux_data* has not been recycled, and only when the z utility fields have not been used for another purpose.

The result might be negative when an edge has been suppressed. Moreover, we can in fact have $miles\_distance(u, v) < 0$ when $miles\_distance(v, u) > 0$, if the distance in question was suppressed by the *max_degree* constraint on u but not on v.

> **long** *miles_distance*(u, v)
> > **Vertex** $*u, *v$;
>
> {
> > **return** $d(u\rightarrow index\_no, v\rightarrow index\_no)$;
>
> }

21.   ⟨ gb_miles.h   1 ⟩ +≡
> **extern long** *miles_distance*( );

GB_PLANE

Important: Before reading GB_PLANE, please read or at least skim the program for GB_MILES.

1. Introduction. This GraphBase module contains the *plane* subroutine, which constructs undirected planar graphs from vertices located randomly in a rectangle, as well as the *plane_miles* routine, which constructs planar graphs based on the mileage and coordinate data in `miles.dat`. Both routines use a general-purpose *delaunay* subroutine, which computes the Delaunay triangulation of a given set of points.

#define *plane_miles*   p_miles   /* abbreviation for Procrustean external linkage */

⟨ gb_plane.h 1 ⟩ ≡
  #define *plane_miles*   p_miles
  extern Graph **plane*( );
  extern Graph **plane_miles*( );
  extern void *delaunay*( );

See also sections 2 and 7.

2. The subroutine call *plane*(n, x_range, y_range, extend, prob, seed) constructs a planar graph whose vertices have integer coordinates uniformly distributed in the rectangle

$$\{ (x,y) \mid 0 \le x < x\_range,\ 0 \le y < y\_range \}.$$

The values of *x_range* and *y_range* must be at most $2^{14} = 16384$; the latter value is the default, which is substituted if *x_range* or *y_range* is given as zero. If *extend* ≡ 0, the graph will have n vertices; otherwise it will have $n + 1$ vertices, where the $(n+1)$st is assigned the coordinates $(-1, -1)$ and may be regarded as a point at ∞. Some of the n finite vertices might have identical coordinates, particularly if the point density $n/(x\_range * y\_range)$ is not very small.

The subroutine works by first constructing the Delaunay triangulation of the points, then discarding each edge of the resulting graph with probability $prob/65536$. Thus, for example, if *prob* is zero the full Delaunay triangulation will be returned; if $prob \equiv 32768$, about half of the Delaunay edges will remain. Each finite edge is assigned a length equal to the Euclidean distance between points, multiplied by 2^{10} and rounded to the nearest integer. If $extend \ne 0$, the Delaunay triangulation will also contain edges between ∞ and all points of the convex hull; such edges, if not discarded, are assigned length 2^{28}, otherwise known as INFTY.

If $extend \ne 0$ and $prob \equiv 0$, the graph will have $n + 1$ vertices and $3(n - 1)$ edges; this is the maximum number of edges that a planar graph on $n + 1$ vertices can have. In such a case the average degree of a vertex will be $6(n - 1)/(n + 1)$, slightly less than 6; hence, if $prob \equiv 32768$, the average degree of a vertex will usually be near 3.

As with all other GraphBase routines that rely on random numbers, different values of *seed* will produce different graphs, in a machine-independent fashion that is reproducible on many different computers. Any *seed* value between 0 and $2^{31} - 1$ is permissible.

#define INFTY   #10000000 L   /* "infinite" length */

⟨ gb_plane.h 1 ⟩ +≡
  #define INFTY   #10000000 L

3. If the *plane* routine encounters a problem, it returns Λ (NULL), after putting a code number into the external variable *panic_code*. This code number identifies the type of failure. Otherwise *plane* returns a pointer to the newly created graph, which will be represented with the data structures explained in GB_GRAPH. (The external variable *panic_code* is itself defined in GB_GRAPH.)

#define *panic(c)* { *panic_code* = *c*; *gb_trouble_code* = 0; **return** Λ; }

4. Here is the overall shape of the C file `gb_plane.c` :

#include `"gb_flip.h"`   /* we will use the GB_FLIP routines for random numbers */
#include `"gb_graph.h"`   /* we will use the GB_GRAPH data structures */
#include `"gb_miles.h"`   /* and we might use GB_MILES for mileage data */
#include `"gb_io.h"`   /* and GB_MILES uses GB_IO, which has *str_buf* */
  ⟨ Preprocessor definitions ⟩

  ⟨ Type declarations 25 ⟩
  ⟨ Global variables 10 ⟩
  ⟨ Subroutines for arithmetic 13 ⟩
  ⟨ Other subroutines 12 ⟩
  ⟨ The *delaunay* routine 9 ⟩
  ⟨ The *plane* routine 5 ⟩
  ⟨ The *plane_miles* routine 41 ⟩

---

delaunay: **void** ( ), §9.
extend: **unsigned long**, §41.
gb_trouble_code: **long**,
  GB_GRAPH §14.
Graph = **struct**, GB_GRAPH §20.

n: **unsigned long**, §5.
panic_code: **long**, GB_GRAPH §5.
plane: **Graph** *( ), §5.
prob: **unsigned long**, §41.

seed: **long**, §41.
str_buf: **char** [ ], GB_IO §26.
x_range: **unsigned long**, §5.
y_range: **unsigned long**, §5.

5. ⟨The *plane* routine 5⟩ ≡

    Graph *∗plane*(*n, x_range, y_range, extend, prob, seed*)

        unsigned long *n*;   /∗ number of vertices desired ∗/

        unsigned long *x_range, y_range*;   /∗ upper bounds on rectangular coordinates ∗/

        unsigned long *extend*;   /∗ should a point at infinity be included? ∗/

        unsigned long *prob*;   /∗ probability of rejecting a Delaunay edge ∗/

        long *seed*;   /∗ random number seed ∗/

  { **Graph** *∗new_graph*;   /∗ the graph constructed by *plane* ∗/

    register Vertex *∗v*;   /∗ the current vertex of interest ∗/

    register long *k*;   /∗ the canonical all-purpose index ∗/

    gb_init_rand(*seed*);

    if (*x_range* > 16384 ∨ *y_range* > 16384) *panic*(*bad_specs*);   /∗ range too large ∗/

    if (*n* < 2) *panic*(*very_bad_specs*);   /∗ don't make *n* so small, you fool ∗/

    if (*x_range* ≡ 0) *x_range* = 16384;   /∗ default ∗/

    if (*y_range* ≡ 0) *y_range* = 16384;   /∗ default ∗/

    ⟨Set up a graph with *n* uniformly distributed vertices 6⟩;

    ⟨Compute the Delaunay triangulation and run through the Delaunay edges; reject them with

        probability *prob*/65536, otherwise append them with their Euclidean length 11⟩;

    if (*gb_trouble_code*) {

      gb_recycle(*new_graph*);

      panic(*alloc_fault*);   /∗ oops, we ran out of memory somewhere back there ∗/

    }

    if (*extend*) *new_graph*⇀*n*++;   /∗ make the "infinite" vertex legitimate ∗/

    return *new_graph*;

  }

This code is used in section 4.

6. The coordinates are placed into utility fields *x_coord* and *y_coord*. A random ID number is also stored in utility field *z_coord*; this number is used by the *delaunay* subroutine to break ties when points are equal or collinear or cocircular. No two vertices have the same ID number. (The header file `gb_miles.h` defines *x_coord*, *y_coord*, and *index_no* to be *x.I*, *y.I*, and *z.I* respectively.)

#define *z_coord* *z.I*

⟨Set up a graph with *n* uniformly distributed vertices 6⟩ ≡

    if (*extend*) *extra_n*++;   /∗ allocate one more vertex than usual ∗/

    new_graph = *gb_new_graph*(*n*);

    if (*new_graph* ≡ Λ) *panic*(*no_room*);   /∗ out of memory before we're even started ∗/

    sprintf(*new_graph*⇀*id*, "plane(%lu,%lu,%lu,%lu,%lu,%ld)", *n, x_range, y_range, extend, prob, seed*);

    strcpy(*new_graph*⇀*util_types*, "ZZZIIIZZZZZZZZ");

    for (*k* = 0, *v* = *new_graph*⇀*vertices*; *k* < *n*; *k*++, *v*++) {

      v⇀*x_coord* = *gb_unif_rand*(*x_range*);

      v⇀*y_coord* = *gb_unif_rand*(*y_range*);

      v⇀*z_coord* = ((**long**) (*gb_next_rand*()/*n*)) ∗ *n* + *k*;

      sprintf(*str_buf*, "%ld", *k*); *v*⇀*name* = *gb_save_string*(*str_buf*);

    }

    if (*extend*) {

      v⇀*name* = *gb_save_string*("INF");

$$v \rightarrow x\_coord = v \rightarrow y\_coord = v \rightarrow z\_coord = -1;$$

$$extra\_n --;$$

}

This code is used in section 5.

7. ⟨ gb_plane.h 1 ⟩ +≡

#define *x_coord*   *x.I*

#define *y_coord*   *y.I*

#define *z_coord*   *z.I*

8. Delaunay triangulation.   The Delaunay triangulation of a set of vertices in the plane consists of all line segments uv such that there exists a circle passing through u and v containing no other vertices. Equivalently, uv is a Delaunay edge if and only if the Voronoi regions for u and v are adjacent; the Voronoi region of a vertex u is the polygon with the property that all points inside it are closer to u than to any other vertex. In this sense, we can say that Delaunay edges connect vertices with their "neighbors."

The definitions in the previous paragraph assume that no two vertices are equal, that no three vertices lie on a straight line, and that no four vertices lie on a circle. If those nondegeneracy conditions aren't satisfied, we can perturb the points very slightly so that the assumptions do hold.

Another way to characterize the Delaunay triangulation is to consider what happens when we map a given set of points onto the unit sphere via stereographic projection: Point (x, y) is mapped to

$$\left(2x/(r^2 + 1), 2y/(r^2 + 1), (r^2 - 1)/(r^2 + 1)\right),$$

where $r^2 = x^2 + y^2$. If we now extend the configuration by adding $(0, 0, 1)$, which is the limiting point on the sphere when r approaches infinity, the Delaunay edges of the original points turn out to be edges of the polytope defined by the mapped points. This polytope, which is the 3-dimensional convex hull of $n + 1$ points on the sphere, also has edges from $(0, 0, 1)$ to the mapped points that correspond to the 2-dimensional convex hull of the original points. Under our assumption of nondegeneracy, the faces of this polytope are all triangles; hence its edges are said to form a triangulation.

A self-contained presentation of all the relevant theory, together with an exposition and proof of correctness of the algorithm below, can be found in the author's monograph *Axioms and Hulls*, Lecture Notes in Computer Science **606** (Springer-Verlag, 1992). For further references, see Franz Aurenhammer, *ACM Computing Surveys* **23** (1991), 345–405.

9.   The *delaunay* procedure, which finds the Delaunay triangulation of a given set of vertices, is the key ingredient in *gb_plane*'s algorithms for generating planar graphs. The given vertices should appear in a GraphBase graph g whose edges, if any, are ignored by *delaunay*. The coordinates of each vertex appear in utility fields *x_coord* and *y_coord*, which must be nonnegative and less than $2^{14} = 16384$. The utility fields *z_coord* must contain unique ID numbers, distinct for every vertex, so that the algorithm can break ties in cases of degeneracy. (Note: These assumptions about the input data are the responsibility of the calling procedure; *delaunay* does not double-check them. If they are violated, catastrophic failure is possible.)

Instead of returning the Delaunay triangulation as a graph, *delaunay* communicates its answer implicitly by performing the procedure call $f(u, v)$ on every pair of vertices u and v joined by a Delaunay edge. Here f is a procedure supplied as a parameter; u and v are either pointers to vertices or Λ (i.e., NULL), where Λ denotes the vertex "∞." As remarked above, edges run between ∞ and all vertices on the convex hull of the given points. The graph of all edges, including the infinite edges, is planar.

For example, if the vertex at infinity is being ignored, the user can declare

> **void** *ins_finite*(u, v)
> **Vertex** *$\ast u$, $\ast v$;*
> { **if** $(u \wedge v)$ *gb_new_edge*$(u, v, 1_{\mathrm{L}})$; }

Then the procedure call $delaunay(g, ins\_finite)$ will add all the finite Delaunay edges to the current graph g, giving them all length 1.

If *delaunay* is unable to allocate enough storage to do its work, it will set *gb_trouble_code* nonzero and there will be no edges in the triangulation.

⟨ The *delaunay* routine 9 ⟩ ≡
    void $delaunay(g, f)$
        Graph $*g$;   /* vertices in the plane */
        void $(*f)()$;   /* procedure that absorbs the triangulated edges */
    { ⟨ Local variables for *delaunay* 26 ⟩;
       ⟨ Find the Delaunay triangulation of g, or return with *gb_trouble_code* nonzero if out of memory 34 ⟩;
       ⟨ Call $f(u, v)$ for each Delaunay edge uv 28 ⟩;
       $gb\_free(working\_storage)$;
    }
This code is used in section 4.

10. The procedure passed to *delaunay* will communicate with *plane* via global variables called *gprob* and *inf_vertex*.

⟨ Global variables 10 ⟩ ≡
    static unsigned long *gprob*;   /* copy of the *prob* parameter */
    static Vertex $*inf\_vertex$;   /* pointer to the vertex ∞, or Λ */
This code is used in section 4.

11. ⟨ Compute the Delaunay triangulation and run through the Delaunay edges; reject them with probability $prob/65536$, otherwise append them with their Euclidean length 11 ⟩ ≡
    $gprob = prob$;
    if $(extend)$ $inf\_vertex = new\_graph{\rightarrow}vertices + n$;
    else $inf\_vertex = \Lambda$;
    $delaunay(new\_graph, new\_euclid\_edge)$;
This code is used in section 5.

---

extend: **unsigned long**, §5.	**Graph** = **struct**, GB_GRAPH §20.	**Vertex** = **struct**, GB_GRAPH §9.
gb_free: **void** ( ), GB_GRAPH §16.	*n*: **unsigned long**, §5.	*vertices*: **Vertex** *, GB_GRAPH §20.
gb_new_edge: **void** ( ),	*new_euclid_edge*: **void** ( ), §12.	*working_storage*: **Area**, §30.
GB_GRAPH §31.	*new_graph*: **Graph** *, §5.	$x\_coord = x.I$, GB_MILES §15.
gb_trouble_code: **long**,	*plane*: **Graph** *( ), §5.	$y\_coord = y.I$, GB_MILES §15.
GB_GRAPH §14.	*prob*: **unsigned long**, §5.	$z\_coord = z.I$, §7.

12. ⟨ Other subroutines 12 ⟩ ≡

```
void new_euclid_edge(u, v)
     Vertex *u, *v;
{ register long dx, dy;
  if ((gb_next_rand( ) ≫ 15) ≥ gprob) {
    if (u) {
      if (v) {
        dx = u→x_coord − v→x_coord;
        dy = u→y_coord − v→y_coord;
        gb_new_edge(u, v, int_sqrt(dx * dx + dy * dy));
      } else if (inf_vertex) gb_new_edge(u, inf_vertex, INFTY);
    } else if (inf_vertex) gb_new_edge(inf_vertex, v, INFTY);
  }
}
```

See also sections 20, 21, 40, and 44.

This code is used in section 4.

13. Arithmetic.   Before we lunge into the world of geometric algorithms, let's build up some confidence by polishing off some subroutines that will be needed to ensure correct results. We assume that **long** integers are less than 2^{31}.

First is a routine to calculate $s = \lfloor 2^{10}\sqrt{x} + \frac{1}{2} \rfloor$, the nearest integer to 2^{10} times the square root of a given nonnegative integer x. If $x > 0$, this is the unique integer such that $2^{20}x - s \le s^2 < 2^{20}x + s$.

The following routine appears to work by magic, but the mystery goes away when one considers the invariant relations

$$m = \lfloor 2^{2k-21} \rfloor, \qquad 0 < y = \lfloor 2^{20-2k}x \rfloor - s^2 + s \le q = 2s.$$

(Exception: We might actually have $y = 0$ for a short time when $q = 2$.)

⟨ Subroutines for arithmetic 13 ⟩ ≡
```
  long int_sqrt(x)
        long x;
  { register long y, m, q = 2;
    long k;

    if (x ≤ 0) return 0;
    for (k = 25, m = #20000000; x < m; k−−, m ≫= 2) ;      /* find the range */
    if (x ≥ m + m)  y = 1;
    else  y = 0;
    do ⟨ Decrease k by 1, maintaining the invariant relations between x, y, m, and q 14 ⟩ while (k);
    return q ≫ 1;
  }
```
See also sections 15 and 24.

This code is used in section 4.

14.   ⟨ Decrease k by 1, maintaining the invariant relations between x, y, m, and q 14 ⟩ ≡
```
  {
    if (x & m) y += y + 1;
    else  y += y;
    m ≫= 1;
    if (x & m) y += y − q + 1;
    else  y += y − q;
    q += q;
    if (y > q)  y −= q, q += 2;
    else if (y ≤ 0)  q −= 2, y += q;
    m ≫= 1;
    k−−;
  }
```
This code is used in section 13.

---

gb_new_edge: **void** (),   *gprob*: **static unsigned long**, §10.   **Vertex** = **struct**, GB_GRAPH §9.
  GB_GRAPH §31.   *inf_vertex*: **static Vertex** *, §10.   *x_coord* = *x.I*, GB_MILES §15.
gb_next_rand = macro (),   INFTY = #10000000 L, §2.   *y_coord* = *y.I*, GB_MILES §15.
  GB_FLIP §6.

15. We are going to need multiple-precision arithmetic in order to calculate certain geometric predicates properly, but it turns out that we do not need to implement general-purpose subroutines for bignums. It suffices to have a single special-purpose routine called $sign\_test(x1, x2, x3, y1, y2, y3)$, which computes a single-precision integer having the same sign as the dot product

$$x1 * y1 + x2 * y2 + x3 * y3$$

when we have $-2^{29} < x1, x2, x3 < 2^{29}$ and $0 \leq y1, y2, y3 < 2^{29}$.

⟨Subroutines for arithmetic 13⟩ +≡
  long $sign\_test(x1, x2, x3, y1, y2, y3)$
      long $x1$, $x2$, $x3$, $y1$, $y2$, $y3$;
  { **long** $s1$, $s2$, $s3$;   /* signs of individual terms */
    long a, b, c;   /* components of a redundant representation of the dot product */
    register long t;   /* temporary register for swapping */
    ⟨Determine the signs of the terms 16⟩;
    ⟨If the answer is obvious, return it without further ado; otherwise, arrange things so that $x3 * y3$
        has the opposite sign to $x1 * y1 + x2 * y2$ 17⟩;
    ⟨Compute a redundant representation of $x1 * y1 + x2 * y2 + x3 * y3$ 18⟩;
    ⟨Return the sign of the redundant representation 19⟩;
  }

16. ⟨Determine the signs of the terms 16⟩ ≡
  if $(x1 \equiv 0 \vee y1 \equiv 0)$ $s1 = 0$;
  else {
    if $(x1 > 0)$ $s1 = 1$;
    else $x1 = -x1, s1 = -1$;
  }
  if $(x2 \equiv 0 \vee y2 \equiv 0)$ $s2 = 0$;
  else {
    if $(x2 > 0)$ $s2 = 1$;
    else $x2 = -x2, s2 = -1$;
  }
  if $(x3 \equiv 0 \vee y3 \equiv 0)$ $s3 = 0$;
  else {
    if $(x3 > 0)$ $s3 = 1$;
    else $x3 = -x3, s3 = -1$;
  }

This code is used in section 15.

17. The answer is obvious unless one of the terms is positive and one of the terms is negative.

⟨If the answer is obvious, return it without further ado; otherwise, arrange things so that $x3 * y3$ has
    the opposite sign to $x1 * y1 + x2 * y2$ 17⟩ ≡
  if $((s1 \geq 0 \wedge s2 \geq 0 \wedge s3 \geq 0) \vee (s1 \leq 0 \wedge s2 \leq 0 \wedge s3 \leq 0))$ **return** $(s1 + s2 + s3)$;
  if $(s3 \equiv 0 \vee s3 \equiv s1)$ {
    $t = s3$; $s3 = s2$; $s2 = t$;
    $t = x3$; $x3 = x2$; $x2 = t$;
    $t = y3$; $y3 = y2$; $y2 = t$;

```
  } else if (s3 ≡ s2) {
    t = s3; s3 = s1; s1 = t;
    t = x3; x3 = x1; x1 = t;
    t = y3; y3 = y1; y1 = t;
  }
```

This code is used in section 15.

18. We make use of a redundant representation $2^{28}a + 2^{14}b + c$, which can be computed by brute force. (Everything is understood to be multiplied by $-s3$.)

⟨ Compute a redundant representation of $x1 * y1 + x2 * y2 + x3 * y3$ 18 ⟩ ≡

```
  { register long lx, rx, ly, ry;
    lx = x1 / #4000;  rx = x1 % #4000;      /* split off the least significant 14 bits */
    ly = y1 / #4000;  ry = y1 % #4000;
    a = lx * ly;  b = lx * ry + ly * rx;  c = rx * ry;
    lx = x2 / #4000;  rx = x2 % #4000;
    ly = y2 / #4000;  ry = y2 % #4000;
    a += lx * ly;  b += lx * ry + ly * rx;  c += rx * ry;
    lx = x3 / #4000;  rx = x3 % #4000;
    ly = y3 / #4000;  ry = y3 % #4000;
    a -= lx * ly;  b -= lx * ry + ly * rx;  c -= rx * ry;
  }
```

This code is used in section 15.

19. Here we use the fact that $|c| < 2^{29}$.

⟨ Return the sign of the redundant representation 19 ⟩ ≡

```
  if (a ≡ 0) goto ez;
  if (a < 0) a = -a, b = -b, c = -c, s3 = -s3;
  while (c < 0) {
    a--;  c += #10000000;
    if (a ≡ 0) goto ez;
  }
  if (b ≥ 0) return -s3;      /* the answer is clear when a > 0 ∧ b ≥ 0 ∧ c ≥ 0 */
  b = -b;
  a -= b/#4000;
  if (a > 0) return -s3;
  if (a ≤ -2) return s3;
  return -s3 * ((a * #4000 - b % #4000) * #4000 + c);
ez: if (b ≥ #8000) return -s3;
  if (b ≤ -#8000) return s3;
  return -s3 * (b * #4000 + c);
```

This code is used in section 15.

20. Determinants. The *delaunay* routine bases all of its decisions on two geometric predicates, which depend on whether certain determinants are positive or negative.

The first predicate, $ccw(u, v, w)$, is true if and only if the three points (u, v, w) have a counterclockwise orientation. This means that if we draw the unique circle through those points, and if we travel along that circle in the counterclockwise direction starting at u, we will encounter v before w.

It turns out that that $ccw(u, v, w)$ holds if and only if the determinant

$$\begin{vmatrix} x_u & y_u & 1 \\ x_v & y_v & 1 \\ x_w & y_w & 1 \end{vmatrix} = \begin{vmatrix} x_u - x_w & y_u - y_w \\ x_v - x_w & y_v - y_w \end{vmatrix}$$

is positive. The evaluation must be exact; if the answer is zero, a special tie-breaking rule must be used because the three points were collinear. The tie-breaking rule is tricky (and necessarily so, according to the theory in *Axioms and Hulls*).

Integer evaluation of that determinant will not cause **long** integer overflow, because we have assumed that all x and y coordinates lie between 0 and $2^{14} - 1$, inclusive. In fact, we could go up to $2^{15} - 1$ without risking overflow; but the limitation to 14 bits will be helpful when we consider a more complicated determinant below.

⟨ Other subroutines 12 ⟩ +≡

```
long ccw(u, v, w)
      Vertex *u, *v, *w;
{ register long wx = w→x_coord, wy = w→y_coord;     /* xw, yw */
  register long det = (u→x_coord − wx) * (v→y_coord − wy) − (u→y_coord − wy) * (v→x_coord − wx);
  Vertex *t;

  if (det ≡ 0) {
    det = 1;
    if (u→z_coord > v→z_coord) {
      t = u;  u = v;  v = t;  det = −det;
    }
    if (v→z_coord > w→z_coord) {
      t = v;  v = w;  w = t;  det = −det;
    }
    if (u→z_coord > v→z_coord) {
      t = u;  u = v;  v = t;  det = −det;
    }
    if (u→x_coord > v→x_coord ∨ (u→x_coord ≡ v→x_coord ∧
            (u→y_coord > v→y_coord ∨ (u→y_coord ≡ v→y_coord ∧
            (w→x_coord > u→x_coord ∨ (w→x_coord ≡ u→x_coord ∧ w→y_coord ≥ u→y_coord))))))
      det = −det;
  }
  return (det > 0);
}
```

21. The other geometric predicate, *incircle*(t, u, v, w), is true if and only if point t lies outside the circle passing through u, v, and w, assuming that $ccw(u, v, w)$ holds. This predicate makes us work harder, because it is equivalent to the sign of a 4×4 determinant that requires twice as much precision:

$$\begin{vmatrix} x_t & y_t & x_t^2 + y_t^2 & 1 \\ x_u & y_u & x_u^2 + y_u^2 & 1 \\ x_v & y_v & x_v^2 + y_v^2 & 1 \\ x_w & y_w & x_w^2 + y_w^2 & 1 \end{vmatrix} = \begin{vmatrix} x_t - x_w & y_t - y_w & (x_t - x_w)^2 + (y_t - y_w)^2 \\ x_u - x_w & y_u - y_w & (x_u - x_w)^2 + (y_u - y_w)^2 \\ x_v - x_w & y_v - y_w & (x_v - x_w)^2 + (y_v - y_w)^2 \end{vmatrix}.$$

The sign can, however, be deduced by the *sign_test* subroutine we had the foresight to provide earlier.

⟨ Other subroutines 12 ⟩ +≡
  long *incircle*(t, u, v, w)
    Vertex $*t, *u, *v, *w$;
  { **register long** $wx = w \rightarrow x\_coord$, $wy = w \rightarrow y\_coord$;   /* x_w, y_w */
    long $tx = t \rightarrow x\_coord - wx$, $ty = t \rightarrow y\_coord - wy$;   /* $x_t - x_w, y_t - y_w$ */
    long $ux = u \rightarrow x\_coord - wx$, $uy = u \rightarrow y\_coord - wy$;   /* $x_u - x_w, y_u - y_w$ */
    long $vx = v \rightarrow x\_coord - wx$, $vy = v \rightarrow y\_coord - wy$;   /* $x_v - x_w, y_v - y_w$ */
    register long $det = sign\_test(tx * uy - ty * ux, ux * vy - uy * vx, vx * ty - vy * tx,$
       $vx * vx + vy * vy, tx * tx + ty * ty, ux * ux + uy * uy)$;
  Vertex $*s$;
  if $(det \equiv 0)$ {
    ⟨ Sort (t, u, v, w) by ID number 22 ⟩;
    ⟨ Remove incircle degeneracy 23 ⟩;
  }
  return $(det > 0)$;
  }

---

delaunay: **void** ( ), §9.   **Vertex** = **struct**, GB_GRAPH §9.   *y_coord* = *y.I*, GB_MILES §15.
sign_test: **long** ( ), §15.   *x_coord* = *x.I*, GB_MILES §15.   *z_coord* = *z.I*, §7.

22.   ⟨ Sort (t, u, v, w) by ID number 22 ⟩ ≡
   $det = 1$;
  if $(t{\rightarrow}z\_coord > u{\rightarrow}z\_coord)$ {
    $s = t$; $t = u$; $u = s$; $det = -det$;
  }
  if $(v{\rightarrow}z\_coord > w{\rightarrow}z\_coord)$ {
    $s = v$; $v = w$; $w = s$; $det = -det$;
  }
  if $(t{\rightarrow}z\_coord > v{\rightarrow}z\_coord)$ {
    $s = t$; $t = v$; $v = s$; $det = -det$;
  }
  if $(u{\rightarrow}z\_coord > w{\rightarrow}z\_coord)$ {
    $s = u$; $u = w$; $w = s$; $det = -det$;
  }
  if $(u{\rightarrow}z\_coord > v{\rightarrow}z\_coord)$ {
    $s = u$; $u = v$; $v = s$; $det = -det$;
  }
This code is used in section 21.

23.   By slightly perturbing the points, we can always make them nondegenerate, although the details are complicated. A sequence of 12 steps, involving up to four auxiliary functions

$$f(t, u, v, w) = \begin{vmatrix} x_t - x_v & (x_t - x_w)^2 + (y_t - y_w)^2 - (x_v - x_w)^2 - (y_v - y_w)^2 \\ x_u - x_v & (x_u - x_w)^2 + (y_u - y_w)^2 - (x_v - x_w)^2 - (y_v - y_w)^2 \end{vmatrix},$$

$$g(t, u, v, w) = \begin{vmatrix} y_t - y_v & (x_t - x_w)^2 + (y_t - y_w)^2 - (x_v - x_w)^2 - (y_v - y_w)^2 \\ y_u - y_v & (x_u - x_w)^2 + (y_u - y_w)^2 - (x_v - x_w)^2 - (y_v - y_w)^2 \end{vmatrix},$$

$$h(t, u, v, w) = (x_u - x_t)(y_v - y_w),$$

$$j(t, u, v, w) = (x_u - x_v)^2 + (y_u - y_w)^2 - (x_t - x_v)^2 - (y_t - y_w)^2,$$

does the trick, as explained in *Axioms and Hulls*.
⟨ Remove incircle degeneracy 23 ⟩ ≡
  { **long** dd;
    if $((dd = \mathit{ff}(t, u, v, w)) < 0 \lor (dd \equiv 0 \land$
        $((dd = \mathit{gg}(t, u, v, w)) < 0 \lor (dd \equiv 0 \land$
        $((dd = \mathit{ff}(u, t, w, v)) < 0 \lor (dd \equiv 0 \land$
        $((dd = \mathit{gg}(u, t, w, v)) < 0 \lor (dd \equiv 0 \land$
        $((dd = \mathit{ff}(v, w, t, u)) < 0 \lor (dd \equiv 0 \land$
        $((dd = \mathit{gg}(v, w, t, u)) < 0 \lor (dd \equiv 0 \land$
        $((dd = \mathit{hh}(t, u, v, w)) < 0 \lor (dd \equiv 0 \land$
        $((dd = \mathit{jj}(t, u, v, w)) < 0 \lor (dd \equiv 0 \land$
        $((dd = \mathit{hh}(v, t, u, w)) < 0 \lor (dd \equiv 0 \land$
        $((dd = \mathit{jj}(v, t, u, w)) < 0 \lor (dd \equiv 0 \land \mathit{jj}(t, w, u, v) < 0)))))))))))))))))))$ $det = -det$;
  }
This code is used in section 21.

24.   ⟨ Subroutines for arithmetic 13 ⟩ +≡

 long $ff(t, u, v, w)$
  Vertex $*t, *u, *v, *w;$
 { **register long** $wx = w\text{→}x\_coord,\ wy = w\text{→}y\_coord;$ /* x_w, y_w */
  long $tx = t\text{→}x\_coord - wx,\ ty = t\text{→}y\_coord - wy;$ /* $x_t - x_w,\ y_t - y_w$ */
  long $ux = u\text{→}x\_coord - wx,\ uy = u\text{→}y\_coord - wy;$ /* $x_u - x_w,\ y_u - y_w$ */
  long $vx = v\text{→}x\_coord - wx,\ vy = v\text{→}y\_coord - wy;$ /* $x_v - x_w,\ y_v - y_w$ */
  return $sign\_test(ux - tx, vx - ux, tx - vx, vx * vx + vy * vy, tx * tx + ty * ty, ux * ux + uy * uy);$
 }

 long $gg(t, u, v, w)$
  Vertex $*t, *u, *v, *w;$
 { **register long** $wx = w\text{→}x\_coord,\ wy = w\text{→}y\_coord;$ /* x_w, y_w */
  long $tx = t\text{→}x\_coord - wx,\ ty = t\text{→}y\_coord - wy;$ /* $x_t - x_w,\ y_t - y_w$ */
  long $ux = u\text{→}x\_coord - wx,\ uy = u\text{→}y\_coord - wy;$ /* $x_u - x_w,\ y_u - y_w$ */
  long $vx = v\text{→}x\_coord - wx,\ vy = v\text{→}y\_coord - wy;$ /* $x_v - x_w,\ y_v - y_w$ */
  return $sign\_test(uy - ty, vy - uy, ty - vy, vx * vx + vy * vy, tx * tx + ty * ty, ux * ux + uy * uy);$
 }

 long $hh(t, u, v, w)$
  Vertex $*t, *u, *v, *w;$
 {
  return $(u\text{→}x\_coord - t\text{→}x\_coord) * (v\text{→}y\_coord - w\text{→}y\_coord);$
 }

 long $jj(t, u, v, w)$
  Vertex $*t, *u, *v, *w;$
 { **register long** $vx = v\text{→}x\_coord,\ wy = w\text{→}y\_coord;$
  return $(u\text{→}x\_coord - vx) * (u\text{→}x\_coord - vx) + (u\text{→}y\_coord - wy) * (u\text{→}y\_coord - wy)$
   $- (t\text{→}x\_coord - vx) * (t\text{→}x\_coord - vx) - (t\text{→}y\_coord - wy) * (t\text{→}y\_coord - wy);$
 }

---

det: **register long**, §21. u: **Vertex** *, §21. $x\_coord = x.I$, GB_MILES §15.
s: **Vertex** *, §21. v: **Vertex** *, §21. $y\_coord = y.I$, GB_MILES §15.
sign_test: **long** ( ), §15.   **Vertex** = **struct**, GB_GRAPH §9. $z\_coord = z.I$, §7.
t: **Vertex** *, §21. w: **Vertex** *, §21.

25. Delaunay data structures.   Now we have the primitive predicates we need, and we can get on with the geometric aspects of *delaunay*. As mentioned above, each vertex is represented by two coordinates and an ID number, stored in the utility fields *x_coord*, *y_coord*, and *z_coord*.

Each edge of the current triangulation is represented by two arcs pointing in opposite directions; the two arcs are called *mates*. Each arc conceptually has a triangle on its left and a mate on its right.

An **arc** record differs from an **Arc**; it has three fields:

vert is the vertex this arc leads to, or Λ if that vertex is ∞;

next is the next arc having the same triangle at the left;

inst is the branch node that points to the triangle at the left, as explained below.

If p points to an arc, then $p\text{-}next\text{-}next\text{-}next \equiv p$, because a triangle is bounded by three arcs. We also have $p\text{-}next\text{-}inst \equiv p\text{-}inst$, for all arcs p.

⟨ Type declarations 25 ⟩ ≡
> **typedef struct a_struct** {
>> **Vertex** *∗vert*;   /∗ *v*, if this arc goes from *u* to *v* ∗/
>> **struct a_struct** *∗next*;   /∗ the arc from *v* that shares a triangle with this one ∗/
>> **struct n_struct** *∗inst*;   /∗ instruction to change when the triangle is modified ∗/
> } **arc**;

See also section 29.

This code is used in section 4.

26.   Storage is allocated in such a way that, if p and q point respectively to an arc and its mate, then $p + q = \&arc\_block[0] + \&arc\_block[m-1]$, where m is the total number of arc records allocated in the *arc_block* array. This convention saves us one pointer field in each arc.

When setting q to the mate of p, we need to do the calculation cautiously using an auxiliary register, because the constant $\&arc\_block[0] + \&arc\_block[m-1]$ might be too large to evaluate without integer overflow on some systems.

#define *mate*(*a, b*)
> {   /∗ given *a*, set *b* to its mate ∗/
>> *reg = max_arc* − (**siz_t**) *a*;
>> *b* = (**arc** ∗) (*reg* + *min_arc*);
> }

⟨ Local variables for *delaunay* 26 ⟩ ≡
> **register siz_t** *reg*;   /∗ used while computing mates ∗/
> **siz_t** *min_arc, max_arc*;   /∗ $\&arc\_block[0]$, $\&arc\_block[m-1]$ ∗/
> **arc** *∗next_arc*;   /∗ the first arc record that hasn't yet been used ∗/

See also sections 30 and 32.

This code is used in section 9.

27.   ⟨ Initialize the array of arcs 27 ⟩ ≡
> *next_arc = gb_typed_alloc*(6 ∗ *g*-*n* − 6, **arc**, *working_storage*);
> **if** (*next_arc* ≡ Λ) **return**;   /∗ *gb_trouble_code* is nonzero ∗/
> *min_arc* = (**siz_t**) *next_arc*;
> *max_arc* = (**siz_t**) (*next_arc* + (6 ∗ *g*-*n* − 7));

This code is used in section 31.

28. ⟨ Call $f(u,v)$ for each Delaunay edge uv 28 ⟩ ≡
  $a = (\textbf{arc} *) \; min\_arc;$
  $b = (\textbf{arc} *) \; max\_arc;$
  for (; $a < next\_arc;$ $a{+}{+}, b{-}{-}$) $(*f)(a{\rightarrow}vert, b{\rightarrow}vert);$
This code is used in section 9.

29. The last and probably most crucial component of the data structure is the collection of *branch nodes*, which will be linked together into a binary tree. Given a new vertex w, we will ascertain what triangle it belongs to by starting at the root of this tree and executing a sequence of instructions, each of which has the form 'if w lies to the right of the straight line from u to v then go to α else go to β', where α and β are nodes that continue the search. This process continues until we reach a terminal node, which says 'congratulations, you're done, w is in triangle such-and-such'. The terminal node points to one of the three arcs bounding that triangle. If a vertex of the triangle is ∞, the terminal node points to the arc whose *vert* pointer is Λ.

⟨ Type declarations 25 ⟩ +≡
  typedef struct n_struct {
    Vertex *$*u$;   /* first vertex, or Λ if this is a terminal node */
    Vertex *$*v$;   /* second vertex, or pointer to the triangle corresponding to a terminal node */
    struct n_struct *$*l$;   /* go here if w lies to the left of uv */
    struct n_struct *$*r$;   /* go here if w lies to the right of uv */
  } **node**;

---

a: **arc** *, §32.   GB_GRAPH §14.   GB_GRAPH §34.
Arc = **struct**, GB_GRAPH §10.   *gb_typed_alloc* = macro ( ),   **Vertex** = **struct**, GB_GRAPH §9.
b: **arc** *, §32.   GB_GRAPH §11.   *working_storage*: **Area**, §30.
delaunay: **void** ( ), §9.   *n*: **long**, GB_GRAPH §20.   *x_coord* = *x.I*, GB_MILES §15.
f: **void** (*)( ), §9.   *p*: **Vertex** *, §40.   *y_coord* = *y.I*, GB_MILES §15.
g: **Graph** *, §9.   **siz_t** = **unsigned long**,   *z_coord* = *z.I*, §7.
gb_trouble_code: **long**,

30. The search tree just described is actually a dag (a directed acyclic graph), because it has overlapping subtrees. As the algorithm proceeds, the dag gets bigger and bigger, since the number of triangles keeps growing. Instructions are never deleted; we just extend the dag by substituting new branches for nodes that once were terminal.

The expected number of nodes in this dag is $O(n)$ when there are n vertices, if we input the vertices in random order. But it can be as high as order n^2 in the worst case. So our program will allocate blocks of nodes dynamically instead of assuming a maximum size.

```
#define nodes_per_block  127      /* on most computers we want it ≡ 15 (mod 16) */
#define new_node(x)
          if (next_node ≡ max_node) {
            x = gb_typed_alloc(nodes_per_block, node, working_storage);
            if (x ≡ Λ) {
              gb_free(working_storage);      /* release delaunay's auxiliary memory */
              return;      /* gb_trouble_code is nonzero */
            }
            next_node = x + 1;
            max_node = x + nodes_per_block;
          } else  x = next_node++;
#define terminal_node(x, p)
          { new_node(x);      /* allocate a new node */
            x⃗v = (Vertex *) (p);      /* make it point to a given arc from the triangle */
          }      /* note that x⃗u ≡ Λ, representing a terminal node */
```

⟨ Local variables for *delaunay* 26 ⟩ +≡
 node *next_node;   /* the first yet-unused node slot in the current block of nodes */
 node *max_node;   /* address of nonexistent node following the current block of nodes */
 node root_node;   /* start here to locate a vertex in its triangle */
 Area working_storage;   /* where *delaunay* builds its triangulation */

31. The algorithm begins with a trivial triangulation that contains only the first two vertices, together with two "triangles" extending to infinity at their left and right.

⟨ Initialize the data structures 31 ⟩ ≡
 next_node = max_node = Λ;
 init_area(working_storage);
 ⟨ Initialize the array of arcs 27 ⟩;
 u = g⃗vertices;
 v = u + 1;
 ⟨ Make two "triangles" for u, v, and ∞ 33 ⟩;
This code is used in section 34.

32. We'll need a bunch of local variables to do elementary operations on data structures.

⟨ Local variables for *delaunay* 26 ⟩ +≡
 Vertex *p, *q, *r, *s, *t, *tp, *tpp, *u, *v;
 arc *a, *aa, *b, *c, *d, *e;
 node *x, *y, *yp, *ypp;

33. ⟨ Make two "triangles" for u, v, and ∞ 33 ⟩ ≡
 root_node.u = u;

$root\_node.v = v$;
$a = next\_arc$;
$terminal\_node(x, a + 1)$;
$root\_node.l = x$;
$a{\rightarrow}vert = v$; $a{\rightarrow}next = a + 1$; $a{\rightarrow}inst = x$;
$(a + 1){\rightarrow}next = a + 2$; $(a + 1){\rightarrow}inst = x$;   /* $(a + 1){\rightarrow}vert = \Lambda$, representing ∞ */
$(a + 2){\rightarrow}vert = u$; $(a + 2){\rightarrow}next = a$; $(a + 2){\rightarrow}inst = x$;
$mate(a, b)$;
$terminal\_node(x, b - 2)$;
$root\_node.r = x$;
$b{\rightarrow}vert = u$; $b{\rightarrow}next = b - 2$; $b{\rightarrow}inst = x$;
$(b - 2){\rightarrow}next = b - 1$; $(b - 2){\rightarrow}inst = x$;   /* $(b - 2){\rightarrow}vert = \Lambda$, representing ∞ */
$(b - 1){\rightarrow}vert = v$; $(b - 1){\rightarrow}next = b$; $(b - 1){\rightarrow}inst = x$;
$next\_arc \mathrel{+}= 3$;

This code is used in section 31.

34. Delaunay updating. The main loop of the algorithm updates the data structure incrementally by adding one new vertex at a time. The new vertex will always be connected by an edge (i.e., by two arcs) to each of the vertices of the triangle that previously enclosed it. It might also deserve to be connected to other nearby vertices.

⟨Find the Delaunay triangulation of g, or return with *gb_trouble_code* nonzero if out of memory 34⟩ ≡
  if $(g\text{-}n < 2)$ **return;**   /* no edges unless there are at least 2 vertices */
  ⟨Initialize the data structures 31⟩;
  for $(p = g\text{-}vertices + 2;\ p < g\text{-}vertices + g\text{-}n;\ p\text{++})$ {
    ⟨Find an arc a on the boundary of the triangle containing p 35⟩;
    ⟨Divide the triangle left of a into three triangles surrounding p 36⟩;
    ⟨Explore the triangles surrounding p, "flipping" their neighbors until all triangles that should
        touch p are found 39⟩;
  }

This code is used in section 9.

35. We have set up the branch nodes so that they solve the triangle location problem.

⟨Find an arc a on the boundary of the triangle containing p 35⟩ ≡
  $x = \&root\_node;$
  do {
    if $(ccw(x\text{-}u, x\text{-}v, p))\ x = x\text{-}l;$
    else $x = x\text{-}r;$
  } **while** $(x\text{-}u);$
  $a = (\textbf{arc} *)\ x\text{-}v;$ /* terminal node points to the arc we want */

This code is used in section 34.

36. Subdividing a triangle is an easy exercise in data structure manipulation, except that we must do something special when one of the vertices is infinite. Let's look carefully at what needs to be done.

Suppose the triangle containing p has the vertices q, r, and s in counterclockwise order. Let x be the terminal node that points to the triangle $\triangle qrs$. We want to change x so that we will be able to locate a future point of $\triangle qrs$ within either $\triangle pqr$, $\triangle prs$, or $\triangle psq$.

If q, r, and s are finite, we will change x and add five new nodes as follows:

$$x: \text{ if left of } rp, \text{ go to } x'', \text{ else go to } x';$$
$$x': \text{ if left of } sp, \text{ go to } y, \text{ else go to } y';$$
$$x'': \text{ if left of } qp, \text{ go to } y', \text{ else go to } y'';$$
$$y: \text{ you're in } \triangle prs;$$
$$y': \text{ you're in } \triangle psq;$$
$$y'': \text{ you're in } \triangle pqr.$$

But if, say, $q = \infty$, such instructions make no sense, because there are lines in all directions that run from ∞ to any point. In such a case we use "wedges" instead of triangles, as explained below.

At the beginning of the following code, we have $x \equiv a\text{-}inst$.

⟨Divide the triangle left of a into three triangles surrounding p 36⟩ ≡
  $b = a\text{-}next;\ c = b\text{-}next;$

$q = a \rightarrow vert; \quad r = b \rightarrow vert; \quad s = c \rightarrow vert;$

⟨ Create new terminal nodes y, yp, ypp, and new arcs pointing to them 37 ⟩;

if $(q \equiv \Lambda)$ ⟨ Compile instructions to update convex hull 38 ⟩

else { **register node** *xp;

$\quad x \rightarrow u = r; \quad x \rightarrow v = p;$

$\quad new\_node(xp);$

$\quad xp \rightarrow u = q; \quad xp \rightarrow v = p; \quad xp \rightarrow l = yp; \quad xp \rightarrow r = ypp; \quad\quad /*$ instruction x'' above $*/$

$\quad x \rightarrow l = xp;$

$\quad new\_node(xp);$

$\quad xp \rightarrow u = s; \quad xp \rightarrow v = p; \quad xp \rightarrow l = y; \quad xp \rightarrow r = yp; \quad\quad /*$ instruction x' above $*/$

$\quad x \rightarrow r = xp;$

}

This code is used in section 34.

37. The only subtle point here is that $q = a \rightarrow vert$ might be Λ. A terminal node must point to the proper arc of an infinite triangle.

⟨ Create new terminal nodes y, yp, ypp, and new arcs pointing to them 37 ⟩ $\equiv$

$\quad terminal\_node(yp, a); \quad terminal\_node(ypp, next\_arc); \quad terminal\_node(y, c);$

$\quad c \rightarrow inst = y; \quad a \rightarrow inst = yp; \quad b \rightarrow inst = ypp;$

$\quad mate(next\_arc, e);$

$\quad a \rightarrow next = e; \quad b \rightarrow next = e - 1; \quad c \rightarrow next = e - 2;$

$\quad next\_arc \rightarrow vert = q; \quad next\_arc \rightarrow next = b; \quad next\_arc \rightarrow inst = ypp;$

$\quad (next\_arc + 1) \rightarrow vert = r; \quad (next\_arc + 1) \rightarrow next = c; \quad (next\_arc + 1) \rightarrow inst = y;$

$\quad (next\_arc + 2) \rightarrow vert = s; \quad (next\_arc + 2) \rightarrow next = a; \quad (next\_arc + 2) \rightarrow inst = yp;$

$\quad e \rightarrow vert = (e - 1) \rightarrow vert = (e - 2) \rightarrow vert = p;$

$\quad e \rightarrow next = next\_arc + 2; \quad (e - 1) \rightarrow next = next\_arc; \quad (e - 2) \rightarrow next = next\_arc + 1;$

$\quad e \rightarrow inst = yp; \quad (e - 1) \rightarrow inst = ypp; \quad (e - 2) \rightarrow inst = y;$

$\quad next\_arc \mathrel{+}= 3;$

This code is used in section 36.

---

a: **arc** *, §32.

arc = **struct a_struct**, §25.

b: **arc** *, §32.

c: **arc** *, §32.

ccw: **long** ( ), §20.

e: **arc** *, §32.

g: **Graph** *, §9.

gb_trouble_code: **long**, GB_GRAPH §14.

inst: **struct n_struct** *, §25.

l: **struct n_struct** *, §29.

mate = macro ( ), §26.

n: **long**, GB_GRAPH §20.

new_node = macro ( ), §30.

next: **struct a_struct** *, §25.

next_arc: **arc** *, §26.

node = **struct n_struct**, §29.

p: **Vertex** *, §32.

q: **Vertex** *, §32.

r: **struct n_struct** *, §29.

r: **Vertex** *, §32.

root_node: **node**, §30.

s: **Vertex** *, §32.

terminal_node = macro ( ), §30.

u: **Vertex** *, §29.

v: **Vertex** *, §29.

vert: **Vertex** *, §25.

vertices: **Vertex** *, GB_GRAPH §20.

x: **node** *, §32.

y: **node** *, §32.

yp: **node** *, §32.

ypp: **node** *, §32.

38. Outside of the current convex hull, we have "wedges" instead of triangles. Wedges are exterior angles whose points lie outside an edge rs of the convex hull, but not outside the next edge on the other side of point r. When a new point lies in such a wedge, we have to see if it also lies outside the edges st, tu, etc., in the clockwise direction, in which case the convex hull loses points s, t, etc., and we must update the new wedges accordingly.

This was the hardest part of the program to prove correct; a complete proof can be found in *Axioms and Hulls*.

⟨ Compile instructions to update convex hull 38 ⟩ ≡
```
{ register node *xp;
    x⃗u = r;  x⃗v = p;  x⃗l = ypp;
    new_node(xp);
    xp⃗u = s;  xp⃗v = p;  xp⃗l = y;  xp⃗r = yp;
    x⃗r = xp;
    mate(a, aa);  d = aa⃗next;  t = d⃗vert;
    while (t ≠ r ∧ (ccw(p, s, t))) {  register node *xpp;
        terminal_node(xpp, d);
        xp⃗r = d⃗inst;
        xp = d⃗inst;
        xp⃗u = t;  xp⃗v = p;  xp⃗l = xpp;  xp⃗r = yp;
        flip(a, aa, d, s, Λ, t, p, xpp, yp);
        a = aa⃗next;  mate(a, aa);  d = aa⃗next;
        s = t;  t = d⃗vert;
        yp⃗v = (Vertex *) a;
    }
    terminal_node(xp, d⃗next);
    x = d⃗inst;  x⃗u = s;  x⃗v = p;  x⃗l = xp;  x⃗r = yp;
    d⃗inst = xp;  d⃗next⃗inst = xp;  d⃗next⃗next⃗inst = xp;
    r = s;      /* this value of r shortens the exploration step that follows */
}
```
This code is used in section 36.

39. The updating process finishes by walking around the triangles that surround p, making sure that none of them are adjacent to triangles containing p in their circumcircle. (Such triangles are no longer in the Delaunay triangulation, by definition.)

⟨ Explore the triangles surrounding p, "flipping" their neighbors until all triangles that should touch p are found 39 ⟩ ≡
```
while (1) {
    mate(c, d);  e = d⃗next;
    t = d⃗vert;  tp = c⃗vert;  tpp = e⃗vert;
    if (tpp ∧ incircle(tpp, tp, t, p)) {      /* triangle tt″t′ no longer Delaunay */
        register node *xp, *xpp;
        terminal_node(xp, e);  terminal_node(xpp, d);
        x = c⃗inst;  x⃗u = tpp;  x⃗v = p;  x⃗l = xp;  x⃗r = xpp;
        x = d⃗inst;  x⃗u = tpp;  x⃗v = p;  x⃗l = xp;  x⃗r = xpp;
        flip(c, d, e, t, tp, tpp, p, xp, xpp);
        c = e;
```

```
    } else if (tp ≡ r) break;
    else {
      mate(c→next, aa);
      c = aa→next;
    }
  }
}
```
This code is used in section 34.

40.   Here d is the mate of c, $e = d\text{-}next$, $t = d\text{-}vert$, $tp = c\text{-}vert$, and $tpp = e\text{-}vert$. The triangles $\Delta tt'p$ and $\Delta t'tt''$ to the left and right of arc c are being replaced in the current triangulation by $\Delta ptt''$ and $\Delta t''t'p$, corresponding to terminal nodes xp and xpp. (The values of t and tp are not actually used, so some optimization is possible.)

⟨ Other subroutines 12 ⟩ +≡
```
  void flip(c, d, e, t, tp, tpp, p, xp, xpp)
    arc *c, *d, *e;
    Vertex *t, *tp, *tpp, *p;
    node *xp, *xpp;
  { register arc *ep = e→next, *cp = c→next, *cpp = cp→next;

    e→next = c;  c→next = cpp;  cpp→next = e;
    e→inst = c→inst = cpp→inst = xp;
    c→vert = p;
    d→next = ep;  ep→next = cp;  cp→next = d;
    d→inst = ep→inst = cp→inst = xpp;
    d→vert = tpp;
  }
```

---

a: **arc** *, §32.	*mate* = macro ( ), §26.	*tp*: **Vertex** *, §32.
aa: **arc** *, §32.	*new_node* = macro ( ), §30.	*tpp*: **Vertex** *, §32.
arc = **struct a_struct**, §25.	*next*: **struct a_struct** *, §25.	*u*: **Vertex** *, §29.
c: **arc** *, §32.	**node** = **struct n_struct**, §29.	*v*: **Vertex** *, §29.
ccw: **long** ( ), §20.	*p*: **Vertex** *, §32.	*vert*: **Vertex** *, §25.
d: **arc** *, §32.	*r*: **Vertex** *, §32.	**Vertex** = **struct**, GB_GRAPH §9.
e: **arc** *, §32.	*r*: **struct n_struct** *, §29.	*x*: **node** *, §32.
incircle: **long** ( ), §21.	*s*: **Vertex** *, §32.	*y*: **node** *, §32.
inst: **struct n_struct** *, §25.	*t*: **Vertex** *, §32.	*yp*: **node** *, §32.
l: **struct n_struct** *, §29.	*terminal_node* = macro ( ), §30.	*ypp*: **node** *, §32.

41. Use of mileage data. The *delaunay* routine is now complete, and the only missing piece of code is the promised routine that generates planar graphs based on data from the real world.

The subroutine call *plane_miles*(n, *north_weight*, *west_weight*, *pop_weight*, *extend*, *prob*, *seed*) will construct a planar graph with $\min(128, n)$ vertices, where the vertices are exactly the same as the cities produced by the subroutine call *miles*(n, *north_weight*, *west_weight*, *pop_weight*, $0, 0$, *seed*). (As explained in module GB_MILES, the weight parameters *north_weight*, *west_weight*, and *pop_weight* are used to rank the cities by location and/or population.) The edges of the new graph are obtained by first constructing the Delaunay triangulation of those cities, based on a simple projection onto the plane using their latitude and longitude, then discarding each Delaunay edge with probability *prob*/65536. The length of each surviving edge is the same as the mileage between cities that would appear in the complete graph produced by *miles*.

If *extend* $\neq 0$, an additional vertex representing ∞ is also included. The Delaunay triangulation includes edges of length INFTY connecting this vertex with all cities on the convex hull; these edges, like the others, are subject to being discarded with probability *prob*/65536. (See the description of *plane* for further comments about using *prob* to control the sparseness of the graph.)

The weight parameters must satisfy

$$|north\_weight| \leq 100{,}000, \quad |west\_weight| \leq 100{,}000, \quad |pop\_weight| \leq 100.$$

Vertices of the graph will appear in order of decreasing weight. The *seed* parameter defines the pseudo-random numbers used wherever a "random" choice between equal-weight vertices needs to be made, or when deciding whether to discard a Delaunay edge.

⟨ The *plane_miles* routine 41 ⟩ ≡

```
Graph *plane_miles(n, north_weight, west_weight, pop_weight, extend, prob, seed)
    unsigned long n;        /* number of vertices desired */
    long north_weight;      /* coefficient of latitude in the weight function */
    long west_weight;       /* coefficient of longitude in the weight function */
    long pop_weight;        /* coefficient of population in the weight function */
    unsigned long extend;   /* should a point at infinity be included? */
    unsigned long prob;     /* probability of rejecting a Delaunay edge */
    long seed;        /* random number seed */
{ Graph *new_graph;        /* the graph constructed by plane_miles */

  ⟨ Use miles to set up the vertices of a graph 42 ⟩;
  ⟨ Compute the Delaunay triangulation and run through the Delaunay edges; reject them with
        probability prob/65536, otherwise append them with the road length in miles 43 ⟩;
  if (gb_trouble_code) {
    gb_recycle(new_graph);
    panic(alloc_fault);        /* oops, we ran out of memory somewhere back there */
  }
  gb_free(new_graph⁀aux_data);   /* recycle special memory used by miles */
  if (extend) new_graph⁀n++;     /* make the "infinite" vertex legitimate */
  return new_graph;
}
```

This code is used in section 4.

42. By setting the *max_distance* parameter to 1, we cause *miles* to produce a graph having the desired vertices but no edges. The vertices of this graph will have appropriate coordinate fields *x_coord*, *y_coord*, and *z_coord*.

⟨ Use *miles* to set up the vertices of a graph 42 ⟩ ≡
  if (*extend*) *extra_n* ++; /* allocate one more vertex than usual */
  if ($n \equiv 0 \lor n >$ MAX_N) $n =$ MAX_N; /* compute true number of vertices */
  new_graph = *miles*(*n, north_weight, west_weight, pop_weight*, $1_L, 0_L$, *seed*);
  if (*new_graph* $\equiv \Lambda$) **return** Λ; /* *panic_code* has been set by *miles* */
  sprintf(*new_graph*→*id*, "plane_miles(%lu,%ld,%ld,%ld,%lu,%ld)", *n, north_weight, west_weight,*
    pop_weight, extend, prob, seed);
  if (*extend*) *extra_n* −−; /* restore *extra_n* to its previous value */

This code is used in section 41.

43. ⟨ Compute the Delaunay triangulation and run through the Delaunay edges; reject them with
  probability *prob*/65536, otherwise append them with the road length in miles 43 ⟩ ≡
gprob = *prob*;
if (*extend*) {
 inf_vertex = *new_graph*→*vertices* + *new_graph*→*n*;
 inf_vertex→*name* = *gb_save_string*("INF");
 inf_vertex→*x_coord* = *inf_vertex*→*y_coord* = *inf_vertex*→*z_coord* = −1;
} **else** *inf_vertex* = Λ;
delaunay(*new_graph*, *new_mile_edge*);

This code is used in section 41.

44. The mileages will all have been negated by *miles*, so we make them positive again.

⟨ Other subroutines 12 ⟩ +≡
 void *new_mile_edge*(*u, v*)
   Vertex *∗u, ∗v*;
 {
  if ((*gb_next_rand*() $\gg$ 15) $\geq$ *gprob*) {
   if (*u*) {
    if (*v*) *gb_new_edge*(*u, v*, −*miles_distance*(*u, v*));
    else if (*inf_vertex*) *gb_new_edge*(*u, inf_vertex*, INFTY);
   } **else if** (*inf_vertex*) *gb_new_edge*(*inf_vertex, v*, INFTY);
  }
 }

---

alloc_fault = −1, GB_GRAPH §7.
aux_data: **Area**, GB_GRAPH §20.
delaunay: **void** ( ), §9.
extra_n: **long**, GB_GRAPH §24.
gb_free: **void** ( ), GB_GRAPH §16.
gb_new_edge: **void** ( ),
 GB_GRAPH §31.
gb_next_rand = macro ( ),
 GB_FLIP §6.
gb_recycle: **void** ( ), GB_GRAPH §40.
gb_save_string: **char** *( ),
 GB_GRAPH §35.
gb_trouble_code: **long**,

GB_GRAPH §14.
gprob: **static unsigned long**, §10.
Graph = **struct**, GB_GRAPH §20.
id: **char** [], GB_GRAPH §20.
inf_vertex: **static Vertex** *, §10.
INFTY = #10000000 L, §2.
max_distance: **unsigned long**,
 GB_MILES §5.
MAX_N = 128, GB_MILES §2.
miles: **Graph** *( ), GB_MILES §5.
miles_distance: **long** ( ),
 GB_MILES §20.

n: **long**, GB_GRAPH §20.
name: **char** *, GB_GRAPH §9.
new_mile_edge: **void** ( ), §44.
panic = macro ( ), §3.
panic_code: **long**, GB_GRAPH §5.
plane: **Graph** *( ), §5.
sprintf: **int** ( ), <stdio.h>.
Vertex = **struct**, GB_GRAPH §9.
vertices: **Vertex** *, GB_GRAPH §20.
x_coord = *x.I*, GB_MILES §15.
y_coord = *y.I*, GB_MILES §15.
z_coord = *z.I*, §7.

GB_RAMAN

Important: Before reading GB_RAMAN, please read or at least skim the program for GB_GRAPH.

1. Introduction. This GraphBase module contains the *raman* subroutine, which creates a family of "Ramanujan graphs" based on a theory developed by Alexander Lubotzky, Ralph Phillips, and Peter Sarnak [see *Combinatorica* **8** (1988), 261–277].

Ramanujan graphs are defined by the following properties: They are connected, undirected graphs in which every vertex has degree k, and every eigenvalue of the adjacency matrix is either $\pm k$ or has absolute value $\leq 2\sqrt{k-1}$. Such graphs are known to have good expansion properties, small diameter, and relatively small independent sets; they cannot be colored with fewer than $k/(2\sqrt{k-1})$ colors unless they are bipartite. The particular examples of Ramanujan graphs constructed here are based on interesting properties of quaternions with integer coefficients.

An example of the use of this procedure can be found in the demo program called GIRTH.

⟨ gb_raman.h 1 ⟩ ≡
  extern Graph *raman*( );

2. The subroutine call *raman*($p, q, type, reduce$) constructs an undirected graph in which each vertex has degree $p + 1$. The number of vertices is $q + 1$ if *type* = 1, or $\frac{1}{2}q(q+1)$ if *type* = 2, or $\frac{1}{2}(q-1)q(q+1)$ if *type* = 3, or $(q-1)q(q+1)$ if *type* = 4. The graph will be bipartite if and only if it has type 4. Parameters p and q must be distinct prime numbers, and q must be odd. Furthermore there are additional restrictions: If $p = 2$, the other parameter q must satisfy $q \bmod 8 \in \{1, 3\}$ and $q \bmod 13 \in 1, 3, 4, 9, 10, 12$; this rules out about one fourth of all primes. Moreover, if *type* = 3 the value of p must be a quadratic residue modulo q; in other words, there must be an integer x such that $x^2 \equiv p \pmod{q}$. If *type* = 4, the value of p must not be a quadratic residue.

If you specify *type* = 0, the procedure will choose the largest permissible type (either 3 or 4); the value of the type selected will appear as part of the string placed in the resulting graph's *id* field. For example, if *type* = 0, $p = 2$, and $q = 43$, a type 4 graph will be generated, because 2 is not a quadratic residue modulo 43. This graph will have $44 \times 43 \times 42 = 79464$ vertices, each of degree 3. (Notice that graphs of types 3 and 4 can be quite large even when q is rather small.)

The largest permissible value of q is 46337; this is the largest prime whose square is less than 2^{31}. Of course you would use it only for a graph of type 1.

If *reduce* is nonzero, loops and multiple edges will be suppressed. In this case the degrees of some vertices might turn out to be less than $p + 1$, in spite of what was said above.

Although type 4 graphs are bipartite, the vertices are not separated into two blocks as in other bipartite graphs produced by GraphBase routines.

All edges of the graphs have length 1.

3. If the *raman* routine encounters a problem, it returns Λ (NULL), after putting a code number into the external variable *panic_code*. This code number identifies the type of failure. Otherwise *raman* returns a pointer to the newly created graph, which will be represented with the data structures explained in GB_GRAPH. (The external variable *panic_code* is itself defined in GB_GRAPH.)

#define *panic*(c) { *panic_code* = c; *gb_trouble_code* = 0; **return** Λ; }

#define *dead_panic*(*c*)
        { *gb_free*(*working_storage*); *panic*(*c*); }
#define *late_panic*(*c*)
        { *gb_recycle*(*new_graph*); *dead_panic*(*c*); }

4.   The C file **gb_raman.c** has the following general shape:

#include "gb_graph.h"   /* we will use the GB_GRAPH data structures */
  ⟨ Preprocessor definitions ⟩
  ⟨ Type declarations 18 ⟩
  ⟨ Private variables and routines 6 ⟩

  Graph *∗raman*(*p, q, type, reduce*)
        long *p*;   /* one less than the desired degree; must be prime */
        long *q*;   /* size parameter; must be prime and properly related to *type* */
        unsigned long *type*;   /* selector between different possible constructions */
        unsigned long *reduce*;   /* if nonzero, multiple edges and self-loops won't occur */
  { ⟨ Local variables 5 ⟩
     ⟨ Prepare tables for doing arithmetic modulo *q* 7 ⟩;
     ⟨ Choose or verify the *type*, and determine the number *n* of vertices 12 ⟩;
     ⟨ Set up a graph with *n* vertices, and assign vertex labels 13 ⟩;
     ⟨ Compute *p* + 1 generators that will define the graph's edges 19 ⟩;
     ⟨ Append the edges 26 ⟩;
     if (*gb_trouble_code*) *late_panic*(*alloc_fault*);
           /* oops, we ran out of memory somewhere back there */
     gb_free(*working_storage*);
     return *new_graph*;
  }

5.   ⟨ Local variables 5 ⟩ ≡
  Graph *∗new_graph*;   /* the graph constructed by *raman* */
  Area *working_storage*;   /* place for auxiliary tables */

See also section 9.

This code is used in section 4.

---

alloc_fault = −1, GB_GRAPH §7.
Area, GB_GRAPH §12.
gb_free: **void** ( ), GB_GRAPH §16.
gb_recycle: **void** ( ), GB_GRAPH §40.

gb_trouble_code: **long**,
    GB_GRAPH §14.
Graph = **struct**, GB_GRAPH §20.

id: **char** [], GB_GRAPH §20.
n: **long**, §9.
panic_code: **long**, GB_GRAPH §5.

6. Brute force number theory. Instead of using routines like Euclid's algorithm to compute inverses and square roots modulo q, we have plenty of time to build complete tables, since q is smaller than the number of vertices we will be generating.

We will make three tables: $q\_sqr[k]$ will contain k^2 modulo q; $q\_sqrt[k]$ will contain one of the values of $\sqrt{k}$ if k is a quadratic residue; and $q\_inv[k]$ will contain the multiplicative inverse of k.

⟨ Private variables and routines 6 ⟩ ≡
    static long $*q\_sqr$;   /* squares */
    static long $*q\_sqrt$;   /* square roots (or -1 if not a quadratic residue) */
    static long $*q\_inv$;   /* reciprocals */
See also sections 15, 20, 22, and 30.

This code is used in section 4.

7. ⟨ Prepare tables for doing arithmetic modulo q 7 ⟩ ≡
    if $(q < 3 \vee q > 46337)$ *panic*(*very_bad_specs*);   /* q is way too small or way too big */
    if $(p < 2)$ *panic*(*very_bad_specs* $+ 1$);   /* p is way too small */
    init_area(*working_storage*);
    $q\_sqr =$ *gb_typed_alloc*$(3 * q,$ **long**, *working_storage*$)$;
    if $(q\_sqr \equiv 0)$ *panic*(*no_room* $+ 1$);
    $q\_sqrt = q\_sqr + q$;
    $q\_inv = q\_sqrt + q$;   /* note that *gb_alloc* has initialized everything to zero */
    ⟨ Compute the q_sqr and q_sqrt tables 8 ⟩;
    ⟨ Find a primitive root a, modulo q, and its inverse aa 10 ⟩;
    ⟨ Compute the q_inv table 11 ⟩;
This code is used in section 4.

8. ⟨ Compute the q_sqr and q_sqrt tables 8 ⟩ ≡
    for $(a = 1;\ a < q;\ a{+}{+})$ $q\_sqrt[a] = -1$;
    for $(a = 1, aa = 1;\ a < q;\ aa = (aa + a + a + 1)\,\%\,q, a{+}{+})$ {
        $q\_sqr[a] = aa$;
        $q\_sqrt[aa] = q - a$;   /* the smaller square root will survive */
        $q\_inv[aa] = -1$;   /* we make $q\_inv[aa]$ nonzero when aa can't be a primitive root */
    }
This code is used in section 7.

9. ⟨ Local variables 5 ⟩ +≡
    register long $a,\ aa,\ k$;   /* primary indices in loops */
    long $b,\ bb,\ c,\ cc,\ d,\ dd$;   /* secondary indices */
    long n;   /* the number of vertices */
    long n_factor;   /* either $\frac{1}{2}(q-1)$ (type 3) or $q - 1$ (type 4) */
    register Vertex $*v$;   /* the current vertex of interest */

10. Here we implicitly test that q is prime, by finding a primitive root whose powers generate everything. If q is not prime, its smallest divisor will cause the inner loop in this step to terminate with $k \geq q$, because no power of that divisor will be congruent to 1.

⟨ Find a primitive root a, modulo q, and its inverse aa 10 ⟩ ≡
    for $(a = 2;\ ;\ a{+}{+})$
        if $(q\_inv[a] \equiv 0)$ {
            for $(b = a, k = 1;\ b \neq 1 \wedge k < q;\ aa = b, b = (a * b)\,\%\,q, k{+}{+})$ $q\_inv[b] = -1$;

```
    if (k ≥ q) dead_panic(bad_specs + 1);     /* q is not prime */
    if (k ≡ q − 1) break;      /* good, a is the primitive root we seek */
  }
```

This code is used in section 7.

11. As soon as we have discovered a primitive root, it is easy to generate all the inverses. (We could also generate the discrete logarithms if we had a need for them.)

We set $q\_inv[0] = q$; this will be our internal representation of ∞.

⟨ Compute the q_inv table 11 ⟩ ≡
```
    for (b = a, bb = aa; b ≠ bb; b = (a ∗ b) % q, bb = (aa ∗ bb) % q) q_inv[b] = bb, q_inv[bb] = b;
    q_inv[1] = 1;
    q_inv[b] = b;      /* at this point b must equal q − 1 */
    q_inv[0] = q;
```

This code is used in section 7.

12. The conditions we stated for validity of q when $p = 2$ are equivalent to the existence of $\sqrt{-2}$ and $\sqrt{13}$ modulo q, according to the law of quadratic reciprocity (see, for example, *Fundamental Algorithms*, exercise 1.2.4–47).

⟨ Choose or verify the *type*, and determine the number n of vertices 12 ⟩ ≡
```
    if (p ≡ 2) {
      if (q_sqrt[13 % q] < 0 ∨ q_sqrt[q − 2] < 0) dead_panic(bad_specs + 2);
          /* improper prime to go with p = 2 */
    }
    if ((a = p % q) ≡ 0) dead_panic(bad_specs + 3);     /* p divisible by q */
    if (type ≡ 0) type = (q_sqrt[a] > 0 ? 3 : 4);
    n_factor = (type ≡ 3 ? (q − 1)/2 : q − 1);
    switch (type) {
    case 1: n = q + 1; break;
    case 2: n = q ∗ (q + 1)/2; break;
    default:
      if ((q_sqrt[a] > 0 ∧ type ≠ 3) ∨ (q_sqrt[a] < 0 ∧ type ≠ 4))
        dead_panic(bad_specs + 4);     /* wrong type for p modulo q */
      if (q > 1289) dead_panic(bad_specs + 5);      /* way too big for types 3, 4 */
      n = n_factor ∗ q ∗ (q + 1);
      break;
    }
    if (p ≥ (long) (#3fffffff/n)) dead_panic(bad_specs + 6);     /* (p + 1)n ≥ 2³⁰ */
```

This code is used in section 4.

---

bad_specs = 30, GB_GRAPH §7.
dead_panic = macro ( ), §3.
gb_alloc: **char** ∗( ), GB_GRAPH §13.
gb_typed_alloc = macro ( ),
  GB_GRAPH §11.
init_area = macro ( ), GB_GRAPH §12.
no_room = 1, GB_GRAPH §7.
p: **long**, §4.
panic = macro ( ), §3.
q: **long**, §4.
type: **unsigned long**, §4.
Vertex = **struct**, GB_GRAPH §9.
very_bad_specs = 40, GB_GRAPH §7.
working_storage: **Area**, §5.

13. The vertices.   Graphs of type 1 have vertices from the set $\{0, 1, \ldots, q-1, \infty\}$, namely the integers modulo q with an additional "infinite" element thrown in. The idea is to operate on these quantities by adding constants, and/or multiplying by constants, and/or taking reciprocals, modulo q.

Graphs of type 2 have vertices that are unordered pairs of distinct elements from that same $(q+1)$-element set.

Graphs of types 3 and 4 have vertices that are 2×2 matrices having nonzero determinants modulo q. The determinants of type 3 matrices are, in fact, nonzero quadratic residues. We consider two matrices to be equivalent if one is obtained from the other by multiplying all entries by a constant (modulo q); therefore we will normalize all the matrices so that the second row is either $(0, 1)$ or has the form $(1, x)$ for some x. The total number of equivalence classes of type 4 matrices obtainable in this way is $(q+1)q(q-1)$, because we can choose the second row in $q+1$ ways, after which there are two cases: Either the second row is $(0, 1)$, and we can select the upper right corner element arbitrarily and choose the upper left corner element nonzero; or the second row is $(1, x)$, and we can select the upper left corner element arbitrarily and then choose an upper right corner element to make the determinant nonzero. For type 3 the counting is similar, except that "nonzero" becomes "nonzero quadratic residue," hence there are exactly half as many choices.

It is easy to verify that the equivalence classes of matrices that correspond to vertices in these graphs of types 3 and 4 are closed under matrix multiplication. Therefore the vertices can be regarded as the elements of finite groups. The type 3 group for a given q is often called the linear fractional group $LF(2, \mathbf{F}_q)$, or the projective special linear group $PSL(2, \mathbf{F}_q)$, or the linear simple group $L_2(q)$; it can also be regarded as the group of 2×2 matrices with determinant 1 (mod q), when the matrix A is considered equivalent to $-A$. (This group is a simple group for all primes $q > 2$.) The type 4 group is officially known as the projective general linear group of degree 2 over the field of q elements, $PGL(2, \mathbf{F}_q)$.

⟨ Set up a graph with n vertices, and assign vertex labels 13 ⟩ ≡

```
new_graph = gb_new_graph(n);
if (new_graph ≡ Λ) dead_panic(no_room);      /* out of memory before we try to add edges */
sprintf(new_graph→id, "raman(%ld,%ld,%lu,%lu)", p, q, type, reduce);
strcpy(new_graph→util_types, "ZZZIIZIZZZZZZZ");
v = new_graph→vertices;
switch (type) {
case 1: ⟨ Assign labels from the set {0, 1, ..., q − 1, ∞} 14 ⟩; break;
case 2: ⟨ Assign labels for pairs of distinct elements 16 ⟩; break;
default: ⟨ Assign projective matrix labels 17 ⟩; break;
}
```

This code is used in section 4.

14.   Type 1 graphs are the easiest to label. We store a serial number in utility field $x.I$, using q to represent ∞.

⟨ Assign labels from the set $\{0, 1, \ldots, q-1, \infty\}$ 14 ⟩ ≡

```
new_graph→util_types[4] = 'Z';
for (a = 0; a < q; a++) {
  sprintf(name_buf, "%ld", a);
```

$v\text{-}name = gb\_save\_string(name\_buf);$

$v\text{-}x.I = a;$

$v\text{++};$

}

$v\text{-}name = gb\_save\_string(\text{"INF"});$

$v\text{-}x.I = q;$

$v\text{++};$

This code is used in section 13.

15. ⟨ Private variables and routines 6 ⟩ +≡

static char $name\_buf[\,] = \text{"(1111,1111;1,1111)"};$ /* place to form vertex names */

16. The type 2 labels run from $\{0,1\}$ to $\{q-1,\infty\}$; we put the coefficients into $x.I$ and $y.I$, where they might prove useful in some applications.

⟨ Assign labels for pairs of distinct elements 16 ⟩ ≡

  for $(a = 0;\ a < q;\ a\text{++})$

    for $(aa = a + 1;\ aa \le q;\ aa\text{++})$ {

      if $(aa \equiv q)\ sprintf(name\_buf, \text{"\{\%ld,INF\}"}, a);$

      else $sprintf(name\_buf, \text{"\{\%ld,\%ld\}"}, a, aa);$

      $v\text{-}name = gb\_save\_string(name\_buf);$

      $v\text{-}x.I = a;\ \ v\text{-}y.I = aa;$

      $v\text{++};$

    }

This code is used in section 13.

---

a: **register long**, §9.

aa: **register long**, §9.

dead_panic = macro ( ), §3.

gb_new_graph: **Graph** *( ), GB_GRAPH §23.

gb_save_string: **char** *( ), GB_GRAPH §35.

I: **long**, GB_GRAPH §8.

id: **char** [ ], GB_GRAPH §20.

n: **long**, §9.

name: **char** *, GB_GRAPH §9.

new_graph: **Graph** *, §5.

no_room = 1, GB_GRAPH §7.

p: **long**, §4.

q: **long**, §4.

reduce: **unsigned long**, §4.

sprintf: **int** ( ), <stdio.h>.

strcpy: **char** *( ), <string.h>.

type: **unsigned long**, §4.

util_types: **char** [ ], GB_GRAPH §20.

v: **register Vertex** *, §9.

vertices: **Vertex** *, GB_GRAPH §20.

x: **util**, GB_GRAPH §9.

y: **util**, GB_GRAPH §9.

17. For graphs of types 3 and 4, we set the $x.I$ and $y.I$ fields to the elements of the first row of the matrix, and we set the $z.I$ field equal to the ratio of the elements of the second row (again with q representing ∞).

The vertices in this case consist of $q(q+1)$ blocks of vertices having a given second row and a given element in the upper left or upper right position. Within each block of vertices, the determinants are respectively congruent modulo q to 1^2, 2^2, $\ldots$, $(\frac{q-1}{2})^2$ in the case of type 3 graphs, or to $1, 2, \ldots, q-1$ in the case of type 4.

⟨ Assign projective matrix labels 17 ⟩ ≡

```
new_graph→util_types[5] = 'I';
for (c = 0; c ≤ q; c++)
  for (b = 0; b < q; b++)
    for (a = 1; a ≤ n_factor; a++) {
      v→z.I = c;
      if (c ≡ q) {      /* second row of matrix is (0,1) */
        v→y.I = b;
        v→x.I = (type ≡ 3 ? q_sqr[a] : a);      /* determinant is a² or a */
        sprintf(name_buf, "(%ld,%ld;0,1)", v→x.I, b);
      } else {      /* second row of matrix is (1,c) */
        v→x.I = b;
        v→y.I = (b * c + q − (type ≡ 3 ? q_sqr[a] : a)) % q;
        sprintf(name_buf, "(%ld,%ld;1,%ld)", b, v→y.I, c);
      }      /* determinant is a² or a */
      v→name = gb_save_string(name_buf);
      v++;
    }
```

This code is used in section 13.

18.   Group generators.   We will define a set of $p + 1$ permutations $\{\pi_0, \pi_1, \ldots, \pi_p\}$ of the vertices, such that the arcs of our graph will go from v to $v\pi_k$ for $0 \le k \le p$. Thus, each path in the graph will be defined by a product of permutations; the cycles of the graph will correspond to vertices that are left fixed by a product of permutations. The graph will be undirected, because the inverse of each π_k will also be one of the permutations of the generating set.

In fact, each permutation π_k will be defined by a 2×2 matrix. For graphs of types 3 and 4, the permutations will therefore correspond to certain vertices, and the vertex $v\pi_k$ will simply be the product of matrix v by matrix π_k.

For graphs of type 1, the permutations will be defined by linear fractional transformations, which are mappings of the form

$$v \longmapsto \frac{av + b}{cv + d} \bmod q \,.$$

This transformation applies to all $v \in \{0, 1, \ldots, q - 1, \infty\}$, under the usual conventions that $x/0 = \infty$ when $x \ne 0$ and $(x\infty + x')/(y\infty + y') = x/y$. The composition of two such transformations is again a linear fractional transformation, corresponding to the product of the two associated matrices $\left(\begin{smallmatrix} a & b \\ c & d \end{smallmatrix}\right)$.

Graphs of type 2 will be handled just like graphs of type 1, except that we will compute the images of two distinct points $v = \{v_1, v_2\}$ under the linear fractional transformation. The two images will be distinct, because the transformation is invertible.

When $p = 2$, a special set of three generating matrices π_0, π_1, π_2 can be shown to define Ramanujan graphs; these matrices are described below. Otherwise p is odd, and the generators are based on the theory of integral quaternions. Integral quaternions are quadruples of the form $\alpha = a_0 + a_1 i + a_2 j + a_3 k$, where a_0, a_1, a_2, and a_3 are integers; we multiply them by using the associative but noncommutative multiplication rules $i^2 = j^2 = k^2 = ijk = -1$. If we write $\alpha = a + A$, where a is the "scalar" a_0 and A is the "vector" $a_1 i + a_2 j + a_3 k$, the product of quaternions $\alpha = a + A$ and $\beta = b + B$ can be expressed as

$$(a + A)(b + B) = ab - A \cdot B + aB + bA + A \times B \,,$$

where $A \cdot B$ and $A \times B$ are the usual dot product and cross product of vectors. The conjugate of $\alpha = a + A$ is $\overline{\alpha} = a - A$, and we have $\alpha\overline{\alpha} = a_0^2 + a_1^2 + a_2^2 + a_3^2$. This important quantity is called $N(\alpha)$, the norm of α. It is not difficult to verify that $N(\alpha\beta) = N(\alpha)N(\beta)$, because of the basic identity $\overline{\alpha\beta} = \overline{\beta}\,\overline{\alpha}$ and the fact that $\alpha x = x\alpha$ when x is scalar.

Integral quaternions have a beautiful theory; for example, there is a nice variant of Euclid's algorithm by which we can compute the greatest common left divisor of any two integral quaternions. This algorithm makes it possible to prove that integral quaternions whose coefficients are relatively prime can be uniquely factored into quaternions whose norm is prime. However, the details of that theory are beyond the scope of this documentation. It will suffice for our purposes to observe that we can use quaternions to define the finite groups $PSL(2, \mathbf{F}_q)$ and $PGL(2, \mathbf{F}_q)$ in a different way from the definitions given earlier: Suppose we consider two quaternions to be equivalent if their coefficients are equal modulo q, or if one is a nonzero scalar multiple of the other (modulo q). Thus, for example, if $q = 3$ we consider $1 + 4i - j$ to be equivalent to $1 + i + 2j$, and also equivalent to $2 + 2i + j$. It turns out that there are exactly $(q+1)q(q-1)$ such equivalence classes, and they form a group under quaternion multiplication that is the same as the projective group of 2×2 matrices under matrix multiplication, modulo q. One way to prove this is by means of the one-to-one correspondence

$$a_0 + a_1 i + a_2 j + a_3 k \longleftrightarrow \begin{pmatrix} a_0 + a_1 g + a_3 h & a_2 + a_3 g - a_1 h \\ -a_2 + a_3 g - a_1 h & a_0 - a_1 g - a_3 h \end{pmatrix},$$

where g and h are integers with $g^2 + h^2 \equiv -1 \pmod{q}$.

Jacobi proved that the number of ways to represent any odd number p as a sum of four squares $a_0^2 + a_1^2 + a_2^2 + a_3^2$ is 8 times the sum of divisors of p. [This fact appears in the concluding sentence of his monumental work *Fundamenta Nova Theoriæ Functionum Ellipticorum*, Königsberg, 1829.] In particular, when p is prime, the number of such representations is $8(p+1)$; in other words, there are exactly $8(p+1)$ quaternions $\alpha = a_0 + a_1 i + a_2 j + a_3 k$ with $N(\alpha) = p$. These quaternions form $p + 1$ equivalence classes under multiplication by the eight "unit quaternions" $\{\pm 1, \pm i, \pm j, \pm k\}$. We will select one element from each equivalence class, and the resulting $p + 1$ quaternions will correspond to $p + 1$ matrices, which will generate the $p + 1$ arcs leading from each vertex in the graphs to be constructed.

⟨ Type declarations 18 ⟩ ≡
    typedef struct {
        long *a0*, *a1*, *a2*, *a3*;   /* coefficients of a quaternion */
        unsigned long *bar*;   /* the index of the inverse (conjugate) quaternion */
    } **quaternion**;
This code is used in section 4.

19. A global variable *gen_count* will be declared below, indicating the number of generators found so far. When p isn't prime, we will find more than $p + 1$ solutions, so we allocate an extra slot in the *gen* table to hold a possible overflow entry.

⟨ Compute $p + 1$ generators that will define the graph's edges 19 ⟩ ≡
    gen = *gb_typed_alloc*($p + 2$, **quaternion**, *working_storage*);
    if (*gen* ≡ Λ) *late_panic*(*no_room* + 2);   /* not enough memory */
    gen_count = 0; *max_gen_count* = $p + 1$;
    if ($p \equiv 2$) ⟨ Fill the *gen* table with special generators 25 ⟩
    else ⟨ Fill the *gen* table with representatives of all quaternions having norm p 21 ⟩;
    if (*gen_count* ≠ *max_gen_count*) *late_panic*(*bad_specs* + 7);   /* p is not prime */
This code is used in section 4.

20. ⟨ Private variables and routines 6 ⟩ +≡
    static quaternion **gen*;   /* table of the $p + 1$ generators */

---

bad_specs = 30, GB_GRAPH §7.
gb_typed_alloc = **macro** ( ),
  GB_GRAPH §11.
gen_count: **static unsigned long**,

§22.
late_panic = **macro** ( ), §3.
max_gen_count: **static unsigned
  long**, §22.

no_room = 1, GB_GRAPH §7.
p: **long**, §4.
q: **long**, §4.
working_storage: **Area**, §5.

21. As mentioned above, quaternions of norm p come in sets of 8, differing from each other only by unit multiples; we need to choose one of the 8. Suppose $a_0^2 + a_1^2 + a_2^2 + a_3^2 = p$. If $p \bmod 4 = 1$, exactly one of the a's will be odd; so we call it a_0 and assign it a positive sign. When $p \bmod 4 = 3$, exactly one of the a's will be even; we call it a_0, and if it is nonzero we make it positive. If $a_0 = 0$, we make sure that one of the others—say the rightmost appearance of the largest one—is positive. In this way we obtain a unique representative from each set of 8 equivalent quaternions.

For example, the four quaternions of norm 3 are $\pm i \pm j + k$; the six of norm 5 are $1 \pm 2i$, $1 \pm 2j$, $1 \pm 2k$.

In the program here we generate solutions to $a^2 + b^2 + c^2 + d^2 = p$ when $a \not\equiv b \equiv c \equiv d \pmod 2$ and $b \leq c \leq d$. The variables aa, bb, and cc hold the respective values $p - a^2 - b^2 - c^2 - d^2$, $p - a^2 - 3b^2$, and $p - a^2 - 2c^2$. The **for** statements use the fact that a^2 increases by $4(a + 1)$ when a increases by 2.

⟨ Fill the *gen* table with representatives of all quaternions having norm p 21 ⟩ ≡
```
{ long sa, sb;        /* p - a², p - a² - b² */
  long pp = (p ≫ 1) & 1;      /* 0 if p mod 4 = 1,  1 if p mod 4 = 3 */
  for (a = 1 - pp, sa = p - a;  sa > 0;  sa -= (a + 1) ≪ 2, a += 2)
    for (b = pp, sb = sa - b, bb = sb - b - b;  bb ≥ 0;  bb -= 12 * (b + 1), sb -= (b + 1) ≪ 2, b += 2)
      for (c = b, cc = bb;  cc ≥ 0;  cc -= (c + 1) ≪ 3, c += 2)
        for (d = c, aa = cc;  aa ≥ 0;  aa -= (d + 1) ≪ 2, d += 2)
          if (aa ≡ 0) ⟨ Deposit the quaternions associated with a + bi + cj + dk 23 ⟩;
  ⟨ Change the gen table to matrix format 24 ⟩;
}
```
This code is used in section 19.

22. If $a > 0$ and $0 < b < c < d$, we obtain 48 different classes of quaternions having the same norm by permuting $\{b, c, d\}$ in six ways and attaching signs to each permutation in eight ways. This happens, for example, when $p = 71$ and $(a, b, c, d) = (6, 1, 3, 5)$. Fewer quaternions arise when $a = 0$ or $0 = b$ or $b = c$ or $c = d$.

The inverse of the matrix corresponding to a quaternion is the matrix corresponding to the conjugate quaternion. Therefore a generating matrix π_k will be its own inverse if and only if it comes from a quaternion with $a = 0$.

It is convenient to have a subroutine that deposits a new quaternion and its conjugate into the table of generators.

⟨ Private variables and routines 6 ⟩ +≡
```
static unsigned long gen_count;      /* the next available quaternion slot */
static unsigned long max_gen_count;      /* p + 1, stored as a global variable */
static void deposit(a, b, c, d)
    long a, b, c, d;      /* a solution to a² + b² + c² + d² = p */
{
  if (gen_count ≥ max_gen_count)      /* oops, we already found p + 1 solutions */
    gen_count = max_gen_count + 1;      /* this will happen only if p isn't prime */
  else {
    gen[gen_count].a0 = gen[gen_count + 1].a0 = a;
    gen[gen_count].a1 = b; gen[gen_count + 1].a1 = -b;
```

$$gen[gen\_count].a2 = c; \quad gen[gen\_count + 1].a2 = -c;$$
$$gen[gen\_count].a3 = d; \quad gen[gen\_count + 1].a3 = -d;$$

if (a) {
    $gen[gen\_count].bar = gen\_count + 1;$
    $gen[gen\_count + 1].bar = gen\_count;$
    $gen\_count \mathrel{+}= 2;$
} **else** {
    $gen[gen\_count].bar = gen\_count;$
    $gen\_count \mathbin{+\!+};$
}
}
}

23. ⟨ Deposit the quaternions associated with $a + bi + cj + dk$ 23 ⟩ ≡
{
  $deposit(a, b, c, d);$
  if (b) {
    $deposit(a, -b, c, d); \quad deposit(a, -b, -c, d);$
  }
  if (c) $deposit(a, b, -c, d);$
  if $(b < c)$ {
    $deposit(a, c, b, d); \quad deposit(a, -c, b, d); \quad deposit(a, c, d, b); \quad deposit(a, -c, d, b);$
    if (b) {
      $deposit(a, c, -b, d); \quad deposit(a, -c, -b, d); \quad deposit(a, c, d, -b); \quad deposit(a, -c, d, -b);$
    }
  }
  if $(c < d)$ {
    $deposit(a, b, d, c); \quad deposit(a, d, b, c);$
    if (b) {
      $deposit(a, -b, d, c); \quad deposit(a, -b, d, -c); \quad deposit(a, d, -b, c); \quad deposit(a, d, -b, -c);$
    }
    if (c) {
      $deposit(a, b, d, -c); \quad deposit(a, d, b, -c);$
    }
    if $(b < c)$ {
      $deposit(a, d, c, b); \quad deposit(a, d, -c, b);$
      if (b) {
        $deposit(a, d, c, -b); \quad deposit(a, d, -c, -b);$
      }
    }
  }
}

This code is used in section 21.

---

a: **register long**, §9.
$a0$: **long**, §18.
$a1$: **long**, §18.
$a2$: **long**, §18.
$a3$: **long**, §18.

aa: **register long**, §9.
b: **long**, §9.
bar: **unsigned long**, §18.
bb: **long**, §9.
c: **long**, §9.

cc: **long**, §9.
d: **long**, §9.
gen: **static quaternion** *, §20.
p: **long**, §4.

24. Once we've found the generators in quaternion form, we want to convert them to 2×2 matrices, using the correspondence mentioned earlier:

$$a_0 + a_1 i + a_2 j + a_3 k \longleftrightarrow \begin{pmatrix} a_0 + a_1 g + a_3 h & a_2 + a_3 g - a_1 h \\ -a_2 + a_3 g - a_1 h & a_0 - a_1 g - a_3 h \end{pmatrix},$$

where g and h are integers with $g^2 + h^2 \equiv -1 \pmod{q}$. Appropriate values for g and h can always be found by the formulas

$$g = \sqrt{k} \quad \text{and} \quad h \Rightarrow \sqrt{q - 1 - k},$$

where k is the largest quadratic residue modulo q. For if $q - 1$ is not a quadratic residue, and if $k + 1$ isn't a residue either, then $q - 1 - k$ must be a quadratic residue because it is congruent to the product $(q - 1)(k + 1)$ of nonresidues. (We will have $h = 0$ if and only if $q \bmod 4 = 1$; $h = 1$ if and only if $q \bmod 8 = 3$; $h = \sqrt{2}$ if and only if $q \bmod 24 = 7$ or 15; etc.)

⟨ Change the *gen* table to matrix format 24 ⟩ ≡

```
{ register long g, h;
  long a00, a01, a10, a11;      /* entries of 2 × 2 matrix */
  for (k = q - 1; q_sqrt[k] < 0; k--) ;    /* find the largest quadratic residue, k */
  g = q_sqrt[k]; h = q_sqrt[q - 1 - k];
  for (k = p; k >= 0; k--) {
    a00 = (gen[k].a0 + g * gen[k].a1 + h * gen[k].a3) % q;
    if (a00 < 0) a00 += q;
    a11 = (gen[k].a0 - g * gen[k].a1 - h * gen[k].a3) % q;
    if (a11 < 0) a11 += q;
    a01 = (gen[k].a2 + g * gen[k].a3 - h * gen[k].a1) % q;
    if (a01 < 0) a01 += q;
    a10 = (-gen[k].a2 + g * gen[k].a3 - h * gen[k].a1) % q;
    if (a10 < 0) a10 += q;
    gen[k].a0 = a00; gen[k].a1 = a01; gen[k].a2 = a10; gen[k].a3 = a11;
  }
}
```

This code is used in section 21.

25. When $p = 2$, the following three appropriate generating matrices have been found by Patrick Chiu:

$$\begin{pmatrix} 1 & 0 \\ 0 & -1 \end{pmatrix}, \quad \begin{pmatrix} 2 + s & t \\ t & 2 - s \end{pmatrix}, \quad \text{and} \quad \begin{pmatrix} 2 - s & -t \\ -t & 2 + s \end{pmatrix},$$

where $s^2 \equiv -2$ and $t^2 \equiv -26 \pmod{q}$. The determinants of these matrices are respectively -1, 32, and 32; the product of the second and third matrices is 32 times the identity matrix. Notice that when 2 is a quadratic residue (this happens when $q = 8k + 1$), the determinants are all quadratic residues, so we get a graph of type 3. When 2 is a quadratic nonresidue (which happens when $q = 8k + 3$), the determinants are all nonresidues, so we get a graph of type 4.

⟨ Fill the *gen* table with special generators 25 ⟩ ≡

```
{ long s = q_sqrt[q − 2], t = (q_sqrt[13 % q] * s) % q;
    gen[0].a0 = 1; gen[0].a1 = gen[0].a2 = 0; gen[0].a3 = q − 1; gen[0].bar = 0;
    gen[1].a0 = gen[2].a3 = (2 + s) % q;
    gen[1].a1 = gen[1].a2 = t;
    gen[2].a1 = gen[2].a2 = q − t;
    gen[1].a3 = gen[2].a0 = (q + 2 − s) % q;
    gen[1].bar = 2; gen[2].bar = 1;
    gen_count = 3;
}
```

This code is used in section 19.

$a0$: **long**, §18.
$a1$: **long**, §18.
$a2$: **long**, §18.
$a3$: **long**, §18.

bar: **unsigned long**, §18.
gen: **static quaternion** *, §20.
gen_count: **static unsigned long**,
        §22.

k: **register long**, §9.
p: **long**, §4.
q: **long**, §4.
q_sqrt: **static long** *, §6.

26. Constructing the edges. The remaining task is to use the permutations defined by the *gen* table to create the arcs of the graph and their inverses.

The *ref* fields in each arc will refer to the permutation leading to the arc. In most cases each vertex v will have degree exactly $p + 1$, and the edges emanating from it will appear in a linked list having the respective *ref* fields $0, 1, \ldots, p$ in order. However, if *reduce* is nonzero, self-loops and multiple edges will be eliminated, so the degree might be less than $p + 1$; in this case the *ref* fields will still be in ascending order, but some generators won't be referenced.

There is a subtle case where *reduce* = 0 but the degree of a vertex might actually be greater than $p + 1$. We want the graph g generated by *raman* to satisfy the conventions for undirected graphs stated in GB_GRAPH; therefore, if any of the generating permutations has a fixed point, we will create two arcs for that fixed point, and the corresponding vertex v will have an edge running to itself. Since each edge consists of two arcs, such an edge will produce two consecutive entries in the list $v\text{-}arcs$. If the generating permutation happens to be its own inverse, there will be two consecutive entries with the same *ref* field; this means there will be more than $p + 1$ entries in $v\text{-}arcs$, and the total number of arcs $g\text{-}m$ will exceed $(p + 1)n$. Self-inverse generating permutations arise only when $p = 2$ or when p is expressible as a sum of three odd squares (hence $p \bmod 8 = 3$); and such permutations will have fixed points only when *type* < 3. Therefore this anomaly does not arise often. But it does occur, for example, in the smallest graph generated by *raman*, namely when $p = 2$, $q = 3$, and *type* = 1, when there are 4 vertices and 14 (not 12) arcs.

#define *ref* *a.I*   /* the *ref* field of an arc refers to its permutation number */
⟨ Append the edges 26 ⟩ ≡
  for $(k = p;\ k \geq 0;\ k\text{--})$ { **long** *kk*;
    if $((kk = gen[k].bar) \leq k)$ /* we assume that $kk = k$ or $kk = k - 1$ */
      for $(v = new\_graph\text{-}vertices;\ v < new\_graph\text{-}vertices + n;\ v\text{++})$ {
        register Vertex *$*u$;

        ⟨ Compute the image, u, of v under the permutation defined by *gen*[k] 27 ⟩;
        if $(u \equiv v)$ {
          if $(\neg reduce)$ {
            $gb\_new\_edge(v, v, 1_L)$;
            $v\text{-}arcs\text{-}ref = kk$; $(v\text{-}arcs + 1)\text{-}ref = k$;
              /* see the remarks above regarding the case $kk = k$ */
          }
        } **else** { **register Arc** *$*ap$;
          if $(u\text{-}arcs \wedge u\text{-}arcs\text{-}ref \equiv kk)$ **continue**;
              /* $kk = k$ and we've already done this two-cycle */
          else if $(reduce)$
            for $(ap = v\text{-}arcs;\ ap;\ ap = ap\text{-}next)$
              if $(ap\text{-}tip \equiv u)$ **goto** *done*;   /* there's already an edge between u and v */
          $gb\_new\_edge(v, u, 1_L)$;
          $v\text{-}arcs\text{-}ref = k$; $u\text{-}arcs\text{-}ref = kk$;
          if $((ap = v\text{-}arcs\text{-}next) \neq \Lambda \wedge ap\text{-}ref \equiv kk)$ {
            $v\text{-}arcs\text{-}next = ap\text{-}next$; $ap\text{-}next = v\text{-}arcs$; $v\text{-}arcs = ap$;
          }   /* now the $v\text{-}arcs$ list has *ref* fields in order again */
        done: ;
      }

```
        }
    }
```
This code is used in section 4.

27. For graphs of types 3 and 4, our job is to compute a 2×2 matrix product, reduce it modulo q, and find the appropriate equivalence class u.

$\langle$ Compute the image, u, of v under the permutation defined by $gen[k]$ 27 $\rangle \equiv$
  if $(type < 3)$
    $\langle$ Compute the image, u, of v under the linear fractional transformation defined by $gen[k]$ 31 $\rangle$
  else $\{$ **long** $a00 = gen[k].a0$, $a01 = gen[k].a1$, $a10 = gen[k].a2$, $a11 = gen[k].a3$;
    $a = v{\rightarrow}x.I$; $b = v{\rightarrow}y.I$;
    if $(v{\rightarrow}z.I \equiv q)$ $c = 0, d = 1$;
    else $c = 1, d = v{\rightarrow}z.I$;
    $\langle$ Compute the matrix product $(aa, bb;\ cc, dd) = (a, b;\ c, d) * (a00, a01;\ a10, a11)$ 28 $\rangle$;
    $a = (cc\ ?\ q\_inv[cc] : q\_inv[dd])$;   /* now a is a normalization factor */
    $d = (a * dd)\ \%\ q$; $c = (a * cc)\ \%\ q$; $b = (a * bb)\ \%\ q$; $a = (a * aa)\ \%\ q$;
    $\langle$ Set u to the vertex whose label is $(a, b;\ c, d)$ 29 $\rangle$;
    $\}$
This code is used in section 26.

28. $\langle$ Compute the matrix product $(aa, bb;\ cc, dd) = (a, b;\ c, d) * (a00, a01;\ a10, a11)$ 28 $\rangle \equiv$
  $aa = (a * a00 + b * a10)\ \%\ q$;
  $bb = (a * a01 + b * a11)\ \%\ q$;
  $cc = (c * a00 + d * a10)\ \%\ q$;
  $dd = (c * a01 + d * a11)\ \%\ q$;
This code is used in section 27.

29. $\langle$ Set u to the vertex whose label is $(a, b;\ c, d)$ 29 $\rangle \equiv$
  if $(c \equiv 0)$ $d = q, aa = a$;
  else $\{$
    $aa = (a * d - b)\ \%\ q$;
    if $(aa < 0)$ $aa\ \mathrel{+}= q$;
    $b = a$;
  $\}$ /* now aa is the determinant of the matrix */
  $u = new\_graph{\rightarrow}vertices + ((d * q + b) * n\_factor + (type \equiv 3\ ?\ q\_sqrt[aa] : aa) - 1)$;
This code is used in section 27.

---

a: **register long**, §9.	d: **long**, §9.	q: **long**, §4.
a: **util**, GB_GRAPH §10.	dd: **long**, §9.	q_inv: **static long** *, §6.
$a0$: **long**, §18.	gb_new_edge: **void** ( ),	q_sqrt: **static long** *, §6.
$a1$: **long**, §18.	GB_GRAPH §31.	$raman$: **Graph** *( ), §4.
$a2$: **long**, §18.	gen: **static quaternion** *, §20.	$reduce$: **unsigned long**, §4.
$a3$: **long**, §18.	I: **long**, GB_GRAPH §8.	tip: **Vertex** *, GB_GRAPH §10.
aa: **register long**, §9.	k: **register long**, §9.	$type$: **unsigned long**, §4.
Arc = **struct**, GB_GRAPH §10.	m: **long**, GB_GRAPH §20.	v: **register Vertex** *, §9.
$arcs$: **Arc** *, GB_GRAPH §9.	n: **long**, §9.	**Vertex** = **struct**, GB_GRAPH §9.
b: **long**, §9.	n_factor: **long**, §9.	$vertices$: **Vertex** *, GB_GRAPH §20.
bar: **unsigned long**, §18.	new_graph: **Graph** *, §5.	x: **util**, GB_GRAPH §9.
bb: **long**, §9.	$next$: **Arc** *, GB_GRAPH §10.	y: **util**, GB_GRAPH §9.
c: **long**, §9.	p: **long**, §4.	z: **util**, GB_GRAPH §9.
cc: **long**, §9.		

30. Linear fractional transformations. Given a nonsingular 2×2 matrix $\left(\begin{smallmatrix} a & b \\ c & d \end{smallmatrix} \right)$, the linear fractional transformation $z \mapsto (az + b)/(cz + d)$ is defined modulo q by the following subroutine. We assume that the matrix $\left(\begin{smallmatrix} a & b \\ c & d \end{smallmatrix} \right)$ appears in row k of the *gen* table.

⟨ Private variables and routines 6 ⟩ +≡

```
    static long lin_frac(a, k)
        long a;        /* the number being transformed; q represents ∞ */
        long k;        /* index into gen table */
{   register long q = q_inv[0];        /* the modulus */
    long a00 = gen[k].a0, a01 = gen[k].a1, a10 = gen[k].a2, a11 = gen[k].a3;
        /* the coefficients */
    register long num, den;        /* numerator and denominator */
    if (a ≡ q)  num = a00, den = a10;
    else  num = (a00 * a + a01) % q, den = (a10 * a + a11) % q;
    if (den ≡ 0)  return q;
    else return (num * q_inv[den]) % q;
}
```

31. We are computing the same values of *lin_frac* over and over again in type 2 graphs, but the author was too lazy to optimize this.

⟨ Compute the image, u, of v under the linear fractional transformation defined by *gen*[k] 31 ⟩ ≡

```
    if (type ≡ 1)  u = new_graph→vertices + lin_frac(v→x.I, k);
    else {
        a = lin_frac(v→x.I, k);  aa = lin_frac(v→y.I, k);
        u = new_graph→vertices + (a < aa ? (a * (2 * q − 1 − a))/2 + aa − 1 : (aa * (2 * q − 1 − aa))/2 + a − 1);
    }
```

This code is used in section 27.

GB_RAND _____

Important: Before reading GB_RAND, please read or at least skim the program for GB_GRAPH.

1. Random graphs. This GraphBase module provides two external subroutines called *random_graph* and *random_bigraph*, which generate graphs in which the arcs or edges have been selected "at random." A third subroutine, *random_lengths*, randomizes the lengths of the arcs of a given graph. The performance of algorithms on such graphs can fruitfully be compared to their performance on the nonrandom graphs generated by other GraphBase routines.

Before reading this code, the reader should be familiar with the basic data structures and conventions described in GB_GRAPH. The routines in GB_GRAPH are loaded together with all GraphBase applications, and the programs below are typical illustrations of how to use them.

```
#define random_graph    r_graph      /* abbreviations for Procrustean external linkage */
#define random_bigraph  r_bigraph
#define random_lengths  r_lengths
```

⟨ gb_rand.h 1 ⟩ ≡
```
#define random_graph  r_graph        /* users of GB_RAND should include this header info */
#define random_bigraph  r_bigraph
#define random_lengths  r_lengths
  extern Graph *random_graph( );
  extern Graph *random_bigraph( );
  extern long random_lengths( );
```

2. Here is an overview of the file `gb_rand.c`, the C code for the routines in question.
```
#include "gb_graph.h"     /* this header file teaches C about GraphBase */
#include "gb_flip.h"      /* we will use the GB_FLIP routines for random numbers */
  ⟨ Preprocessor definitions ⟩
  ⟨ Private declarations 8 ⟩
  ⟨ Internal functions 18 ⟩
  ⟨ External functions 5 ⟩
```

3. The procedure *random_graph*($n, m, multi, self, directed, dist\_from, dist\_to, min\_len, max\_len, seed$) is designed to produce a pseudo-random graph with n vertices and m arcs or edges, using pseudo-random numbers that depend on *seed* in a system-independent fashion. The remaining parameters specify a variety of options:

> *multi* $\neq 0$ permits duplicate arcs;
> *self* $\neq 0$ permits self-loops (arcs from a vertex to itself);
> *directed* $\neq 0$ makes the graph directed; otherwise each arc becomes an undirected edge;
> *dist_from* and *dist_to* specify probability distributions on the arcs;
> *min_len* and *max_len* bound the arc lengths, which will be uniformly distributed between
>> these limits.

If *dist_from* or *dist_to* are Λ, the probability distribution is uniform over vertices; otherwise the *dist* parameter points to an array of n nonnegative integers that sum to 2^{30}, specifying the respective probabilities (times 2^{30}) that each given vertex will appear as the source or destination of the random arcs.

A special option *multi* = −1 is provided. This acts exactly like *multi* = 1, except that arcs are not physically duplicated in computer memory—they are replaced by a single arc whose length is the minimum of all arcs having a common source and destination.

The vertices are named simply "0", "1", "2", and so on.

4. Examples: *random_graph*(1000, 5000, 0, 0, 0, Λ, Λ, 1, 1, 0) creates a random undirected graph with 1000 vertices and 5000 edges (hence 10000 arcs) of length 1, having no duplicate edges or self-loops. There are $\binom{1000}{2} = 499500$ possible undirected edges on 1000 vertices; hence there are exactly $\binom{499500}{5000}$ possible graphs meeting these specifications. Every such graph would be equally likely, if *random_graph* had access to an ideal source of random numbers. The GraphBase programs are designed to be system-independent, so that identical graphs will be obtained by everybody who asks for *random_graph*(1000, 5000, 0, 0, 0, Λ, Λ, 1, 1, 0). Equivalent experiments on algorithms for graph manipulation can therefore be performed by researchers in different parts of the world.

The subroutine call *random_graph*(1000, 5000, 0, 0, 0, Λ, Λ, 1, 1, s) will produce different graphs when the random seed s varies; however, the graph for any particular value of s will be the same on all computers. The seed value can be any integer in the range $0 \le s < 2^{31}$.

To get a random directed graph, allowing self-loops and repeated arcs, and with a uniform distribution on vertices, ask for

$$random\_graph(n, m, 1, 1, 1, Λ, Λ, 1, 1, s).$$

Each of the m arcs of that digraph has probability $1/n^2$ of being from u to v, for all u and v. If self-loops are disallowed (by changing '1, 1, 1' to '1, 0, 1'), each arc has probability $1/(n^2 - n)$ of being from u to v, for all $u \ne v$.

To get a random directed graph in which vertex k is twice as likely as vertex $k + 1$ to be the source of an arc but only half as likely to be the destination of an arc, for all k, try

$$random\_graph(25, m, 1, 1, 1, d0, d1, 0, 255, s)$$

where the arrays *d0* and *d1* have the static declarations

> **long** $d0[31] = \{^\#20000000, {}^\#10000000, \dots, 4, 2, 1, 1\};$
> **long** $d1[31] = \{1, 1, 2, 4, \dots, {}^\#10000000, {}^\#20000000\};$

then about 1/4 of the arcs will run from 0 to 30, while arcs from 30 to 0 will be extremely rare (occurring with probability 2^{-60}). Incidentally, the arc lengths in this example will be random bytes, uniformly distributed between 0 and 255, because *min_len* = 0 and *max_len* = 255.

If we forbid repeated arcs in this example, by setting *multi* = 0, the effect is to discard all arcs having the same source and destination as a previous arc, regardless of length. In such a case m had better not be too large, because the algorithm will keep going until it has found m distinct arcs, and many arcs are quite rare indeed; they will probably not be found until hundreds of centuries have elapsed.

A random bipartite graph can also be obtained as a special case of *random_graph*; this case is explained below.

Semantics: If *multi* = *directed* = 0 and *self* $\ne$ 0, we have an undirected graph without duplicate edges but with self-loops permitted. A self-loop then consists of two identical self-arcs, in spite of the fact that *multi* = 0.

5. If the *random_graph* routine encounters a problem, it returns Λ, after putting a code number into the external variable *panic_code*. This code number identifies the type of failure. Otherwise *random_graph* returns a pointer to the newly created graph and leaves *panic_code* unchanged. The *gb_trouble_code* will be cleared to zero after *random_graph* has acted.

#define *panic(c)* { *panic_code = c*; *gb_trouble_code = 0*; **return** Λ; }

⟨ External functions 5 ⟩ ≡
    Graph **random_graph*(*n, m, multi, self, directed, dist_from, dist_to, min_len, max_len, seed*)
        unsigned long *n*;   /* number of vertices desired */
        unsigned long *m*;   /* number of arcs or edges desired */
        long *multi*;   /* allow duplicate arcs? */
        long *self*;   /* allow self loops? */
        long *directed*;   /* directed graph? */
        long **dist_from*;   /* distribution of arc sources */
        long **dist_to*;   /* distribution of of arc destinations */
        long *min_len, max_len*;   /* bounds on random lengths */
        long *seed*;   /* random number seed */
    { ⟨ Local variables 6 ⟩
      if ($n \equiv 0$) *panic(bad_specs)*;   /* we gotta have a vertex */
      if (*min_len > max_len*) *panic(very_bad_specs)*;   /* what are you trying to do? */
      if (((**unsigned long**) (*max_len*)) − ((**unsigned long**) (*min_len*)) ≥ ((**unsigned long**)
            #80000000)) *panic(bad_specs + 1)*;   /* too much range */
      ⟨ Check the distribution parameters 11 ⟩;
      gb_init_rand(seed);
      ⟨ Create a graph with *n* vertices and no arcs 7 ⟩;
      ⟨ Build tables for nonuniform distributions, if needed 13 ⟩;
      for (*mm = m*; *mm*; *mm* −−) ⟨ Add a random arc or a random edge 9 ⟩;
    trouble:
      if (*gb_trouble_code*) {
        gb_recycle(new_graph);
        panic(alloc_fault);   /* oops, we ran out of memory somewhere back there */
      }
      gb_free(new_graph→aux_data);
      return *new_graph*;
    }

See also sections 22 and 24.

This code is used in section 2.

6. ⟨ Local variables 6 ⟩ ≡
    Graph **new_graph*;   /* the graph constructed by *random_graph* */
    long *mm*;   /* the number of arcs or edges we still need to generate */
    register long *k*;   /* vertex being processed */

See also section 12.

This code is used in section 5.

---

alloc_fault = −1, GB_GRAPH §7.
aux_data: **Area**, GB_GRAPH §20.
bad_specs = 30, GB_GRAPH §7.
gb_free: **void** ( ), GB_GRAPH §16.

gb_init_rand: **void** ( ), GB_FLIP §8.
gb_recycle: **void** ( ), GB_GRAPH §40.
gb_trouble_code: **long**,
  GB_GRAPH §14.

Graph = **struct**, GB_GRAPH §20.
panic_code: **long**, GB_GRAPH §5.
very_bad_specs = 40, GB_GRAPH §7.

7. **#define** *dist_code*(*x*) (*x* ? "dist" : "0")

⟨ Create a graph with *n* vertices and no arcs 7 ⟩ ≡
  new_graph = *gb_new_graph*(*n*);
  if (*new_graph* ≡ Λ) *panic*(*no_room*);   /* out of memory before we're even started */
  for (*k* = 0; *k* < *n*; *k*++) {
    sprintf(*name_buffer*, "%ld", *k*);
    (*new_graph*→*vertices* + *k*)→*name* = *gb_save_string*(*name_buffer*);
  }
  sprintf(*new_graph*→*id*, "random_graph(%lu,%lu,%d,%d,%d,%s,%s,%ld,%ld,%ld)",
      n, *m*, *multi* > 0 ? 1 : *multi* < 0 ? −1 : 0, *self* ? 1 : 0, *directed* ? 1 : 0,
      dist_code(*dist_from*), *dist_code*(*dist_to*), *min_len*, *max_len*, *seed*);

This code is used in section 5.

8. ⟨ Private declarations 8 ⟩ ≡
  static char *name_buffer*[ ] = "9999999999";

See also sections 14, 17, and 25.

This code is used in section 2.

9. **#define** *rand_len* (*min_len* ≡ *max_len* ? *min_len* : *min_len* + *gb_unif_rand*(*max_len* − *min_len*))

⟨ Add a random arc or a random edge 9 ⟩ ≡
  { **register Vertex** *u, *v;
  repeat:
    if (*dist_from*) ⟨ Generate a random vertex *u* according to *dist_from* 15 ⟩
    else *u* = *new_graph*→*vertices* + *gb_unif_rand*(*n*);
    if (*dist_to*) ⟨ Generate a random vertex *v* according to *dist_to* 16 ⟩
    else *v* = *new_graph*→*vertices* + *gb_unif_rand*(*n*);
    if (*u* ≡ *v* ∧ ¬*self*) **goto** *repeat*;
    if (*multi* ≤ 0) ⟨ Search for duplicate arcs or edges; **goto** *repeat* or *done* if found 10 ⟩;
    if (*directed*) *gb_new_arc*(*u*, *v*, *rand_len*);
    else *gb_new_edge*(*u*, *v*, *rand_len*);
  done: ;
  }

This code is used in section 5.

10. When we decrease the length of an existing edge, we use the fact that its two arcs are adjacent in memory. If *u* ≡ *v* in this case, we encounter the first of two mated arcs before seeing the second; hence the mate of the arc we find is in location *a* + 1 when *u* ≤ *v*, and in location *a* − 1 when *u* > *v*.

We must exit to location *trouble* if memory has been exhausted; otherwise there is a danger of an infinite loop, with *dummy_arc*→*next* = *dummy_arc*.

⟨ Search for duplicate arcs or edges; **goto** *repeat* or *done* if found 10 ⟩ ≡
  if (*gb_trouble_code*) **goto** *trouble*;
  else { **register Arc** *a;
    long *len*;   /* length of new arc or edge being combined with previous */
    for (*a* = *u*→*arcs*; *a*; *a* = *a*→*next*)
      if (*a*→*tip* ≡ *v*)
        if (*multi* ≡ 0) **goto** *repeat*;   /* reject a duplicate arc */

```
      else {        /* multi < 0 */
         len = rand_len;
         if (len < a⃗len) {
            a⃗len = len;
            if (¬directed) {
               if (u ≤ v) (a + 1)⃗len = len;
               else (a − 1)⃗len = len;
            }
         }
         goto done;
      }
   }
```

This code is used in section 9.

Arc = **struct**, GB_GRAPH §10.
arcs: **Arc** ∗, GB_GRAPH §9.
directed: **long**, §5.
dist_from: **long** ∗, §5.
dist_to: **long** ∗, §5.
dummy_arc: **static Arc** [],
   GB_GRAPH §28.
gb_new_arc: **void** ( ),
   GB_GRAPH §30.
gb_new_edge: **void** ( ),
   GB_GRAPH §31.
gb_new_graph: **Graph** ∗( ),
   GB_GRAPH §23.

gb_save_string: **char** ∗( ),
   GB_GRAPH §35.
gb_trouble_code: **long**,
   GB_GRAPH §14.
gb_unif_rand: **long** ( ), GB_FLIP §12.
id: **char** [], GB_GRAPH §20.
k: **register long**, §6.
m: **unsigned long**, §5.
max_len: **long**, §5.
min_len: **long**, §5.
multi: **long**, §5.
n: **unsigned long**, §5.

name: **char** ∗, GB_GRAPH §9.
new_graph: **Graph** ∗, §6.
next: **Arc** ∗, GB_GRAPH §10.
no_room = 1, GB_GRAPH §7.
panic = macro ( ), §5.
seed: **long**, §5.
self: **long**, §5.
sprintf: **int** ( ), <stdio.h>.
tip: **Vertex** ∗, GB_GRAPH §10.
trouble: label, §5.
Vertex = **struct**, GB_GRAPH §9.
vertices: **Vertex** ∗, GB_GRAPH §20.

11.   Nonuniform random number generation.   The *random_graph* procedure is complete except for the parts that handle general distributions *dist_from* and *dist_to*. Before attempting to generate those distributions, we had better check them to make sure that the specifications are well formed; otherwise disaster might ensue later. This part of the program is easy.

⟨ Check the distribution parameters 11 ⟩ ≡

```
{ register long acc;       /* sum of probabilities */
  register long *p;        /* pointer to current probability of interest */
  if (dist_from) {
    for (acc = 0, p = dist_from;  p < dist_from + n;  p++) {
      if (*p < 0) panic(invalid_operand);      /* dist_from contains a negative entry */
      if (*p > #40000000 − acc) panic(invalid_operand + 1);      /* probability too high */
      acc += *p;
    }
    if (acc ≠ #40000000) panic(invalid_operand + 2);      /* dist_from table doesn't sum to 2³⁰ */
  }
  if (dist_to) {
    for (acc = 0, p = dist_to;  p < dist_to + n;  p++) {
      if (*p < 0) panic(invalid_operand + 5);      /* dist_to contains a negative entry */
      if (*p > #40000000 − acc) panic(invalid_operand + 6);      /* probability too high */
      acc += *p;
    }
    if (acc ≠ #40000000) panic(invalid_operand + 7);      /* dist_to table doesn't sum to 2³⁰ */
  }
}
```

This code is used in section 5.

12.   We generate nonuniform distributions by using Walker's alias method (see, for example, *Seminumerical Algorithms*, second edition, exercise 3.4.1–7). Walker's method involves setting up "magic" tables of length nn, where nn is the smallest power of 2 that is $\geq n$.

⟨ Local variables 6 ⟩ +≡

```
long nn = 1;       /* this will be increased to 2^⌈lg n⌉ */
long kk = 31;      /* this will be decreased to 31 − ⌈lg n⌉ */
magic_entry *from_table, *to_table;      /* alias tables. */
```

13.   ⟨ Build tables for nonuniform distributions, if needed 13 ⟩ ≡

```
{
  if (dist_from) { while (nn < n) nn += nn, kk −−;
    from_table = walker(n, nn, dist_from, new_graph);
  }
  if (dist_to) { while (nn < n) nn += nn, kk −−;
    to_table = walker(n, nn, dist_to, new_graph);
  }
  if (gb_trouble_code) { gb_recycle(new_graph);
    panic(alloc_fault);      /* oops, we ran out of memory somewhere back there */
  }
}
```

This code is used in section 5.

14. ⟨Private declarations 8⟩ +≡
   typedef struct {
      long *prob*;   /* a probability, multiplied by 2^{31} and translated */
      long *inx*;   /* index that might be selected */
   } **magic_entry**;

15. Once the magic tables have been set up, we can generate nonuniform vertices by using the following code:

⟨Generate a random vertex *u* according to *dist_from* 15⟩ ≡
   { **register magic_entry** **magic*;
      register long *uu = gb_next_rand*( );   /* uniform random number */

      k = uu ≫ kk;
      magic = from_table + k;
      if (*uu ≤ magic→prob*) *u = new_graph→vertices + k*;
      else *u = new_graph→vertices + magic→inx*;
   }

This code is used in section 9.

16. ⟨Generate a random vertex *v* according to *dist_to* 16⟩ ≡
   { **register magic_entry** **magic*;
      register long *uu = gb_next_rand*( );   /* uniform random number */

      k = uu ≫ kk;
      magic = to_table + k;
      if (*uu ≤ magic→prob*) *v = new_graph→vertices + k*;
      else *v = new_graph→vertices + magic→inx*;
   }

This code is used in section 9.

---

alloc_fault = −1, GB_GRAPH §7.
dist_from: **long** *, §5.
dist_to: **long** *, §5.
gb_next_rand = macro ( ),
   GB_FLIP §6.
gb_recycle: **void** ( ), GB_GRAPH §40.
gb_trouble_code: **long**,

GB_GRAPH §14.
invalid_operand = 60, GB_GRAPH §7.
k: **register long**, §6.
n: **unsigned long**, §5.
new_graph: **Graph** *, §6.
panic = macro ( ), §5.

random_graph: **Graph** *( ), §5.
u: **register Vertex** *, §9.
v: **register Vertex** *, §9.
vertices: **Vertex** *, GB_GRAPH §20.
walker: **static magic_entry** *( ),
   §18.

17. So all we have to do is set up those magic tables. If uu is a uniform random integer between 0 and $2^{31} - 1$, the index $k = uu \gg kk$ is a uniform random integer between 0 and $nn - 1$, because of the relation between nn and kk. Once k is computed, the code above selects vertex k with probability $(p + 1 - (k \ll kk))/2^{31}$, where $p = magic\text{-}prob$ and $magic$ is the kth element of the magic table; otherwise the code selects vertex $magic\text{-}inx$. The trick is to set things up so that each vertex is selected with the proper overall probability.

Let's imagine that the given distribution vector has length nn, instead of n, by extending it if necessary with zeroes. Then the average entry among these nn integers is exactly $t = 2^{30}/nn$. If some entry, say entry i, exceeds t, there must be another entry that's less than t, say entry j. We can set the jth entry of the magic table so that its *prob* field selects vertex j with the correct probability, and so that its *inx* field equals i. Then we are selecting vertex i with a certain residual probability; so we subtract that residual from i's present probability, and repeat the process with vertex j eliminated. The average of the remaining entries is still t, so we can repeat this procedure until all remaining entries are exactly equal to t. The rest is easy.

During the calculation, we maintain two linked lists of $(prob, inx)$ pairs. The hi list contains entries with $prob > t$, and the lo list contains the rest. During this part of the computation we call these list elements 'nodes', and we use the field names key and j instead of $prob$ and inx.

⟨ Private declarations 8 ⟩ +≡
```
typedef struct node_struct {
    long key;      /* a numeric quantity */
    struct node_struct *link;      /* the next node on the list */
    long j;      /* a vertex number to be selected with probability key/2³⁰ */
} node;
static Area temp_nodes;      /* nodes will be allocated in this area */
static node *base_node;      /* beginning of a block of nodes */
```

18. ⟨ Internal functions 18 ⟩ ≡
```
static magic_entry *walker(n, nn, dist, g)
        long n;      /* length of dist vector */
        long nn;      /* 2^⌈lg n⌉ */
        register long *dist;      /* start of distribution table, which sums to 2³⁰ */
        Graph *g;      /* tables will be allocated for this graph's vertices */
{ magic_entry *table;      /* this will be the magic table we compute */
    long t;      /* average key value */
    node *hi = Λ, *lo = Λ;      /* nodes not yet included in magic table */
    register node *p, *q;      /* pointer variables for list manipulation */

    base_node = gb_typed_alloc(nn, node, temp_nodes);
    table = gb_typed_alloc(nn, magic_entry, g-aux_data);
    if (¬gb_trouble_code) {
        ⟨ Initialize the hi and lo lists 19 ⟩;
        while (hi) ⟨ Remove a lo element and match it with a hi element; deduct the residual
                probability from that hi element 20 ⟩;
        while (lo) ⟨ Remove a lo element of key value t 21 ⟩;
    }
    gb_free(temp_nodes);
    return table;      /* if gb_trouble_code is nonzero, the table is empty */
```

```
}
```
This code is used in section 2.

19. ⟨ Initialize the *hi* and *lo* lists 19 ⟩ ≡
```
  t = #40000000/nn;        /* this division is exact */
  p = base_node;
  while (nn > n) { p⃗key = 0; p⃗link = lo; p⃗j = −−nn; lo = p++; }
  for (dist = dist + n − 1; n > 0; dist −−, p++) {
    p⃗key = *dist; p⃗j = −−n;
    if (*dist > t) p⃗link = hi, hi = p; else p⃗link = lo, lo = p;
  }
```
This code is used in section 18.

20. When we change the scale factor from 2^{30} to 2^{31}, we need to be careful lest integer overflow occur. The introduction of register x into this code removes the risk.

⟨ Remove a *lo* element and match it with a *hi* element; deduct the residual probability from that *hi* element 20 ⟩ ≡
```
  { register magic_entry *r;
    register long x;

    p = hi, hi = p⃗link;
    q = lo, lo = q⃗link;
    r = table + q⃗j;
    x = t * q⃗j + q⃗key − 1;
    r⃗prob = x + x + 1;
    r⃗inx = p⃗j;        /* we have just given q⃗key units of probability to vertex q⃗j, and t − q⃗key units
                          to vertex p⃗j */
    if ((p⃗key −= t − q⃗key) > t) p⃗link = hi, hi = p;
    else p⃗link = lo, lo = p;
  }
```
This code is used in section 18.

21. When all remaining entries have the average probability, the *inx* component need not be set, because it will never be used.

⟨ Remove a *lo* element of *key* value t 21 ⟩ ≡
```
  { register magic_entry *r;
    register long x;

    q = lo, lo = q⃗link; r = table + q⃗j; x = t * q⃗j + t − 1;
    r⃗prob = x + x + 1;        /* that's t units of probability for vertex q⃗j */
  }
```
This code is used in section 18.

Area, GB_GRAPH §12.
aux_data: **Area**, GB_GRAPH §20.
gb_free: **void** ( ), GB_GRAPH §16.
gb_trouble_code: **long**,
   GB_GRAPH §14.
gb_typed_alloc = macro ( ),

GB_GRAPH §11.
Graph = **struct**, GB_GRAPH §20.
inx: **long**, §14.
k: **register long**, §6.
kk: **long**, §12.
magic: **register magic_entry** *,

§16.
magic_entry = **struct**, §14.
n: **unsigned long**, §5.
nn: **long**, §12.
prob: **long**, §14.
uu: **register long**, §16.

22. Random bipartite graphs.   The procedure call

$$random\_bigraph(n1, n2, m, multi, dist1, dist2, min\_len, max\_len, seed)$$

is designed to produce a pseudo-random bipartite graph with $n1$ vertices in one part and $n2$ in the other, having m edges. The remaining parameters $multi$, $dist1$, $dist2$, min_len, max_len, and $seed$ have the same meaning as the analogous parameters of $random\_graph$.

In fact, $random\_bigraph$ works by reducing its parameters to a special case of $random\_graph$. Almost all that needs to be done is to pad $dist1$ with $n2$ trailing zeroes and $dist2$ with $n1$ leading zeroes. The only slightly tricky part occurs when $dist1$ and/or $dist2$ are null, since non-null distribution vectors summing exactly to 2^{30} must then be fabricated.

⟨ External functions 5 ⟩ +≡
```
  Graph *random_bigraph(n1, n2, m, multi, dist1, dist2, min_len, max_len, seed)
      unsigned long n1, n2;     /* number of vertices desired in each part */
      unsigned long m;     /* number of edges desired */
      long multi;     /* allow duplicate edges? */
      long *dist1, *dist2;     /* distribution of edge endpoints */
      long min_len, max_len;     /* bounds on random lengths */
      long seed;     /* random number seed */
  { unsigned long n = n1 + n2;     /* total number of vertices */
    Area new_dists;
    long *dist_from, *dist_to;
    Graph *new_graph;

    init_area(new_dists);
    if (n1 ≡ 0 ∨ n2 ≡ 0) panic(bad_specs);     /* illegal options */
    if (min_len > max_len) panic(very_bad_specs);     /* what are you trying to do? */
    if (((unsigned long) (max_len)) − ((unsigned long) (min_len)) ≥ ((unsigned long)
            #80000000)) panic(bad_specs + 1);     /* too much range */
    dist_from = gb_typed_alloc(n, long, new_dists);
    dist_to = gb_typed_alloc(n, long, new_dists);
    if (gb_trouble_code) {
      gb_free(new_dists);
      panic(no_room + 2);     /* no room for auxiliary distribution tables */
    }
    ⟨ Compute the entries of dist_from and dist_to 23 ⟩;
    new_graph = random_graph(n, m, multi, 0_L, 0_L, dist_from, dist_to, min_len, max_len, seed);
    sprintf(new_graph→id, "random_bigraph(%lu,%lu,%lu,%d,%s,%s,%ld,%ld,%ld)",
        n1, n2, m, multi > 0 ? 1 : multi < 0 ? −1 : 0, dist_code(dist1), dist_code(dist2),
        min_len, max_len, seed);
    mark_bipartite(new_graph, n1);
    gb_free(new_dists);
    return new_graph;
  }
```

23.   The relevant identity we need here is the replicative law for the floor function:

$$\left\lfloor \frac{x}{n} \right\rfloor + \left\lfloor \frac{x+1}{n} \right\rfloor + \cdots + \left\lfloor \frac{x+n-1}{n} \right\rfloor = \lfloor x \rfloor .$$

⟨ Compute the entries of *dist_from* and *dist_to* 23 ⟩ ≡
  { **register long** *p, *q;   /* traversers of the dists */
    register long k;   /* vertex count */
    · p = dist1;
    q = dist_from;
    if (p)
      while (p < dist1 + n1) *q++ = *p++;
    else
      for (k = 0; k < n1; k++) *q++ = (#40000000 + k)/n1;
    p = dist2;
    q = dist_to + n1;
    if (p)
      while (p < dist2 + n2) *q++ = *p++;
    else
      for (k = 0; k < n2; k++) *q++ = (#40000000 + k)/n2;
  }

This code is used in section 22.

---

24. Random lengths. The subroutine call

$$random\_lengths\,(g, directed, min\_len, max\_len, dist, seed)$$

takes an existing graph and assigns new lengths to each of its arcs. If $dist = \Lambda$, the lengths will be uniformly distributed between min_len and max_len inclusive; otherwise $dist$ should be a probability distribution vector of length $max\_len - min\_len + 1$, like those in *random_graph*.

If $directed = 0$, pairs of arcs $u \to v$ and $v \to u$ will be regarded as a single edge, both arcs receiving the same length.

The procedure returns a nonzero value if something goes wrong; in that case, graph g will not have been changed.

Alias tables for generating nonuniform random lengths will survive in $g\text{-}aux\_data$.

⟨ External functions 5 ⟩ +≡
```
long random_lengths (g, directed, min_len, max_len, dist, seed)
     Graph *g;      /* graph whose lengths will be randomized */
     long directed;      /* is it directed? */
     long min_len, max_len;      /* bounds on random lengths */
     long *dist;      /* distribution of lengths */
     long seed;      /* random number seed */
{ register Vertex *u, *v;      /* current vertices of interest */
  register Arc *a;      /* current arc of interest */
  long nn = 1, kk = 31;      /* variables for nonuniform generation */
  magic_entry *dist_table;      /* alias table for nonuniform generation */
  if (g ≡ Λ) return missing_operand;      /* where is g? */
  gb_init_rand (seed);
  if (min_len > max_len) return very_bad_specs;      /* what are you trying to do? */
  if (((unsigned long) (max_len)) − ((unsigned long) (min_len)) ≥ ((unsigned long)
          #80000000)) return bad_specs;      /* too much range */
  ⟨ Check dist for validity, and set up the dist_table 26 ⟩;
  sprintf (buffer, ",%d,%ld,%ld,%s,%ld)", directed ? 1 : 0, min_len, max_len, dist_code (dist), seed);
  make_compound_id (g, "random_lengths(", g, buffer);
  ⟨ Run through all arcs and assign new lengths 27 ⟩;
  return 0;
}
```

25. ⟨ Private declarations 8 ⟩ +≡
```
static char buffer[] = "1,-1000000001,-1000000000,dist,1000000000)";
```

26. ⟨ Check dist for validity, and set up the dist_table 26 ⟩ ≡
```
if (dist) { register long acc;      /* sum of probabilities */
  register long *p;      /* pointer to current probability of interest */
  register long n = max_len − min_len + 1;
  for (acc = 0, p = dist; p < dist + n; p++) {
    if (*p < 0) return −1;      /* negative probability */
    if (*p > #40000000 − acc) return 1;      /* probability too high */
    acc += *p;
  }
```

```
    if (acc ≠ #40000000) return 2;        /* probabilities don't sum to 1 */
    while (nn < n) nn += nn, kk --;
    dist_table = walker(n, nn, dist, g);
    if (gb_trouble_code) {
      gb_trouble_code = 0;
      return alloc_fault;        /* not enough room to generate the magic tables */
    }
  }
}
```
This code is used in section 24.

27. ⟨Run through all arcs and assign new lengths 27⟩ ≡
```
  for (u = g⃗vertices; u < g⃗vertices + g⃗n; u++)
    for (a = u⃗arcs; a; a = a⃗next) {
      v = a⃗tip;
      if (directed ≡ 0 ∧ u > v) a⃗len = (a − 1)⃗len;
      else { register long len;        /* a random length */
        if (dist ≡ 0) len = rand_len;
        else { long uu = gb_next_rand();
          long k = uu ≫ kk;
          magic_entry *magic = dist_table + k;
          if (uu ≤ magic⃗prob) len = min_len + k;
          else len = min_len + magic⃗inx;
        }
        a⃗len = len;
        if (directed ≡ 0 ∧ u ≡ v ∧ a⃗next ≡ a + 1) (++a)⃗len = len;
      }
    }
```
This code is used in section 24.

---

alloc_fault = −1, GB_GRAPH §7.
Arc = **struct**, GB_GRAPH §10.
arcs: **Arc** *, GB_GRAPH §9.
aux_data: **Area**, GB_GRAPH §20.
bad_specs = 30, GB_GRAPH §7.
dist_code = macro ( ), §7.
gb_init_rand: **void** ( ), GB_FLIP §8.
gb_next_rand = macro ( ),
  GB_FLIP §6.
gb_trouble_code: **long**,
  GB_GRAPH §14.

Graph = **struct**, GB_GRAPH §20.
inx: **long**, §14.
len: **long**, GB_GRAPH §10.
magic_entry = **struct**, §14.
make_compound_id: **void** ( ),
  GB_GRAPH §26.
missing_operand = 50,
  GB_GRAPH §7.
n: **long**, GB_GRAPH §20.
next: **Arc** *, GB_GRAPH §10.

prob: **long**, §14.
rand_len = macro, §9.
random_graph: **Graph** *( ), §5.
sprintf: **int** ( ), <stdio.h>.
tip: **Vertex** *, GB_GRAPH §10.
Vertex = **struct**, GB_GRAPH §9.
vertices: **Vertex** *, GB_GRAPH §20.
very_bad_specs = 40, GB_GRAPH §7.
walker: **static magic_entry** *( ),
  §18.

GB_ROGET

Important: Before reading GB_ROGET, please read or at least skim the programs for GB_GRAPH and GB_IO.

1. Introduction. This GraphBase module contains the *roget* subroutine, which creates a family of graphs based on Roget's Thesaurus. An example of the use of this procedure can be found in the demo program ROGET_COMPONENTS.

⟨ gb_roget.h 1 ⟩ ≡
  extern Graph *∗roget*( );

See also section 12.

2. The subroutine call *roget*(*n*, *min_distance*, *prob*, *seed*) constructs a graph based on the information in `roget.dat`. Each vertex of the graph corresponds to one of the 1022 categories in the 1879 edition of Peter Mark Roget's *Thesaurus of English Words and Phrases*, edited by John Lewis Roget. An arc goes from one category to another if Roget gave a reference to the latter among the words and phrases of the former, or if the two categories were directly related to each other by their positions in Roget's book. For example, the vertex for category 312 ('ascent') has arcs to the vertices for categories 224 ('obliquity'), 313 ('descent'), and 316 ('leap'), because Roget gave explicit cross-references from 312 to 224 and 316, and because category 312 was implicitly paired with 313 in his scheme.

The constructed graph will have $\min(n, 1022)$ vertices; however, the default value $n = 1022$ is substituted when $n = 0$. If n is less than 1022, the n categories will be selected at random, and all arcs to unselected categories will be omitted. Arcs will also be omitted if they correspond to categories whose numbers differ by less than *min_distance*. For example, if $min\_distance > 1$, the arc between categories 312 and 313 will not be included. (Roget sometimes formed clusters of three interrelated categories; to avoid cross-references within all such clusters, you can set $min\_distance = 3$.)

If $prob > 0$, arcs that would ordinarily be included in the graph are rejected with probability $prob/65536$. This provides a way to obtain sparser graphs.

The vertices will appear in random order. However, all "randomness" in GraphBase graphs is reproducible; it depends only on the value of a given *seed*, which can be any nonnegative integer less than 2^{31}. For example, everyone who asks for *roget*(1000, 3, 32768, 50) will obtain exactly the same graph, regardless of their computer system.

Changing the value of *prob* will affect only the arcs of the generated graph; it will change neither the choice of vertices nor the vertex order.

#define MAX_N 1022   /∗ the number of categories in Roget's book ∗/

3. If the *roget* routine encounters a problem, it returns Λ (NULL), after putting a code number into the external variable *panic_code*. This code number identifies the type of failure. Otherwise *roget* returns a pointer to the newly created graph, which will be represented with the data structures explained in GB_GRAPH. (The external variable *panic_code* is itself defined in GB_GRAPH.)

#define *panic*(*c*) { *panic_code* = *c*; *gb_trouble_code* = 0; **return** Λ; }

4. The C file `gb_roget.c` has the following general shape:

#**include** `"gb_io.h"`   /∗ we will use the GB_IO routines for input ∗/
#**include** `"gb_flip.h"`   /∗ we will use the GB_FLIP routines for random numbers ∗/
#**include** `"gb_graph.h"`   /∗ and we will use the GB_GRAPH data structures ∗/
  ⟨ Preprocessor definitions ⟩
  ⟨ Private variables 7 ⟩

  Graph ∗*roget*(*n*, *min_distance*, *prob*, *seed*)
      unsigned long *n*;   /∗ number of vertices desired ∗/
      unsigned long *min_distance*;   /∗ smallest inter-category distance allowed in an arc ∗/
      unsigned long *prob*;   /∗ 65536 times the probability of rejecting an arc ∗/
      long *seed*;   /∗ random number seed ∗/
 { ⟨ Local variables 5 ⟩
   gb_init_rand(*seed*);
   if (*n* ≡ 0 ∨ *n* > `MAX_N`) *n* = `MAX_N`;
   ⟨ Set up a graph with *n* vertices 6 ⟩;
   ⟨ Determine the *n* categories to use in the graph 8 ⟩;
   ⟨ Input `roget.dat` and build the graph 10 ⟩;
   if (*gb_trouble_code*) {
     gb_recycle(*new_graph*);
     panic(*alloc_fault*);   /∗ oops, we ran out of memory somewhere back there ∗/
   }
   return *new_graph*;
 }

5. ⟨ Local variables 5 ⟩ ≡
  Graph ∗*new_graph*;   /∗ the graph constructed by *roget* ∗/
See also section 9.

This code is used in section 4.

---

alloc_fault = −1, GB_GRAPH §7.   *gb_trouble_code*: **long**,   **Graph** = **struct**, GB_GRAPH §20.
gb_init_rand: **void** ( ), GB_FLIP §8.   GB_GRAPH §14.   *panic_code*: **long**, GB_GRAPH §5.
gb_recycle: **void** ( ), GB_GRAPH §40.

6. Vertices.

⟨ Set up a graph with n vertices 6 ⟩ ≡
   new_graph = *gb_new_graph*(n);
   if (*new_graph* ≡ Λ) *panic*(*no_room*);   /* out of memory before we're even started */
   sprintf(*new_graph*→*id*, "roget(%lu,%lu,%lu,%ld)", n, *min_distance*, *prob*, *seed*);
   strcpy(*new_graph*→*util_types*, "IZZZZZZZZZZZZZ");

This code is used in section 4.

7. The first nontrivial thing we need to do is find a random selection and permutation of n vertices. We will compute a *mapping* table such that *mapping*[k] is non-Λ for exactly n randomly selected category numbers k. Moreover, these non-Λ values will be a random permutation of the vertices of the graph.

⟨ Private variables 7 ⟩ ≡
   Vertex *mapping[MAX_N + 1];   /* the vertex corresponding to a given category */
   long *cats*[MAX_N];   /* table of category numbers that have not yet been used */
This code is used in section 4.

8. During the loop on v in this step, k is the number of categories whose *mapping* value is still Λ. The first k entries of *cats* will contain those category numbers in some order.

⟨ Determine the n categories to use in the graph 8 ⟩ ≡
   for ($k = 0$; $k <$ MAX_N; k++) *cats*[k] = $k + 1$, *mapping*[$k + 1$] = Λ;
   for (v = *new_graph*→*vertices* + $n - 1$; $v \geq$ *new_graph*→*vertices*; v−−) {
      j = *gb_unif_rand*(k);
      mapping[*cats*[j]] = v;
      cats[j] = *cats*[−−k];
   }
This code is used in section 4.

9. ⟨ Local variables 5 ⟩ +≡
   register long j, k;   /* all-purpose indices */
   register Vertex *v;   /* current vertex */

10.   Arcs.   The data in `roget.dat` appears in 1022 lines, one for each category. For example, the line

312ascent:224 313 316

specifies the arcs from category 312 as explained earlier. First comes the category number, then the category name, then a colon, then zero or more numbers specifying arcs to other categories; the numbers are separated by spaces.

Some categories have too many arcs to fit on a single line; the data for these categories can be found on two lines, the first line ending with a backslash and the second line beginning with a space.

⟨ Input `roget.dat` and build the graph 10 ⟩ ≡
  if (*gb_open*("`roget.dat`") ≠ 0) *panic*(*early_data_fault*);
    /∗ couldn't open "`roget.dat`" using GraphBase conventions ∗/
  for (*k* = 1; ¬*gb_eof*( ); *k*++)
    ⟨ Read the data for category *k*, and put it in the graph if it has been selected 11 ⟩;
  if (*gb_close*( ) ≠ 0) *panic*(*late_data_fault*);
    /∗ something's wrong with "`roget.dat`"; see *io_errors* ∗/
  if (*k* ≠ `MAX_N` + 1) *panic*(*impossible*);   /∗ we don't have the right value of `MAX_N` ∗/

This code is used in section 4.

---

early_data_fault = 10, GB_GRAPH §7.
gb_close: **long** ( ), GB_IO §39.
gb_eof: **long** ( ), GB_IO §20.
gb_new_graph: **Graph** ∗( ),
  GB_GRAPH §23.
gb_open: **long** ( ), GB_IO §32.
gb_unif_rand: **long** ( ), GB_FLIP §12.
id: **char** [ ], GB_GRAPH §20.

impossible = 90, GB_GRAPH §7.
io_errors: **long**, GB_IO §5.
late_data_fault = 11, GB_GRAPH §7.
`MAX_N` = 1022, §2.
min_distance: **unsigned long**, §4.
n: **unsigned long**, §4.
new_graph: **Graph** ∗, §5.
no_room = 1, GB_GRAPH §7.

panic = macro ( ), §3.
prob: **unsigned long**, §4.
seed: **long**, §4.
sprintf: **int** ( ), <stdio.h>.
strcpy: **char** ∗( ), <string.h>.
util_types: **char** [ ], GB_GRAPH §20.
Vertex = **struct**, GB_GRAPH §9.
vertices: **Vertex** ∗, GB_GRAPH §20.

11. We check that the data isn't garbled, except that we don't bother to look at unselected categories.

The original category number is stored in vertex utility field *cat_no*, in case anybody wants to see it.

#define *cat_no* *u.I*   /* utility field *u* of each vertex holds the category number */

⟨ Read the data for category k, and put it in the graph if it has been selected 11 ⟩ ≡

```
{
    if (mapping[k]) {       /* yes, this category has been selected */
        if (gb_number(10) ≠ k) panic(syntax_error);       /* out of synch */
        (void) gb_string(str_buf, ':');
        if (gb_char() ≠ ':') panic(syntax_error + 1);      /* no colon found */
        v = mapping[k];
        v⃗name = gb_save_string(str_buf);
        v⃗cat_no = k;
        ⟨ Add arcs from v for every category that's both listed on the line and selected 13 ⟩;
    } else ⟨ Skip past the data for one category 14 ⟩;
}
```

This code is used in section 10.

12. ⟨ gb_roget.h 1 ⟩ +≡

#define *cat_no* *u.I*   /* definition of *cat_no* is repeated in the header file */

13.   **#define** *iabs*(*x*)   ((*x*) < 0 ? −(*x*) : (*x*))

⟨ Add arcs from v for every category that's both listed on the line and selected 13 ⟩ ≡

```
    j = gb_number(10);
    if (j ≡ 0) goto done;     /* some categories lead to no arcs at all */
    while (1) {
        if (j > MAX_N) panic(syntax_error + 2);      /* category code out of range */
        if (mapping[j] ∧ iabs(j − k) ≥ min_distance ∧ (prob ≡ 0 ∨ ((gb_next_rand() ≫ 15) ≥ prob)))
            gb_new_arc(v, mapping[j], 1 L);
        switch (gb_char()) {
        case '\\': gb_newline();
            if (gb_char() ≠ '␣') panic(syntax_error + 3);      /* space should begin a continuation line */
                /* fall through to the space case */
        case '␣': j = gb_number(10); break;
        case '\n': goto done;
        default: panic(syntax_error + 4);      /* illegal character following category number */
        }
    }
done: gb_newline();
```

This code is used in section 11.

14. We want to call *gb_newline*( ) twice if the current line ends with a backslash; otherwise we want to call it just once. There's an obvious way to do that, and there's also a faster and trickier way. The author apologizes here for succumbing to some old-fashioned impulses. (Recall that *gb_string* returns the location just following the '\0' it places at the end of a scanned string.)

⟨ Skip past the data for one category 14 ⟩ ≡

```
{
    if (*(gb_string(str_buf, '\n') − 2) ≡ '\\') gb_newline( );
        /* the first line ended with backslash */
    gb_newline( );
}
```

This code is used in section 11.

---

gb_char: **char** ( ), GB_IO §22.
gb_new_arc: **void** ( ),
   GB_GRAPH §30.
gb_newline: **void** ( ), GB_IO §18.
gb_next_rand = macro ( ),
   GB_FLIP §6.
gb_number: **unsigned long** ( ),
   GB_IO §24.

gb_save_string: **char** *( ),
   GB_GRAPH §35.
gb_string: **char** *( ), GB_IO §26.
l: **long**, GB_GRAPH §8.
j: **register long**, §9.
k: **register long**, §9.
mapping: **Vertex** *[], §7.
MAX_N = 1022, §2.

min_distance: **unsigned long**, §4.
name: **char** *, GB_GRAPH §9.
panic = macro ( ), §3.
prob: **unsigned long**, §4.
str_buf: **char** [], GB_IO §26.
syntax_error = 20, GB_GRAPH §7.
u: **util**, GB_GRAPH §9.
v: **register Vertex** *, §9.

GB_SAVE

Important: Before reading GB_SAVE, please read or at least skim the programs for GB_GRAPH and GB_IO.

1. Introduction. This GraphBase module contains the code for two special utility routines, *save_graph* and *restore_graph*, which convert graphs back and forth between the internal representation that is described in GB_GRAPH and a symbolic file format that is described below. Researchers can use these routines to transmit graphs between computers in a machine-independent way, or to use GraphBase graphs with other graph manipulation software that supports the same symbolic format.

All kinds of tricks are possible in the C language, so it is easy to abuse the GraphBase conventions and to create data structures that make sense only on a particular machine. But if users follow the recommended ground rules, *save_graph* will be able to transform their graphs into files that any other GraphBase installation will be able to read with *restore_graph*. The graphs created on remote machines will then be semantically equivalent to the originals.

Restrictions: Strings must contain only standard printable characters, not including \ or " or newline, and must be at most 4095 characters long; the $g\text{-}id$ string should be at most 154 characters long. All pointers to vertices and arcs must be confined to blocks within the $g\text{-}data$ area; blocks within $g\text{-}aux\_data$ are not saved or restored. Storage blocks in $g\text{-}data$ must be "pure"; that is, each block must be entirely devoted either to **Vertex** records, or to **Arc** records, or to characters of strings. The *save_graph* procedure places all **Vertex** records into a single **Vertex** block and all **Arc** records into a single **Arc** block, preserving the relative order of the original records where possible; but it does not preserve the relative order of string data in memory. For example, if $u\text{-}name$ and $v\text{-}name$ point to the same memory location in the saved graph, they will point to different memory locations (representing equal strings) in the restored graph. All utility fields must conform to the conventions of the graph's *util_types* string; the G option, which leads to graphs within graphs, is not permitted in that string.

```
#define MAX_SV_STRING 4095    /* longest strings supported */
#define MAX_SV_ID 154    /* longest id supported, is less than ID_FIELD_SIZE */
```

⟨ gb_save.h 1 ⟩ ≡
  extern long *save_graph*( );
  extern Graph **restore_graph*( );

2. Here is an overview of the C code, gb_save.c, for this module:

```
#include "gb_io.h"      /* we use the input/output conventions of GB_IO */
#include "gb_graph.h"      /* and, of course, the data structures of GB_GRAPH */
```
  ⟨ Preprocessor definitions ⟩
  ⟨ Type declarations 21 ⟩
  ⟨ Private variables 8 ⟩
  ⟨ Private functions 7 ⟩
  ⟨ External functions 4 ⟩

3. External representation of graphs. The internal representation of graphs has been described in GB_GRAPH. We now need to supplement that description by devising an alternative format suitable for human-and-machine-readable files.

The following somewhat contrived example illustrates the simple conventions that we shall follow:

```
* GraphBase graph (util_types IZAZZZZVZZZZSZ,3V,4A)
"somewhat_contrived_example(3.14159265358979323846264338327\
95028841971693993751058209749445923078164062862089986828)",1,
3,"pi"
* Vertices
"look",A0,15,A1
"feel",0,-9,A1
"",0,0,0
* Arcs
V0,A2,3,V1
V1,0,5,0
V1,0,-8,1
0,0,0,0
* Checksum 271828
```

The first line specifies the 14 characters of *util_types* and the total number of **Vertex** and **Arc** records; in this case there are 3 vertices and 4 arcs. The next line or lines specify the *id*, *n*, and *m* fields of the **Graph** record, together with any utility fields that are not being ignored. In this case, the *id* is a rather long string; a string may be broken into parts by ending the initial parts with a backslash, so that no line of the file has more than 79 characters. The last six characters of *util_types* refer to the utility fields of the **Graph** record, and in this case they are ZZZZSZ; so all utility fields are ignored except the second-to-last, *yy*, which is of type string. The *restore_graph* routine will construct a **Graph** record g from this example in which $g\text{-}n = 1$, $g\text{-}m = 3$, and $g\text{-}yy.S = $ "pi".

Notice that the individual field values for a record are separated by commas. If a line ends with a comma, the following line contains additional fields of the same record.

After the **Graph** record fields have been specified, there's a special line '*⊔Vertices', after which we learn the fields of each vertex in turn. First comes the *name* field, then the *arcs* field, and then any non-ignored utility fields. In this example the *util_types* for **Vertex** records are IZAZZZ, so the utility field values are $u.I$ and $w.A$. Let v point to the first **Vertex** record (which incidentally is also pointed to by $g\text{-}vertices$), and let a point to the first **Arc** record. Then in this example we will have $v\text{-}name = $ "look", $v\text{-}arcs = a$, $v\text{-}u.I = 15$, and $v\text{-}w.A = (a+1)$.

After the **Vertex** records comes a special line '*⊔Arcs', followed by the fields of each **Arc** record in an entirely analogous way. First comes the *tip* field, then the *next* field, then the *len*, and finally the utility fields (if any). In this example the *util_types* for **Arc** utility fields are ZV; hence field a is ignored, and field b is a pointer to a **Vertex**. We will have $a\text{-}tip = v$, $a\text{-}next = (a+2)$, $a\text{-}len = 3$, and $a\text{-}b.V = (v+1)$.

The null pointer Λ is denoted by 0. Furthermore, a **Vertex** pointer is allowed to have the special value 1, because of conventions explained in GB_GATES. (This special value appears in the fourth field of the third arc in the example above.) The *restore_graph* procedure does not allow **Vertex** pointers to take on constant values greater than 1, nor does it permit the value '1' where an **Arc** pointer ought to be.

There should be exactly as many **Vertex** and **Arc** specifications as indicated after the utility types at the beginning of the file. The final **Arc** should then be followed by a special checksum line, which must contain either a number consistent with the data on all the previous lines or a negative value (which is not checked). All information after the checksum line is ignored.

Users should not edit the files produced by *save_graph*, because an incorrect checksum is liable to ruin everything. However, additional lines beginning with '*' may be placed as comments at the very beginning of the file; such lines are immune to checksumming.

4. We can establish these conventions firmly in mind by writing the *restore_graph* routine before we write *save_graph*. The subroutine call *restore_graph*(`"foo.gb"`) produces a pointer to the graph defined in file `"foo.gb"`, or a null pointer in case that file is unreadable or incorrect. In the latter case, *panic_code* indicates the problem.

⟨ External functions 4 ⟩ ≡
   Graph *∗restore_graph*(*f*)
      char *∗f*;   /∗ the file name ∗/
  { **Graph** *∗g* = Λ;   /∗ the graph being restored ∗/
   register char *∗p*;   /∗ register for string manipulation ∗/
   long *m*;   /∗ the number of **Arc** records to allocate ∗/
   long *n*;   /∗ the number of **Vertex** records to allocate ∗/

   ⟨ Open the file and parse the first line; **goto** *sorry* if there's trouble 5 ⟩;
   ⟨ Create the **Graph** record *g* and fill in its fields 6 ⟩;
   ⟨ Fill in the fields of all **Vertex** records 16 ⟩;
   ⟨ Fill in the fields of all **Arc** records 17 ⟩;
   ⟨ Check the checksum and close the file 18 ⟩;
   return *g*;
  sorry: *gb_raw_close*( ); *gb_recycle*(*g*); **return** Λ;
  }

See also section 20.

This code is used in section 2.

A: **Arc** *∗*, GB_GRAPH §8.
Arc = **struct**, GB_GRAPH §10.
arcs: **Arc** *∗*, GB_GRAPH §9.
b: **util**, GB_GRAPH §10.
gb_raw_close: **long** ( ), GB_IO §42.
gb_recycle: **void** ( ), GB_GRAPH §40.
Graph = **struct**, GB_GRAPH §20.
I: **long**, GB_GRAPH §8.
id: **char** [ ], GB_GRAPH §20.

len: **long**, GB_GRAPH §10.
m: **long**, GB_GRAPH §20.
n: **long**, GB_GRAPH §20.
name: **char** *∗*, GB_GRAPH §9.
next: **Arc** *∗*, GB_GRAPH §10.
panic_code: **long**, GB_GRAPH §5.
S: **char** *∗*, GB_GRAPH §8.
save_graph: **long** ( ), §20.

tip: **Vertex** *∗*, GB_GRAPH §10.
u: **util**, GB_GRAPH §9.
util_types: **char** [ ], GB_GRAPH §20.
V: **Vertex** *∗*, GB_GRAPH §8.
Vertex = **struct**, GB_GRAPH §9.
vertices: **Vertex** *∗*, GB_GRAPH §20.
w: **util**, GB_GRAPH §9.
yy: **util**, GB_GRAPH §20.

5. As mentioned above, users can add comment lines at the beginning of the file, if they put a
* at the beginning of every such line. But the line that precedes the data proper must adhere to
strict standards.

#define *panic*(*c*) { *panic_code* = *c*; **goto** *sorry*; }

⟨ Open the file and parse the first line; **goto** *sorry* if there's trouble 5 ⟩ ≡
  gb_raw_open(*f*);
  if (*io_errors*) *panic*(*early_data_fault*);   /* can't open the file */
  while (1) {
    gb_string(*str_buf*, ')');
    if (*sscanf*(*str_buf*, "*␣GraphBase␣graph␣(util_types␣%14[ZIVSA],%1dV,%1dA", *str_buf* + 80, &*n*,
      &*m*) ≡ 3 ∧ *strlen*(*str_buf* + 80) ≡ 14) **break**;
    if (*str_buf*[0] ≠ '*') *panic*(*syntax_error*);   /* first line is unreadable */
  }

This code is used in section 4.

6. The previous code has placed the graph's *util_types* into location *str_buf* + 80 and verified
that it contains precisely 14 characters, all belonging to the set {Z, I, V, S, A}.

⟨ Create the **Graph** record *g* and fill in its fields 6 ⟩ ≡
  g = *gb_new_graph*(0_L);
  if (*g* ≡ Λ) *panic*(*no_room*);   /* out of memory before we're even started */
  gb_free(*g*→*data*);
  g→*vertices* = *verts* = *gb_typed_alloc*(*n* ≡ 0 ? 1 : *n*, **Vertex**, *g*→*data*);
  last_vert = *verts* + *n*;
  arcs = *gb_typed_alloc*(*m* ≡ 0 ? 1 : *m*, **Arc**, *g*→*data*);
  last_arc = *arcs* + *m*;
  if (*gb_trouble_code*) *panic*(*no_room* + 1);   /* not enough room for vertices and arcs */
  strcpy(*g*→*util_types*, *str_buf* + 80);
  gb_newline();
  if (*gb_char*() ≠ '"') *panic*(*syntax_error* + 1);   /* missing quotes before graph *id* string */
  p = *gb_string*(*g*→*id*, '"');
  if (*(*p* − 2) ≡ '\n' ∧ *(*p* − 3) ≡ '\\' ∧ *p* > *g*→*id* + 2) {
    gb_newline();
    gb_string(*p* − 3, '"');
  }
  if (*gb_char*() ≠ '"') *panic*(*syntax_error* + 2);   /* missing quotes after graph *id* string */
  ⟨ Fill in *g*→*n*, *g*→*m*, and *g*'s utility fields 15 ⟩;

This code is used in section 4.

7. The *util_types* and *id* fields are slightly different from other string fields, because we store
them directly in the **Graph** record instead of storing a pointer. The other fields to be filled by
restore_graph can all be done by a macro called *fillin*, which invokes a subroutine called *fill_field*.
The first parameter to *fillin* is the address of a field in a record; the second parameter is one of
the codes {Z, I, V, S, A}. A global variable *comma_expected* is nonzero when this field is not the
first in its record.

  The value returned by *fill_field* is nonzero if something goes wrong.

  We assume here that a utility field takes exactly as much space as a field of any of its constituent
types.

#define *fillin*(*l*, *t*)   **if** (*fill_field*((**util** ∗) &(*l*), *t*)) **goto** *sorry*

⟨ Private functions 7 ⟩ ≡
    static long *fill_field*(*l*, *t*)
        util ∗*l*;   /∗ location of field to be filled in ∗/
        char *t*;   /∗ its type code ∗/
  { **register char** *c*;   /∗ character just read ∗/
    if (*t* ≠ 'Z' ∧ *comma_expected*) {
      if (*gb_char*( ) ≠ ',') **return** (*panic_code* = *syntax_error* − 1);   /∗ missing comma ∗/
      if (*gb_char*( ) ≡ '\n') *gb_newline*( );
      else *gb_backup*( );
    }
    else *comma_expected* = 1;
    c = *gb_char*( );
    switch (*t*) {
    case 'I': ⟨ Fill in a numeric field 9 ⟩;
    case 'V': ⟨ Fill in a vertex pointer 10 ⟩;
    case 'S': ⟨ Fill in a string pointer 12 ⟩;
    case 'A': ⟨ Fill in an arc pointer 11 ⟩;
    default: *gb_backup*( ); **break**;
    }
    return *panic_code*;
  }

See also sections 14, 25, 35, 36, 37, and 39.

This code is used in section 2.

8. Some of the communication between *restore_graph* and *fillin* is best done via global variables.

⟨ Private variables 8 ⟩ ≡
    static long *comma_expected*;   /∗ should *fillin* look for a comma? ∗/
    static Vertex ∗*verts*;   /∗ beginning of the block of **Vertex** records ∗/
    static Vertex ∗*last_vert*;   /∗ end of the block of **Vertex** records ∗/
    static Arc ∗*arcs*;   /∗ beginning of the block of **Arc** records ∗/
    static Arc ∗*last_arc*;   /∗ end of the block of **Arc** records ∗/

See also sections 13, 19, 22, and 34.

This code is used in section 2.

Arc = **struct**, GB_GRAPH §10.
data: **Area**, GB_GRAPH §20.
early_data_fault = 10, GB_GRAPH §7.
f: **char** ∗, §4.
g: **Graph** ∗, §4.
gb_backup: **void** ( ), GB_IO §22.
gb_char: **char** ( ), GB_IO §22.
gb_free: **void** ( ), GB_GRAPH §16.
gb_new_graph: **Graph** ∗( ),
    GB_GRAPH §23.
gb_newline: **void** ( ), GB_IO §18.
gb_raw_open: **void** ( ), GB_IO §30.

gb_string: **char** ∗( ), GB_IO §26.
gb_trouble_code: **long**,
    GB_GRAPH §14.
gb_typed_alloc = macro ( ),
    GB_GRAPH §11.
Graph = **struct**, GB_GRAPH §20.
id: **char** [ ], GB_GRAPH §20.
io_errors: **long**, GB_IO §5.
m: **long**, §4.
n: **long**, §4.
no_room = 1, GB_GRAPH §7.
p: **register char** ∗, §4.

panic_code: **long**, GB_GRAPH §5.
restore_graph: **Graph** ∗( ), §4.
sorry: label, §4.
sscanf: **int** ( ), <stdio.h>.
str_buf: **char** [ ], GB_IO §26.
strcpy: **char** ∗( ), <string.h>.
strlen: **int** ( ), <string.h>.
syntax_error = 20, GB_GRAPH §7.
util = **union**, GB_GRAPH §8.
util_types: **char** [ ], GB_GRAPH §20.
Vertex = **struct**, GB_GRAPH §9.
vertices: **Vertex** ∗, GB_GRAPH §20.

9. ⟨ Fill in a numeric field 9 ⟩ ≡
  if $(c \equiv$ '-'$)$ $l{\to}I = -gb\_number(10);$
  else {
    $gb\_backup(\,);$
    $l{\to}I = gb\_number(10);$
  }
  break;

This code is used in section 7.

10. ⟨ Fill in a vertex pointer 10 ⟩ ≡
  if $(c \equiv$ 'V'$)$ {
    $l{\to}V = verts + gb\_number(10);$
    if $(l{\to}V \geq last\_vert \vee l{\to}V < verts)$ $panic\_code = syntax\_error - 2;$ /∗ vertex address too big ∗/
  } **else if** $(c \equiv$ '0' $\vee c \equiv$ '1'$)$ $l{\to}I = c -$ '0';
  else $panic\_code = syntax\_error - 3;$ /∗ vertex numeric address illegal ∗/
  break;

This code is used in section 7.

11. ⟨ Fill in an arc pointer 11 ⟩ ≡
  if $(c \equiv$ 'A'$)$ {
    $l{\to}A = arcs + gb\_number(10);$
    if $(l{\to}A \geq last\_arc \vee l{\to}A < arcs)$ $panic\_code = syntax\_error - 4;$ /∗ arc address too big ∗/
  } **else if** $(c \equiv$ '0'$)$ $l{\to}A = \Lambda;$
  else $panic\_code = syntax\_error - 5;$ /∗ arc numeric address illegal ∗/
  break;

This code is used in section 7.

12. We can restore a string slightly longer than the strings we can save.
⟨ Fill in a string pointer 12 ⟩ ≡
  if $(c \neq$ '"'$)$ $panic\_code = syntax\_error - 6;$ /∗ missing quotes at beginning of string ∗/
  else { **register char** ∗p;
    $p = gb\_string(item\_buf,$ '"'$);$
    while $(\ast(p-2) \equiv$ '\n' $\wedge \ast(p-3) \equiv$ '\\' $\wedge p > item\_buf + 2 \wedge p \leq buffer)$ {
      $gb\_newline(\,);$
      $p = gb\_string(p-3,$ '"'$);$ /∗ splice a broken string together ∗/
    }
    if $(gb\_char(\,) \neq$ '"'$)$ $panic\_code = syntax\_error - 7;$ /∗ missing quotes at end of string ∗/
    else if $(item\_buf[0] \equiv$ '\0'$)$ $l{\to}S = null\_string;$
    else $l{\to}S = gb\_save\_string(item\_buf);$
  }
  break;

This code is used in section 7.

13. ⟨ Private variables 8 ⟩ +≡
  static char $item\_buf$[MAX_SV_STRING + 3];   /∗ an item to be output ∗/
  static char $buffer$[81];   /∗ a line of output ∗/
    /∗ NB: $buffer$ must immediately follow $item\_buf$ ∗/

14. When all fields of a record have been filled in, we call *finish_record* and hope that it returns 0.

⟨ Private functions 7 ⟩ +≡
  static long *finish_record*( )
  {
    if (*gb_char*( ) ≠ '\n') **return** (*panic_code* = *syntax_error* − 8);   /∗ garbage present ∗/
    gb_newline( );
    comma_expected = 0;
    return 0;
  }

15. ⟨ Fill in *g→n*, *g→m*, and *g*'s utility fields 15 ⟩ ≡
  panic_code = 0;
  comma_expected = 1;
  fillin(*g→n*, 'I');
  fillin(*g→m*, 'I');
  fillin(*g→uu*, *g→util_types*[8]);
  fillin(*g→vv*, *g→util_types*[9]);
  fillin(*g→ww*, *g→util_types*[10]);
  fillin(*g→xx*, *g→util_types*[11]);
  fillin(*g→yy*, *g→util_types*[12]);
  fillin(*g→zz*, *g→util_types*[13]);
  if (*finish_record*( )) **goto** *sorry*;
This code is used in section 6.

A: **Arc** ∗, GB_GRAPH §8.
arcs: **static Arc** ∗, §8.
c: **register char**, §7.
comma_expected: **static long**, §8.
fillin = macro ( ), §7.
g: **Graph** ∗, §4.
gb_backup: **void** ( ), GB_IO §22.
gb_char: **char** ( ), GB_IO §22.
gb_newline: **void** ( ), GB_IO §18.
gb_number: **unsigned long** ( ),
  GB_IO §24.
gb_save_string: **char** ∗( ),

GB_GRAPH §35.
gb_string: **char** ∗( ), GB_IO §26.
I: **long**, GB_GRAPH §8.
l: **util** ∗, §7.
last_arc: **static Arc** ∗, §8.
last_vert: **static Vertex** ∗, §8.
m: **long**, GB_GRAPH §20.
MAX_SV_STRING = 4095, §1.
n: **long**, GB_GRAPH §20.
null_string: **char** [ ], GB_GRAPH §24.
panic_code: **long**, GB_GRAPH §5.
S: **char** ∗, GB_GRAPH §8.

sorry: label, §4.
syntax_error = 20, GB_GRAPH §7.
util_types: **char** [ ], GB_GRAPH §20.
uu: **util**, GB_GRAPH §20.
V: **Vertex** ∗, GB_GRAPH §8.
verts: **static Vertex** ∗, §8.
vv: **util**, GB_GRAPH §20.
ww: **util**, GB_GRAPH §20.
xx: **util**, GB_GRAPH §20.
yy: **util**, GB_GRAPH §20.
zz: **util**, GB_GRAPH §20.

16. The rest is easy.

⟨ Fill in the fields of all **Vertex** records 16 ⟩ ≡

```
{ register Vertex *v;
    gb_string(str_buf, '\n');
    if (strcmp(str_buf, "*␣Vertices") ≠ 0)
        panic(syntax_error + 3);      /* introductory line for vertices is missing */
    gb_newline();
    for (v = verts; v < last_vert; v++) {
        fillin(v→name, 'S');
        fillin(v→arcs, 'A');
        fillin(v→u, g→util_types[0]);
        fillin(v→v, g→util_types[1]);
        fillin(v→w, g→util_types[2]);
        fillin(v→x, g→util_types[3]);
        fillin(v→y, g→util_types[4]);
        fillin(v→z, g→util_types[5]);
        if (finish_record()) goto sorry;
    }
}
```

This code is used in section 4.

17. ⟨ Fill in the fields of all **Arc** records 17 ⟩ ≡

```
{ register Arc *a;
    gb_string(str_buf, '\n');
    if (strcmp(str_buf, "*␣Arcs") ≠ 0) panic(syntax_error + 4);
            /* introductory line for arcs is missing */
    gb_newline();
    for (a = arcs; a < last_arc; a++) {
        fillin(a→tip, 'V');
        fillin(a→next, 'A');
        fillin(a→len, 'I');
        fillin(a→a, g→util_types[6]);
        fillin(a→b, g→util_types[7]);
        if (finish_record()) goto sorry;
    }
}
```

This code is used in section 4.

18. ⟨ Check the checksum and close the file 18 ⟩ ≡

```
{ long s;
    gb_string(str_buf, '\n');
    if (sscanf(str_buf, "*␣Checksum␣%ld", &s) ≠ 1) panic(syntax_error + 5);
            /* checksum line is missing */
    if (gb_raw_close() ≠ s ∧ s ≥ 0) panic(late_data_fault);      /* checksum does not match */
}
```

This code is used in section 4.

19. Saving a graph. Now that we know how to restore a graph, once it has been saved, we are ready to write the *save_graph* routine.

Users say *save_graph*(g, "foo.gb"); our job is to create a file "foo.gb" from which the subroutine call *restore_graph*("foo.gb") will be able to reconstruct a graph equivalent to g, assuming that g meets the restrictions stated earlier. If nothing goes wrong, *save_graph* should return the value zero. Otherwise it should return an encoded trouble report.

We will set things up so that *save_graph* produces a syntactically correct file "foo.gb" in almost every case, with explicit error indications written at the end of the file whenever certain aspects of the given graph have had to be changed. The value -1 will be returned if $g \equiv \Lambda$; the value -2 will be returned if $g \neq \Lambda$ but the file "foo.gb" could not be opened for output; in other cases a file "foo.gb" will be created.

Here is a list of things that might go wrong, and the corresponding corrective actions to be taken in each case, assuming that *save_graph* does create a file:

#define *bad_type_code*   #1   /* illegal character, is changed to 'Z' */
#define *string_too_long*   #2   /* extralong string, is truncated */
#define *addr_not_in_data_area* #4   /* address out of range, is changed to Λ */
#define *addr_in_mixed_block*   #8   /* address not in pure block, is Λified */
#define *bad_string_char*   #10   /* illegal string character, is changed to '?' */
#define *ignored_data*   #20   /* nonzero value in 'Z' format, is not output */

⟨ Private variables 8 ⟩ +≡
  static long *anomalies*;   /* problems accumulated by *save_graph* */
  static FILE *save_file;   /* the file being written */

20. ⟨External functions 4⟩ +≡

  long *save_graph*(*g*, *f*)

      Graph **g*;   /* graph to be saved */

      char **f*;   /* name of the file to be created */

  { ⟨Local variables for *save_graph* 24⟩;

    if (*g* ≡ Λ ∨ *g*→*vertices* ≡ Λ) **return** −1;   /* where is *g*? */

    save_file = *fopen*(*f*, "w");

    if (¬*save_file*) **return** −2;   /* oops, the operating system won't cooperate */

    anomalies = 0;

    ⟨Figure out the extent of *g*'s internal records 27⟩;

    ⟨Translate *g* into external format 30⟩;

    ⟨Make notes at the end of the file about any changes that were necessary 46⟩;

    fclose(*save_file*);

    gb_free(*working_storage*);

    return *anomalies*;

  }

21. The main difficulty faced by *save_graph* is the problem of translating vertex and arc pointers into symbolic form. A graph's vertices usually appear in a single block, *g*→*vertices*, but its arcs usually appear in separate blocks that were created whenever the *gb_new_arc* routine needed more space. Other blocks, created by *gb_save_string*, are usually also present in the *g*→*data* area. We need to classify the various data blocks. We also want to be able to handle graphs that have been created with homegrown methods of memory allocation, because GraphBase structures need not conform to the conventions of *gb_new_arc* and *gb_save_string*.

A simple data structure based on **block_rep** records will facilitate our task. Each **block_rep** will be set up to contain the information we need to know about a particular block of data accessible from *g*→*data*. Such blocks are classified into four categories, identified by the *cat* field in a **block_rep**:

#define *unk* 0   /* *cat* value for blocks of unknown nature */

#define *ark* 1   /* *cat* value for blocks assumed to hold **Arc** records */

#define *vrt* 2   /* *cat* value for blocks assumed to hold **Vertex** records */

#define *mxt* 3   /* *cat* value for blocks being used for more than one purpose */

⟨Type declarations 21⟩ ≡

  typedef struct {

    char **start_addr*;   /* starting address of a data block */

    char **end_addr*;   /* ending address of a data block */

    long *offset*;   /* index number of first record in the block, if known */

    long *cat*;   /* *cat* code for the block */

    long *expl*;   /* have we finished exploring this block? */

  } **block_rep**;

This code is used in section 2.

22. The **block_rep** records don't need to be linked together in any fancy way, because there usually aren't very many of them. We will simply create an array, organized in decreasing order of *start_addr* and *end_addr*, with a dummy record standing as a sentinel at the end.

A system-dependent change might be necessary in the following code, if pointer values can be longer than 32 bits.

⟨ Private variables 8 ⟩ +≡
   static block_rep *blocks*;   /* beginning of table of block representatives */
   static Area *working_storage*;

23. Initially we set the *end_addr* field to the location following a block's data area. Later we will change it as explained below.

   The code in this section uses the fact that all bits of storage blocks are zero until set nonzero. In particular, the *cat* field of each **block_rep** will initially be *unk*, and the *expl* will be zero; the *start_addr* and *end_addr* of the sentinel record will be zero.

⟨ Initialize the *blocks* array 23 ⟩ ≡
   { **Area** t;   /* variable that runs through g→*data* */
      for ($*t = *(g$→*data*), *block_count* = 0; $*t$; $*t = (*t)$→*next*) *block_count*++;
      blocks = *gb_typed_alloc*(*block_count* + 1, **block_rep**, *working_storage*);
      for ($*t = *(g$→*data*), *block_count* = 0; $*t$; $*t = (*t)$→*next*, *block_count*++) {
         cur_block = *blocks* + *block_count*;
         while (*cur_block* > *blocks* ∧ (*cur_block* − 1)→*start_addr* < (*t)→*first*) {
            cur_block→*start_addr* = (*cur_block* − 1)→*start_addr*;
            cur_block→*end_addr* = (*cur_block* − 1)→*end_addr*;
            cur_block −−;
         }
         cur_block→*start_addr* = (*t)→*first*;
         cur_block→*end_addr* = (**char** *) $*t$;
      }
   }

This code is used in section 27.

24. ⟨ Local variables for *save_graph* 24 ⟩ ≡
   register block_rep *cur_block*;   /* the current block of interest */
   long *block_count*;   /* how many blocks have we processed? */

See also section 31.

This code is used in section 20.

25. The *save_graph* routine makes two passes over the graph. The goal of the first pass is reconnaissance: We try to see where everything is, and we prune off parts that don't conform to the restrictions. When we get to the second pass, our task will then be almost trivial. We will be able to march through the known territory and spew out a copy of what we encounter. (Items that are "pruned" are not actually removed from g itself, only from the portion of g that is saved.)

The first pass is essentially a sequence of calls of the *lookup* macro, which looks at one field of one record and notes whether the existence of this field extends the known boundaries of the graph. The *lookup* macro is a shorthand notation for calling the *classify* subroutine. We make the same assumption about field sizes as the *fill_field* routine did above.

```
#define  lookup(l, t)   classify((util *) &(l), t)     /* explore field l of type t */
⟨ Private functions 7 ⟩ +≡
  static void classify(l, t)
      util *l;      /* location of field to be classified */
      char t;       /* its type code, from the set {Z, I, V, S, A} */
  { register block_rep *cur_block;
    register char *loc;
    register long tcat;       /* category corresponding to t */
    register long tsize;      /* record size corresponding to t */

    switch (t) {
    default: return;
    case 'V':
      if (l→I ≡ 1) return;
      tcat = vrt;  tsize = sizeof(Vertex);
      break;
    case 'A': tcat = ark;  tsize = sizeof(Arc);
      break;
    }
    if (l→I ≡ 0) return;
    ⟨ Classify a pointer variable 26 ⟩;
  }
```

26. Here we know that l points to a **Vertex** or to an **Arc**, according as *tcat* is *vrt* or *ark*. We need to check that this doesn't violate any assumptions about all such pointers lying in pure blocks within the g→*data* area.

```
⟨ Classify a pointer variable 26 ⟩ ≡
  loc = (char *) l→V;
  for (cur_block = blocks; cur_block→start_addr > loc; cur_block ++) ;
  if (loc < cur_block→end_addr) {
    if ((loc − cur_block→start_addr) % tsize ≠ 0 ∨ loc + tsize > cur_block→end_addr)
      cur_block→cat = mxt;
    if (cur_block→cat ≡ unk) cur_block→cat = tcat;
    else if (cur_block→cat ≠ tcat) cur_block→cat = mxt;
  }
```

This code is used in section 25.

27.   We go through the list of blocks repeatedly until we reach a stable situation in which every *vrt* or *ark* block has been explored.

⟨ Figure out the extent of *g*'s internal records 27 ⟩ ≡

```
{ long activity;
    ⟨ Initialize the blocks array 23 ⟩;
    lookup(g→vertices, 'V');
    lookup(g→uu, g→util_types[8]);
    lookup(g→vv, g→util_types[9]);
    lookup(g→ww, g→util_types[10]);
    lookup(g→xx, g→util_types[11]);
    lookup(g→yy, g→util_types[12]);
    lookup(g→zz, g→util_types[13]);
    do { activity = 0;
        for (cur_block = blocks; cur_block→end_addr; cur_block ++) {
            if (cur_block→cat ≡ vrt ∧ ¬cur_block→expl) ⟨ Explore a block of supposed vertex records 28 ⟩
            else if (cur_block→cat ≡ ark ∧ ¬cur_block→expl) ⟨ Explore a block of supposed arc records 29 ⟩
            else continue;
            cur_block→expl = activity = 1;
        }
    } while (activity);
}
```

This code is used in section 20.

Arc = **struct**, GB_GRAPH §10.
ark = 1, §21.
block_rep = **struct**, §21.
blocks: **static block_rep** ∗, §22.
cat: **long**, §21.
cur_block: **register block_rep** ∗,
  §24.
data: **Area**, GB_GRAPH §20.
end_addr: **char** ∗, §21.
expl: **long**, §21.

fill_field: **static long** ( ), §7.
g: **Graph** ∗, §20.
l: **long**, GB_GRAPH §8.
mxt = 3, §21.
save_graph: **long** ( ), §20.
start_addr: **char** ∗, §21.
unk = 0, §21.
util = **union**, GB_GRAPH §8.
util_types: **char** [], GB_GRAPH §20.
uu: **util**, GB_GRAPH §20.

V: **Vertex** ∗, GB_GRAPH §8.
Vertex = **struct**, GB_GRAPH §9.
vertices: **Vertex** ∗, GB_GRAPH §20.
vrt = 2, §21.
vv: **util**, GB_GRAPH §20.
ww: **util**, GB_GRAPH §20.
xx: **util**, GB_GRAPH §20.
yy: **util**, GB_GRAPH §20.
zz: **util**, GB_GRAPH §20.

28. While we are exploring a block, the *lookup* routine might classify a previously explored block (or even the current block) as *mxt*. Therefore some data we assumed would be accessible will actually be removed from the graph; contradictions that arose might no longer exist. But we plunge ahead anyway, because we aren't going to try especially hard to "save" portions of graphs that violate our ground rules.

⟨ Explore a block of supposed vertex records 28 ⟩ ≡
    { **register Vertex** *v;

    for (v = (**Vertex** *) cur_block→start_addr;
            (**char** *) (v + 1) ≤ cur_block→end_addr ∧ cur_block→cat ≡ vrt; v++) {
        lookup(v→arcs, 'A');
        lookup(v→u, g→util_types[0]);
        lookup(v→v, g→util_types[1]);
        lookup(v→w, g→util_types[2]);
        lookup(v→x, g→util_types[3]);
        lookup(v→y, g→util_types[4]);
        lookup(v→z, g→util_types[5]);
        }
    }

This code is used in section 27.

29. ⟨ Explore a block of supposed arc records 29 ⟩ ≡
    { **register Arc** *a;

    for (a = (**Arc** *) cur_block→start_addr;
            (**char** *) (a + 1) ≤ cur_block→end_addr ∧ cur_block→cat ≡ ark; a++) {
        lookup(a→tip, 'V');
        lookup(a→next, 'A');
        lookup(a→a, g→util_types[6]);
        lookup(a→b, g→util_types[7]);
        }
    }

This code is used in section 27.

30. OK, the first pass is complete. And the second pass is routine:

⟨ Translate g into external format 30 ⟩ ≡
    ⟨ Orient the *blocks* table for translation 32 ⟩;
    ⟨ Initialize the output buffer mechanism and output the first line 38 ⟩;
    ⟨ Translate the **Graph** record 41 ⟩;
    ⟨ Translate the **Vertex** records 42 ⟩;
    ⟨ Translate the **Arc** records 44 ⟩;
    ⟨ Output the checksum line 45 ⟩;

This code is used in section 20.

31. During this pass we decrease the *end_addr* field of a **block_rep**, so that it points to the first byte of the final record in a *vrt* or *ark* block.

The variables m and n are set to the number of arc records and vertex records, respectively.

⟨ Local variables for *save_graph* 24 ⟩ +≡
  long m;   /* total number of **Arc** records to be translated */
  long n;   /* total number of **Vertex** records to be translated */
  register long s;   /* accumulator register for arithmetic calculations */

32. One tricky point needs to be observed, in the unusual case that there are two or more blocks of **Vertex** records: The base block *g⃗vertices* must come first in the final ordering. (This is the only exception to the rule that **Vertex** and **Arc** records each retain their relative order with respect to less-than and greater-than.)

⟨ Orient the *blocks* table for translation 32 ⟩ ≡
  $m = 0$; ⟨ Set n to the size of the block that starts with *g⃗vertices* 33 ⟩;
  for ($cur\_block = blocks + block\_count - 1$; $cur\_block \geq blocks$; cur_block −−) {
    if ($cur\_block{\rightarrow}cat \equiv vrt$) {
      $s = (cur\_block{\rightarrow}end\_addr - cur\_block{\rightarrow}start\_addr)/\textbf{sizeof}(\textbf{Vertex})$;
      $cur\_block{\rightarrow}end\_addr = cur\_block{\rightarrow}start\_addr + ((s - 1) * \textbf{sizeof}(\textbf{Vertex}))$;
      if ($cur\_block{\rightarrow}start\_addr \neq (\textbf{char} *) g{\rightarrow}vertices$) {
        $cur\_block{\rightarrow}offset = n$; $n \mathrel{+}= s$;
      }   /* otherwise *cur_block⃗offset* remains zero */
    } **else if** ($cur\_block{\rightarrow}cat \equiv ark$) {
      $s = (cur\_block{\rightarrow}end\_addr - cur\_block{\rightarrow}start\_addr)/\textbf{sizeof}(\textbf{Arc})$;
      $cur\_block{\rightarrow}end\_addr = cur\_block{\rightarrow}start\_addr + ((s - 1) * \textbf{sizeof}(\textbf{Arc}))$;
      $cur\_block{\rightarrow}offset = m$;
      $m \mathrel{+}= s$;
    }
  }

This code is used in section 30.

33. ⟨ Set n to the size of the block that starts with *g⃗vertices* 33 ⟩ ≡
  $n = 0$;
  for ($cur\_block = blocks + block\_count - 1$; $cur\_block \geq blocks$; cur_block −−)
    if ($cur\_block{\rightarrow}start\_addr \equiv (\textbf{char} *) g{\rightarrow}vertices$) {
      $n = (cur\_block{\rightarrow}end\_addr - cur\_block{\rightarrow}start\_addr)/\textbf{sizeof}(\textbf{Vertex})$;
      break;
    }

This code is used in section 32.

---

a: **util**, GB_GRAPH §10.
Arc = **struct**, GB_GRAPH §10.
arcs: **Arc** *, GB_GRAPH §9.
ark = 1, §21.
b: **util**, GB_GRAPH §10.
block_count: **long**, §24.
block_rep = **struct**, §21.
blocks: **static block_rep** *, §22.
cat: **long**, §21.
cur_block: **register block_rep** *, §24.

end_addr: **char** *, §21.
g: **Graph** *, §20.
Graph = **struct**, GB_GRAPH §20.
lookup = macro ( ), §25.
mxt = 3, §21.
next: **Arc** *, GB_GRAPH §10.
offset: **long**, §21.
save_graph: **long** ( ), §20.
start_addr: **char** *, §21.
tip: **Vertex** *, GB_GRAPH §10.

u: **util**, GB_GRAPH §9.
util_types: **char** [ ], GB_GRAPH §20.
v: **util**, GB_GRAPH §9.
Vertex = **struct**, GB_GRAPH §9.
vertices: **Vertex** *, GB_GRAPH §20.
vrt = 2, §21.
w: **util**, GB_GRAPH §9.
x: **util**, GB_GRAPH §9.
y: **util**, GB_GRAPH §9.
z: **util**, GB_GRAPH §9.

34. We will store material to be output in the *buffer* array, so that we can compute the correct checksum.

⟨ Private variables 8 ⟩ +≡
   static char *buf_ptr*;   /∗ the first unfilled position in *buffer* ∗/
   static long *magic*;   /∗ the checksum ∗/

35. ⟨ Private functions 7 ⟩ +≡
  static void *flushout*( )   /∗ output the buffer to *save_file* ∗/
  {
    ∗*buf_ptr*++ = '\n';
    ∗*buf_ptr* = '\0';
    magic = *new_checksum*(*buffer*, *magic*);
    fputs(*buffer*, *save_file*);
    buf_ptr = *buffer*;
  }

36. If a supposed string pointer is zero, we output the null string. (This case arises when a string field has not been initialized, for example in vertices and arcs that have been allocated but not used.)

⟨ Private functions 7 ⟩ +≡
  static void *prepare_string*(*s*)
      char ∗*s*;   /∗ string that is moved to *item_buf* ∗/
  { **register char** ∗*p*, ∗*q*;
    item_buf[0] = '"';
    p = &*item_buf*[1];
    if (*s* ≡ 0) **goto** *sready*;
    for (*q* = *s*; ∗*q* ∧ *p* ≤ &*item_buf*[MAX_SV_STRING]; *q*++, *p*++)
      if (∗*q* ≡ '"' ∨ ∗*q* ≡ '\n' ∨ ∗*q* ≡ '\\' ∨ *imap_ord*(∗*q*) ≡ *unexpected_char*) {
        anomalies |= *bad_string_char*;
        ∗*p* = '?';
      } **else** ∗*p* = ∗*q*;
    if (∗*q*) *anomalies* |= *string_too_long*;
  sready: ∗*p* = '"';
    ∗(*p* + 1) = '\0';
  }

37. The main idea of this part of the program is to format an item into *item_buf*, then move it to *buffer*, making sure that there is always room for a comma.

#**define** *append_comma* ∗*buf_ptr*++ = ','
⟨ Private functions 7 ⟩ +≡
  static void *move_item*( )
  { **register long** *l* = *strlen*(*item_buf*);
    if (*buf_ptr* + *l* > &*buffer*[78]) {
      if (*l* ≤ 78) *flushout*( );
      else { **register char** ∗*p* = *item_buf*;
        if (*buf_ptr* > &*buffer*[77]) *flushout*( );   /∗ no room for initial " ∗/
        do {

```
        for ( ; buf_ptr < &buffer[78]; buf_ptr ++, p++, l--) *buf_ptr = *p;
        *buf_ptr ++ = '\\';
        flushout( );
      }  while (l > 78);
      strcpy(buffer, p);
      buf_ptr = &buffer[l];
      return;
    }
  }
  strcpy(buf_ptr, item_buf);
  buf_ptr += l;
}
```

38. ⟨Initialize the output buffer mechanism and output the first line 38⟩ ≡

```
  buf_ptr = buffer;
  magic = 0;
  fputs("*␣GraphBase␣graph␣(util_types␣", save_file);
  { register char *p;
    for (p = g⃗util_types; p < g⃗util_types + 14; p++)
      if (*p ≡ 'Z' ∨ *p ≡ 'I' ∨ *p ≡ 'V' ∨ *p ≡ 'S' ∨ *p ≡ 'A') fputc(*p, save_file);
      else fputc('Z', save_file);
  }
  fprintf(save_file, ",%ldV,%ldA)\n", n, m);
```

This code is used in section 30.

---

anomalies: **static long**, §19.
bad_string_char = #10, §19.
buffer: **static char** [], §13.
fprintf: **int** ( ), <stdio.h>.
fputc: **int** ( ), <stdio.h>.
fputs: **int** ( ), <stdio.h>.
g: **Graph** *, §20.

imap_ord: **long** ( ), GB_IO §12.
item_buf: **static char** [], §13.
m: **long**, §31.
MAX_SV_STRING = 4095, §1.
n: **long**, §31.
new_checksum: **long** ( ), GB_IO §17.

save_file: **static FILE** *, §19.
strcpy: **char** *( ), <string.h>.
string_too_long = #2, §19.
strlen: **int** ( ), <string.h>.
unexpected_char = 127, GB_IO §11.
util_types: **char** [], GB_GRAPH §20.

39. A macro called *trans*, which is sort of an inverse to *fillin*, takes care of the main work in the second pass.

#define *trans*(*l, t*)   *translate_field*((**util** ∗) &(*l*), *t*)
⟨ Private functions 7 ⟩ +≡
  static void *translate_field*(*l, t*)
      util ∗*l*;   /∗ address of field to be output in symbolic form ∗/
      char *t*;   /∗ type of formatting desired ∗/
  { **register block_rep** ∗*cur_block*;
   register char ∗*loc*;
   register long *tcat*;   /∗ category corresponding to *t* ∗/
   register long *tsize*;   /∗ record size corresponding to *t* ∗/

   if (*comma_expected*) *append_comma*;
   else *comma_expected* = 1;
   switch (*t*) {
   default: *anomalies* |= *bad_type_code*;   /∗ fall through to case Z ∗/
   case 'Z': *buf_ptr* −−;   /∗ forget spurious comma ∗/
     if (*l*→*I*) *anomalies* |= *ignored_data*;
     return;
   case 'I': *numeric*: *sprintf*(*item_buf*, "%1d", *l*→*I*); **goto** *ready*;
   case 'S': *prepare_string*(*l*→*S*); **goto** *ready*;
   case 'V':
     if (*l*→*I* ≡ 1) **goto** *numeric*;
     tcat = *vrt*; *tsize* = **sizeof**(**Vertex**); **break**;
   case 'A': *tcat* = *ark*; *tsize* = **sizeof**(**Arc**); **break**;
   }
  ⟨ Translate a pointer variable 40 ⟩;
  ready: *move_item*( );
  }

40. ⟨ Translate a pointer variable 40 ⟩ ≡
  loc = (**char** ∗) *l*→*V*;
  item_buf[0] = '0'; *item_buf*[1] = '\0';   /∗ Λ will be the default ∗/
  if (*loc* ≡ Λ) **goto** *ready*;
  for (*cur_block* = *blocks*; *cur_block*→*start_addr* > *loc*; *cur_block* ++) ;
  if (*loc* > *cur_block*→*end_addr*) {
   anomalies |= *addr_not_in_data_area*;
   goto *ready*;
  }
  if (*cur_block*→*cat* ≠ *tcat* ∨ (*loc* − *cur_block*→*start_addr*) % *tsize* ≠ 0) {
   anomalies |= *addr_in_mixed_block*;
   goto *ready*;
  }
  sprintf(*item_buf*, "%c%1d", *t*, *cur_block*→*offset* + ((*loc* − *cur_block*→*start_addr*)/*tsize*));
This code is used in section 39.

41. ⟨ Translate the **Graph** record 41 ⟩ ≡
  prepare_string(*g*→*id*);
  if (*strlen*(*g*→*id*) > MAX_SV_ID) {

$strcpy(item\_buf + \texttt{MAX\_SV\_ID} + 1, \texttt{"\\""});$
    $anomalies \mathrel{|=} string\_too\_long;$
  }
  $move\_item();$
  $comma\_expected = 1;$
  $trans(g\text{-}n, \texttt{'I'});$
  $trans(g\text{-}m, \texttt{'I'});$
  $trans(g\text{-}uu, g\text{-}util\_types[8]);$
  $trans(g\text{-}vv, g\text{-}util\_types[9]);$
  $trans(g\text{-}ww, g\text{-}util\_types[10]);$
  $trans(g\text{-}xx, g\text{-}util\_types[11]);$
  $trans(g\text{-}yy, g\text{-}util\_types[12]);$
  $trans(g\text{-}zz, g\text{-}util\_types[13]);$
  $flushout();$

This code is used in section 30.

42. ⟨ Translate the **Vertex** records 42 ⟩ ≡
  { **register Vertex** *v*;

  $fputs(\texttt{"*\_Vertices\\n"}, save\_file);$
  for $(cur\_block = blocks + block\_count - 1; \; cur\_block \geq blocks; \; cur\_block--)$
    if $(cur\_block\text{-}cat \equiv vrt \land cur\_block\text{-}offset \equiv 0)$ ⟨Translate all **Vertex** records in *cur_block* 43⟩;
  for $(cur\_block = blocks + block\_count - 1; \; cur\_block \geq blocks; \; cur\_block--)$
    if $(cur\_block\text{-}cat \equiv vrt \land cur\_block\text{-}offset \neq 0)$ ⟨Translate all **Vertex** records in *cur_block* 43⟩;
  }

This code is used in section 30.

---

43. ⟨ Translate all **Vertex** records in *cur_block* 43 ⟩ ≡

  for $(v = (\textbf{Vertex} *) \; cur\_block \rightarrow start\_addr; \; v \le (\textbf{Vertex} *) \; cur\_block \rightarrow end\_addr; \; v\texttt{++})$ {

    comma_expected = 0;

    trans(*v*→*name*, 'S');

    trans(*v*→*arcs*, 'A');

    trans(*v*→*u*, *g*→*util_types*[0]);

    trans(*v*→*v*, *g*→*util_types*[1]);

    trans(*v*→*w*, *g*→*util_types*[2]);

    trans(*v*→*x*, *g*→*util_types*[3]);

    trans(*v*→*y*, *g*→*util_types*[4]);

    trans(*v*→*z*, *g*→*util_types*[5]);

    flushout( );

  }

This code is used in section 42 (twice).

44. ⟨ Translate the **Arc** records 44 ⟩ ≡

  { **register Arc** **a*;

  fputs("*␣Arcs\n", *save_file*);

  for (*cur_block* = *blocks* + *block_count* − 1; *cur_block* ≥ *blocks*; *cur_block* −−)

    if (*cur_block*→*cat* ≡ *ark*)

      for $(a = (\textbf{Arc} *) \; cur\_block \rightarrow start\_addr; \; a \le (\textbf{Arc} *) \; cur\_block \rightarrow end\_addr; \; a\texttt{++})$ {

        comma_expected = 0;

        trans(*a*→*tip*, 'V');

        trans(*a*→*next*, 'A');

        trans(*a*→*len*, 'I');

        trans(*a*→*a*, *g*→*util_types*[6]);

        trans(*a*→*b*, *g*→*util_types*[7]);

        flushout( );

      }

  }

This code is used in section 30.

45. ⟨ Output the checksum line 45 ⟩ ≡

  fprintf(*save_file*, "*␣Checksum␣%ld\n", *magic*);

This code is used in section 30.

46. ⟨ Make notes at the end of the file about any changes that were necessary 46 ⟩ ≡

  if (*anomalies*) {

    fputs(">␣WARNING:␣I␣had␣trouble␣making␣this␣file␣from␣the␣given␣graph!\n", *save_file*);

    if (*anomalies* & *bad_type_code*)

      fputs(">>␣The␣original␣util_types␣had␣to␣be␣corrected.\n", *save_file*);

    if (*anomalies* & *ignored_data*)

      fputs(">>␣Some␣data␣suppressed␣by␣Z␣format␣was␣actually␣nonzero.\n", *save_file*);

    if (*anomalies* & *string_too_long*)

      fputs(">>␣At␣least␣one␣long␣string␣had␣to␣be␣truncated.\n", *save_file*);

    if (*anomalies* & *bad_string_char*)

      fputs(">>␣At␣least␣one␣string␣character␣had␣to␣be␣changed␣to␣'?'.\n", *save_file*);

    if (*anomalies* & *addr_not_in_data_area*)

      fputs(">>␣At␣least␣one␣pointer␣led␣out␣of␣the␣data␣area.\n", *save_file*);

```
    if (anomalies & addr_in_mixed_block)
        fputs(">>␣At␣least␣one␣data␣block␣had␣an␣illegal␣mixture␣of␣records.\n", save_file);
    if (anomalies & (addr_not_in_data_area + addr_in_mixed_block))
        fputs(">>␣␣(Pointers␣to␣improper␣data␣have␣been␣changed␣to␣0.)\n", save_file);
    fputs(">␣You␣should␣be␣able␣to␣read␣this␣file␣with␣restore_graph,\n", save_file);
    fputs(">␣but␣the␣graph␣you␣get␣won't␣be␣exactly␣like␣the␣original.\n", save_file);
}
```

This code is used in section 20.

---

a: **util**, GB_GRAPH §10.
addr_in_mixed_block = #8, §19.
addr_not_in_data_area = #4, §19.
anomalies: **static long**, §19.
Arc = **struct**, GB_GRAPH §10.
arcs: **Arc** *, GB_GRAPH §9.
ark = 1, §21.
b: **util**, GB_GRAPH §10.
bad_string_char = #10, §19.
bad_type_code = #1, §19.
block_count: **long**, §24.
blocks: **static block_rep** *, §22.
cat: **long**, §21.
comma_expected: **static long**, §8.

cur_block: **register block_rep** *,
  §24.
end_addr: **char** *, §21.
flushout: **static void** ( ), §35.
fprintf: **int** ( ), <stdio.h>.
fputs: **int** ( ), <stdio.h>.
g: **Graph** *, §20.
ignored_data = #20, §19.
len: **long**, GB_GRAPH §10.
magic: **static long**, §34.
name: **char** *, GB_GRAPH §9.
next: **Arc** *, GB_GRAPH §10.
save_file: **static FILE** *, §19.

start_addr: **char** *, §21.
string_too_long = #2, §19.
tip: **Vertex** *, GB_GRAPH §10.
trans = macro ( ), §39.
u: **util**, GB_GRAPH §9.
util_types: **char** [], GB_GRAPH §20.
v: **register Vertex** *, §42.
v: **util**, GB_GRAPH §9.
Vertex = **struct**, GB_GRAPH §9.
w: **util**, GB_GRAPH §9.
x: **util**, GB_GRAPH §9.
y: **util**, GB_GRAPH §9.
z: **util**, GB_GRAPH §9.

GB_SORT

1. Introduction. This short GraphBase module provides a simple utility routine called *gb_linksort*, which is used in many of the other programs.

```
#include <stdio.h>      /* the NULL pointer (Λ) is defined here */
#include "gb_flip.h"     /* we need to use the random number generator */
⟨ Preprocessor definitions ⟩
⟨ Declarations 2 ⟩
⟨ The gb_linksort routine 5 ⟩
```

2. Most of the graphs obtained from GraphBase data are parameterized, so that different effects can be obtained easily from the same underlying body of information. In many cases the desired graph is determined by selecting the "heaviest" vertices according to some notion of "weight," and/or by taking a random sample of vertices. For example, the GraphBase routine *words*(n, wt_vector, wt_threshold, seed) creates a graph based on the n most common words of English, where common-ness is determined by a given weight vector. When several words have equal weight, we want to choose between them at random. In particular, this means that we can obtain a completely random choice of words if the weight vector assigns the same weight to each word.

The *gb_linksort* routine is a convenient tool for this purpose. It takes a given linked list of nodes and shuffles their link fields so that the nodes can be read in decreasing order of weight, and so that equal-weight nodes appear in random order. *Note: The random number generator of* GB_FLIP *must be initialized before gb_linksort is called.*

The nodes sorted by *gb_linksort* can be records of any structure type, provided only that the first field is '**long** *key*' and the second field is '**struct** *this_struct_type *link*'. Further fields are not examined. The **node** type defined in this section is the simplest possible example of such a structure.

Sorting is done by means of the *key* fields, which must each contain nonnegative integers less than 2^{31}.

After sorting is complete, the data will appear in 128 linked lists: *gb_sorted*[127], *gb_sorted*[126], ..., *gb_sorted*[0]. To examine the nodes in decreasing order of weight, one can read through these lists with a routine such as

```
{
    int j;
    node *p;
    for (j = 127; j ≥ 0; j--)
        for (p = (node *) gb_sorted[j]; p; p = p‑link)
            look_at(p);
}
```

All nodes whose keys are in the range $j \cdot 2^{24} \le key < (j+1) \cdot 2^{24}$ will appear in list *gb_sorted*[j]. Therefore the results will all be found in the single list *gb_sorted*[0], if all the keys are strictly less than 2^{24}.

⟨ Declarations 2 ⟩ ≡
  typedef struct node_struct {
    long *key*;   /* a numeric quantity, assumed nonnegative */
    struct node_struct *link;   /* the next node on a list */
  } **node**;   /* applications of *gb_linksort* may have other fields after *link* */
See also section 4.

This code is used in section 1.

3. In the header file, *gb_sorted* is declared to be an array of pointers to **char**, since nodes may have different types in different applications. User programs should cast *gb_sorted* to the appropriate type as in the example above.

⟨ **gb_sort.h** 3 ⟩ ≡
  extern void *gb_linksort*( );   /* procedure to sort a linked list */
  extern char *gb_sorted[ ];   /* the results of *gb_linksort* */

4. Six passes of a radix sort, using radix 256, will accomplish the desired objective rather quickly. (See, for example, Algorithm 5.2.5R in *Sorting and Searching*.) The first two passes use random numbers instead of looking at the key fields, thereby effectively extending the keys so that nodes with equal keys will appear in reasonably random order.

  We move the nodes back and forth between two arrays of lists: the external array *gb_sorted* and a private array called *alt_sorted*.

⟨ Declarations 2 ⟩ +≡
  node *gb_sorted[256];   /* external bank of lists, for even-numbered passes */
  static node *alt_sorted[256];   /* internal bank of lists, for odd-numbered passes */

5. So here we go with six passes over the data.

⟨ The *gb_linksort* routine 5 ⟩ ≡
  void *gb_linksort*(*l*)
     node *l;
  { **register long** *k*;   /* index to destination list */
    register node **pp;   /* current place in list of pointers */
    register node *p, *q;   /* pointers for list manipulation */
    ⟨ Partition the given list into 256 random sublists *alt_sorted* 6 ⟩;
    ⟨ Partition the *alt_sorted* lists into 256 random sublists *gb_sorted* 7 ⟩;
    ⟨ Partition the *gb_sorted* lists into *alt_sorted* by low-order byte 8 ⟩;
    ⟨ Partition the *alt_sorted* lists into *gb_sorted* by second-lowest byte 9 ⟩;
    ⟨ Partition the *gb_sorted* lists into *alt_sorted* by second-highest byte 10 ⟩;
    ⟨ Partition the *alt_sorted* lists into *gb_sorted* by high-order byte 11 ⟩;
  }
This code is used in section 1.

6. ⟨ Partition the given list into 256 random sublists *alt_sorted* 6 ⟩ ≡

   for (*pp* = *alt_sorted* + 255; *pp* ≥ *alt_sorted*; *pp* −−) *pp = Λ;

      /∗ empty all the destination lists ∗/

   for (*p* = *l*; *p*; *p* = *q*) {

      k = *gb_next_rand*( ) ≫ 23;   /∗ extract the eight most significant bits ∗/

      q = *p→link*;

      p→link = *alt_sorted*[*k*];

      alt_sorted[*k*] = *p*;

   }

This code is used in section 5.

7. ⟨ Partition the *alt_sorted* lists into 256 random sublists *gb_sorted* 7 ⟩ ≡

   for (*pp* = *gb_sorted* + 255; *pp* ≥ *gb_sorted*; *pp* −−) *pp = Λ;   /∗ empty all the destination lists ∗/

   for (*pp* = *alt_sorted* + 255; *pp* ≥ *alt_sorted*; *pp* −−)

      for (*p* = *pp; *p*; *p* = *q*) {

        k = *gb_next_rand*( ) ≫ 23;   /∗ extract the eight most significant bits ∗/

        q = *p→link*;

        p→link = *gb_sorted*[*k*];

        gb_sorted[*k*] = *p*;

      }

This code is used in section 5.

8. ⟨ Partition the *gb_sorted* lists into *alt_sorted* by low-order byte 8 ⟩ ≡

   for (*pp* = *alt_sorted* + 255; *pp* ≥ *alt_sorted*; *pp* −−) *pp = Λ;

      /∗ empty all the destination lists ∗/

   for (*pp* = *gb_sorted* + 255; *pp* ≥ *gb_sorted*; *pp* −−)

      for (*p* = *pp; *p*; *p* = *q*) {

        k = *p→key* & #ff;   /∗ extract the eight least significant bits ∗/

        q = *p→link*;

        p→link = *alt_sorted*[*k*];

        alt_sorted[*k*] = *p*;

      }

This code is used in section 5.

9. Here we must read from *alt_sorted* from 0 to 255, not from 255 to 0, to get the desired final order. (Each pass reverses the order of the lists; it's tricky, but it works.)

⟨ Partition the *alt_sorted* lists into *gb_sorted* by second-lowest byte 9 ⟩ ≡

   for (*pp* = *gb_sorted* + 255; *pp* ≥ *gb_sorted*; *pp* −−) *pp = Λ;   /∗ empty all the destination lists ∗/

   for (*pp* = *alt_sorted*; *pp* < *alt_sorted* + 256; *pp* ++)

      for (*p* = *pp; *p*; *p* = *q*) {

        k = (*p→key* ≫ 8) & #ff;   /∗ extract the next eight bits ∗/

        q = *p→link*;

        p→link = *gb_sorted*[*k*];

        gb_sorted[*k*] = *p*;

      }

This code is used in section 5.

10. ⟨ Partition the *gb_sorted* lists into *alt_sorted* by second-highest byte 10 ⟩ ≡
   for (*pp* = *alt_sorted* + 255; *pp* ≥ *alt_sorted*; *pp* −−) *pp* = Λ;
      /∗ empty all the destination lists ∗/
   for (*pp* = *gb_sorted* + 255; *pp* ≥ *gb_sorted*; *pp* −−)
     for (*p* = ∗*pp*; *p*; *p* = *q*) {
       k = (*p⟶key* ≫ 16) & #ff;   /∗ extract the next eight bits ∗/
       q = *p⟶link*;
       p⟶link = *alt_sorted*[*k*];
       alt_sorted[*k*] = *p*;
     }
This code is used in section 5.

11. The most significant bits will lie between 0 and 127, because we assumed that the keys are nonnegative and less than 2^{31}. (A similar routine would be able to sort signed integers, or unsigned long integers, but the C code would not then be portable.)

⟨ Partition the *alt_sorted* lists into *gb_sorted* by high-order byte 11 ⟩ ≡
   for (*pp* = *gb_sorted* + 255; *pp* ≥ *gb_sorted*; *pp* −−) *pp* = Λ;   /∗ empty all the destination lists ∗/
   for (*pp* = *alt_sorted*; *pp* < *alt_sorted* + 256; *pp* ++)
     for (*p* = ∗*pp*; *p*; *p* = *q*) {
       k = (*p⟶key* ≫ 24) & #ff;   /∗ extract the most significant bits ∗/
       q = *p⟶link*;
       p⟶link = *gb_sorted*[*k*];
       gb_sorted[*k*] = *p*;
     }
This code is used in section 5.

---

alt_sorted: **static node** ∗[ ], §4.
gb_next_rand = macro ( ),
   GB_FLIP §6.
gb_sorted: **node** ∗[ ], §4.

k: **register long**, §5.
key: **long**, §2.
l: **node** ∗, §5.
link: **struct node_struct** ∗, §2.

p: **register node** ∗, §5.
pp: **register node** ∗∗, §5.
q: **register node** ∗, §5.

GB_WORDS

Important: Before reading GB_WORDS, please read or at least skim the programs for GB_GRAPH and GB_IO.

1. Introduction. This GraphBase module provides two external subroutines:

> *words*, a routine that creates a graph based on five-letter words;
> *find_word*, a routine that looks for a given vertex in such a graph.

Examples of the use of these routines can be found in two demo programs, WORD_COMPONENTS and LADDERS.

⟨ gb_words.h 1 ⟩ ≡
  extern Graph *words( );
  extern Vertex *find_word( );

See also section 26.

2. The subroutine call $words(n, wt\_vector, wt\_threshold, seed)$ constructs a graph based on the five-letter words in `words.dat`. Each vertex of the graph corresponds to a single five-letter word. Two words are adjacent in the graph if they are the same except in one letter position. For example, 'words' is adjacent to other words such as 'cords', 'wards', 'woods', 'worms', and 'wordy'.

The constructed graph has at most n vertices; indeed, it has exactly n vertices if there are enough qualifying words. A word qualifies if its "weight" is $wt\_threshold$ or more, when weights are computed from a table pointed to by wt_vector according to rules described below. (If parameter wt_vector is Λ, i.e., NULL, default weights are used.) The fourth parameter, $seed$, is the seed of a random number generator.

All words of `words.dat` will be sorted by weight. The first vertex of the graph will be the word of largest weight, the second vertex will have second-largest weight, and so on. Words of equal weight will appear in pseudo-random order, as determined by the value of $seed$ in a system-independent fashion. The first n words in order of decreasing weight are chosen to be vertices of the graph. However, if fewer than n words have weight $\geq wt\_threshold$, the graph will contain only the words that qualify. In such cases the graph will have fewer than n vertices—possibly none at all.

Exception: The special case $n = 0$ is equivalent to the case when n has been set to the highest possible value. It causes all qualifying words to appear.

3. Every word in `words.dat` has been classified as 'common' (*), 'advanced' (+), or 'unusual' (␣). Each word has also been assigned seven frequency counts $c_1, \ldots, c_7$, separated by commas; these counts show how often the word has occurred in different publication contexts:

> c_1 times in the American Heritage Intermediate Corpus of elementary school material;
> c_2 times in the Brown Corpus of reading material from America;
> c_3 times in the Lancaster-Oslo/Bergen Corpus of reading material from Britain;
> c_4 times in the Melbourne-Surrey Corpus of newspaper material from Australia;
> c_5 times in the Revised Standard Version of the Bible;
> c_6 times in *The TEXbook* and *The METAFONTbook* by D. E. Knuth;
> c_7 times in *Concrete Mathematics* by Graham, Knuth, and Patashnik.

For example, one of the entries in `words.dat` is

$$\texttt{happy*774,92,121,2,26,8,1}$$

indicating a common word with $c_1 = 774, \ldots, c_7 = 1$.

Parameter *wt_vector* points to an array of nine integers $(a, b, w_1, \ldots, w_7)$. The weight of each word is computed from these nine numbers by using the formula

$$c_1 w_1 + \cdots + c_7 w_7 + \begin{cases} a, & \text{if the word is 'common';} \\ b, & \text{if the word is 'advanced';} \\ 0, & \text{if the word is 'unusual'.} \end{cases}$$

The components of *wt_vector* must be chosen so that

$$\max(|a|, |b|) + C_1|w_1| + \cdots + C_7|w_7| < 2^{30},$$

where C_j is the maximum value of c_j in the file; this restriction ensures that the *words* procedure will produce the same results on all computer systems.

4. The maximum frequency counts actually present are $C_1 = 15194$, $C_2 = 3560$, $C_3 = 4467$, $C_4 = 460$, $C_5 = 6976$, $C_6 = 756$, and $C_7 = 362$; these can be found in the entries for the common words 'shall', 'there', 'which', and 'would'.

The default weights are $a = 100$, $b = 10$, $c_1 = 4$, $c_2 = c_3 = 2$, $c_4 = c_5 = c_6 = c_7 = 1$.

File `words.dat` contains 5757 words, of which 3300 are 'common', 1194 are 'advanced', and 1263 are 'unusual'. Included among the unusual words are 891 having $c_1 = \cdots = c_7 = 0$; such words will always have weight zero, regardless of the weight vector parameter.

⟨ Private variables 4 ⟩ ≡
    static long *max_c*[ ] = {15194, 3560, 4467, 460, 6976, 756, 362};   /* maximum counts C_j */
    static long *default_wt_vector*[ ] = {100, 10, 4, 2, 2, 1, 1, 1, 1};   /* use this if *wt_vector* = Λ */
See also sections 17 and 25.

This code is used in section 7.

---

find_word: **Vertex** *( ), §30.   *seed*: **long**, §7.   *wt_threshold*: **long**, §7.
Graph = **struct**, GB_GRAPH §20.   **Vertex** = **struct**, GB_GRAPH §9.   *wt_vector*: **long** [ ], §7.
n: **unsigned long**, §7.   *words*: **Graph** *( ), §7.

5. Examples: If you call $words(2000, \Lambda, 0, 0)$, you get a graph with 2000 of the most common five-letter words of English, using the default weights. The GraphBase programs are designed to be system-independent, so that identical graphs will be obtained by everybody who asks for $words(2000, \Lambda, 0, 0)$. Equivalent experiments on algorithms for graph manipulation can therefore be performed by researchers in different parts of the world.

The subroutine call $words(2000, \Lambda, 0, s)$ will produce slightly different graphs when the random seed s varies, because some words have equal weight. However, the graph for any particular value of s will be the same on all computers. The seed value can be any integer in the range $0 \le s < 2^{31}$.

Suppose you call $words(0, w, 1, 0)$, with w defined by the C declaration

$$\textbf{long } w[9] = \{1\};$$

this means that $a = 1$ and $b = w_1 = \cdots = w_7 = 0$. Therefore you'll get a graph containing only the 3300 'common' words. Similarly, it's possible to obtain only the $3300 + 1194 = 4494$ non-'unusual' words, by specifying the weight vector

$$\textbf{long } w[9] = \{1, 1\};$$

this makes $a = b = 1$ and $w_1 = \cdots = w_7 = 0$. In both of these examples, the qualifying words all have weight 1, so the vertices of the graph will appear in pseudo-random order.

If w points to an array of nine 0's, the call $words(n, w, 0, s)$ gives a random sample of n words, depending on s in a system-independent fashion.

If the entries of the weight vector are all nonnegative, and if the weight threshold is zero, every word of `words.dat` will qualify. Thus you will obtain a graph with $\min(n, 5757)$ vertices.

If w points to an array with *negative* weights, the call $words(n, w, -{}^{\#}\texttt{7fffffff}, 0)$ selects n of the *least* common words in `words.dat`.

6. If the *words* routine encounters a problem, it returns Λ, after putting a code number into the external variable *panic_code*. This code number identifies the type of failure. Otherwise *words* returns a pointer to the newly created graph, which will be represented with the data structures explained in GB_GRAPH. (The external variable *panic_code* is itself defined in GB_GRAPH.)

#define $panic(c)$ { $gb\_free(node\_blocks)$;
$\qquad\qquad\qquad panic\_code = c$; $gb\_trouble\_code = 0$; **return** Λ; }

7. Now let's get going on the program. The C file `gb_words.c` begins as follows:

```
#include "gb_io.h"       /* we will use the GB_IO routines for input */
#include "gb_flip.h"       /* we will use the GB_FLIP routines for random numbers */
#include "gb_graph.h"       /* we will use the GB_GRAPH data structures */
#include "gb_sort.h"       /* and gb_linksort for sorting */
```
⟨ Preprocessor definitions ⟩

⟨ Type declarations 15 ⟩
⟨ Private variables 4 ⟩
⟨ Private functions 10 ⟩

Graph *$*words(n, wt\_vector, wt\_threshold, seed)$*
    unsigned long n;   /* maximum number of vertices desired */
    long $wt\_vector[\,]$;   /* pointer to array of weights */
    long $wt\_threshold$;   /* minimum qualifying weight */
    long $seed$;   /* random number seed */
{ ⟨ Local variables 8 ⟩
  $gb\_init\_rand(seed)$;
  ⟨ Check that wt_vector is valid 9 ⟩;
  ⟨ Input the qualifying words to a linked list, computing their weights 18 ⟩;
  ⟨ Sort and output the words, determining adjacencies 22 ⟩;
  if $(gb\_trouble\_code)$ {
    $gb\_recycle(new\_graph)$;
    $panic(alloc\_fault)$;   /* oops, we ran out of memory somewhere back there */
  }
  return new_graph;
}

8. ⟨ Local variables 8 ⟩ ≡
  Graph *new_graph;   /* the graph constructed by *words* */
See also sections 14, 16, and 24.
This code is used in section 7.

$alloc\_fault = -1$, GB_GRAPH §7.
gb_free: **void** ( ), GB_GRAPH §16.
gb_init_rand: **void** ( ), GB_FLIP §8.
$gb\_linksort$: **void** ( ), GB_SORT §5.

$gb\_recycle$: **void** ( ), GB_GRAPH §40.
$gb\_trouble\_code$: **long**, GB_GRAPH §14.

Graph = **struct**, GB_GRAPH §20.
$node\_blocks$: **Area**, §17.
$panic\_code$: **long**, GB_GRAPH §5.

9. Validating the weights. The first job that *words* needs to tackle is comparatively trivial: We want to verify the condition

$$\max(|a|, |b|) + C_1|w_1| + \cdots + C_7|w_7| < 2^{30}. \qquad (*)$$

This proves to be an interesting exercise in "portable C programming," because we don't want to risk integer overflow. Our approach is to do the calculation first in floating point arithmetic, thereby ruling out cases that are clearly unacceptable. Once that test is passed, we can safely test the condition with ordinary integer arithmetic. Floating point arithmetic is system dependent, but we use it carefully so that system-independent results are obtained.

⟨ Check that *wt_vector* is valid 9 ⟩ ≡
 if (¬*wt_vector*) *wt_vector* = *default_wt_vector*;
 else { **register double** *flacc*;
  register long *∗p*, *∗q*;
  register long *acc*;
  ⟨ Use floating point arithmetic to check that *wt_vector* isn't totally off base 11 ⟩;
  ⟨ Use integer arithmetic to check that *wt_vector* is truly OK 12 ⟩;
 }

This code is used in section 7.

10. The floating-point calculations are facilitated by a routine that converts an integer to its absolute value, expressed as a **double**:

⟨ Private functions 10 ⟩ ≡
 static double *flabs*(*x*)
   long *x*;
 { **if** (*x* ≥ 0) **return** (**double**) *x*;
  return −((**double**) *x*);
 }

See also section 13.

This code is used in section 7.

11. Although floating point arithmetic is system dependent, we can certainly assume that at least 16 bits of precision are used. This implies that the difference between *flabs*(*x*) and |*x*| must be less than 2^{14}. Also, if *x* and *y* are nonnegative values less than 2^{31}, the difference between their floating-point sum and their true sum must be less than 2^{14}.

The floating point calculations in the following test will never reject a valid weight vector. For if condition (*) holds, the floating-point value of max(*flabs*(*a*), *flabs*(*b*)) + C_1 ∗ *flabs*(*w_1*) + ⋯ + C_7 ∗ *flabs*(*w_7*) will be less than $2^{30} + (8 + C_1 + \cdots + C_7)2^{14}$, which is less than $2^{30} + 2^{29}$.

⟨ Use floating point arithmetic to check that *wt_vector* isn't totally off base 11 ⟩ ≡
 p = *wt_vector*;
 flacc = *flabs*(*∗p*++);
 if (*flacc* < *flabs*(*∗p*)) *flacc* = *flabs*(*∗p*); /∗ now *flacc* = max(|*a*|, |*b*|) ∗/
 for (*q* = &*max_c*[0]; *q* < &*max_c*[7]; *q*++) *flacc* += *∗q* ∗ *flabs*(*∗*++*p*);
 if (*flacc* ≥ (**double**) #60000000) /∗ this constant is 6 × 2^{28} = $2^{30} + 2^{29}$ ∗/
  panic(*very_bad_specs*); /∗ whoa; the weight vector is way too big ∗/

This code is used in section 9.

12. Conversely, if the floating point test just made is passed, the true value of the sum will be less than $2^{30} + 2^{29} + 2^{29} = 2^{31}$; hence integer overflow will never occur when we make the following more refined test:

⟨ Use integer arithmetic to check that *wt_vector* is truly OK 12 ⟩ ≡
```
  p = wt_vector;
  acc = iabs(*p++);
  if (acc < iabs(*p))  acc = iabs(*p);       /* now acc = max(|a|, |b|) */
  for (q = &max_c[0]; q < &max_c[7]; q++)  acc += *q * iabs(*++p);
  if (acc ≥ #40000000)  panic(bad_specs);        /* the weight vector is a bit too big */
```
This code is used in section 9.

13. ⟨ Private functions 10 ⟩ +≡
```
  static long iabs(x)
        long x;
  { if (x ≥ 0) return (long) x;
    return −((long) x);
  }
```

bad_specs = 30, GB_GRAPH §7.   *max_c*: **static long** [], §4.   *words*: **Graph** *( ), §7.
default_wt_vector: **static long** [],   *panic* = macro ( ), §6.   *wt_vector*: **long** [], §7.
   §4.   *very_bad_specs* = 40, GB_GRAPH §7.

14. The input phase. Now we're ready to read `words.dat`.

⟨ Local variables 8 ⟩ +≡
    register long *wt*;   /∗ the weight of the current word ∗/
    char *word*[5];   /∗ the current five-letter word ∗/
    long *nn* = 0;   /∗ the number of qualifying words found so far ∗/

15. As we read the words, we will form a linked list of nodes containing each qualifying word and its weight, using the memory management routines of GB_GRAPH to allocate space for 111 nodes at a time. These nodes should be returned to available memory later, so we will keep them in a separate area under local control.

The nodes start out with *key* and *link* fields, as required by the *gb_linksort* routine, which we'll use to sort by weight. The sort key must be nonnegative; we obtain it by adding 2^{30} to the weight.

#define *nodes_per_block*   111
⟨ Type declarations 15 ⟩ ≡
    typedef struct node_struct {
        long *key*;   /∗ the sort key (weight plus 2^{30}) ∗/
        struct node_struct ∗*link*;   /∗ links the nodes together ∗/
        char *wd*[5];   /∗ five-letter word (which typically consumes eight bytes, too bad) ∗/
    } **node**;
See also section 23.
This code is used in section 7.

16. ⟨ Local variables 8 ⟩ +≡
    node ∗*next_node*;   /∗ the next node available for allocation ∗/
    node ∗*bad_node*;   /∗ if *next_node* = *bad_node*, the node isn't really there ∗/
    node ∗*stack_ptr*;   /∗ the most recently created node ∗/
    node ∗*cur_node*;   /∗ current node being created or examined ∗/

17. ⟨ Private variables 4 ⟩ +≡
    Area *node_blocks*;   /∗ the memory area for blocks of nodes ∗/

18. ⟨ Input the qualifying words to a linked list, computing their weights 18 ⟩ ≡
    next_node = *bad_node* = *stack_ptr* = Λ;
    if (*gb_open*("words.dat") ≠ 0) *panic*(*early_data_fault*);
          /∗ couldn't open "words.dat" using GraphBase conventions; *io_errors* tells why ∗/
    do ⟨ Read one word, and put it on the stack if it qualifies 19 ⟩   **while** (¬*gb_eof*( ));
    if (*gb_close*( ) ≠ 0) *panic*(*late_data_fault*);
          /∗ something's wrong with "words.dat"; see *io_errors* ∗/
This code is used in section 7.

19. ⟨ Read one word, and put it on the stack if it qualifies 19 ⟩ ≡
    { **register long** *j*;   /∗ position in *word* ∗/
        for (*j* = 0; *j* < 5; *j*++) *word*[*j*] = *gb_char*( );
        ⟨ Compute the weight *wt* 21 ⟩;
        if (*wt* ≥ *wt_threshold*) {   /∗ it qualifies ∗/
        ⟨ Install *word* and *wt* in a new node 20 ⟩;
        nn ++;

```
    }
  gb_newline( );
}
```

This code is used in section 18.

20.   **#define** *copy5*(*y, x*)
            { *(*y*) = *(*x*); *((*y*) + 1) = *((*x*) + 1); *((*y*) + 2) = *((*x*) + 2);
            *((*y*) + 3) = *((*x*) + 3); *((*y*) + 4) = *((*x*) + 4); }

⟨ Install *word* and *wt* in a new node 20 ⟩ ≡
  if (*next_node* ≡ *bad_node*) {
    cur_node = *gb_typed_alloc*(*nodes_per_block*, **node**, *node_blocks*);
    if (*cur_node* ≡ Λ) *panic*(*no_room* + 1);   /* out of memory already */
    next_node = *cur_node* + 1; *bad_node* = *cur_node* + *nodes_per_block*;
  } **else** *cur_node* = *next_node*++;
  cur_node→*key* = *wt* + #40000000; *cur_node*→*link* = *stack_ptr*; *copy5*(*cur_node*→*wd*, *word*);
  stack_ptr = *cur_node*;

This code is used in section 19.

21.   Recall that *gb_number*( ) returns 0, without giving an error, if no digit is present in the current position of the file being read. This implies that the `words.dat` file need not include zero counts explicitly. Furthermore, we can arrange things so that trailing zero counts are unnecessary; commas can be omitted if all counts following them on the current line are zero.

⟨ Compute the weight *wt* 21 ⟩ ≡
  { **register long** *p, *q;   /* pointers to C_j and w_j */
    register long *c*;   /* current count */

    switch (*gb_char*( )) {
    case '*': *wt* = *wt_vector*[0]; **break**;   /* 'common' word */
    case '+': *wt* = *wt_vector*[1]; **break**;   /* 'advanced' word */
    case '␣': **case** '\n': *wt* = 0; **break**;   /* 'unusual' word */
    default: *panic*(*syntax_error*);   /* unknown type of word */
    }
    p = &*max_c*[0]; *q* = &*wt_vector*[2];
    do { **if** (*p* ≡ &*max_c*[7]) *panic*(*syntax_error* + 1);   /* too many counts */
      c = *gb_number*(10);
      if (*c* > **p*++) *panic*(*syntax_error* + 2);   /* count too large */
      wt += *c* * *q*++;
    } **while** (*gb_char*( ) ≡ ',');
  }

This code is used in section 19.

---

22. The output phase. Once the input phase has examined all of `words.dat`, we are left with a stack of nn nodes containing the qualifying words, starting at *stack_ptr*.

The next step is to call *gb_linksort*, which takes the qualifying words and distributes them into the 128 lists *gb_sorted*[j], for $0 \le j < 128$. We can then access the words in order of decreasing weight by reading through these lists, starting with *gb_sorted*[127] and ending with *gb_sorted*[0]. (See the documentation of *gb_linksort* in the GB_SORT module.)

The output phase therefore has the following general outline:

⟨ Sort and output the words, determining adjacencies 22 ⟩ ≡
  gb_linksort(*stack_ptr*);
  ⟨ Allocate storage for the new graph; adjust n if it is zero or too large 27 ⟩;
  if (*gb_trouble_code* ≡ 0 ∧ n) {
    register long j;   /* runs through sorted lists */
    register node *p;   /* the current node being output */

    $nn = n$;
    for ($j = 127$; $j \ge 0$; j−−)
      for ($p = $ (**node** *) *gb_sorted*[j]; p; $p = p{\rightarrow}link$) {
        ⟨ Add the word $p{\rightarrow}wd$ to the graph 28 ⟩;
        if (−−nn ≡ 0) **goto** *done*;
      }
  }
done: *gb_free*(*node_blocks*);

This code is used in section 7.

23. The only slightly unusual data structure needed is a set of five hash tables, one for each of the strings of four letters obtained by suppressing a single letter of a five-letter word. For example, a word like 'words' will lead to entries for '␣ords', 'w␣rds, 'wo␣ds', 'wor␣s', and 'word␣', one in each of the hash tables.

#define *hash_prime* 6997   /* a prime number larger than the total number of words */
⟨ Type declarations 15 ⟩ +≡
  typedef Vertex *****hash_table**[*hash_prime*];

24. ⟨ Local variables 8 ⟩ +≡
  Vertex *****cur_vertex*;   /* the current vertex being created or examined */
  char *****next_string*;   /* where we'll store the next five-letter word */

25. ⟨ Private variables 4 ⟩ +≡
  static hash_table *****htab*;   /* five dynamically allocated hash tables */

26. The weight of each word will be stored in the utility field *u.I* of its **Vertex** record. The position in which adjacent words differ will be stored in utility field *a.I* of the **Arc** records between them.

#define *weight* *u.I*   /* weighted frequencies */
#define *loc*   *a.I*   /* index of difference (0, 1, 2, 3, or 4) */

⟨ gb_words.h 1 ⟩ +≡
  #define *weight* *u.I*   /* repeat the definitions in the header file */
  #define *loc*   *a.I*

27. ⟨ Allocate storage for the new graph; adjust n if it is zero or too large 27 ⟩ ≡
　　if $(n \equiv 0 \lor nn < n)$ $n = nn$;
　　new_graph $= gb\_new\_graph(n)$;
　　if $(new\_graph \equiv \Lambda)$ *panic*(*no_room*);　　　/* out of memory before we're even started */
　　if $(wt\_vector \equiv default\_wt\_vector)$
　　　　sprintf(*new_graph*→*id*, "words(%lu,0,%ld,%ld)", n, *wt_threshold*, *seed*);
　　else *sprintf*(*new_graph*→*id*, "words(%lu,{%ld,%ld,%ld,%ld,%ld,%ld,%ld,%ld,%ld},%ld,%ld)", n,
　　　　　　wt_vector[0], *wt_vector*[1], *wt_vector*[2], *wt_vector*[3], *wt_vector*[4], *wt_vector*[5], *wt_vector*[6],
　　　　　　wt_vector[7], *wt_vector*[8], *wt_threshold*, *seed*);
　　strcpy(*new_graph*→*util_types*, "IZZZZZIZZZZZZZ");
　　cur_vertex = *new_graph*→*vertices*;
　　next_string = *gb_typed_alloc*(6 * n, **char**, *new_graph*→*data*);
　　htab = *gb_typed_alloc*(5, **hash_table**, *new_graph*→*aux_data*);
This code is used in section 22.

28. ⟨ Add the word p→*wd* to the graph 28 ⟩ ≡
　　{ **register char** *q;　　/* the new word */

　　q = *cur_vertex*→*name* = *next_string*;
　　next_string += 6;
　　copy5(q, p→*wd*);
　　cur_vertex→*weight* = p→*key* − #40000000;
　　⟨ Add edges for all previous words r that nearly match q 29 ⟩;
　　cur_vertex ++;
　　}
This code is used in section 22.

---

a: **util**, GB_GRAPH §10.
Arc = **struct**, GB_GRAPH §10.
aux_data: **Area**, GB_GRAPH §20.
copy5 = macro ( ), §20.
data: **Area**, GB_GRAPH §20.
default_wt_vector: **static long** [],
　§4.
gb_free: **void** ( ), GB_GRAPH §16.
gb_linksort: **void** ( ), GB_SORT §5.
gb_new_graph: **Graph** *( ),
　GB_GRAPH §23.
gb_sorted: **char** *[], GB_SORT §4.
gb_trouble_code: **long**,
　GB_GRAPH §14.

gb_typed_alloc = macro ( ),
　GB_GRAPH §11.
l: **long**, GB_GRAPH §8.
id: **char** [], GB_GRAPH §20.
key: **long**, §15.
link: **struct node_struct** *, §15.
n: **unsigned long**, §7.
name: **char** *, GB_GRAPH §9.
new_graph: **Graph** *, §8.
nn: **long**, §14.
no_room = 1, GB_GRAPH §7.
node = **struct node_struct**, §15.
node_blocks: **Area**, §17.

panic = macro ( ), §6.
r: **register char** *, §30.
seed: **long**, §7.
sprintf: **int** ( ), <stdio.h>.
stack_ptr: **node** *, §16.
strcpy: **char** *( ), <string.h>.
u: **util**, GB_GRAPH §9.
util_types: **char** [], GB_GRAPH §20.
Vertex = **struct**, GB_GRAPH §9.
vertices: **Vertex** *, GB_GRAPH §20.
wd: **char** [], §15.
wt_threshold: **long**, §7.
wt_vector: **long** [], §7.

29. The length of each edge in a *words* graph is set to 1; the calling routine can change it later if desired.

#define *mtch*(*i*) $(*(q + i) \equiv *(r + i))$
#define *match*(*a*, *b*, *c*, *d*) $(mtch(a) \wedge mtch(b) \wedge mtch(c) \wedge mtch(d))$
#define *store_loc_of_diff*(*k*) *cur_vertex*→*arcs*→*loc* = (*cur_vertex*→*arcs* − 1)→*loc* = *k*
#define *ch*(*q*) $((\textbf{long}) *(q))$
#define *hdown*(*k*) $h \equiv htab[k] ? h = htab[k + 1] - 1 : h--$

⟨ Add edges for all previous words *r* that nearly match *q* 29 ⟩ ≡
  { **register char** **r*;   /* previous word possibly adjacent to *q* */
    register Vertex ***h*;   /* hash address for linear probing */
    register long *raw_hash*;   /* five-letter hash code before remaindering */

    $raw\_hash = ((((((ch(q) \ll 5) + ch(q + 1)) \ll 5) + ch(q + 2)) \ll 5) + ch(q + 3)) \ll 5) + ch(q + 4);$
    for $(h = htab[0] + (raw\_hash - (ch(q) \ll 20)) \% hash\_prime; *h; hdown(0))$ {
      $r = (*h)$→*name*;
      if $(match(1, 2, 3, 4))$ *gb_new_edge*(*cur_vertex*, **h*, 1 ₗ), *store_loc_of_diff*(0);
    }
    *h = *cur_vertex*;
    for $(h = htab[1] + (raw\_hash - (ch(q + 1) \ll 15)) \% hash\_prime; *h; hdown(1))$ {
      $r = (*h)$→*name*;
      if $(match(0, 2, 3, 4))$ *gb_new_edge*(*cur_vertex*, **h*, 1 ₗ), *store_loc_of_diff*(1);
    }
    *h = *cur_vertex*;
    for $(h = htab[2] + (raw\_hash - (ch(q + 2) \ll 10)) \% hash\_prime; *h; hdown(2))$ {
      $r = (*h)$→*name*;
      if $(match(0, 1, 3, 4))$ *gb_new_edge*(*cur_vertex*, **h*, 1 ₗ), *store_loc_of_diff*(2);
    }
    *h = *cur_vertex*;
    for $(h = htab[3] + (raw\_hash - (ch(q + 3) \ll 5)) \% hash\_prime; *h; hdown(3))$ {
      $r = (*h)$→*name*;
      if $(match(0, 1, 2, 4))$ *gb_new_edge*(*cur_vertex*, **h*, 1 ₗ), *store_loc_of_diff*(3);
    }
    *h = *cur_vertex*;
    for $(h = htab[4] + (raw\_hash - ch(q + 4)) \% hash\_prime; *h; hdown(4))$ {
      $r = (*h)$→*name*;
      if $(match(0, 1, 2, 3))$ *gb_new_edge*(*cur_vertex*, **h*, 1 ₗ), *store_loc_of_diff*(4);
    }
    *h = *cur_vertex*;
  }

This code is used in section 28.

30. Finding a word. After *words* has created a graph *g*, the user can remove the hash tables by calling *gb_free*(*g*→*aux_data*). But if the hash tables have not been removed, another procedure can be used to find vertices that match or nearly match a given word.

The subroutine call *find_word*(*q*, *f*) will return a pointer to a vertex that matches a given five-letter word *q*, if that word is in the graph; otherwise, it returns Λ (i.e., NULL), after calling *f*(*v*) for each vertex *v* whose word matches *q* in all but one letter position.

```
Vertex *find_word(q, f)
    char *q;
    void (*f)();        /* *f should take one argument, of type Vertex *, or f should be Λ */
{ register char *r;      /* previous word possibly adjacent to q */
  register Vertex **h;       /* hash address for linear probing */
  register long raw_hash;        /* five-letter hash code before remaindering */
  raw_hash = (((((((ch(q) ≪ 5) + ch(q + 1)) ≪ 5) + ch(q + 2)) ≪ 5) + ch(q + 3)) ≪ 5) + ch(q + 4);
  for (h = htab[0] + (raw_hash − (ch(q) ≪ 20)) % hash_prime; *h; hdown(0)) {
    r = (*h)→name;
    if (mtch(0) ∧ match(1, 2, 3, 4)) return *h;
  }
  ⟨Invoke f on every vertex that is adjacent to word q 31⟩;
  return Λ;
}
```

31. ⟨Invoke *f* on every vertex that is adjacent to word *q* 31⟩ ≡

```
if (f) {
  for (h = htab[0] + (raw_hash − (ch(q) ≪ 20)) % hash_prime; *h; hdown(0)) {
    r = (*h)→name; if (match(1, 2, 3, 4)) (*f)(*h);
  }
  for (h = htab[1] + (raw_hash − (ch(q + 1) ≪ 15)) % hash_prime; *h; hdown(1)) {
    r = (*h)→name; if (match(0, 2, 3, 4)) (*f)(*h);
  }
  for (h = htab[2] + (raw_hash − (ch(q + 2) ≪ 10)) % hash_prime; *h; hdown(2)) {
    r = (*h)→name; if (match(0, 1, 3, 4)) (*f)(*h);
  }
  for (h = htab[3] + (raw_hash − (ch(q + 3) ≪ 5)) % hash_prime; *h; hdown(3)) {
    r = (*h)→name; if (match(0, 1, 2, 4)) (*f)(*h);
  }
  for (h = htab[4] + (raw_hash − ch(q + 4)) % hash_prime; *h; hdown(4)) {
    r = (*h)→name; if (match(0, 1, 2, 3)) (*f)(*h);
  }
}
```

This code is used in section 30.

---

arcs: **Arc** *, GB_GRAPH §9.
aux_data: **Area**, GB_GRAPH §20.
cur_vertex: **Vertex** *, §24.
gb_free: **void** ( ), GB_GRAPH §16.
gb_new_edge: **void** ( ),

GB_GRAPH §31.
hash_prime = 6997, §23.
htab: **static hash_table** *, §25.
loc: = *a*.I, §26.

name: **char** *, GB_GRAPH §9.
q: **register char** *, §28.
Vertex = **struct**, GB_GRAPH §9.
words: **Graph** *( ), §7.

GIRTH _____

Important: Before reading GIRTH, please read or at least skim the program for GB_RAMAN.

1. Introduction. This demonstration program uses graphs constructed by the *raman* procedure in the GB_RAMAN module to produce an interactive program called `girth`, which computes the girth and diameter of a class of Ramanujan graphs.

The girth of a graph is the length of its shortest cycle; the diameter is the maximum length of a shortest path between two vertices. A Ramanujan graph is a connected, undirected graph in which every vertex has degree $p + 1$, with the property that every eigenvalue of its adjacency matrix is either $\pm(p + 1)$ or has absolute value $\leq 2\sqrt{p}$.

Exact values for the girth are of interest because the bipartite graphs produced by *raman* apparently have larger girth than any other known family of regular graphs, even if we consider graphs whose existence is known only by nonconstructive methods, except for the cubic "sextet" graphs of Biggs, Hoare, and Weiss [*Combinatorica* **3** (1983), 153–165; **4** (1984), 241–245].

Exact values for the diameter are of interest because the diameter of any Ramanujan graph is at most twice the minimum possible diameter of any regular graph.

The program will prompt you for two numbers, p and q. These should be distinct prime numbers, not too large, with $q > 2$. A graph is constructed in which each vertex has degree $p+1$. The number of vertices is $(q^3 - q)/2$ if p is a quadratic residue modulo q, or $q^3 - q$ if p is not a quadratic residue. In the latter case, the graph is bipartite and it is known to have rather large girth.

If $p = 2$, the value of q is further restricted to be of the form $104k + (1, 3, 9, 17, 25, 27, 35, 43, 49, 51, 75, 81)$. This means that the only feasible values of q to go with $p = 2$ are probably 3, 17, and 43; the next case, $q = 107$, would generate a bipartite graph with 1,224,936 vertices and 3,675,808 arcs, thus requiring approximately 113 megabytes of memory (not to mention a nontrivial amount of computer time). If you want to compute the girth and diameter of Ramanujan graphs for large p and/or q, much better methods are available based on number theory; the present program is merely a demonstration of how to interface with the output of *raman*. Incidentally, the graph for $p = 2$ and $q = 43$ turns out to have 79464 vertices, girth 20, and diameter 22.

The program will examine the graph and compute its girth and its diameter, then will prompt you for another choice of p and q.

2. Here is the general layout of this program, as seen by the C compiler:

```
#include "gb_graph.h"    /* the standard GraphBase data structures */
#include "gb_raman.h"    /* Ramanujan graph generator */
⟨Preprocessor definitions⟩
⟨Global variables 3⟩
main()
{
    printf("This program explores the girth and diameter of Ramanujan graphs.\n");
    printf("The bipartite graphs have q^3-q vertices, and the non-bipartite\n");
    printf("graphs have half that number. Each vertex has degree p+1.\n");
    printf("Both p and q should be odd prime numbers;\n");
    printf("  or you can try p = 2 with q = 17 or 43.\n");
```

```
      while (1) {
         ⟨Prompt the user for p and q; break if unsuccessful 4⟩;
         g = raman(p, q, 0_L, 0_L);
         if (g ≡ Λ) ⟨Explain that the graph could not be constructed 5⟩
         else {
            ⟨Print the theoretical bounds on girth and diameter of g 10⟩;
            ⟨Compute and print the true girth and diameter of g 12⟩;
            gb_recycle(g);
         }
      }
      return 0;      /∗ normal exit ∗/
   }
```

3.   ⟨Global variables 3⟩ ≡
   Graph *∗g;*   /∗ the current Ramanujan graph ∗/
   long *p;*   /∗ the branching factor (degree minus one) ∗/
   long *q;*   /∗ cube root of the graph size ∗/
   char *buffer*[16];   /∗ place to collect what the user types ∗/

See also section 11.

This code is used in section 2.

4.   **#define** *prompt*(s)
             { *printf*(s); *fflush*(stdout);   /∗ make sure the user sees the prompt ∗/
               if (*fgets*(buffer, 15, stdin) ≡ Λ) **break**; }
⟨Prompt the user for p and q; **break** if unsuccessful 4⟩ ≡
   prompt("\nChoose␣a␣branching␣factor,␣p:␣");
   if (*sscanf*(buffer, "%ld", &p) ≠ 1) **break**;
   prompt("OK,␣now␣choose␣the␣cube␣root␣of␣graph␣size,␣q:␣");
   if (*sscanf*(buffer, "%ld", &q) ≠ 1) **break**;

This code is used in section 2.

5.   ⟨Explain that the graph could not be constructed 5⟩ ≡
   printf("␣Sorry,␣I␣couldn't␣make␣that␣graph␣(%s).\n",
         panic_code ≡ *very_bad_specs* ? "q␣is␣out␣of␣range" :
         panic_code ≡ *very_bad_specs* + 1 ? "p␣is␣out␣of␣range" :
         panic_code ≡ *bad_specs* + 5 ? "q␣is␣too␣big" :
         panic_code ≡ *bad_specs* + 6 ? "p␣is␣too␣big" :
         panic_code ≡ *bad_specs* + 1 ? "q␣isn't␣prime" :
         panic_code ≡ *bad_specs* + 7 ? "p␣isn't␣prime" :
         panic_code ≡ *bad_specs* + 3 ? "p␣is␣a␣multiple␣of␣q" :
         panic_code ≡ *bad_specs* + 2 ? "q␣isn't␣compatible␣with␣p=2" :
         "not␣enough␣memory");

This code is used in section 2.

---

bad_specs = 30, GB_GRAPH §7.
fflush: **int** ( ), <stdio.h>.
fgets: **char** ∗( ), <stdio.h>.
gb_recycle: **void** ( ), GB_GRAPH §40.

Graph = **struct**, GB_GRAPH §20.
panic_code: **long**, GB_GRAPH §5.
printf: **int** ( ), <stdio.h>.
raman: **Graph** ∗( ), GB_RAMAN §4.

sscanf: **int** ( ), <stdio.h>.
stdin: **FILE** ∗, <stdio.h>.
stdout: **FILE** ∗, <stdio.h>.
very_bad_specs = 40, GB_GRAPH §7.

6. Bounds. The theory of Ramanujan graphs allows us to predict the girth and diameter to within a factor of 2 or so.

In the first place, we can easily derive an upper bound on the girth and a lower bound on the diameter, valid for any regular graph of degree $p+1$. Such a graph has at most $(p+1)p^{k-1}$ points at distance k from any given vertex; this implies a lower bound on the diameter d:

$$1 + (p+1) + (p+1)p + (p+1)p^2 + \cdots + (p+1)p^{d-1} \geq n.$$

Similarly, if the girth g is odd, say $g = 2k + 1$, the points at distance $\leq k$ from any vertex must be distinct, so we have

$$1 + (p+1) + (p+1)p + (p+1)p^2 + \cdots + (p+1)p^{k-1} \leq n;$$

and if $g = 2k + 2$, at least p^k further points must exist at distance $k + 1$, because the $(p+1)p^k$ paths of length $k+1$ can end at a particular vertex at most $p+1$ times. Thus

$$1 + (p+1) + (p+1)p + (p+1)p^2 + \cdots + (p+1)p^{k-1} + p^k \leq n$$

when the girth is even.

In the following code we let $pp = p^{dl}$ and $s = 1 + (p+1) + \cdots + (p+1)p^{dl}$.

⟨ Compute the "trivial" bounds gu and dl on girth and diameter 6 ⟩ ≡

```
s = p + 2;  dl = 1;  pp = p;  gu = 3;
while (s < n) {
    s += pp;
    if (s ≤ n)  gu++;
    dl++;
    pp *= p;
    s += pp;
    if (s ≤ n)  gu++;
}
```

This code is used in section 10.

7. When $p > 2$, we can use the theory of integral quaternions to derive a lower bound on the girth of the graphs produced by *raman*. A path of length g from a vertex to itself exists if and only if there is an integral quaternion $\alpha = a_0 + a_1 i + a_2 j + a_3 k$ of norm p_g such that the a's are not all multiples of p, while a_1, a_2, and a_3 are multiples of q and $a_0 \not\equiv a_1 \equiv a_2 \equiv a_3$ (mod 2). This means we have integers (a_0, a_1, a_2, a_3) with

$$a_0^2 + a_1^2 + a_2^2 + a_3^2 = p^g,$$

satisfying the stated properties mod q and mod 2. If a_1, a_2, and a_3 are even, they cannot all be zero so we must have $p^g \geq 1 + 4q^2$; if they are odd, we must have $p^g \geq 4 + 3q^2$. (The latter is possible only when g is odd and p mod 4 = 3.) Since n is roughly proportional to q^3, this means g must be at least about $\frac{2}{3} \log_p n$. Thus g isn't too much less than the maximum girth possible in any regular graph, which we have shown is at most about $2 \log_p n$.

When the graph is bipartite we can, in fact, prove that g is approximately $\frac{4}{3}\log_p n$. The bipartite case occurs if and only if p is not a quadratic residue modulo q; hence the number g in the previous paragraph must be even, say $g = 2r$. Then $p^g \bmod 4 = 1$, and a_0 must be odd. The congruence $a_0^2 \equiv p^{2r} \pmod{q^2}$ implies that $a_0 \equiv \pm p^r$, because all numbers relatively prime to q^2 are powers of a primitive root. We can assume without loss of generality that $a_0 = p^r - 2mq^2$, where $0 < m < p^r/q^2$; it follows in particular that $p^r > q^2$. Conversely, if $p^r - q^2$ can be written as a sum of three squares $b_1^2 + b_2^2 + b_3^2$, then $p^{2r} = (p^r - 2q^2)^2 + (2b_1 q)^2 + (2b_2 q)^2 + (2b_3 q)^2$ is a representation of the required type. If $p^r - q^2$ is a positive integer that cannot be represented as a sum of three squares, a well-known theorem of Legendre tells us that $p^r - q^2 = 4^t s$, where $s \equiv 7 \pmod 8$. Since p and q are odd, we have $t \geq 1$; hence $p^r - 2q^2$ is odd. If $p^r - 2q^2$ is a positive odd integer, Legendre's theorem tells us that we can write $2p^r - 4q^2 = b_1^2 + b_2^2 + b_3^2$; hence $p^{2r} = (p^r - 4q^2)^2 + (2b_1 q)^2 + (2b_2 q)^2 + (2b_3 q)^2$. We conclude that the girth is either $2\lceil \log_p q^2 \rceil$ or $2\lceil \log_p 2q^2 \rceil$. (This explicit calculation, which makes our program for calculating the girth unnecessary or at best redundant in the bipartite case, is due to G. A. Margulis and, independently, to Biggs and Boshier [*Journal of Combinatorial Theory* **B49** (1990), 190–194].)

A girth of 1 or 2 can occur, since these graphs might have self-loops or multiple edges if p is sufficiently large.

⟨ Compute a lower bound *gl* on the girth 7 ⟩ ≡

```
if (bipartite) { long b = q * q;
    for (gl = 1, pp = p; pp ≤ b; gl++, pp *= p) ;       /* iterate until p^g > q^2 */
    gl += gl;
} else { long b1 = 1 + 4 * q * 2, b2 = 4 + 3 * q * q;       /* bounds on p^g */
    for (gl = 1, pp = p; pp < b1; gl++, pp *= p) {
        if (pp ≥ b2 ∧ (gl & 1) ∧ (p & 2)) break;
    }
}
```

This code is used in section 10.

---

8. Upper bounds on the diameter of any Ramanujan graph can be derived as shown in the paper by Lubotzky, Phillips, and Sarnak in *Combinatorica* **8** (1988), page 275. (However, a slight correction to their proof is necessary—their parameter l should be odd when x and y lie in different parts of a bipartite graph.) Their argument demonstrates that $p^{(d-1)/2} < 2n$ in the nonbipartite case and $p^{(d-2)/2} < n$ in the bipartite case; therefore we obtain the upper bound $d \leq 2\log_p n + O(1)$, which is about twice the lower bound that holds in an arbitrary regular graph.

⟨ Compute an upper bound du on the diameter 8 ⟩ ≡

```
{ long nn = (bipartite ? n : 2 * n);
    for (du = 0, pp = 1;  pp < nn;  du += 2, pp *= p) ;
    ⟨ Decrease du by 1, if pp/nn ≥ √p̄ 9 ⟩;
    if (bipartite) du++;
}
```

This code is used in section 10.

9. Floating point arithmetic might not be accurate enough for the test required in this section. We avoid it by using an all-integer method analogous to Euclid's algorithm, based on the continued fraction for $\sqrt{p}$ [*Seminumerical Algorithms*, exercise 4.5.3–12]. In the loop here we want to compare nn/pp to $(\sqrt{p} + a)/b$, where $\sqrt{p} + a > b > 0$ and $p - a^2$ is a multiple of b.

⟨ Decrease du by 1, if $pp/nn \geq \sqrt{p}$ 9 ⟩ ≡

```
{ long qq = pp/nn;
    if (qq * qq > p)  du --;
    else if ((qq + 1) * (qq + 1) > p) {     /* qq = ⌊√p̄⌋ */
        long aa = qq, bb = p - aa * aa, parity = 0;
        pp -= qq * nn;
        while (1) {
            long x = (aa + qq)/bb, y = nn - x * pp;
            if (y ≤ 0) break;
            aa = bb * x - aa;      /* now 0 < aa < √p̄ */
            bb = (p - aa * aa)/bb;
            nn = pp; pp = y;
            parity ⊕= 1;
        }
        if (¬parity) du --;
    }
}
```

This code is used in section 8.

10. ⟨ Print the theoretical bounds on girth and diameter of g 10 ⟩ ≡

```
n = g↠n;
if (n ≡ (q + 1) * q * (q − 1))  bipartite = 1;
else  bipartite = 0;
printf("The⎵graph⎵has⎵%ld⎵vertices,⎵each⎵of⎵degree⎵%ld,⎵and⎵it⎵is⎵%sbipartite.\n", n,
    p + 1, bipartite ? "" : "not⎵");
⟨ Compute the "trivial" bounds gu and dl on girth and diameter 6 ⟩;
printf("Any⎵such⎵graph⎵must⎵have⎵diameter⎵>=⎵%ld⎵and⎵girth⎵<=⎵%ld;\n", dl, gu);
```

⟨Compute an upper bound du on the diameter 8⟩;
$printf$("theoretical␣considerations␣tell␣us␣that␣this␣one's␣diameter␣is␣<=␣%ld", du);
if $(p \equiv 2)$ $printf$(".\n");
else {
   ⟨Compute a lower bound gl on the girth 7⟩;
   $printf$(",\nand␣its␣girth␣is␣>=␣%ld.\n", gl);
}
This code is used in section 2.

11.   We had better declare all the variables we've been using so freely.

⟨Global variables 3⟩ +≡
  long gl, gu, dl, du;   /* theoretical bounds */
  long pp;   /* power of p */
  long s;   /* accumulated sum */
  long n;   /* number of vertices */
  char $bipartite$;   /* is the graph bipartite? */

---

g: **Graph** *, §3.
n: **long**, GB_GRAPH §20.
p: **long**, §3.
$printf$: **int** (), <stdio.h>.
q: **long**, §3.

12. Breadth-first search. The graphs produced by *raman* are symmetrical, in the sense that there is an automorphism taking any vertex into any other. Each vertex V and each edge P corresponds to a 2×2 matrix, and the path $P_1 P_2 \ldots P_k$ leading from vertex V to vertex $V P_1 P_2 \ldots P_k$ has the same properties as the path leading from vertex U to vertex $U P_1 P_2 \ldots P_k$. Therefore we can find the girth and the diameter by starting at any vertex v_0.

We compute the number of points at distance k from v_0 for all k, by explicitly forming a linked list of all such points. Utility field *link* is used for the links. The lists terminate with a non-null *sentinel* value, so that we can also use the condition $link \equiv \Lambda$ to tell if a vertex has been encountered before. Another utility field, *dist*, contains the distance from the starting point, and *back* points to a vertex one step closer.

```
#define  link  w.V      /* the field where we store links, initially Λ */
#define  dist  v.I      /* the field where we store distances, initially 0 */
#define  back  u.V      /* the field where we store backpointers, initially Λ */
```

⟨ Compute and print the true girth and diameter of g 12 ⟩ ≡

```
printf("Starting␣at␣any␣given␣vertex,␣there␣are\n");
{ long k;      /* current distance being generated */
  long c;       /* how many we've seen so far at this distance */
  register Vertex *v;      /* current vertex in list at distance k − 1 */
  register Vertex *u;      /* head of list for distance k */
  Vertex *sentinel = g→vertices + n;      /* nonzero link at end of lists */
  long girth = 999;      /* length of smallest cycle found, initially infinite */
  k = 0;
  u = g→vertices;
  u→link = sentinel;
  c = 1;
  while (c) {
    for (v = u, u = sentinel, c = 0, k++; v ≠ sentinel; v = v→link)
      ⟨ Place all vertices adjacent to v onto list u, unless they've been encountered before, increasing
           c whenever the list grows 13 ⟩;
    printf("%8ld␣vertices␣at␣distance␣%ld%s\n", c, k, c > 0 ? "," : ".");
  }
  printf("So␣the␣diameter␣is␣%ld,␣and␣the␣girth␣is␣%ld.\n", k − 1, girth);
}
```

This code is used in section 2.

13. ⟨ Place all vertices adjacent to v onto list u, unless they've been encountered before, increasing c whenever the list grows 13 ⟩ ≡

```
{ register Arc *a;
  for (a = v→arcs; a; a = a→next) { register Vertex *w;      /* vertex adjacent to v */
    w = a→tip;
    if (w→link ≡ Λ) {
      w→link = u;
      w→dist = k;
      w→back = v;
      u = w;
      c++;
```

```
    } else if (w→dist + k < girth ∧ w ≠ v→back)  girth = w→dist + k;
  }
}
```
This code is used in section 12.

LADDERS

Important: Before reading LADDERS, please read or at least skim the programs for GB_WORDS and GB_DIJK.

1. Introduction. This demonstration program uses graphs constructed by the GB_WORDS module to produce an interactive program called `ladders`, which finds shortest paths between two given five-letter words of English.

The program assumes that UNIX conventions are being used. Some code in sections listed under 'UNIX dependencies' in the index might need to change if this program is ported to other operating systems.

To run the program under UNIX, say '`ladders ⟨options⟩`', where ⟨options⟩ consists of zero or more of the following specifications in any order:

`-v`	Verbosely print all words encountered during the shortest-path computation, showing also their distances from the goal word.
`-a`	Use alphabetic distance instead of considering adjacent words to be one unit apart; for example, the alphabetic distance from 'words' to 'woods' is 3, because 'r' is three places from 'o' in the alphabet.
`-f`	Use distance based on frequency (see below), instead of considering adjacent words to be one unit apart. This option is ignored if either `-a` or `-r` has been specified.
`-h`	Use a lower-bound heuristic to shorten the search (see below). This option is ignored if option `-f` has been selected.
`-e`	Echo the input to the output (useful if input comes from a file instead of from the terminal).
`-n⟨number⟩`	Limit the graph to the n most common English words, where n is the given ⟨number⟩.
`-r⟨number⟩`	Limit the graph to ⟨number⟩ randomly selected words. This option is incompatible with `-n`.
`-s⟨number⟩`	Use ⟨number⟩ instead of 0 as the seed for random numbers, to get different random samples or to explore words of equal frequency in a different order.

Option `-f` assigns a cost of 0 to the most common words and a cost of 16 to the least common words; a cost between 0 and 16 is assigned to words of intermediate frequency. The word ladders that are found will then have minimum total cost by this criterion. Experience shows that the `-f` option tends to give the "friendliest," most intuitively appealing ladders.

Option `-h` attempts to focus the search by giving priority to words that are near the goal. (More precisely, it modifies distances between adjacent words by using a heuristic function $hh(v)$, which would be the shortest possible distance between v and the goal if every five-letter combination happened to be an English word.) The GB_DIJK module explains more about such heuristics; this option is most interesting to watch when used in conjunction with `-v`.

2. The program will prompt you for a starting word. If you simply type ⟨return⟩, it exits; otherwise you should enter a five-letter word (with no uppercase letters) before typing ⟨return⟩.

Then the program will prompt you for a goal word. If you simply type ⟨return⟩ at this point, it will go back and ask for a new starting word; otherwise you should specify another five-letter word.

Then the program will find and display an optimal word ladder from the start to the goal, if there is a path from one to the other that changes only one letter at a time.

And then you have a chance to start all over again, with another starting word.

The start and goal words need not be present in the program's graph of "known" words. They are temporarily added to that graph, but removed again whenever new start and goal words are given. If the -f option is being used, the cost of the goal word will be 20 when it is not in the program's dictionary.

3. Here is the general layout of this program, as seen by the C compiler:

```
#include <ctype.h>        /* system file for character types */
#include "gb_graph.h"      /* the standard GraphBase data structures */
#include "gb_words.h"      /* routines for five-letter word graphs */
#include "gb_dijk.h"       /* routines for shortest paths */
  ⟨ Preprocessor definitions ⟩
  ⟨ Global variables 4 ⟩
  ⟨ Subroutines 11 ⟩
main (argc, argv)
      int argc;       /* the number of command-line arguments */
      char *argv[];      /* an array of strings containing those arguments */
{
    ⟨ Scan the command-line options 5 ⟩;
    ⟨ Set up the graph of words 6 ⟩;
    while (1) {
      ⟨ Prompt for starting word and goal word; break if none given 26 ⟩;
      ⟨ Find a minimal ladder from start to goal, if one exists, and print it 13 ⟩;
    }
    return 0;       /* normal exit */
}
```

---

goal: **char** [], §12.   *start*: **char** [], §12.

4. Parsing the options. Let's get the UNIX command-line junk out of the way first, so that we can concentrate on meatier stuff. Our job in this part of the program is to see if the default value zero of external variable *verbose* should change, and/or if the default values of any of the following internal variables should change:

⟨ Global variables 4 ⟩ ≡
    char *alph* = 0;   /* nonzero if the alphabetic distance option is selected */
    char *freq* = 0;   /* nonzero if the frequency-based distance option is selected */
    char *heur* = 0;   /* nonzero if the heuristic search option is selected */
    char *echo* = 0;   /* nonzero if the input-echo option is selected */
    unsigned long *n* = 0;   /* maximum number of words in the graph (0 means infinity) */
    char *rand* = 0;   /* nonzero if we will ignore the weight of words */
    long *seed* = 0;   /* seed for random number generator */

See also sections 7, 12, and 23.

This code is used in section 3.

5. ⟨ Scan the command-line options 5 ⟩ ≡
    while (−−*argc*) {
      if (*strcmp*(*argv*[*argc*], "-v") ≡ 0) *verbose* = 1;
      else if (*strcmp*(*argv*[*argc*], "-a") ≡ 0) *alph* = 1;
      else if (*strcmp*(*argv*[*argc*], "-f") ≡ 0) *freq* = 1;
      else if (*strcmp*(*argv*[*argc*], "-h") ≡ 0) *heur* = 1;
      else if (*strcmp*(*argv*[*argc*], "-e") ≡ 0) *echo* = 1;
      else if (*sscanf*(*argv*[*argc*], "-n%lu", &*n*) ≡ 1) *rand* = 0;
      else if (*sscanf*(*argv*[*argc*], "-r%lu", &*n*) ≡ 1) *rand* = 1;
      else if (*sscanf*(*argv*[*argc*], "-s%ld", &*seed*) ≡ 1) ;
      else {
        fprintf(*stderr*, "Usage:␣%s␣[-v] [-a] [-f] [-h] [-e] [-nN] [-rN] [-sN]\n", *argv*[0]);
        return −2;
      }
    }
    if (*alph* ∨ *rand*) *freq* = 0;
    if (*freq*) *heur* = 0;

This code is used in section 3.

6. Creating the graph.   The GraphBase *words* procedure will produce the five-letter words
we want, organized in a graph structure.

#define *quit_if* (x, c)
        if (x) {
            fprintf (*stderr*, "Sorry,␣I␣couldn't␣build␣a␣dictionary␣(trouble␣code␣%ld)!\n", *c*);
            return *c*;
        }
⟨ Set up the graph of words 6 ⟩ ≡
    $g = words(n, (rand ? zero\_vector : \Lambda), 0_{\text{L}}, seed)$;
    quit_if $(g \equiv \Lambda, panic\_code)$;
    ⟨ Confirm the options selected 8 ⟩;
    ⟨ Modify the edge lengths, if the *alph* or *freq* option was selected 9 ⟩;
    ⟨ Modify the priority queue algorithm, if unequal edge lengths are possible 10 ⟩;
This code is used in section 3.

7.   ⟨ Global variables 4 ⟩ +≡
Graph **g*;   /* graph created by *words* */
long *zero_vector*[9];   /* weights to use when ignoring all frequency information */

8.   The actual number of words might be decreased to the size of the GraphBase dictionary, so
we wait until the graph is generated before confirming the user-selected options.

⟨ Confirm the options selected 8 ⟩ ≡
    if (*verbose*) {
        if (*alph*) *printf* ("(alphabetic␣distance␣selected)\n");
        if (*freq*) *printf* ("(frequency-based␣distances␣selected)\n");
        if (*heur*) *printf* ("(lowerbound␣heuristic␣will␣be␣used␣to␣focus␣the␣search)\n");
        if (*rand*) *printf* ("(random␣selection␣of␣%ld␣words␣with␣seed␣%ld)\n", $g\text{-}n$, *seed*);
        else *printf* ("(the␣graph␣has␣%ld␣words)\n", $g\text{-}n$);
    }
This code is used in section 6.

---

argc: **int**, §3.
argv: **char** *[], §3.
fprintf: **int** ( ), <stdio.h>.
Graph = **struct**, GB_GRAPH §20.

n: **long**, GB_GRAPH §20.
panic_code: **long**, GB_GRAPH §5.
printf: **int** ( ), <stdio.h>.
sscanf: **int** ( ), <stdio.h>.

stderr: **FILE** *, <stdio.h>.
strcmp: **int** ( ), <string.h>.
verbose: **long**, GB_GRAPH §5.
words: **Graph** *( ), GB_WORDS §7.

9. The edges in a *words* graph normally have length 1, so we must change them if the user has selected *alph* or *freq*. The character position in which adjacent words differ is recorded in the *loc* field of each arc. The frequency of a word is stored in the *weight* field of its vertex.

#define $a\_dist(k)$ $(*(p+k) < *(q+k) \mathbin{?} *(q+k) - *(p+k) : *(p+k) - *(q+k))$

⟨ Modify the edge lengths, if the *alph* or *freq* option was selected 9 ⟩ ≡

```
if (alph) { register Vertex *u;
    for (u = g⃗vertices + g⃗n - 1; u ≥ g⃗vertices; u−−) { register Arc *a;
        register char *p = u⃗name;
        for (a = u⃗arcs; a; a = a⃗next) { register char *q = a⃗tip⃗name;
            a⃗len = a_dist(a⃗loc);
        }
    }
} else if (freq) { register Vertex *u;
    for (u = g⃗vertices + g⃗n - 1; u ≥ g⃗vertices; u−−) { register Arc *a;
        for (a = u⃗arcs; a; a = a⃗next) a⃗len = freq_cost(a⃗tip);
    }
}
```

This code is used in section 6.

10. The default priority queue algorithm of *dijkstra* is quite efficient when all edge lengths are 1. Otherwise we change it to the alternative method that works best for edge lengths less than 128.

⟨ Modify the priority queue algorithm, if unequal edge lengths are possible 10 ⟩ ≡

```
if (alph ∨ freq ∨ heur) {
    init_queue = init_128; del_min = del_128;
    enqueue = enq_128; requeue = req_128;
}
```

This code is used in section 6.

11. The frequency has been computed with the default weights explained in the documentation of *words*; it is usually less than 2^{16}. A word whose frequency is 0 costs 16; a word whose frequency is 1 costs 15; a word whose frequency is 2 or 3 costs 14; and the costs keep decreasing by 1 as the frequency doubles, until we reach a cost of 0.

⟨ Subroutines 11 ⟩ ≡

```
long freq_cost(v)
    Vertex *v;
{ register long acc = v⃗weight;      /* the frequency, to be shifted right */
    register long k = 16;
    while (acc) k−−, acc ≫= 1;
    return (k < 0 ? 0 : k);
}
```

See also sections 17, 18, 20, 22, and 27.

This code is used in section 3.

12. Minimal ladders. The guts of this program is a routine to compute shortest paths between two given words, *start* and *goal*.

The *dijkstra* procedure does this, in any graph with nonnegative arc lengths. The only complication we need to deal with here is that *start* and *goal* might not themselves be present in the graph. In that case we want to insert them, albeit temporarily.

The conventions of GB_GRAPH allow us to do the desired augmentation by creating a new graph *gg* whose vertices are borrowed from *g*. The graph *g* has space for two more vertices (actually for four), and any new memory blocks allocated for the additional arcs present in *gg* will be freed later by the operation *gb_recycle(gg)* without confusion.

⟨ Global variables 4 ⟩ +≡
    Graph *∗gg*;   /∗ clone of *g* with possible additional words ∗/
    char *start*[6], *goal*[6];   /∗ **words** dear to the user's **heart**, plus '\0' ∗/
    Vertex *∗uu, ∗vv*;   /∗ start and goal vertices in *gg* ∗/

13. ⟨ Find a minimal ladder from *start* to *goal*, if one exists, and print it 13 ⟩ ≡
    ⟨ Build the amplified graph *gg* 14 ⟩;
    ⟨ Let *dijkstra* do the hard work 21 ⟩;
    ⟨ Print the answer 24 ⟩;
    ⟨ Remove all traces of *gg* 25 ⟩;
This code is used in section 3.

14. ⟨ Build the amplified graph *gg* 14 ⟩ ≡
    gg = gb_new_graph(0_L);
    quit_if(*gg* ≡ Λ, *no_room* + 5);   /∗ out of memory ∗/
    gg→vertices = g→vertices;
    gg→n = g→n;
    ⟨ Put the *start* word into *gg*, and let *uu* point to it 15 ⟩;
    ⟨ Put the *goal* word into *gg*, and let *vv* point to it 16 ⟩;
    if (*gg→n* ≡ *g→n* + 2) ⟨ Check if *start* is adjacent to *goal* 19 ⟩;
    quit_if(*gb_trouble_code, no_room* + 6);   /∗ out of memory ∗/
This code is used in section 13.

---

alph: **char**, §4.
Arc = **struct**, GB_GRAPH §10.
arcs: **Arc** ∗, GB_GRAPH §9.
del_128: **Vertex** ∗( ), GB_DIJK §22.
del_min: **Vertex** ∗(∗)( ),
  GB_DIJK §15.
dijkstra: **long** ( ), GB_DIJK §9.
enq_128: **void** ( ), GB_DIJK §23.
enqueue: **void** (∗)( ), GB_DIJK §15.
freq: **char**, §4.
g: **Graph** ∗, §7.
gb_new_graph: **Graph** ∗( ),

  GB_GRAPH §23.
gb_recycle: **void** ( ), GB_GRAPH §40.
gb_trouble_code: **long**,
  GB_GRAPH §14.
Graph = **struct**, GB_GRAPH §20.
heur: **char**, §4.
init_128: **void** ( ), GB_DIJK §21.
init_queue: **void** (∗)( ),
  GB_DIJK §15.
len: **long**, GB_GRAPH §10.
loc = *a.I*, GB_WORDS §26.
n: **long**, GB_GRAPH §20.

name: **char** ∗, GB_GRAPH §9.
next: **Arc** ∗, GB_GRAPH §10.
no_room = 1, GB_GRAPH §7.
quit_if = macro ( ), §6.
req_128: **void** ( ), GB_DIJK §24.
requeue: **void** (∗)( ), GB_DIJK §15.
tip: **Vertex** ∗, GB_GRAPH §10.
Vertex = **struct**, GB_GRAPH §9.
vertices: **Vertex** ∗, GB_GRAPH §20.
weight = *u.I*, GB_WORDS §26.
words: **Graph** ∗( ), GB_WORDS §7.

15. The *find_word* procedure returns Λ if it can't find the given word in the graph just constructed by *words*. In that case it has applied its second argument to every adjacent word. Hence the program logic here does everything needed to add a new vertex to *gg* when necessary.

⟨ Put the *start* word into *gg*, and let *uu* point to it 15 ⟩ ≡
```
(gg→vertices + gg→n)→name = start;       /* a tentative new vertex */
uu = find_word(start, plant_new_edge);
if (¬uu) uu = gg→vertices + gg→n++;       /* recognize the new vertex and refer to it */
```
This code is used in section 14.

16. ⟨ Put the *goal* word into *gg*, and let *vv* point to it 16 ⟩ ≡
```
if (strncmp(start, goal, 5) ≡ 0) vv = uu;       /* avoid inserting a word twice */
else {
  (gg→vertices + gg→n)→name = goal;       /* a tentative new vertex */
  vv = find_word(goal, plant_new_edge);
  if (¬vv) vv = gg→vertices + gg→n++;       /* recognize the new vertex and refer to it */
}
```
This code is used in section 14.

17. The *alph_dist* subroutine calculates the alphabetic distance between arbitrary five-letter words, whether they are adjacent or not.

⟨ Subroutines 11 ⟩ +≡
```
long alph_dist(p, q)
     register char *p, *q;
{
  return a_dist(0) + a_dist(1) + a_dist(2) + a_dist(3) + a_dist(4);
}
```

18. ⟨ Subroutines 11 ⟩ +≡
```
void plant_new_edge(v)
     Vertex *v;
{ Vertex *u = gg→vertices + gg→n;       /* the new edge runs from u to v */
  gb_new_edge(u, v, 1_L);
  if (alph) u→arcs→len = (u→arcs − 1)→len = alph_dist(u→name, v→name);
  else if (freq) {
    u→arcs→len = 20;       /* adjust the arc length from v to u */
    (u→arcs − 1)→len = freq_cost(v);       /* adjust the arc length from u to v */
  }
}
```

19. There's a bug in the above logic that could be embarrassing, although it will come up only when a user is trying to be clever: The *find_word* routine knows only the words of *g*, so it will fail to make any direct connection between *start* and *goal* if they happen to be adjacent to each other yet not in the original graph. We had better fix this, otherwise the computer will look stupid.

⟨ Check if *start* is adjacent to *goal* 19 ⟩ ≡
```
if (hamm_dist(start, goal) ≡ 1) {
  gg→n−−;       /* temporarily pretend vv hasn't been added yet */
```

```
    plant_new_edge(uu);      /* make vv adjacent to uu */
    gg⁻n++;        /* and recognize it again */
  }
```

This code is used in section 14.

20. The Hamming distance between words is the number of character positions in which they differ.

#define *h_dist*(k)   (*(p + k) ≡ *(q + k) ? 0 : 1)

⟨ Subroutines 11 ⟩ +≡
  long *hamm_dist*(p, q)
     register char *p, *q;
  {
    return *h_dist*(0) + *h_dist*(1) + *h_dist*(2) + *h_dist*(3) + *h_dist*(4);
  }

21. OK, now we've got a graph in which *dijkstra* can operate.

⟨ Let *dijkstra* do the hard work 21 ⟩ ≡
  if (¬*heur*) *min_dist* = *dijkstra*(uu, vv, gg, Λ);
  else if (*alph*) *min_dist* = *dijkstra*(uu, vv, gg, alph_heur);
  else *min_dist* = *dijkstra*(uu, vv, gg, hamm_heur);

This code is used in section 13.

22. ⟨ Subroutines 11 ⟩ +≡
  long *alph_heur*(v)
     Vertex *v;
  { **return** *alph_dist*(v⁻name, goal); }
  long *hamm_heur*(v)
     Vertex *v;
  { **return** *hamm_dist*(v⁻name, goal); }

23. ⟨ Global variables 4 ⟩ +≡
  long *min_dist*;   /* length of the shortest ladder */

24. ⟨ Print the answer 24 ⟩ ≡
  if (*min_dist* < 0) *printf*("Sorry,␣there's␣no␣ladder␣from␣%s␣to␣%s.\n", start, goal);
  else *print_dijkstra_result*(vv);

This code is used in section 13.

---

a_dist = macro ( ), §9.
alph: **char**, §4.
arcs: **Arc** *, GB_GRAPH §9.
dijkstra: **long** ( ), GB_DIJK §9.
find_word: **Vertex** *( ),
  GB_WORDS §30.
freq: **char**, §4.
freq_cost: **long** ( ), §11.
g: **Graph** *, §7.

gb_new_edge: **void** ( ),
  GB_GRAPH §31.
gg: **Graph** *, §12.
goal: **char** [], §12.
heur: **char**, §4.
len: **long**, GB_GRAPH §10.
n: **long**, GB_GRAPH §20.
name: **char** *, GB_GRAPH §9.
print_dijkstra_result: **void** ( ),

GB_DIJK §14.
printf: **int** ( ), <stdio.h>.
start: **char** [], §12.
strncmp: **int** ( ), <string.h>.
uu: **Vertex** *, §12.
Vertex = **struct**, GB_GRAPH §9.
vertices: **Vertex** *, GB_GRAPH §20.
vv: **Vertex** *, §12.
words: **Graph** *( ), GB_WORDS §7.

25. Finally, we have to clean up our tracks. It's easy to remove all arcs from the new vertices of gg to the old vertices of g; it's a bit trickier to remove the arcs from old to new. The loop here will also remove arcs properly between start and goal vertices, if they both belong to gg not g.

⟨ Remove all traces of gg 25 ⟩ ≡
```
for (uu = g→vertices + gg→n − 1; uu ≥ g→vertices + g→n; uu −−) { register Arc *a;
    for (a = uu→arcs; a; a = a→next) {
        vv = a→tip;      /* now vv→arcs ≡ a − 1, since arcs for edges come in pairs */
        vv→arcs = vv→arcs→next;
    }
    uu→arcs = Λ;      /* we needn't clear uu→name */
}
gb_recycle(gg);      /* the gg→data blocks disappear, but g→data remains */
```
This code is used in section 13.

26.   Terminal interaction.   We've finished doing all the interesting things. Only one minor part of the program still remains to be written.

⟨Prompt for starting word and goal word; **break** if none given 26⟩ ≡
    putchar('\n');   /* make a blank line for visual punctuation */
restart: /* if we try to avoid this label, the **break** command will be broken */
    if (*prompt_for_five*("Starting", *start*) ≠ 0) **break**;
    if (*prompt_for_five*("␣␣␣␣Goal", *goal*) ≠ 0) **goto** *restart*;
This code is used in section 3.

27.   ⟨Subroutines 11⟩ +≡
    long *prompt_for_five*(*s, p*)
        char *s;   /* string used in prompt message */
        register char *p;   /* where to put a string typed by the user */
    { **register char** *q;   /* current position to store characters */
      register long *c*;   /* current character of input */

      while (1) {
        printf("%s␣word:␣", *s*);
        fflush(*stdout*);   /* make sure the user sees the prompt */
        q = *p*;
        while (1) {
          c = *getchar*( );
          if (*c* ≡ EOF) **return** −1;   /* end-of-file */
          if (*echo*) *putchar*(*c*);
          if (*c* ≡ '\n') **break**;
          if (¬*islower*(*c*)) *q* = *p* + 5;
          else if (*q* < *p* + 5) *q* = *c*;
          q++;
        }
        if (*q* ≡ *p* + 5) **return** 0;   /* got a good five-letter word */
        if (*q* ≡ *p*) **return** 1;   /* got just ⟨return⟩ */
        printf("(Please␣type␣five␣lowercase␣letters␣and␣RETURN.)\n");
      }
    }

Arc = **struct**, GB_GRAPH §10.
arcs: **Arc** *, GB_GRAPH §9.
data: **Area**, GB_GRAPH §20.
echo: **char**, §4.
EOF = (−1), <stdio.h>.
fflush: **int** ( ), <stdio.h>.
g: **Graph** *, §7.
gb_recycle: **void** ( ), GB_GRAPH §40.

getchar = macro ( ), <stdio.h>.
gg: **Graph** *, §12.
goal: **char** [], §12.
islower: **int** ( ), <ctype.h>.
n: **long**, GB_GRAPH §20.
name: **char** *, GB_GRAPH §9.
next: **Arc** *, GB_GRAPH §10.
printf: **int** ( ), <stdio.h>.

prompt_for_five: **long** ( ), §27.
putchar = macro ( ), <stdio.h>.
start: **char** [], §12.
stdout: **FILE** *, <stdio.h>.
tip: **Vertex** *, GB_GRAPH §10.
uu: **Vertex** *, §12.
vertices: **Vertex** *, GB_GRAPH §20.
vv: **Vertex** *, §12.

MILES_SPAN

Important: Before reading MILES_SPAN, please read or at least skim the program for GB_MILES.

1. Minimum spanning trees. A classic paper by R. L. Graham and Pavol Hell about the history of algorithms to find the minimum-length spanning tree of a graph [*Annals of the History of Computing* **7** (1985), 43–57] describes three main approaches to that problem. Algorithm 1, "two nearest fragments," repeatedly adds a shortest edge that joins two hitherto unconnected fragments of the graph; this algorithm was first published by J. B. Kruskal in 1956. Algorithm 2, "nearest neighbor," repeatedly adds a shortest edge that joins a particular fragment to a vertex not in that fragment; this algorithm was first published by V. Jarník in 1930. Algorithm 3, "all nearest fragments," repeatedly adds to each existing fragment the shortest edge that joins it to another fragment; this method, seemingly the most sophisticated in concept, also turns out to be the oldest, being first published by Otakar Borůvka in 1926.

The present program contains simple implementations of all three approaches, in an attempt to make practical comparisons of how they behave on "realistic" data. One of the main goals of this program is to demonstrate a simple way to make machine-independent comparisons of programs written in C, by counting memory references or "mems." In other words, this program is intended to be read, not just performed.

The author believes that mem counting sheds considerable light on the problem of determining the relative efficiency of competing algorithms for practical problems. He hopes other researchers will enjoy rising to the challenge of devising algorithms that find minimum spanning trees in significantly fewer mem units than the algorithms presented here, on problems of the size considered here.

Indeed, mem counting promises to be significant for combinatorial algorithms of all kinds. The standard graphs available in the Stanford GraphBase should make it possible to carry out a large number of machine-independent experiments concerning the practical efficiency of algorithms that have previously been studied only asymptotically.

2. The graphs we will deal with are produced by the *miles* subroutine, found in the GB_MILES module. As explained there, $miles(n, north\_weight, west\_weight, pop\_weight, 0, max\_degree, seed)$ produces a graph of $n \leq 128$ vertices based on the driving distances between North American cities. By default we take $n = 100$, $north\_weight = west\_weight = pop\_weight = 0$, and $max\_degree = 10$; this gives billions of different sparse graphs, when different $seed$ values are specified, since a different random number seed generally results in the selection of another one of the $\binom{128}{100}$ possible subgraphs.

The default parameters can be changed by specifying options on the command line, at least in a UNIX implementation, thereby obtaining a variety of special effects. For example, the value of n can be raised or lowered and/or the graph can be made more or less sparse. The user can bias the selection by ranking cities according to their population and/or position, if nonzero values are given to any of the parameters $north\_weight$, $west\_weight$, or pop_weight. Command-line options $-n\langle$number$\rangle$, $-N\langle$number$\rangle$, $-W\langle$number$\rangle$, $-P\langle$number$\rangle$, $-d\langle$number$\rangle$, and $-s\langle$number$\rangle$ are used to specify non-default values of the respective quantities n, $north\_weight$, $west\_weight$, pop_weight, max_degree, and $seed$.

If the user specifies a `-r` option, for example by saying 'miles_span -r10', this program will investigate the spanning trees of a series of, say, 10 graphs having consecutive *seed* values. (This option makes sense only if *north_weight* = *west_weight* = *pop_weight* = 0, because *miles* chooses the top *n* cities by weight. The procedure rarely needs to use random numbers to break ties when the weights are nonzero, because cities rarely have exactly the same weight in that case.)

The special command-line option `-g`⟨filename⟩ overrides all others. It substitutes an external graph previously saved by *save_graph* for the graphs produced by *miles*.

Here is the overall layout of this C program:

```
#include "gb_graph.h"      /* the GraphBase data structures */
#include "gb_save.h"       /* restore_graph */
#include "gb_miles.h"      /* the miles routine */
⟨ Preprocessor definitions ⟩
⟨ Global variables 3 ⟩
⟨ Procedures to be declared early 67 ⟩
⟨ Priority queue subroutines 24 ⟩
⟨ Subroutines 7 ⟩;

main(argc, argv)
     int argc;        /* the number of command-line arguments */
     char *argv[ ];     /* an array of strings containing those arguments */
{ unsigned long n = 100;      /* the desired number of vertices */
  unsigned long n_weight = 0;      /* the north_weight parameter */
  unsigned long w_weight = 0;      /* the west_weight parameter */
  unsigned long p_weight = 0;      /* the pop_weight parameter */
  unsigned long d = 10;      /* the max_degree parameter */
  long s = 0;      /* the random number seed */
  unsigned long r = 1;      /* the number of repetitions */
  char *file_name = Λ;      /* external graph to be restored */
  ⟨ Scan the command-line options 4 ⟩;
  while (r−−) {
    if (file_name) g = restore_graph(file_name);
    else  g = miles(n, n_weight, w_weight, p_weight, 0 L, d, s);
    if (g ≡ Λ ∨ g⃗n ≤ 1) {
       fprintf(stderr, "Sorry,␣can't␣create␣the␣graph!␣(error␣code␣%ld)\n", panic_code);
       return −1;      /* error code 0 means the graph is too small */
    }
    ⟨ Report the number of mems needed to compute a minimum spanning tree of g by various
         algorithms 5 ⟩;
    gb_recycle(g);  s++;      /* increase the seed value */
  }
  return 0;      /* normal exit */
}
```

---

fprintf: **int** ( ), <stdio.h>.
g: **Graph** *, §3.
gb_recycle: **void** ( ), GB_GRAPH §40.
max_degree: **unsigned long**,
  GB_MILES §5.
miles: **Graph** *( ), GB_MILES §5.

n: **unsigned long**, GB_MILES §5.
north_weight: **long**, GB_MILES §5.
panic_code: **long**, GB_GRAPH §5.
pop_weight: **long**, GB_MILES §5.
restore_graph: **Graph** *( ),

GB_SAVE §4.
save_graph: **long** ( ), GB_SAVE §20.
seed: **long**, GB_MILES §5.
stderr: **FILE** *, <stdio.h>.
west_weight: **long**, GB_MILES §5.

3. ⟨ Global variables 3 ⟩ ≡

  Graph *g;   /* the graph we will work on */

See also sections 6, 10, 13, 19, 23, 31, 37, 57, and 68.

This code is used in section 2.

4. ⟨ Scan the command-line options 4 ⟩ ≡

  while $(--argc)$ {

    if $(sscanf(argv[argc], \texttt{"-n\%lu"}, \&n) \equiv 1)$;

    else if $(sscanf(argv[argc], \texttt{"-N\%lu"}, \&n\_weight) \equiv 1)$;

    else if $(sscanf(argv[argc], \texttt{"-W\%lu"}, \&w\_weight) \equiv 1)$;

    else if $(sscanf(argv[argc], \texttt{"-P\%lu"}, \&p\_weight) \equiv 1)$;

    else if $(sscanf(argv[argc], \texttt{"-d\%lu"}, \&d) \equiv 1)$;

    else if $(sscanf(argv[argc], \texttt{"-r\%lu"}, \&r) \equiv 1)$;

    else if $(sscanf(argv[argc], \texttt{"-s\%ld"}, \&s) \equiv 1)$;

    else if $(strcmp(argv[argc], \texttt{"-v"}) \equiv 0)$ $verbose = 1;$

    else if $(strncmp(argv[argc], \texttt{"-g"}, 2) \equiv 0)$ $file\_name = argv[argc] + 2;$

    else {

      $fprintf(stderr, \texttt{"Usage:\_\%s\_[-nN][-dN][-rN][-sN][-NN][-WN][-PN][-v][-gfoo]\textbackslash n"}, argv[0]);$

      return $-2;$

    }

  }

  if $(file\_name)$ $r = 1;$

This code is used in section 2.

5. We will try out four basic algorithms that have received prominent attention in the literature. Graham and Hell's Algorithm 1 is represented by the *krusk* procedure, which uses Kruskal's algorithm after the edges have been sorted by length with a radix sort. Their Algorithm 2 is represented by the *jar_pr* procedure, which incorporates a priority queue structure that we implement in two ways, either as a simple binary heap or as a Fibonacci heap. And their Algorithm 3 is represented by the *cher_tar_kar* procedure, which implements a method similar to Borůvka's that was independently discovered by Cheriton and Tarjan and later simplified and refined by Karp and Tarjan.

#define INFINITY **(unsigned long)** -1 /* value returned when there's no spanning tree */

⟨ Report the number of mems needed to compute a minimum spanning tree of g by various algorithms 5 ⟩ ≡

  $printf(\texttt{"The\_graph\_\%s\_has\_\%ld\_edges,\textbackslash n"}, g\text{-}id, g\text{-}m/2);$

  $sp\_length = krusk(g);$

  if $(sp\_length \equiv \text{INFINITY})$ $printf(\texttt{"\_\_and\_it\_isn't\_connected.\textbackslash n"});$

  else $printf(\texttt{"\_\_and\_its\_minimum\_spanning\_tree\_has\_length\_\%ld.\textbackslash n"}, sp\_length);$

  $printf(\texttt{"\_The\_Kruskal/radix-sort\_algorithm\_takes\_\%ld\_mems;\textbackslash n"}, mems);$

  ⟨ Execute $jar\_pr(g)$ with binary heaps as the priority queue algorithm 28 ⟩;

  $printf(\texttt{"\_the\_Jarnik/Prim/binary-heap\_algorithm\_takes\_\%ld\_mems;\textbackslash n"}, mems);$

  ⟨ Allocate additional space needed by the more complex algorithms; or **goto** *done* if there isn't enough room 32 ⟩;

  ⟨ Execute $jar\_pr(g)$ with Fibonacci heaps as the priority queue algorithm 42 ⟩;

  $printf(\texttt{"\_the\_Jarnik/Prim/Fibonacci-heap\_algorithm\_takes\_\%ld\_mems;\textbackslash n"}, mems);$

  if $(sp\_length \neq cher\_tar\_kar(g))$ {

```
  if (gb_trouble_code) printf("␣...oops,␣I've␣run␣out␣of␣memory!\n");
  else printf("␣...oops,␣I've␣got␣a␣bug,␣please␣fix␣fix␣fix\n");
  return −3;
}
printf("␣the␣Cheriton/Tarjan/Karp␣algorithm␣takes␣%ld␣mems.\n\n", mems);
done: ;
```

This code is used in section 2.

6. ⟨ Global variables 3 ⟩ +≡
  unsigned long *sp_length*;   /∗ length of the minimum spanning tree ∗/

7. When the *verbose* switch is nonzero, edges found by the various algorithms will call the
report subroutine.

⟨ Subroutines 7 ⟩ ≡
  report(*u*, *v*, *l*)
      Vertex ∗*u*, ∗*v*;   /∗ adjacent vertices in the minimum spanning tree ∗/
      long *l*;   /∗ the length of the edge between them ∗/
  {
      printf("␣␣%ld␣miles␣between␣%s␣and␣%s␣[%ld␣mems]\n", *l*, *u*→*name*, *v*→*name*, *mems*);
  }

See also sections 14, 20, and 55.

This code is used in section 2.

---

8. Strategies and ground rules. Let us say that a *fragment* is any subtree of a minimum spanning tree. All three algorithms we implement make use of a basic principle first stated in full generality by R. C. Prim in 1957: "If a fragment F does not include all the vertices, and if e is a shortest edge joining F to a vertex not in F, then $F \cup e$ is a fragment." To prove Prim's principle, let T be a minimum spanning tree that contains F but not e. Adding e to T creates a circuit containing some edge $e' \neq e$, where e' runs from a vertex in F to a vertex not in F. Deleting e' from $T \cup e$ produces a spanning tree T' of total length no larger than the total length of T. Hence T' is a minimum spanning tree containing $F \cup e$, QED.

9. The graphs produced by *miles* have special properties, and it is fair game to make use of those properties if we can.

First, the length of each edge is a positive integer less than 2^{12}.

Second, the kth vertex v_k of the graph is represented in C programs by the pointer expression $g\text{-}vertices + k$. If weights have been assigned, these vertices will be in order by weight. For example, if $north\_weight = 1$ but $west\_weight = pop\_weight = 0$, vertex v_0 will be the most northerly city and vertex v_{n-1} will be the most southerly.

Third, the edges accessible from a vertex v appear in a linked list starting at $v\text{-}arcs$. An edge from v to v_j will precede an edge from v to v_k in this list if and only if $j > k$.

Fourth, the vertices have coordinates $v\text{-}x\_coord$ and $v\text{-}y\_coord$ that are correlated with the length of edges between them: The Euclidean distance between the coordinates of two vertices tends to be small if and only if those vertices are connected by a relatively short edge. (This is only a tendency, not a certainty; for example, some cities around Chesapeake Bay are fairly close together as the crow flies, but not within easy driving range of each other.)

Fifth, the edge lengths satisfy the triangle inequality: Whenever three edges form a cycle, the longest is no longer than the sum of the lengths of the two others. (It can be proved that the triangle inequality is of no use in finding minimum spanning trees; we mention it here only to exhibit yet another way in which the data produced by *miles* is known to be nonrandom.)

Our implementation of Kruskal's algorithm will make use of the first property, and it also uses part of the third to avoid considering an edge more than once. We will not exploit the other properties, but a reader who wants to design algorithms that use fewer mems to find minimum spanning trees of these graphs is free to use any idea that helps.

10. Speaking of mems, here are the simple C instrumentation macros that we use to count memory references. The macros are called o, oo, ooo, and $oooo$; hence Jon Bentley has called this a "little oh analysis." Implementors who want to count mems are supposed to say, e.g., 'oo,' just before an assignment statement or boolean expression that makes two references to memory. The C preprocessor will convert this to a statement that increases *mems* by 2 as that statement or expression is evaluated.

The semantics of C tell us that the evaluation of an expression like '$a \wedge (o, a\text{-}len > 10)$' will increment *mems* if and only if the pointer variable a is non-null. Warning: The parentheses are very important in this example, because C's operator $\wedge$ (i.e., **&&**) has higher precedence than comma.

Values of significant variables, like a in the previous example, can be assumed to be in "registers," and no charge is made for arithmetic computations that involve only registers. But the total number of registers in an implementation must be finite and fixed, independent of the problem size.

C does not allow the *o* macros to appear in declarations, so we cannot take full advantage of C's initialization mechanism when we are counting mems. But it's easy to initialize variables in separate statements after the declarations are done.

#define *o*   *mems* ++
#define *oo*   *mems* += 2
#define *ooo*   *mems* += 3
#define *oooo*   *mems* += 4

⟨ Global variables 3 ⟩ +≡
　long *mems*;　　/∗ the number of memory references counted ∗/

11.　Examples of these mem-counting conventions appear throughout the program that follows. Some people will undoubtedly ask why the insertion of macros by hand is being recommended here, when it would be possible to develop a fancy system that counts mems automatically. The author believes that it is best to rely on programmers to introduce *o* and *oo*, etc., by themselves, for several reasons. (1) The macros can be inserted easily and quickly using a text editor. (2) An implementation need not pay for mems that could be avoided by a suitable optimizing compiler or by making the C program text slightly more complex; thus, authors can use their good judgment to keep programs more readable than if the code were overly hand-optimized. (3) The programmer should be able to see exactly where mems are being charged, as an aid to bottleneck elimination. Occurrences of *o* and *oo* make this plain without messing up the program text. (4) An implementation need not be charged for mems that merely provide diagnostic output, or mems that do redundant computations just to double-check the validity of "proven" assertions as a program is being tested.

Computer architecture is converging rapidly these days to the design of machines in which the exact running time of a program depends on complicated interactions between pipelined circuitry and the dynamic properties of cache mapping in a memory hierarchy, not to mention the effects of compilers and operating systems. But a good approximation to running time is usually obtained if we assume that the amount of computation is proportional to the activity of the memory bus between registers and main memory. This approximation is likely to get even better in the future, as RISC computers get faster and faster in comparison to memory devices. Although the mem measure is far from perfect, it appears to be significantly less distorted than any other measurement that can be obtained without considerably more work. An implementation that is designed to use few mems will almost certainly be efficient on today's sequential computers, as well as on the sequential computers we can expect to be built in the foreseeable future. And the converse statement is even more true: An algorithm that runs fast will not consume many mems.

Of course authors are expected to be reasonable and fair when they are competing for minimum-mem prizes. They must be ready to submit their programs to inspection by impartial judges. A good algorithm will not need to abuse the spirit of realistic mem-counting.

Mems can be analyzed theoretically as well as empirically. This means we can attach constants to estimates of running time, instead of always resorting to O notation.

---

arcs: **Arc** ∗, GB_GRAPH §9.
g: **Graph** ∗, §3.
miles: **Graph** ∗( ), GB_MILES §5.

north_weight: **long**, GB_MILES §5.
pop_weight: **long**, GB_MILES §5.
vertices: **Vertex** ∗, GB_GRAPH §20.

west_weight: **long**, GB_MILES §5.
x_coord = *x.I*, GB_MILES §15.
y_coord = *y.I*, GB_MILES §15.

12. Kruskal's algorithm. The first algorithm we shall implement and instrument is the simplest: It considers the edges one by one in order of nondecreasing length, selecting each edge that does not form a cycle with previously selected edges.

We know that the edge lengths are less than 2^{12}, so we can sort them into order with two passes of a 2^6-bucket radix sort. We will arrange to have them appear in the buckets as linked lists of **Arc** records; the two utility fields of an **Arc** will be called *from* and *klink*, respectively.

```
#define from   a.V      /* an edge goes from vertex a→from to vertex a→tip */
#define klink  b.A      /* the next longer edge after a will be a→klink */
```

⟨ Put all the edges into *bucket*[0] through *bucket*[63] 12 ⟩ ≡

```
o, n = g→n;
for (l = 0; l < 64; l++)  oo, aucket[l] = bucket[l] = Λ;
for (o, v = g→vertices;  v < g→vertices + n;  v++)
    for (o, a = v→arcs;  a ∧ (o, a→tip > v);  o, a = a→next) {
        o, a→from = v;
        o, l = a→len & #3f;      /* length mod 64 */
        oo, a→klink = aucket[l];
        o, aucket[l] = a;
    }
for (l = 63;  l ≥ 0;  l−−)
    for (o, a = aucket[l];  a;  ) { register long ll;
        register Arc *aa = a;
        o, a = a→klink;
        o, ll = aa→len ≫ 6;      /* length divided by 64 */
        oo, aa→klink = bucket[ll];
        o, bucket[ll] = aa;
    }
```

This code is used in section 14.

13. ⟨ Global variables 3 ⟩ +≡

```
Arc *aucket[64], *bucket[64];      /* heads of linked lists of arcs */
```

14. Kruskal's algorithm now takes the following form.

⟨ Subroutines 7 ⟩ +≡

```
unsigned long krusk(g)
      Graph *g;
{ ⟨ Local variables for krusk 15 ⟩;
    mems = 0;
    ⟨ Put all the edges into bucket[0] through bucket[63] 12 ⟩;
    if (verbose)  printf("␣␣␣[%ld␣mems␣to␣sort␣the␣edges␣into␣buckets]\n", mems);
    ⟨ Put all the vertices into components by themselves 17 ⟩;
    for (l = 0;  l < 64;  l++)
        for (o, a = bucket[l];  a;  o, a = a→klink) {
            o, u = a→from;
            o, v = a→tip;
            ⟨ If u and v are already in the same component, continue 16 ⟩;
            if (verbose)  report(a→from, a→tip, a→len);
            o, tot_len += a→len;
```

        if $(--\,components \equiv 1)$ **return** *tot_len*;
        ⟨ Merge the components containing *u* and *v* 18 ⟩;
      }
    return INFINITY;   /* the graph wasn't connected */
}

15.   Lest we forget, we'd better declare all the local variables we've been using.

⟨ Local variables for *krusk* 15 ⟩ ≡
    register Arc *a;   /* current edge of interest */
    register long *l*;   /* current bucket of interest */
    register Vertex *u, *v, *w;   /* current vertices of interest */
    unsigned long *tot_len* = 0;   /* total length of edges already chosen */
    long *n*;   /* the number of vertices */
    long *components*;

This code is used in section 14.

16.   The remaining things that *krusk* needs to do are easily recognizable as an application of "equivalence algorithms" or "union/find" data structures. We will use a simple approach whose average running time on random graphs was shown to be linear by Knuth and Schönhage in *Theoretical Computer Science* **6** (1978), 281–315.

    The vertices of each component (that is, of each connected fragment defined by the edges selected so far) will be linked circularly by *clink* pointers. Each vertex also has a *comp* field that points to a unique vertex representing its component. Each component representative also has a *csize* field that tells how many vertices are in the component.

#define *clink*   *z.V*   /* pointer to another vertex in the same component */
#define *comp*   *y.V*   /* pointer to component representative */
#define *csize*   *x.I*   /* size of the component (maintained only for representatives) */

⟨ If *u* and *v* are already in the same component, **continue** 16 ⟩ ≡
    if $(oo, u{\rightarrow}comp \equiv v{\rightarrow}comp)$ **continue**;

This code is used in section 14.

17.   We don't need to charge any mems for fetching *g⃗vertices*, because *krusk* has already referred to it.

⟨ Put all the vertices into components by themselves 17 ⟩ ≡
    for $(v = g{\rightarrow}vertices;\ v < g{\rightarrow}vertices + n;\ v{++})$ {
        $oo, v{\rightarrow}clink = v{\rightarrow}comp = v;\ o, v{\rightarrow}csize = 1;$
    }
    components = *n*;

This code is used in section 14.

---

a: **util**, GB_GRAPH §10.
A: **Arc** *, GB_GRAPH §8.
Arc = **struct**, GB_GRAPH §10.
arcs: **Arc** *, GB_GRAPH §9.
b: **util**, GB_GRAPH §10.
Graph = **struct**, GB_GRAPH §20.
I: **long**, GB_GRAPH §8.
INFINITY = macro, §5.

len: **long**, GB_GRAPH §10.
mems: **long**, §10.
n: **long**, GB_GRAPH §20.
next: **Arc** *, GB_GRAPH §10.
o = macro, §10.
oo = macro, §10.
printf: **int** ( ), <stdio.h>.
report: **int** ( ), §7.

tip: **Vertex** *, GB_GRAPH §10.
V: **Vertex** *, GB_GRAPH §8.
verbose: **long**, GB_GRAPH §5.
Vertex = **struct**, GB_GRAPH §9.
vertices: **Vertex** *, GB_GRAPH §20.
x: **util**, GB_GRAPH §9.
y: **util**, GB_GRAPH §9.
z: **util**, GB_GRAPH §9.

18. The operation of merging two components together requires us to change two *clink* pointers, one *csize* field, and the *comp* fields in each vertex of the smaller component.

Here we charge two mems for the first **if** test, since $u\text{-}csize$ and $v\text{-}csize$ are being fetched from memory. Then we charge only one mem when $u\text{-}csize$ is being updated, since the values being added together have already been fetched. True, the compiler has to be smart to realize that it's safe to add the fetched values $u\text{-}csize + v\text{-}csize$ even though u and v might have been swapped in the meantime; but we are assuming that the compiler is extremely clever. (Otherwise we would have to clutter up our program every time we don't trust the compiler. After all, programs that count mems are intended primarily to be read. They aren't intended for production jobs.)

⟨ Merge the components containing u and v 18 ⟩ ≡

```
u = u-comp;      /* u-comp has already been fetched from memory */
v = v-comp;      /* ditto for v-comp */
if (oo, u-csize < v-csize) {
   w = u;  u = v;  v = w;
}     /* now v's component is smaller than u's (or equally small) */
o, u-csize += v-csize;
o, w = v-clink;
oo, v-clink = u-clink;
o, u-clink = w;
for ( ; ; o, w = w-clink) {
   o, w-comp = u;
   if (w ≡ v) break;
}
```

This code is used in section 14.

19. Jarník and Prim's algorithm. A second approach to minimum spanning trees is also pretty simple, except for one technicality: We want to write it in a sufficiently general manner that different priority queue algorithms can be plugged in. The basic idea is to choose an arbitrary vertex v_0 and connect it to its nearest neighbor v_1, then to connect that fragment to its nearest neighbor v_2, and so on. A priority queue holds all vertices that are adjacent to but not already in the current fragment; the key value stored with each vertex is its distance to the current fragment.

We want the priority queue data structure to support the four operations *init_queue*(d), *enqueue*(v, d), *requeue*(v, d), and *del_min*( ), described in the GB_DIJK module. Dijkstra's algorithm for shortest paths, described there, is remarkably similar to Jarník and Prim's algorithm for minimum spanning trees; in fact, Dijkstra discovered the latter algorithm independently, at the same time as he came up with his procedure for shortest paths.

As in GB_DIJK, we define pointers to priority queue subroutines so that the queueing mechanism can be varied.

#define *dist*   *z.I*   /* this is the key field for vertices in the priority queue */
#define *backlink* *y.V*   /* this vertex is the stated *dist* away */

⟨ Global variables 3 ⟩ +≡
    void (*init_queue*)( );   /* create an empty priority queue */
    void (*enqueue*)( );   /* insert a new element in the priority queue */
    void (*requeue*)( );   /* decrease the key of an element in the queue */
    Vertex *(*del_min*)( );   /* remove an element with smallest key */

---

clink = z.V, §16.
comp = y.V, §16.
csize = x.I, §16.
d: **unsigned long**, §2.
I: **long**, GB_GRAPH §8.

o = macro, §10.
oo = macro, §10.
u: **register Vertex** *, §15.
v: **register Vertex** *, §15.
V: **Vertex** *, GB_GRAPH §8.

Vertex = **struct**, GB_GRAPH §9.
w: **register Vertex** *, §15.
y: **util**, GB_GRAPH §9.
z: **util**, GB_GRAPH §9.

20. The vertices in this algorithm are initially "unseen"; they become "seen" when they enter the priority queue, and finally "known" when they leave it and enter the current fragment. We will put a special constant in the *backlink* field of known vertices. A vertex will be unseen if and only if its *backlink* is Λ.

#define KNOWN (**Vertex** *) 1   /* special *backlink* to mark known vertices */
⟨ Subroutines 7 ⟩ +≡
  unsigned long *jar_pr*(*g*)
      Graph **g*;
  { **register Vertex** **t*;   /* vertex that is just becoming known */
    long *fragment_size*;   /* number of vertices in the tree so far */
    unsigned long *tot_len* = 0;   /* sum of edge lengths in the tree so far */
    mems = 0;
    ⟨ Make *t* = *g*⃗*vertices* the only vertex seen; also make it known 21 ⟩;
    while (*fragment_size* < *g*⃗*n*) {
      ⟨ Put all unseen vertices adjacent to *t* into the queue, and update the distances of the other
          vertices adjacent to *t* 22 ⟩;
      t = (**del_min*)( );
      if (*t* ≡ Λ) **return** INFINITY;   /* the graph is disconnected */
      if (*verbose*) *report*(*t*⃗*backlink*, *t*, *t*⃗*dist*);
      o, *tot_len* += *t*⃗*dist*;
      o, *t*⃗*backlink* = KNOWN;
      fragment_size ++;
    }
    return *tot_len*;
  }

21. Notice that we don't charge any mems for the subroutine call to *init_queue*, except for mems counted in the subroutine itself. What should we charge in general for subroutine linkage when we are counting mems? The parameters to subroutines generally go into registers, and registers are "free"; also, a compiler can often choose to implement a procedure in line, thereby reducing the overhead to zero. Hence, the recommended method for charging mems with respect to subroutines is: Charge nothing if the subroutine is not recursive; otherwise charge twice the number of things that need to be saved on a runtime stack. (The return address is one of the things that needs to be saved.)

⟨ Make *t* = *g*⃗*vertices* the only vertex seen; also make it known 21 ⟩ ≡
  for (*oo*, *t* = *g*⃗*vertices* + *g*⃗*n* − 1; *t* > *g*⃗*vertices*; *t*−−) *o*, *t*⃗*backlink* = Λ;
  o, *t*⃗*backlink* = KNOWN;
  fragment_size = 1;
  (**init_queue*)(0_L);   /* make the priority queue empty */
This code is used in section 20.

22. ⟨ Put all unseen vertices adjacent to *t* into the queue, and update the distances of the other
      vertices adjacent to *t* 22 ⟩ ≡
  { **register Arc** **a*;   /* an arc leading from *t* */
    for (*o*, *a* = *t*⃗*arcs*; *a*; *o*, *a* = *a*⃗*next*) {
      register Vertex **v*;   /* a vertex adjacent to *t* */

```
    o, v = a→tip;
    if (o, v→backlink) {      /* v has already been seen */
      if (v→backlink > KNOWN) {
        if (oo, a→len < v→dist) {
          o, v→backlink = t;
          (*requeue)(v, a→len);      /* we found a better way to get there */
        }
      }
    } else {      /* v hasn't been seen before */
      o, v→backlink = t;
      o, (*enqueue)(v, a→len);
    }
  }
}
```

This code is used in section 20.

Arc = **struct**, GB_GRAPH §10.
arcs: **Arc** ∗, GB_GRAPH §9.
backlink = *y*.*V*, §19.
del_min: **Vertex** ∗(∗)( ), §19.
dist = *z*.*I*, §19.
enqueue: **void** (∗)( ), §19.
Graph = **struct**, GB_GRAPH §20.

INFINITY = macro, §5.
init_queue: **void** (∗)( ), §19.
len: **long**, GB_GRAPH §10.
mems: **long**, §10.
n: **long**, GB_GRAPH §20.
next: **Arc** ∗, GB_GRAPH §10.
o = macro, §10.

oo = macro, §10.
report: **int** ( ), §7.
requeue: **void** (∗)( ), §19.
tip: **Vertex** ∗, GB_GRAPH §10.
verbose: **long**, GB_GRAPH §5.
Vertex = **struct**, GB_GRAPH §9.
vertices: **Vertex** ∗, GB_GRAPH §20.

23. Binary heaps. To complete the *jar_pr* routine, we need to fill in the four priority queue functions. Jarník wrote his original paper before computers were known; Prim and Dijkstra wrote theirs before efficient priority queue algorithms were known. Their original algorithms therefore took $\Theta(n^2)$ steps. Kerschenbaum and Van Slyke pointed out in 1972 that binary heaps could do better. A simplified version of binary heaps (invented by Williams in 1964) is presented here.

A binary heap is an array of n elements, and we need space for it. Fortunately the space is already there; we can use utility field u in each of the vertex records of the graph. Moreover, if *heap_elt*(i) points to vertex v, we will arrange things so that $v \rightarrow$*heap_index* $= i$.

```
#define heap_elt(i) (gv + i)→u.V      /* the ith vertex of the heap; gv = g→vertices */
#define heap_index v.I      /* the v utility field says where a vertex is in the heap */
```
⟨ Global variables 3 ⟩ +≡
```
    Vertex *gv;      /* g→vertices, the base of the heap array */
    long hsize;      /* the number of elements currently in the heap */
```

24. To initialize the heap, we need only initialize two "registers" to known values, so we don't have to charge any mems at all. (In a production implementation, this code would appear in-line as part of the spanning tree algorithm.)

Important Note: This routine refers to the global variable g, which is set in *main* (not in *jar_pr*). Suitable changes need to be made if these binary heap routines are used in other programs.

⟨ Priority queue subroutines 24 ⟩ ≡
```
    void init_heap(d)      /* makes the heap empty */
        long d;
    {
        gv = g→vertices;
        hsize = 0;
    }
```
See also sections 25, 26, 27, 30, 33, 34, 38, 45, 50, 51, 52, and 54.

This code is used in section 2.

25. The key invariant property that makes heaps work is

$$heap\_elt(k/2) \rightarrow dist \leq heap\_elt(k) \rightarrow dist, \qquad \text{for } 1 < k \leq hsize.$$

(A reader who has not seen heap ordering before should stop at this point and study the beautiful consequences of this innocuously simple set of inequalities.) The enqueueing operation turns out to be quite simple:

⟨ Priority queue subroutines 24 ⟩ +≡
```
    void enq_heap(v, d)
        Vertex *v;      /* vertex that is entering the queue */
        long d;      /* its key (aka dist) */
    { register unsigned long k;      /* position of a "hole" in the heap */
      register unsigned long j;      /* the parent of that position */
      register Vertex *u;      /* heap_elt(j) */

      o, v→dist = d;
      k = ++hsize;
      j = k ≫ 1;      /* k/2 */
```

```
while (j > 0 ∧ (oo, (u = heap_elt(j))→dist > d)) {
    o, heap_elt(k) = u;        /* the hole moves to parent position */
    o, u→heap_index = k;
    k = j;
    j = k ≫ 1;
}
o, heap_elt(k) = v;
o, v→heap_index = k;
}
```

26.   And in fact, the general requeueing operation is almost identical to enqueueing. This operation is popularly called "siftup," because the vertex whose key is being reduced may displace its ancestors higher in the heap. We could have implemented enqueueing by first placing the new element at the end of the heap, then requeueing it; that would have cost at most a couple mems more.

⟨ Priority queue subroutines 24 ⟩ +≡

```
void req_heap(v, d)
    Vertex *v;        /* vertex whose key is being reduced */
    long d;        /* its new dist */
{ register unsigned long k;        /* position of a "hole" in the heap */
  register unsigned long j;        /* the parent of that position */
  register Vertex *u;        /* heap_elt(j) */

  o, v→dist = d;
  o, k = v→heap_index;        /* now heap_elt(k) = v */
  j = k ≫ 1;        /* k/2 */
  if (j > 0 ∧ (oo, (u = heap_elt(j))→dist > d)) {        /* change is needed */
    do {
      o, heap_elt(k) = u;        /* the hole moves to parent position */
      o, u→heap_index = k;
      k = j;
      j = k ≫ 1;        /* k/2 */
    } while (j > 0 ∧ (oo, (u = heap_elt(j))→dist > d));
    o, heap_elt(k) = v;
    o, v→heap_index = k;
  }
}
```

$dist = z.I$, §19.
g: **Graph** *, §3.
I: **long**, GB_GRAPH §8.
jar_pr: **unsigned long** ( ), §20.

main: **int** ( ), §2.
o = macro, §10.
oo = macro, §10.
u: **util**, GB_GRAPH §9.

v: **util**, GB_GRAPH §9.
V: **Vertex** *, GB_GRAPH §8.
Vertex = **struct**, GB_GRAPH §9.
vertices: **Vertex** *, GB_GRAPH §20.

27. Finally, the procedure for removing the vertex with smallest key is only a bit more difficult. The vertex to be removed is always *heap_elt*(1). After we delete it, we "sift down" *heap_elt*(*hsize*), until the basic heap inequalities hold once again.

At a crucial point in this process, we have $j\text{-}dist < u\text{-}dist$. We cannot then have $j = hsize + 1$, because the previous steps have made $(hsize + 1)\text{-}dist = u\text{-}dist = d$.

⟨ Priority queue subroutines 24 ⟩ +≡

```
Vertex *del_heap( )
{ Vertex *v;      /* vertex to return */
  register Vertex *u;      /* vertex being sifted down */
  register unsigned long k;      /* hole in the heap */
  register unsigned long j;      /* child of that hole */
  register long d;      /* u⁻dist, the vertex of the vertex being sifted */

  if (hsize ≡ 0) return Λ;
  o, v = heap_elt(1);
  o, u = heap_elt(hsize --);
  o, d = u⁻dist;
  k = 1;
  j = 2;
  while (j ≤ hsize) {
    if (oooo, heap_elt(j)⁻dist > heap_elt(j + 1)⁻dist) j++;
    if (heap_elt(j)⁻dist ≥ d) break;
    o, heap_elt(k) = heap_elt(j);      /* NB: we cannot have j > hsize, see above */
    o, heap_elt(k)⁻heap_index = k;
    k = j;      /* the hole moves to child position */
    j = k ≪ 1;      /* 2k */
  }
  o, heap_elt(k) = u;
  o, u⁻heap_index = k;
  return v;
}
```

28. OK, here's how we plug binary heaps into Jarník/Prim.

⟨ Execute *jar_pr*(*g*) with binary heaps as the priority queue algorithm 28 ⟩ ≡

```
init_queue = init_heap;
enqueue = enq_heap;
requeue = req_heap;
del_min = del_heap;
if (sp_length ≠ jar_pr(g)) {
  printf("␣...oops,␣I've␣got␣a␣bug,␣please␣fix␣fix␣fix\n");
  return −4;
}
```

This code is used in section 5.

29. Fibonacci heaps. The running time of Jarník/Prim with binary heaps, when the algorithm is applied to a connected graph with n vertices and m edges, is $O(m \log n)$, because the total number of operations is $O(m + n) = O(m)$ and each heap operation takes at most $O(\log n)$ time.

Fibonacci heaps were invented by Fredman and Tarjan in 1984, in order to do better than this. The Jarník/Prim algorithm does $O(n)$ enqueueing operations, $O(n)$ delete-min operations, and $O(m)$ requeueing operations; so Fredman and Tarjan designed a data structure that would support requeueing in "constant amortized time." In other words, Fibonacci heaps allow us to do m requeueing operations with a total cost of $O(m)$, even though some of the individual requeueings might take longer. The resulting asymptotic running time is then $O(m + n \log n)$. (This turns out to be optimum within a constant factor, when the same technique is applied to Dijkstra's algorithm for shortest paths. But for minimum spanning trees the Fibonacci method is not always optimum; for example, if $m \approx n \sqrt{\log n}$, the algorithm of Cheriton and Tarjan has slightly better asymptotic behavior, $O(m \log \log n)$.)

Fibonacci heaps are more complex than binary heaps, so we can expect that overhead costs will make them non-competitive unless m and n are quite large. Furthermore, it is not clear that the running time with simple binary heaps will behave as $m \log n$ on realistic data, because $O(m \log n)$ is a worst-case estimate based on rather pessimistic assumptions. (For example, requeueing might rarely require many iterations of the siftup loop.) But it will be instructive to implement Fibonacci heaps as best we can, just to see how good they look in actual practice.

Let us say that the *rank* of a node in a forest is the number of children it has. A Fibonacci heap is an unordered forest of trees in which the key of each node is less than or equal to the key of each child of that node, and in which the following further condition, called property F, also holds: The ranks $\{r_1, r_2, \ldots, r_k\}$ of the children of every node of rank k, when put into nondecreasing order $r_1 \le r_2 \le \cdots \le r_k$, satisfy $r_j \ge j - 2$ for all j.

As a consequence of property F, we can prove by induction that every node of rank k has at least F_{k+2} descendants (including itself). Therefore, for example, we cannot have a node of rank ≥ 30 unless the total size of the forest is at least $F_{32} = 2{,}178{,}309$. We cannot have a node of rank ≥ 46 unless the total size of the forest exceeds 2^{32}.

---

30. We will represent a Fibonacci heap with a rather elaborate data structure, in order to guarantee the efficiency of all the necessary operations. Each node will have four pointers: *parent*, the node's parent (or Λ if the node is a root); *child*, one of the node's children (or undefined if the node has no children); *lsib* and *rsib*, the node's left and right siblings. The children of each node, and the roots of the forest, are doubly linked by *lsib* and *rsib* in circular lists; the nodes in these lists can appear in any convenient order, and the *child* pointer can point to any child.

Besides the four pointers, there is a *rank* field, which tells how many children exist, and a *tag* field, which is either 0 or 1.

Suppose a node has children of ranks $\{r_1, r_2, \ldots, r_k\}$, where $r_1 \leq r_2 \leq \cdots \leq r_k$. We know that $r_j \geq j - 2$ for all j; we say that the node has l *critical* children if there are l cases of equality, where $r_j = j - 2$. Our implementation will guarantee that any node with l critical children will have at least l tagged children of the corresponding ranks. For example, suppose a node has seven children, of respective ranks $\{1, 1, 1, 2, 4, 4, 6\}$. Then it has three critical children, because $r_3 = 1$, $r_4 = 2$, and $r_6 = 4$. In our implementation, at least one of the children of rank 1 will have *tag* = 1, and so will the child of rank 2; so will one of the children of rank 4.

There is an external pointer called *F_heap*, which indicates a node whose key is smallest. (If the heap is empty, *F_heap* is Λ.)

⟨ Priority queue subroutines 24 ⟩ +≡
   void *init_F_heap*(d)
      long d;
  { *F_heap* = Λ; }

31. ⟨ Global variables 3 ⟩ +≡
   Vertex **F_heap*;   /* pointer to the ring of root nodes */

32. We can save a bit of space and time by combining the *rank* and *tag* fields into a single *rank_tag* field, which contains *rank* * 2 + *tag*.

Vertices in GraphBase graphs have six utility fields. That's just enough for *parent*, *child*, *lsib*, *rsib*, *rank_tag*, and the key field *dist*. But unfortunately we also need the *backlink* field, so we are over the limit. That's not really so bad, however; we can set up another array of n records, and point to it. The extra running time needed for indirect pointing does not have to be charged to mems, because a production system involving Fibonacci heaps would simply redefine **Vertex** records to have seven utility fields instead of six. In this way we can simulate the behavior of larger records without changing the basic GraphBase conventions.

We will want an **Arc** record for each vertex in our next algorithm, so we might as well allocate storage for it now even though Fibonacci heaps need only two of the five fields.

#define *newarc*   *u.A*   /* *v*⁓*newarc* points to an **Arc** record associated with *v* */
#define *parent*   *newarc→tip*
#define *child*   *newarc→a.V*
#define *lsib*   *v.V*
#define *rsib*   *w.V*
#define *rank_tag*   *x.I*

⟨ Allocate additional space needed by the more complex algorithms; or **goto** *done* if there isn't enough room 32 ⟩ ≡
  { **register Arc** **aa*;
    register Vertex **uu*;

```
aa = gb_typed_alloc(g⃗n, Arc, g⃗aux_data);
if (aa ≡ Λ) {
    printf("␣and␣there␣isn't␣enough␣space␣to␣try␣the␣other␣methods.\n\n");
    goto done;
}
for (uu = g⃗vertices; uu < g⃗vertices + g⃗n; uu++, aa++) uu⃗newarc = aa;
}
```

This code is used in section 5.

33. The *potential energy* of a Fibonacci heap, as we are representing it, is defined to be the number of trees in the forest plus twice the total number of tagged children. When we operate on a heap, we will store potential energy to be used up later; then it will be possible to do the later operations with only a small incremental cost to the running time. (Potential energy is just a way to prove that the amortized cost is small; it does not appear explicitly in our implementation. It simply explains why the number of mems we compute will always be $O(m + n \log n)$.)

Enqueueing is easy: We simply insert the new element as a new tree in the forest. This costs a constant amount of time, including the cost of one new unit of potential energy for the new tree.

We can assume that *F_heap⃗dist* appears in a register, so we need not charge a mem to fetch it.

⟨ Priority queue subroutines 24 ⟩ +≡
```
void enq_F_heap(v, d)
    Vertex *v;       /* vertex that is entering the queue */
    long d;          /* its key (aka dist) */
{
    o, v⃗dist = d;
    o, v⃗parent = Λ;
    o, v⃗rank_tag = 0;      /* v⃗child need not be set */
    if (F_heap ≡ Λ) {
        oo, F_heap = v⃗lsib = v⃗rsib = v;
    } else { register Vertex *u;
        o, u = F_heap⃗lsib;
        o, v⃗lsib = u;
        o, v⃗rsib = F_heap;
        oo, F_heap⃗lsib = u⃗rsib = v;
        if (F_heap⃗dist > d) F_heap = v;
    }
}
```

---

a: **util**, GB_GRAPH §10.
A: **Arc** *, GB_GRAPH §8.
Arc = struct, GB_GRAPH §10.
aux_data: **Area**, GB_GRAPH §20.
backlink = y.V, §19.
dist = z.I, §19.
done: label, §5.
g: **Graph** *, §3.

gb_typed_alloc = macro ( ),
    GB_GRAPH §11.
I: **long**, GB_GRAPH §8.
n: **long**, GB_GRAPH §20.
o = macro, §10.
oo = macro, §10.
printf: **int** ( ), <stdio.h>.
tip: **Vertex** *, GB_GRAPH §10.

u: **util**, GB_GRAPH §9.
v: **util**, GB_GRAPH §9.
V: **Vertex** *, GB_GRAPH §8.
Vertex = struct, GB_GRAPH §9.
vertices: **Vertex** *, GB_GRAPH §20.
w: **util**, GB_GRAPH §9.
x: **util**, GB_GRAPH §9.

34. Requeueing is of medium difficulty. If the key is being decreased in a root node, or if the decrease doesn't make the key less than the key of its parent, no links need to change (except possibly *F_heap* itself). Otherwise we detach the node and its descendants from its present family and put this former subtree into the forest as a new tree. (One unit of potential energy must be stored with it.)

The rank of the former parent, p, decreases by 1. If p is a root, we're done. Otherwise if p was not tagged, we tag it (and pay for two additional units of energy). Property F still holds, because an untagged node can always admit a decrease in rank. If p was tagged, however, we detach p and its remaining descendants, making it another new tree of the forest, with p no longer tagged. Removing the tag releases enough stored energy to pay for the extra work of moving p. Then we must decrease the rank of p's parent, and so on, until finally we get to a root or to an untagged node. The total net cost is at most three units of energy plus the cost of relinking the original node, so it is $O(1)$.

We needn't clear the tag fields of root nodes, because we never look at them.

⟨ Priority queue subroutines 24 ⟩ +≡

```
void req_F_heap(v, d)
        Vertex *v;        /* vertex whose key is being reduced */
        long d;        /* its new dist */
{ register Vertex *p, *pp;        /* parent and grandparent of v */
  register Vertex *u, *w;        /* other vertices being modified */
  register long r;        /* twice the rank plus the tag */
  o, v⃗dist = d;
  o, p = v⃗parent;
  if (p ≡ Λ) {
    if (F_heap⃗dist > d)  F_heap = v;
  } else if (o, p⃗dist > d)
    while (1) {
      o, r = p⃗rank_tag;
      if (r ≥ 4)        /* v is not an only child */
        ⟨ Remove v from its family 35 ⟩;
      ⟨ Insert v into the forest 36 ⟩;
      o, pp = p⃗parent;
      if (pp ≡ Λ) {        /* the parent of v is a root */
        o, p⃗rank_tag = r - 2;  break;
      }
      if ((r & 1) ≡ 0) {        /* the parent of v is untagged */
        o, p⃗rank_tag = r - 1;  break;        /* now it's tagged */
      } else  o, p⃗rank_tag = r - 2;        /* tagged parent will become a root */
      v = p;  p = pp;
    }
}
```

35. ⟨ Remove v from its family 35 ⟩ ≡
```
{
  o, u = v⃗lsib;
  o, w = v⃗rsib;
  o, u⃗rsib = w;
```

    $o, w \to lsib = u$;
    if $(o, p \to child \equiv v)$ $o, p \to child = w$;
  }
This code is used in section 34.

36. ⟨ Insert v into the forest 36 ⟩ ≡
  $o, v \to parent = \Lambda$;
  $o, u = F\_heap \to lsib$;
  $o, v \to lsib = u$;
  $o, v \to rsib = F\_heap$;
  $oo, F\_heap \to lsib = u \to rsib = v$;
  if $(F\_heap \to dist > d)$ $F\_heap = v$;   /* this can happen only with the original v */
This code is used in section 34.

37.   The *del_min* operation is even more interesting; this, in fact, is where most of the action lies. We know that F_heap points to the vertex v we will be deleting. That's nice, but we need to figure out the new value of F_heap. So we have to look at all the children of v and at all the root nodes in the forest. We have stored up enough potential energy to do that, but we can reclaim the potential only if we rebuild the Fibonacci heap so that the rebuilt version contains relatively few trees.

    The solution is to make sure that the new heap has at most one root of each rank. Whenever we have two tree roots of equal rank, we can make one the child of the other, thus reducing the number of trees by 1. (The new child does not violate Property F, nor is it critical, so we can mark it untagged.) The largest rank is always $O(\log n)$, if there are n nodes altogether, and we can afford to pay $\log n$ units of time for the work that isn't reclaimed from potential energy.

    An array of pointers to roots of known rank is used to help control this part of the process.

⟨ Global variables 3 ⟩ +≡
  Vertex $*new\_roots[46]$;   /* big enough for queues of size 2^{32} */

---

child = macro, §32.
del_min: **Vertex** *(*)( ), §19.
dist = $z.I$, §19.
F_heap: **Vertex** *, §31.

lsib = $v.V$, §32.
o = macro, §10.
oo = macro, §10.
parent = macro, §32.

rank_tag = $x.I$, §32.
rsib = $w.V$, §32.
Vertex = **struct**, GB_GRAPH §9.

38. ⟨ Priority queue subroutines 24 ⟩ +≡

   Vertex *∗del_F_heap*( )

  { **Vertex** *∗final_v = F_heap*;   /∗ the node to return ∗/

    register Vertex *∗t, ∗u, ∗v, ∗w*;   /∗ registers for manipulation of links ∗/

    register long $h = -1$;   /∗ the highest rank present in *new_roots* ∗/

    register long *r*;   /∗ rank of current tree ∗/

    if (*F_heap*) {

      if ($o, F\_heap{\rightarrow}rank\_tag < 2$) $o, v = F\_heap{\rightarrow}rsib$;

      else {

        $o, w = F\_heap{\rightarrow}child$; $o, v = w{\rightarrow}rsib$;

        $oo, w{\rightarrow}rsib = F\_heap{\rightarrow}rsib$;   /∗ link children of deleted node into the list ∗/

        for ($w = v$; $w \neq F\_heap{\rightarrow}rsib$; $o, w = w{\rightarrow}rsib$) $o, w{\rightarrow}parent = \Lambda$;

      }

      while ($v \neq F\_heap$) {

        $o, w = v{\rightarrow}rsib$;

        ⟨ Put the tree rooted at *v* into the *new_roots* forest 39 ⟩;

        $v = w$;

      }

      ⟨ Rebuild *F_heap* from *new_roots* 41 ⟩;

    }

    return *final_v*;

  }

39. The work we do in this step is paid for by the unit of potential energy being freed as *v* leaves the old forest, except for the work of increasing *h*; we charge the latter to the $O(\log n)$ cost of building *new_roots*.

⟨ Put the tree rooted at *v* into the *new_roots* forest 39 ⟩ ≡

  $o, r = v{\rightarrow}rank\_tag \gg 1$;

  while (1) {

    if ($h < r$) {

      do { $h{+}{+}$; $o, new\_roots[h] = (h \equiv r ? v : \Lambda)$; } **while** ($h < r$);

      break;

    }

    if ($o, new\_roots[r] \equiv \Lambda$) {

      $o, new\_roots[r] = v$;

      break;

    }

    $u = new\_roots[r]$;

    $o, new\_roots[r] = \Lambda$;

    if ($oo, u{\rightarrow}dist < v{\rightarrow}dist$) {

      $o, v{\rightarrow}rank\_tag = r \ll 1$;   /∗ *v* is not critical and needn't be tagged ∗/

      $t = u$; $u = v$; $v = t$;

    }

    ⟨ Make *u* a child of *v* 40 ⟩;

    $r{+}{+}$;

  }

  $o, v{\rightarrow}rank\_tag = r \ll 1$;   /∗ every root in *new_roots* is untagged ∗/

This code is used in section 38.

40.   When we get to this step, u and v both have rank r, and $u\text{-}dist \geq v\text{-}dist$; u is untagged.

⟨ Make u a child of v 40 ⟩ ≡
```
  if (r ≡ 0) {
    o, v⁻child = u;
    oo, u⁻lsib = u⁻rsib = u;
  } else {
    o, t = v⁻child;
    oo, u⁻rsib = t⁻rsib;
    o, u⁻lsib = t;
    oo, u⁻rsib⁻lsib = t⁻rsib = u;
  }
  u⁻parent = v;
```
This code is used in section 39.

41.   And now we can breathe easy, because the last step is trivial.

⟨ Rebuild *F_heap* from *new_roots* 41 ⟩ ≡
```
  if (h < 0) F_heap = Λ;
  else { long d;      /* smallest key value seen so far */
    o, u = v = new_roots[h];      /* u and v will point to beginning and end of list, respectively */
    o, d = u⁻dist;
    F_heap = u;
    for (h−−; h ≥ 0; h−−)
      if (o, new_roots[h]) {
        w = new_roots[h];
        o, w⁻lsib = v;
        o, v⁻rsib = w;
        if (o, w⁻dist < d) {
          F_heap = w;
          d = w⁻dist;
        }
        v = w;
      }
    o, v⁻rsib = u;
    o, u⁻lsib = v;
  }
```
This code is used in section 38.

$child$ = macro, §32.
$dist$ = z.I, §19.
F_heap: **Vertex** ∗, §31.
$lsib$ = v.V, §32.

new_roots: **Vertex** ∗[], §37.
o = macro, §10.
oo = macro, §10.
$parent$ = macro, §32.

$rank\_tag$ = x.I, §32.
$rsib$ = w.V, §32.
Vertex = **struct**, GB_GRAPH §9.

42. ⟨ Execute *jar_pr*(*g*) with Fibonacci heaps as the priority queue algorithm 42 ⟩ ≡

    init_queue = *init_F_heap*;

    enqueue = *enq_F_heap*;

    requeue = *req_F_heap*;

    del_min = *del_F_heap*;

    if (*sp_length* ≠ *jar_pr*(*g*)) {

        printf("␣...oops,␣I'␣ve␣got␣a␣bug,␣please␣fix␣fix␣fix\n");

        return −5;

    }

This code is used in section 5.

43. Binomial queues. Jean Vuillemin's "binomial queue" structures [*CACM* **21** (1978), 309–314] provide yet another appealing way to maintain priority queues. A binomial queue is a forest of trees with keys ordered as in Fibonacci heaps, satisfying two conditions that are considerably stronger than the Fibonacci heap property: Each node of rank k has children of respective ranks $\{0, 1, \ldots, k-1\}$; and each root of the forest has a different rank. It follows that each node of rank k has exactly 2^k descendants (including itself), and that a binomial queue of n elements has exactly as many trees as the number n has 1's in binary notation.

We could plug binomial queues into the Jarník/Prim algorithm, but they don't offer advantages over the heap methods already considered because they don't support the requeueing operation as nicely. Binomial queues do, however, permit efficient merging—the operation of combining two priority queues into one—and they achieve this without as much space overhead as Fibonacci heaps. In fact, we can implement binomial queues with only two pointers per node, namely a pointer to the largest child and another to the next sibling. This means we have just enough space in the utility fields of GraphBase **Arc** records to link the arcs that extend out of a spanning tree fragment. The algorithm of Cheriton, Tarjan, and Karp, which we will consider soon, maintains priority queues of arcs, not vertices; and it requires the operation of merging, not requeueing. Therefore binomial queues are well suited to it, and we will prepare ourselves for that algorithm by implementing basic binomial queue procedures.

Incidentally, if you wonder why Vuillemin called his structure a binomial queue, it's because the trees of 2^k elements have many pleasant combinatorial properties, among which is the fact that the number of elements on level l is the binomial coefficient $\binom{k}{l}$. The backtrack tree for subsets of a k-set has the same structure. A picture of a binomial-queue tree with $k = 5$, drawn by Jill C. Knuth, appears as the frontispiece of *The Art of Computer Programming*, facing page 1 of Volume 1.

```
#define qchild  a.A    /* pointer to the arc for largest child of an arc */
#define qsib   b.A    /* pointer to next larger sibling, or from largest to smallest */
```

---

a: **util**, GB_GRAPH §10.
A: **Arc** ∗, GB_GRAPH §8.
Arc = **struct**, GB_GRAPH §10.
b: **util**, GB_GRAPH §10.
del_F_heap: **Vertex** ∗( ), §38.
del_min: **Vertex** ∗(∗)( ), §19.

enq_F_heap: **void** ( ), §33.
enqueue: **void** (∗)( ), §19.
g: **Graph** ∗, §3.
init_F_heap: **void** ( ), §30.
init_queue: **void** (∗)( ), §19.

jar_pr: **unsigned long** ( ), §20.
printf: **int** ( ), <stdio.h>.
req_F_heap: **void** ( ), §34.
requeue: **void** (∗)( ), §19.
sp_length: **unsigned long**, §6.

44. A special header node is used at the head of a binomial queue, to represent the queue itself. The *qsib* field of this node points to the smallest root node in the forest. ("Smallest" means smallest in rank, not in key value.) The header also contains a *qcount* field, which takes the place of *qchild*; the *qcount* is the total number of nodes, so its binary representation characterizes the sizes of the trees accessible from *qsib*.

For example, suppose a queue with header node h contains five elements $\{a, b, c, d, e\}$ whose keys happen to be ordered alphabetically. The first tree might be the single node c; the other tree might be rooted at a, with children e and b. Then we have

$$h \rightarrow qcount = 5, \qquad h \rightarrow qsib = c;$$
$$c \rightarrow qsib = a;$$
$$a \rightarrow qchild = b;$$
$$b \rightarrow qchild = d, \qquad b \rightarrow qsib = e;$$
$$e \rightarrow qsib = b.$$

The other fields $c \rightarrow qchild$, $a \rightarrow qsib$, $e \rightarrow qchild$, $d \rightarrow qsib$, and $d \rightarrow qchild$ are undefined. We can save time by not loading or storing the undefined fields, which make up about 3/8 of the structure.

An empty binomial queue would have $h \rightarrow qcount = 0$ and $h \rightarrow qsib$ undefined.

Like Fibonacci heaps, binomial queues store potential energy: The number of energy units present is simply the number of trees in the forest.

#define *qcount* *a.I*   /* this field takes the place of *qchild* in header nodes */

45. Most of the operations we want to do with binomial queues rely on the following basic subroutine, which merges a forest of m nodes starting at q with a forest of mm nodes starting at qq, putting a pointer to the resulting forest of $m + mm$ nodes into $h \rightarrow qsib$. The amortized running time is $O(\log m)$, independent of mm.

The *len* field, not *dist*, is the key field for this queue, because our nodes in this case are arcs instead of vertices.

⟨ Priority queue subroutines 24 ⟩ +≡

```
qunite(m, q, mm, qq, h)
     register long m, mm;     /* number of nodes in the forests */
     register Arc *q, *qq;     /* binomial trees in the forests, linked by qsib */
     Arc *h;     /* h→qsib will get the result */
{ register Arc *p;     /* tail of the list built so far */
  register long k = 1;     /* size of trees currently being processed */
  p = h;
  while (m) {
    if ((m & k) ≡ 0) {
      if (mm & k) {     /* qq goes into the merged list */
        o, p→qsib = qq; p = qq; mm −= k;
        if (mm) o, qq = qq→qsib;
      }
    } else if ((mm & k) ≡ 0) {     /* q goes into the merged list */
      o, p→qsib = q; p = q; m −= k;
      if (m) o, q = q→qsib;
```

```
    } else ⟨Combine q and qq into a "carry" tree, and continue merging until the carry no longer
            propagates 46⟩;
    k ≪= 1;
  }
  if (mm) o, p⃗qsib = qq;
}
```

46. As we have seen in Fibonacci heaps, two heap-ordered trees can be combined by simply attaching one as a new child of the other. This operation preserves binomial trees. (In fact, if we use Fibonacci heaps without ever doing a requeue operation, the forests that appear after every *del_min* are binomial queues.) The number of trees decreases by 1, so we have a unit of potential energy to pay for this computation.

⟨Combine q and qq into a "carry" tree, and continue merging until the carry no longer propagates 46⟩ ≡

```
  { register Arc *c;        /* the "carry," a tree of size 2k */
    register long key;      /* c⃗len */
    register Arc *r, *rr;   /* remainders of the input lists */
    m −= k; if (m) o, r = q⃗qsib;
    mm −= k; if (mm) o, rr = qq⃗qsib;
    ⟨Set c to the combination of q and qq 47⟩;
    k ≪= 1;  q = r;  qq = rr;
    while ((m | mm) & k) {
      if ((m & k) ≡ 0) ⟨Merge qq into c and advance qq 49⟩
      else {
        ⟨Merge q into c and advance q 48⟩;
        if (mm & k) {
          o, p⃗qsib = qq;  p = qq;  mm −= k;
          if (mm) o, qq = qq⃗qsib;
        }
      }
      k ≪= 1;
    }
    o, p⃗qsib = c;  p = c;
  }
```

This code is used in section 45.

a: **util**, GB_GRAPH §10.   *dist* = *z.I*, §19.   o = macro, §10.
Arc = **struct**, GB_GRAPH §10.   *I*: **long**, GB_GRAPH §8.   *qchild* = *a.A*, §43.
del_min: **Vertex** *(*)( ), §19.   *len*: **long**, GB_GRAPH §10.   *qsib* = *b.A*, §43.

47. ⟨ Set c to the combination of q and qq 47 ⟩ ≡
 if $(oo, q\text{-}len < qq\text{-}len)$ {
  $c = q, key = q\text{-}len;$
  $q = qq;$
 } **else** $c = qq, key = qq\text{-}len;$
 if $(k \equiv 1)$ $o, c\text{-}qchild = q;$
 else {
  $o, qq = c\text{-}qchild;$
  $o, c\text{-}qchild = q;$
  if $(k \equiv 2)$ $o, q\text{-}qsib = qq;$
  else $oo, q\text{-}qsib = qq\text{-}qsib;$
  $o, qq\text{-}qsib = q;$
 }
This code is used in section 46.

48. At this point, $k > 1$.
⟨ Merge q into c and advance q 48 ⟩ ≡
 {
  $m \mathrel{-}= k;$ **if** (m) $o, r = q\text{-}qsib;$
  if $(o, q\text{-}len < key)$ {
   $rr = c;$ $c = q;$ $key = q\text{-}len;$ $q = rr;$
  }
  $o, rr = c\text{-}qchild;$
  $o, c\text{-}qchild = q;$
  if $(k \equiv 2)$ $o, q\text{-}qsib = rr;$
  else $oo, q\text{-}qsib = rr\text{-}qsib;$
  $o, rr\text{-}qsib = q;$
  $q = r;$
 }
This code is used in section 46.

49. ⟨ Merge qq into c and advance qq 49 ⟩ ≡
 {
  $mm \mathrel{-}= k;$ **if** (mm) $o, rr = qq\text{-}qsib;$
  if $(o, qq\text{-}len < key)$ {
   $r = c;$ $c = qq;$ $key = qq\text{-}len;$ $qq = r;$
  }
  $o, r = c\text{-}qchild;$
  $o, c\text{-}qchild = qq;$
  if $(k \equiv 2)$ $o, qq\text{-}qsib = r;$
  else $oo, qq\text{-}qsib = r\text{-}qsib;$
  $o, r\text{-}qsib = qq;$
  $qq = rr;$
 }
This code is used in section 46.

50. OK, now the hard work is done and we can reap the benefits of the basic *qunite* routine. One easy application enqueues a new arc in $O(1)$ amortized time.

⟨ Priority queue subroutines 24 ⟩ +≡
  qenque(*h, a*)
     Arc *∗h*;   /∗ header of a binomial queue ∗/
     Arc *∗a*;   /∗ new element for that queue ∗/
  { **long** *m*;
    o, m = *h⟶qcount*;
    o, h⟶qcount = *m* + 1;
    if (*m* ≡ 0) *o, h⟶qsib* = *a*;
    else *o, qunite*(1$_L$, *a, m, h⟶qsib, h*);
  }

51. Here, similarly, is a routine that merges one binomial queue into another. The amortized running time is proportional to the logarithm of the number of nodes in the smaller queue.

⟨ Priority queue subroutines 24 ⟩ +≡
  qmerge(*h, hh*)
     Arc *∗h*;   /∗ header of binomial queue that will receive the result ∗/
     Arc *∗hh*;   /∗ header of binomial queue that will be absorbed ∗/
  { **long** *m, mm*;
    o, mm = *hh⟶qcount*;
    if (*mm*) {
      o, m = *h⟶qcount*;
      o, h⟶qcount = *m* + *mm*;
      if (*m* ≥ *mm*) *oo, qunite*(*mm, hh⟶qsib, m, h⟶qsib, h*);
      else if (*m* ≡ 0) *oo, h⟶qsib* = *hh⟶qsib*;
      else *oo, qunite*(*m, h⟶qsib, mm, hh⟶qsib, h*);
    }
  }

Arc = **struct**, GB_GRAPH §10.	*mm*: **register long**, §45.	*qq*: **register Arc** ∗, §45.
c: **register Arc** ∗, §46.	*o* = macro, §10.	*qsib* = *b.A*, §43.
k: **register long**, §45.	*oo* = macro, §10.	*qunite*: **int** ( ), §45.
key: **register long**, §46.	*q*: **register Arc** ∗, §45.	*r*: **register Arc** ∗, §46.
len: **long**, GB_GRAPH §10.	*qchild* = *a.A*, §43.	*rr*: **register Arc** ∗, §46.
m: **register long**, §45.	*qcount* = *a.I*, §44.	

52. The other important operation is, of course, deletion of a node with the smallest key. The amortized running time is proportional to the logarithm of the queue size.

⟨ Priority queue subroutines 24 ⟩ +≡

```
Arc *qdel_min(h)
        Arc *h;       /* header of binomial queue */
{ register Arc *p, *pp;      /* current node and its predecessor */
  register Arc *q, *qq;      /* current minimum node and its predecessor */
  register long key;      /* q→len, the smallest key known so far */
  long m;      /* number of nodes in the queue */
  long k;      /* number of nodes in tree q */
  register long mm;      /* number of nodes not yet considered */

  o, m = h→qcount;
  if (m ≡ 0) return Λ;
  o, h→qcount = m − 1;
  ⟨ Find and remove a tree whose root q has the smallest key 53 ⟩;
  if (k > 2) {
    if (k + k ≤ m)  oo, qunite(k − 1, q→qchild→qsib, m − k, h→qsib, h);
    else  oo, qunite(m − k, h→qsib, k − 1, q→qchild→qsib, h);
  } else if (k ≡ 2)  o, qunite(1_L, q→qchild, m − k, h→qsib, h);
  return q;
}
```

53. If the tree with smallest key is the largest in the forest, we don't have to change any links to remove it, because our binomial queue algorithms never look at the last *qsib* pointer.

We use a well-known binary number trick: $m \& (m − 1)$ is the same as m, except that the least significant 1 bit is deleted.

⟨ Find and remove a tree whose root q has the smallest key 53 ⟩ ≡

```
mm = m & (m − 1);
o, q = h→qsib;
k = m − mm;
if (mm) {      /* there's more than one tree */
  p = q;  qq = h;
  o, key = q→len;
  do { long t = mm & (mm − 1);
    pp = p;  o, p = p→qsib;
    if (o, p→len ≤ key) {
      q = p;  qq = pp;  k = mm − t;  key = p→len;
    }
    mm = t;
  } while (mm);
  if (k + k ≤ m)  oo, qq→qsib = q→qsib;      /* remove the tree rooted at q */
}
```

This code is used in section 52.

54. To complete our implementation, here is an algorithm that traverses a binomial queue, "visiting" each node exactly once, destroying the queue as it goes. The total number of mems required is about $1.75m$.

⟨ Priority queue subroutines 24 ⟩ +≡

```
qtraverse(h, visit)
      Arc *h;      /* head of binomial queue to be unraveled */
      void (*visit)();      /* procedure to be invoked on each node */
{ register long m;      /* the number of nodes remaining */
  register Arc *p, *q, *r;      /* current position and neighboring positions */
  o, m = h→qcount;
  p = h;
  while (m) {
    o, p = p→qsib;
    (*visit)(p);
    if (m & 1)  m−−;
    else {
      o, q = p→qchild;
      if (m & 2)  (*visit)(q);
      else {
        o, r = q→qsib;
        if (m & (m − 1))  oo, q→qsib = p→qsib;
        (*visit)(r);
        p = r;
      }
      m −= 2;
    }
  }
}
```

Arc = **struct**, GB_GRAPH §10.   *oo* = macro, §10.   *qsib* = *b.A*, §43.
len: **long**, GB_GRAPH §10.   *qchild* = *a.A*, §43.   *qunite*: **int** ( ), §45.
o = macro, §10.   *qcount* = *a.I*, §44.

55. Cheriton, Tarjan, and Karp's algorithm. The final algorithm we shall consider takes yet another approach to spanning tree minimization. It operates in two distinct stages: Stage 1 creates small fragments of the minimum tree, working locally with the edges that lead out of each fragment instead of dealing with the full set of edges at once as in Kruskal's method. As soon as the number of component fragments has been reduced from n to $\lfloor\sqrt{n}\,\rfloor$, stage 2 begins. Stage 2 runs through the remaining edges and builds a $\lfloor\sqrt{n}\,\rfloor \times \lfloor\sqrt{n}\,\rfloor$ matrix, which represents the problem of finding a minimum spanning tree on the remaining $\lfloor\sqrt{n}\,\rfloor$ components. A simple $O(\sqrt{n}\,)^2 = O(n)$ algorithm then completes the job.

The philosophy underlying stage 1 is that an edge leading out of a vertex in a small component is likely to lead to a vertex in another component, rather than in the same one. Thus each delete-min operation tends to be productive. Karp and Tarjan proved [*Journal of Algorithms* **1** (1980), 374–393] that the average running time on a random graph with n vertices and m edges will be $O(m)$.

The philosophy underlying stage 2 is that the problem on an initially sparse graph eventually reduces to a problem on a smaller but dense graph that is best solved by a different method.

⟨ Subroutines 7 ⟩ +≡
  unsigned long *cher_tar_kar*(*g*)
    Graph **g*;
  { ⟨ Local variables for *cher_tar_kar* 56 ⟩;
    mems = 0;
    ⟨ Do stage 1 of *cher_tar_kar* 58 ⟩;
    if (*verbose*) *printf*("␣␣␣␣[Stage␣1␣has␣used␣%ld␣mems]\n", *mems*);
    ⟨ Do stage 2 of *cher_tar_kar* 64 ⟩;
    return *tot_len*;
  }

56. We say that a fragment is *large* if it contains $\lfloor\sqrt{n+1}+\frac{1}{2}\rfloor$ or more vertices. As soon as a fragment becomes large, stage 1 stops trying to extend it. There cannot be more than $\lfloor\sqrt{n}\,\rfloor$ large fragments, because $(\lfloor\sqrt{n}\,\rfloor+1)\lfloor\sqrt{n+1}+\frac{1}{2}\rfloor > n$. The other fragments are called *small*.

Stage 1 keeps a list of all the small fragments. Initially this list contains n fragments consisting of one vertex each. The algorithm repeatedly looks at the first fragment on its list, and finds the smallest edge leading to another fragment. These two fragments are removed from the list and combined. The resulting fragment is put at the end of the list if it is still small, or put onto another list if it is large.

⟨ Local variables for *cher_tar_kar* 56 ⟩ ≡
  register Vertex **s, *t*;   /* beginning and end of the small list */
  Vertex **large_list*;   /* beginning of the list of large fragments */
  long *frags*;   /* current number of fragments, large and small */
  unsigned long *tot_len* = 0;   /* total length of all edges in fragments */
  register Vertex **u, *v*;   /* registers for list manipulation */
  register Arc **a*;   /* and another */
  register long *j, k*;   /* index registers for stage 2 */
See also section 61.

This code is used in section 55.

57. We need to make *lo_sqrt* global so that the *note_edge* procedure below can access it.

⟨ Global variables 3 ⟩ +≡
    long *lo_sqrt*, *hi_sqrt*;   /* $\lfloor \sqrt{n} \rfloor$ and $\lfloor \sqrt{n+1} + \frac{1}{2} \rfloor$ */

58.   There is a nonobvious way to compute $\lfloor \sqrt{n+1} + \frac{1}{2} \rfloor$ and $\lfloor \sqrt{n} \rfloor$. Since $\sqrt{n}$ is small and arithmetic is mem-free, the author couldn't resist writing the **for** loop shown here. Of course, different ground rules for counting mems would be appropriate if this sort of computing were a critical factor in the running time.

⟨ Do stage 1 of *cher_tar_kar* 58 ⟩ ≡
    o, *frags* = *g*⁀*n*;
    for (*hi_sqrt* = 1; *hi_sqrt* * (*hi_sqrt* + 1) ≤ *frags*; *hi_sqrt* ++) ;
    if (*hi_sqrt* * *hi_sqrt* ≤ *frags*) *lo_sqrt* = *hi_sqrt*;
    else *lo_sqrt* = *hi_sqrt* − 1;
    large_list = Λ;
    ⟨ Create the small list 59 ⟩;
    while (*frags* > *lo_sqrt*) {
        ⟨ Combine the first fragment on the small list with its nearest neighbor 60 ⟩;
        frags −−;
    }
This code is used in section 55.

Arc = **struct**, GB_GRAPH §10.   *n*: **long**, GB_GRAPH §20.   *printf*: **int** ( ), <stdio.h>.
Graph = **struct**, GB_GRAPH §20.   *note_edge*: **void** ( ), §67.   *verbose*: **long**, GB_GRAPH §5.
mems: **long**, §10.   *o* = macro, §10.   **Vertex** = **struct**, GB_GRAPH §9.

59. To represent fragments, we will use several utility fields already defined above. The *lsib* and *rsib* pointers are used between fragments in the small list, which is doubly linked; *s* points to the first small fragment, $s\rightarrow rsib$ to the next, ..., $t\rightarrow lsib$ to the second-from-last, and *t* to the last. The pointer fields $s\rightarrow lsib$ and $t\rightarrow rsib$ are undefined. The *large_list* is singly linked via *rsib* pointers, terminating with Λ.

The *csize* field of each fragment tells how many vertices it contains.

The *comp* field of each vertex is Λ if this vertex represents a fragment (i.e., if this vertex is in the small list or *large_list*); otherwise it points to another vertex that is closer to the fragment representative.

Finally, the *pq* pointer of each fragment points to the header node of its priority queue, which is a binomial queue containing all unlooked-at arcs that originate from vertices in the fragment. This pointer is identical to the *newarc* pointer already set up. In a production implementation, we wouldn't need *pq* as a separate field; it would be part of a vertex record. So we do not pay any mems for referring to it.

#define *pq* *newarc*

⟨ Create the small list 59 ⟩ ≡
```
  o, s = g⃗vertices;
  for (v = s;  v < s + frags;  v++) {
    if (v > s) {
      o, v⃗lsib = v − 1;  o, (v − 1)⃗rsib = v;
    }
    o, v⃗comp = Λ;  o, v⃗csize = 1;
    o, v⃗pq⃗qcount = 0;      /* the binomial queue is initially empty */
    for (o, a = v⃗arcs;  a;  o, a = a⃗next)  qenque(v⃗pq, a);
  }
  t = v − 1;
```
This code is used in section 58.

60. ⟨ Combine the first fragment on the small list with its nearest neighbor 60 ⟩ ≡
```
  v = s;  o, s = s⃗rsib;      /* remove v from small list */
  do {
    a = qdel_min(v⃗pq);
    if (a ≡ Λ) return INFINITY;      /* the graph isn't connected */
    o, u = a⃗tip;
    while (o, u⃗comp)  u = u⃗comp;      /* find the fragment pointed to */
  } while (u ≡ v);      /* repeat until a new fragment is found */
  if (verbose) ⟨ Report the new edge verbosely 63 ⟩;
  o, tot_len += a⃗len;  o, v⃗comp = u;
  qmerge(u⃗pq, v⃗pq);
  o, old_size = u⃗csize;  o, new_size = old_size + v⃗csize;  o, u⃗csize = new_size;
  ⟨ Move u to the proper list position 62 ⟩;
```
This code is used in section 58.

61. ⟨ Local variables for *cher_tar_kar* 56 ⟩ +≡
```
  long old_size, new_size;      /* size of fragment u, before and after */
```

62.   Here is a fussy part of the program. We have just merged the small fragment v into another fragment u. If u was already large, there's nothing to do (except to check if the small list has just become empty). Otherwise we need to move u to the end of the small list, or we need to put it onto the large list. All these cases are special, if we want to avoid unnecessary memory references; so let's hope we get them right.

⟨ Move u to the proper list position 62 ⟩ ≡

```
if (old_size ≥ hi_sqrt) {       /* u was large */
  if (t ≡ v) s = Λ;       /* small list just became empty */
} else if (new_size < hi_sqrt) {       /* u was and still is small */
  if (u ≡ t) goto fin;       /* u is already where we want it */
  if (u ≡ s) o, s = u⃗rsib;       /* remove u from front */
  else {
    ooo, u⃗rsib⃗lsib = u⃗lsib;       /* detach u from middle */
    o, u⃗lsib⃗rsib = u⃗rsib;       /* do you follow the mem-counting here? */
  }
  o, t⃗rsib = u;       /* insert u at the end */
  o, u⃗lsib = t;  t = u;
} else {       /* u has just become large */
  if (u ≡ t) {
    if (u ≡ s) goto fin;       /* well, keep it small, we're done anyway */
    o, t = u⃗lsib;       /* remove u from end */
  } else if (u ≡ s) o, s = u⃗rsib;       /* remove u from front */
  else {
    ooo, u⃗rsib⃗lsib = u⃗lsib;       /* detach u from middle */
    o, u⃗lsib⃗rsib = u⃗rsib;
  }
  o, u⃗rsib = large_list;  large_list = u;       /* make u large */
}
fin: ;
```

This code is used in section 60.

63.   We don't have room in our binomial queues to keep track of both endpoints of the arcs. But the arcs occur in pairs, and by looking at the address of a we can tell whether the matching arc is $a + 1$ or $a - 1$. (See the explanation in GB_GRAPH.)

⟨ Report the new edge verbosely 63 ⟩ ≡

```
report((edge_trick & (siz_t) a ? a − 1 : a + 1)⃗tip, a⃗tip, a⃗len);
```

This code is used in sections 60 and 70.

---

a: **register Arc** *, §56.	len: **long**, GB_GRAPH §10.	$rsib = w.V$, §32.
$arcs$: **Arc** *, GB_GRAPH §9.	$lsib = v.V$, §32.	s: **register Vertex** *, §56.
$cher\_tar\_kar$: **unsigned long** ( ), §55.	$newarc = u.A$, §32.	**siz_t** = **unsigned long**, GB_GRAPH §34.
	$next$: **Arc** *, GB_GRAPH §10.	
$comp = y.V$, §16.	o = macro, §10.	t: **register Vertex** *, §56.
$csize = x.I$, §16.	ooo = macro, §10.	tip: **Vertex** *, GB_GRAPH §10.
$edge\_trick$: **siz_t**, GB_GRAPH §32.	$qcount = a.I$, §44.	tot_len: **unsigned long**, §56.
$frags$: **long**, §56.	$qdel\_min$: **Arc** *( ), §52.	u: **register Vertex** *, §56.
g: **Graph** *, §55.	$qenque$: **int** ( ), §50.	v: **register Vertex** *, §56.
hi_sqrt: **long**, §57.	$qmerge$: **int** ( ), §51.	$verbose$: **long**, GB_GRAPH §5.
INFINITY = macro, §5.	$report$: **int** ( ), §7.	$vertices$: **Vertex** *, GB_GRAPH §20.
$large\_list$: **Vertex** *, §56.		

64. Cheriton, Tarjan, and Karp's algorithm (continued). And now for the second part of the algorithm. Here we need to find room for a $\lfloor \sqrt{n} \rfloor \times \lfloor \sqrt{n} \rfloor$ matrix of edge lengths; we will use random access into the z utility fields of vertex records, since these haven't been used for anything yet by *cher_tar_kar*. We can also use the v utility fields to record the arcs that are the source of the best lengths, since this was the *lsib* field (no longer needed). The program doesn't count mems for updating that field, since it considers its goal to be simply the calculation of minimum spanning tree length; the actual edges of the minimum spanning tree are computed only for *verbose* mode. (We want to see how competitive *cher_tar_kar* is when we streamline it as much as possible.)

In stage 2, the vertices will be assigned integer index numbers between 0 and $\lfloor \sqrt{n} \rfloor - 1$. We'll put this into the *csize* field, which is no longer needed, and call it *findex*.

#define *findex* *csize*
#define *matx*(j, k) $(gv + ((j) * lo\_sqrt + (k))) \rightarrow z.I$ /* distance between fragments j and k */
#define *matx_arc*(j, k) $(gv + ((j) * lo\_sqrt + (k))) \rightarrow v.A$ /* arc corresponding to *matx*(j, k) */
#define INF 30000   /* upper bound on all edge lengths */
⟨ Do stage 2 of *cher_tar_kar* 64 ⟩ ≡
  $gv = g \rightarrow vertices$;   /* the global variable gv helps access auxiliary memory */
  ⟨ Map all vertices to their index numbers 65 ⟩;
  ⟨ Create the reduced matrix by running through all remaining edges 66 ⟩;
  ⟨ Execute Prim's algorithm on the reduced matrix 69 ⟩;
This code is used in section 55.

65. The vertex-mapping algorithm is $O(n)$ because each non-null *comp* link is examined at most three times. We set the *comp* field to null as an indication that *findex* has been set.
⟨ Map all vertices to their index numbers 65 ⟩ ≡
  if $(s \equiv \Lambda)$ $s = large\_list$;
  else $t \rightarrow rsib = large\_list$;
  for $(k = 0, v = s;\ v;\ o, v = v \rightarrow rsib, k{+}{+})$ $o, v \rightarrow findex = k$;
  for $(v = g \rightarrow vertices;\ v < g \rightarrow vertices + g \rightarrow n;\ v{+}{+})$
    if $(o, v \rightarrow comp)$ {
      for $(t = v \rightarrow comp;\ o, t \rightarrow comp;\ t = t \rightarrow comp)$;
      $o, k = t \rightarrow findex$;
      for $(t = v;\ o, u = t \rightarrow comp;\ t = u) \cdot$ {
        $o, t \rightarrow comp = \Lambda$;
        $o, t \rightarrow findex = k$;
      }
    }
  }
This code is used in section 64.

66. ⟨ Create the reduced matrix by running through all remaining edges 66 ⟩ ≡
  for $(j = 0;\ j < lo\_sqrt;\ j{+}{+})$
    for $(k = 0;\ k < lo\_sqrt;\ k{+}{+})$ $o, matx(j, k) = $ INF;
  for $(kk = 0;\ s;\ o, s = s \rightarrow rsib, kk{+}{+})$ $qtraverse(s \rightarrow pq, note\_edge)$;
This code is used in section 64.

67. The *note_edge* procedure "visits" every edge in the binomial queues traversed by *qtraverse* in the preceding code. Global variable kk, which would be a global register in a production version, is the index of the fragment from which this arc emanates.

⟨ Procedures to be declared early 67 ⟩ ≡

```
void note_edge(a)
    Arc *a;
{ register long k;
    o, k = a→tip→findex;
    if (k ≡ kk) return;
    if (oo, a→len < matx(kk, k)) {
        o, matx(kk, k) = a→len;
        o, matx(k, kk) = a→len;
        matx_arc(kk, k) = matx_arc(k, kk) = a;
    }
}
```

This code is used in section 2.

68. As we work on the final subproblem of size $\lfloor \sqrt{n} \rfloor \times \lfloor \sqrt{n} \rfloor$, we'll have a short vector that tells us the distance to each fragment that hasn't yet been joined up with fragment 0. The vector has -1 in positions that already have been joined up. In a production version, we could keep this in row 0 of *matx*.

⟨ Global variables 3 ⟩ +≡

```
long kk;          /* current fragment */
long distance[100];      /* distances to at most ⌊√n⌋ unhit fragments */
Arc *dist_arc[100];      /* the corresponding arcs, for verbose mode */
```

A: **Arc** *, GB_GRAPH §8.
Arc = **struct**, GB_GRAPH §10.
cher_tar_kar: **unsigned long** ( ),
  §55.
comp = *y.V*, §16.
csize = *x.I*, §16.
g: **Graph** *, §55.
gv: **Vertex** *, §23.
I: **long**, GB_GRAPH §8.
j: **register long**, §56.

k: **register long**, §56.
large_list: **Vertex** *, §56.
len: **long**, GB_GRAPH §10.
lo_sqrt: **long**, §57.
lsib = *v.V*, §32.
n: **long**, GB_GRAPH §20.
o = macro, §10.
oo = macro, §10.
pq = *u.A*, §59.
qtraverse: **int** ( ), §54.

rsib = *w.V*, §32.
s: **register Vertex** *, §56.
t: **register Vertex** *, §56.
tip: **Vertex** *, GB_GRAPH §10.
u: **register Vertex** *, §56.
v: **register Vertex** *, §56.
v: **util**, GB_GRAPH §9.
verbose: **long**, GB_GRAPH §5.
vertices: **Vertex** *, GB_GRAPH §20.
z: **util**, GB_GRAPH §9.

69.   The last step, as suggested by Prim, repeatedly updates the distance table against each row
of the matrix as it is encountered. This is the algorithm of choice to find the minimum spanning
tree of a complete graph.

⟨ Execute Prim's algorithm on the reduced matrix 69 ⟩ ≡
```
  { long d;      /* shortest entry seen so far in distance vector */
    o, distance[0] = −1;
    d = INF;
    for (k = 1;  k < lo_sqrt;  k++)  {
      o, distance[k] = matx(0, k);
      dist_arc[k] = matx_arc(0, k);
      if (distance[k] < d)  d = distance[k], j = k;
    }
    while (frags > 1) ⟨ Connect fragment 0 with fragment j, since j is the column achieving the
          smallest distance, d; also compute j and d for the next round 70 ⟩;
  }
```
This code is used in section 64.

70.   ⟨ Connect fragment 0 with fragment j, since j is the column achieving the smallest distance, d;
       also compute j and d for the next round 70 ⟩ ≡
```
  {
    if (d ≡ INF) return INFINITY;       /* the graph isn't connected */
    o, distance[j] = −1;       /* fragment j now will join up with fragment 0 */
    tot_len += d;
    if (verbose)  {
      a = dist_arc[j];
      ⟨ Report the new edge verbosely 63 ⟩;
    }
    frags −−;
    d = INF;
    for (k = 1;  k < lo_sqrt;  k++)
      if (o, distance[k] ≥ 0)  {
        if (o, matx(j, k) < distance[k])  {
          o, distance[k] = matx(j, k);
          dist_arc[k] = matx_arc(j, k);
        }
        if (distance[k] < d)  d = distance[k], kk = k;
      }
    j = kk;
  }
```
This code is used in section 69.

71. Conclusions. The winning algorithm, of the four methods considered here, on problems of the size considered here, with respect mem counting, is clearly Jarník/Prim with binary heaps. Second is Kruskal with radix sorting, on sparse graphs, but the Fibonacci heap method beats it on dense graphs. Procedure *cher_tar_kar* never comes close, although every step it takes seems to be reasonably sensible and efficient, and although the implementation above gives it the benefit of every doubt when counting its mems. It apparently loses because it more or less gives up a factor of 2 by dealing with each edge twice; the other methods put very little effort into discarding an arc whose mate has already been processed.

But it is important to realize that mem counting is not the whole story. Further tests were made on a Sun SPARCstation 2, in order to measure the true running times when all the complications of pipelining, caching, and compiler optimization are taken into account. These runs showed that Kruskal's algorithm was actually best, at least on the particular system tested:

optimization level	-g	-O2	-O3	mems
Kruskal/radix	132	111	111	8379
Jarník/Prim/binary	307	226	212	7972
Jarník/Prim/Fibonacci	432	350	333	11519
Cheriton/Tarjan/Karp	686	509	492	17090

(Times are shown in seconds per 100,000 runs with the default graph $miles(1000, 0, 0, 0, 10, 0)$. Optimization level -O4 gave the same results as -O3. Optimization does not change the mem count.) Thus the Kruskal procedure used only about 160 nanoseconds per mem, without optimization, and about 130 with; the others used about 380 to 400 ns/mem without optimization, 270 to 300 with. The mem measure gave consistent readings for the three "sophisticated" data structures, but the "naïve" Kruskal method blended better with hardware. The complete graph $miles(100, 0, 0, 0, 0, 99, 0)$, obtained by specifying option -d100, gave somewhat different statistics:

optimization level	-g	-O2	-O3	mems
Kruskal/radix	1846	1787	1810	63795
Jarník/Prim/binary	2246	1958	1845	50594
Jarník/Prim/Fibonacci	2675	2377	2248	58544
Cheriton/Tarjan/Karp	8881	6964	6909	165749

Now the identical machine instructions took significantly longer per mem—presumably because of cache misses, although the frequency of conditional jump instructions might also be a factor. Careful analyses of these phenomena should be instructive. Future computers are expected to be more nearly limited by memory speed; therefore the running time per mem is likely to become more uniform between methods, although cache performance will probably always be a factor.

The *krusk* procedure might go even faster if it were given a streamlined union/find algorithm. Or would such "streamlining" negate some of its present efficiency?

---

a: **register Arc** *, §56.
cher_tar_kar: **unsigned long** ( ),
    §55.
dist_arc: **Arc** *[], §68.
distance: **long** [], §68.
frags: **long**, §56.
INF = 30000, §64.

INFINITY = macro, §5.
j: **register long**, §56.
k: **register long**, §56.
kk: **long**, §68.
krusk: **unsigned long** ( ), §14.
lo_sqrt: **long**, §57.

matx = macro ( ), §64.
matx_arc = macro ( ), §64.
miles: **Graph** *( ), GB_MILES §5.
o = macro, §10.
tot_len: **unsigned long**, §56.
verbose: **long**, GB_GRAPH §5.

MULTIPLY

Important: Before reading MULTIPLY, please read or at least skim the program for GB_GATES.

1. Introduction. This demonstration program uses graphs constructed by the *prod* procedure in the GB_GATES module to produce an interactive program called `multiply`, which multiplies and divides small numbers the slow way—by simulating the behavior of a logical circuit, one gate at a time.

The program assumes that UNIX conventions are being used. Some code in sections listed under 'UNIX dependencies' in the index might need to change if this program is ported to other operating systems.

To run the program under UNIX, say '`multiply` m n [*seed*]', where m and n are the sizes of the numbers to be multiplied, in bits, and where *seed* is given if and only if you want the multiplier to be a special-purpose circuit for multiplying a given m-bit number by a randomly chosen n-bit constant.

The program will prompt you for two numbers (or for just one, if the random constant option has been selected), and it will use the gate network to compute their product. Then it will ask for more input, and so on.

2. Here is the general layout of this program, as seen by the C compiler:

```
#include "gb_graph.h"      /* the standard GraphBase data structures */
#include "gb_gates.h"      /* routines for gate graphs */
⟨ Preprocessor definitions ⟩
⟨ Global variables 4 ⟩
⟨ Handy subroutines 10 ⟩
main(argc, argv)
     int argc;       /* the number of command-line arguments */
     char *argv[];      /* an array of strings containing those arguments */
{
  ⟨ Declare variables that ought to be in registers 5 ⟩;
  ⟨ Obtain m, n, and optional seed from the command line 6 ⟩;
  ⟨ Make sure m and n are valid; generate the prod graph g 3 ⟩;
  if (seed < 0)     /* no seed given */
     printf("Here␣I␣am,␣ready␣to␣multiply␣%ld-bit␣numbers␣by␣%ld-bit␣numbers.\n", m, n);
  else {
     g = partial_gates(g, m, 0L, seed, buffer);
     if (g) {
       ⟨ Set y to the decimal value of the second input 9 ⟩;
       printf("OK,␣I'm␣ready␣to␣multiply␣any␣%ld-bit␣number␣by␣%s.\n", m, y);
     } else {     /* there was enough memory to make the original g, but not enough to reduce it;
              this probably can't happen, but who knows? */
       printf("Sorry,␣I␣couldn't␣process␣the␣graph␣(trouble␣code␣%ld)!\n", panic_code);
       return -9;
     }
  }
  printf("(I'm␣simulating␣a␣logic␣circuit␣with␣%ld␣gates,␣depth␣%ld.)\n", g-n, depth(g));
```

```
    while (1) {
        ⟨Prompt for one or two numbers; break if unsuccessful 7⟩;
        ⟨Use the network to compute the product 11⟩;
        printf("%sx%s=%s%s.\n", x, y, (strlen(x) + strlen(y) > 35 ? "\n␣" : ""), z);
    }
    return 0;        /* normal exit */
}
```

3. ⟨Make sure m and n are valid; generate the *prod* graph g 3⟩ ≡

```
    if (m < 2)  m = 2;
    if (n < 2)  n = 2;
    if (m > 999 ∨ n > 999) {
        printf("Sorry,␣I'm␣set␣up␣only␣for␣precision␣less␣than␣1000␣bits.\n");
        return −1;
    }
    if ((g = prod(m, n)) ≡ Λ) {
        printf("Sorry,␣I␣couldn't␣generate␣the␣graph␣(not␣enough␣memory␣for␣%s)!\n", panic_code ≡
            no_room ? "the␣gates" : panic_code ≡ alloc_fault ? "the␣wires" : "local␣optimization");
        return −3;
    }
```

This code is used in section 2.

4. To figure the maximum length of strings x and y, we note that $2^{999} \approx 5.4 \times 10^{300}$.

⟨Global variables 4⟩ ≡

```
    Graph *g;        /* graph that defines a logical network for multiplication */
    long m, n;       /* length of binary numbers to be multiplied */
    long seed;       /* optional seed value, or −1 */
    char x[302], y[302], z[603];        /* input and output numbers, as decimal strings */
    char buffer[2000];        /* workspace for communication between routines */
```

This code is used in section 2.

5. ⟨Declare variables that ought to be in registers 5⟩ ≡

```
    register char *p, *q, *r;        /* pointers for string manipulation */
    register long a, b;        /* amounts being carried over while doing radix conversion */
```

This code is used in section 2.

---

alloc_fault = −1, GB_GRAPH §7.   *no_room* = 1, GB_GRAPH §7.   *printf*: **int** ( ), <stdio.h>.
depth: **long** ( ), §13.   *panic_code*: **long**, GB_GRAPH §5.   *prod*: **Graph** *( ), GB_GATES §66.
Graph = **struct**, GB_GRAPH §20.   *partial_gates*: **Graph** *( ),   *strlen*: **int** ( ), <string.h>.
n: **long**, GB_GRAPH §20.   GB_GATES §84.

6. ⟨ Obtain m, n, and optional *seed* from the command line 6 ⟩ ≡

    if $(argc < 3 \lor argc > 4 \lor sscanf(argv[1], \texttt{"\%ld"}, \&m) \neq 1 \lor sscanf(argv[2], \texttt{"\%ld"}, \&n) \neq 1)$ {

      $fprintf(stderr, \texttt{"Usage:}␣\texttt{\%s}␣\texttt{m}␣\texttt{n}␣\texttt{[seed]\textbackslash n"}, argv[0]);$

      return -2;

    }

    if $(m < 0)$ $m = -m;$ /∗ maybe the user attached '-' to the argument ∗/

    if $(n < 0)$ $n = -n;$

    $seed = -1;$

    if $(argc \equiv 4 \land sscanf(argv[3], \texttt{"\%ld"}, \&seed) \equiv 1 \land seed < 0)$ $seed = -seed;$

This code is used in section 2.

7. This program may not be user-friendly, but at least it is polite.

#define $prompt(s)$
        { $printf(s);$ $fflush(stdout);$ /∗ make sure the user sees the prompt ∗/
          if $(fgets(buffer, 999, stdin) \equiv \Lambda)$ **break**; }

#define $retry(s, t)$
        { $printf(s);$ **goto** $t;$ }

⟨ Prompt for one or two numbers; **break** if unsuccessful 7 ⟩ ≡

$step1$: $prompt(\texttt{"\textbackslash nNumber,}␣\texttt{please?}␣\texttt{"});$

    for $(p = buffer;$ $*p \equiv \texttt{'0'};$ $p\mathord{+}\mathord{+})$;   /∗ bypass leading zeroes ∗/

    if $(*p \equiv \texttt{'\textbackslash n'})$ {

      if $(p > buffer)$ $p\mathord{-}\mathord{-};$ /∗ zero is acceptable ∗/

      else break;   /∗ empty input terminates the run ∗/

    }

    for $(q = p;$ $*q \geq \texttt{'0'} \land *q \leq \texttt{'9'};$ $q\mathord{+}\mathord{+})$;   /∗ check for digits ∗/

    if $(*q \neq \texttt{'\textbackslash n'})$

      $retry(\texttt{"Excuse}␣\texttt{me...}␣\texttt{I'm}␣\texttt{looking}␣\texttt{for}␣\texttt{a}␣\texttt{nonnegative}␣\texttt{sequence}␣\texttt{of}␣\texttt{decimal}␣\texttt{digits."}, step1);$

    $*q = 0;$

    if $(strlen(p) > 301)$ $retry(\texttt{"Sorry,}␣\texttt{that's}␣\texttt{too}␣\texttt{big."}, step1);$

    $strcpy(x, p);$

    if $(seed < 0)$ {

      ⟨ Do the same thing for y instead of x 8 ⟩;

    }

This code is used in section 2.

8. ⟨ Do the same thing for y instead of x 8 ⟩ ≡

$step2$: $prompt(\texttt{"Another?}␣\texttt{"});$

    for $(p = buffer;$ $*p \equiv \texttt{'0'};$ $p\mathord{+}\mathord{+})$;   /∗ bypass leading zeroes ∗/

    if $(*p \equiv \texttt{'\textbackslash n'})$ {

      if $(p > buffer)$ $p\mathord{-}\mathord{-};$ /∗ zero is acceptable ∗/

      else break;   /∗ empty input terminates the run ∗/

    }

    for $(q = p;$ $*q \geq \texttt{'0'} \land *q \leq \texttt{'9'};$ $q\mathord{+}\mathord{+})$;   /∗ check for digits ∗/

    if $(*q \neq \texttt{'\textbackslash n'})$

      $retry(\texttt{"Excuse}␣\texttt{me...}␣\texttt{I'm}␣\texttt{looking}␣\texttt{for}␣\texttt{a}␣\texttt{nonnegative}␣\texttt{sequence}␣\texttt{of}␣\texttt{decimal}␣\texttt{digits."}, step2);$

    $*q = 0;$

    if $(strlen(p) > 301)$ $retry(\texttt{"Sorry,}␣\texttt{that's}␣\texttt{too}␣\texttt{big."}, step2);$

$strcpy(y, p);$

This code is used in section 7.

9. The binary value chosen at random by *partial_gates* appears as a string of 0s and 1s in *buffer*, in little-endian order. We compute the corresponding decimal value by repeated doubling.

If the value turns out to be zero, the whole network will have collapsed. Otherwise, however, the m inputs from the first operand will all remain present, because they all affect the output.

$\langle$ Set y to the decimal value of the second input $9\,\rangle \equiv$

```
*y = '0'; *(y + 1) = 0;      /* now y is "0" */
for (r = buffer + strlen(buffer) − 1; r ≥ buffer; r−−) {
    /* we will set y = 2y + t where t is the next bit, *r */
    if (*y ≥ '5') a = 0, p = y;
    else a = *y − '0', p = y + 1;
    for (q = y; *p; a = b, p++, q++) {
        if (*p ≥ '5') {
            b = *p − '5';
            *q = 2 * a + '1';
        } else {
            b = *p − '0';
            *q = 2 * a + '0';
        }
    }
    if (*r ≡ '1') *q = 2 * a + '1';
    else *q = 2 * a + '0';
    *++q = 0;      /* terminate the string */
}
if (strcmp(y, "0") ≡ 0) {
    printf("Please␣try␣another␣seed␣value;␣%d␣makes␣the␣answer␣zero!\n", seed);
    return (−5);
}
```

This code is used in section 2.

---

a: **register long**, §5.
$argc$: **int**, §2.
$argv$: **char** *[], §2.
b: **register long**, §5.
$buffer$: **char** [], §4.
$fflush$: **int** ( ), <stdio.h>.
$fgets$: **char** *( ), <stdio.h>.
$fprintf$: **int** ( ), <stdio.h>.
m: **long**, §4.

n: **long**, §4.
p: **register char** *, §5.
$partial\_gates$: **Graph** *( ),
  GB_GATES §84.
$printf$: **int** ( ), <stdio.h>.
q: **register char** *, §5.
r: **register char** *, §5.
$seed$: **long**, §4.
$sscanf$: **int** ( ), <stdio.h>.

$stderr$: **FILE** *, <stdio.h>.
$stdin$: **FILE** *, <stdio.h>.
$stdout$: **FILE** *, <stdio.h>.
$strcmp$: **int** ( ), <string.h>.
$strcpy$: **char** *( ), <string.h>.
$strlen$: **int** ( ), <string.h>.
x: **char** [], §4.
y: **char** [], §4.

10. **Using the network.** The reader of the code in the previous section will have noticed that we are representing high-precision decimal numbers as strings. We might as well do that, since the only operations we need to perform on them are input, output, doubling, and halving. In fact, arithmetic on strings is kind of fun, if you like that sort of thing.

Here is a subroutine that converts a decimal string to a binary string. The decimal string is big-endian as usual, but the binary string is little-endian. The decimal string is decimated in the process; it should end up empty, unless the original value was too big.

⟨Handy subroutines 10⟩ ≡

```
decimal_to_binary(x, s, n)
     char *x;      /* decimal string */
     char *s;      /* binary string */
     long n;       /* length of s */
{ register long k;
  register char *p, *q;      /* pointers for string manipulation */
  register long r;      /* remainder */

  for (k = 0; k < n; k++, s++) {
    if (*x ≡ 0) *s = '0';
    else {      /* we will divide x by 2 */
      if (*x > '1') p = x, r = 0;
      else  p = x + 1, r = *x - '0';
      for (q = x; *p; p++, q++) {
        r = 10 * r + *p - '0';
        *q = (r ≫ 1) + '0';
        r = r & 1;
      }
      *q = 0;      /* terminate string x */
      *s = '0' + r;
    }
  }
  *s = 0;      /* terminate the output string */
}
```

See also section 13.

This code is used in section 2.

11. ⟨Use the network to compute the product 11⟩ ≡

```
strcpy(z, x);
decimal_to_binary(z, buffer, m);
if (*z) {
  printf("(Sorry,␣%s␣has␣more␣than␣%ld␣bits.)\n", x, m); continue;
}
if (seed < 0) {
  strcpy(z, y);
  decimal_to_binary(z, buffer + m, n);
  if (*z) {
    printf("(Sorry,␣%s␣has␣more␣than␣%ld␣bits.)\n", y, n); continue;
  }
}
```

```
if (gate_eval(g, buffer, buffer) < 0) {
    printf("???␣An␣internal␣error␣occurred!");
    return 666;       /* this can't happen */
}
```
⟨ Convert the binary number in *buffer* to the decimal string *z* 12 ⟩;
This code is used in section 2.

12.   The remaining task is almost identical to what we needed to do when computing the value of *y* after a random seed was specified. But this time the binary number in *buffer* is big-endian.

⟨ Convert the binary number in *buffer* to the decimal string *z* 12 ⟩ ≡
```
*z = '0'; *(z + 1) = 0;
for (r = buffer; *r; r++) {       /* we'll set z = 2z + t where t is the next bit, *r */
    if (*z ≥ '5') a = 0, p = z;
    else a = *z − '0', p = z + 1;
    for (q = z; *p; a = b, p++, q++) {
        if (*p ≥ '5') {
            b = *p − '5';
            *q = 2 * a + '1';
        } else {
            b = *p − '0';
            *q = 2 * a + '0';
        }
    }
    if (*r ≡ '1') *q = 2 * a + '1';
    else *q = 2 * a + '0';
    *++q = 0;       /* terminate the string */
}
```
This code is used in section 11.

---

a: **register long**, §5.
b: **register long**, §5.
buffer: **char** [], §4.
g: **Graph** ∗, §4.
gate_eval: **long** ( ), GB_GATES §3.

m: **long**, §4.
n: **long**, §4.
p: **register char** ∗, §5.
printf: **int** ( ), <stdio.h>.
q: **register char** ∗, §5.

r: **register char** ∗, §5.
seed: **long**, §4.
strcpy: **char** ∗( ), <string.h>.
y: **char** [], §4.
z: **char** [], §4.

13. Calculating the depth. The depth of a gate network produced by GB_GATES is easily discovered by making one pass over the vertices. An input gate or a constant has depth 0; every other gate has depth one greater than the maximum of its inputs.

This routine is more general than it needs to be for the circuits output by *prod*. The result of a latch is considered to have depth 0.

Utility field $u.I$ is set to the depth of each individual gate.

#define dp $u.I$

⟨ Handy subroutines 10 ⟩ +≡
    long *depth*(*g*)
        Graph **g*;   /* graph with gates as vertices */
    { **register Vertex** **v*;   /* the current vertex of interest */
      register Arc **a*;   /* the current arc of interest */
      long *d*;   /* depth of current vertex */
      if (¬*g*) **return** −1;   /* no graph supplied! */
      for (*v* = *g*⇥*vertices*; *v* < *g*⇥*vertices* + *g*⇥*n*; *v*++) {
        switch (*v*⇥*typ*) {   /* branch on type of gate */
        case 'I': **case** 'L': **case** 'C': *v*⇥*dp* = 0; **break**;
        default: ⟨ Set *d* to the maximum depth of an operand of *v* 14 ⟩;
            v⇥*dp* = 1 + *d*;
        }
      }
      ⟨ Set *d* to the maximum depth of an output of *g* 15 ⟩;
      return *d*;
    }

14. ⟨ Set *d* to the maximum depth of an operand of *v* 14 ⟩ ≡
    d = 0;
    for (*a* = *v*⇥*arcs*; *a*; *a* = *a*⇥*next*)
        if (*a*⇥*tip*⇥*dp* > *d*) *d* = *a*⇥*tip*⇥*dp*;
This code is used in section 13.

15. ⟨ Set *d* to the maximum depth of an output of *g* 15 ⟩ ≡
    d = 0;
    for (*a* = *g*⇥*outs*; *a*; *a* = *a*⇥*next*)
        if (¬*is_boolean*(*a*⇥*tip*) ∧ *a*⇥*tip*⇥*dp* > *d*) *d* = *a*⇥*tip*⇥*dp*;
This code is used in section 13.

QUEEN

1. Queen moves. This is a short demonstration of how to generate and traverse graphs with the Stanford GraphBase. It creates a graph with 12 vertices, representing the cells of a 3×4 rectangular board; two cells are considered adjacent if you can get from one to another by a queen move. Then it prints a description of the vertices and their neighbors, on the standard output file.

An ASCII file called `queen.gb` is also produced. Other programs can obtain a copy of the queen graph by calling *restore_graph*(`"queen.gb"`). You might find it interesting to compare the output of QUEEN with the contents of `queen.gb`; the former is intended to be readable by human beings, the latter by computers.

```
#include "gb_graph.h"      /* we use the GB_GRAPH data structures */
#include "gb_basic.h"      /* we test the basic graph operations */
#include "gb_save.h"       /* and we save our results in ASCII format */
  main()
  { Graph *g, *gg, *ggg;
    g = board(3 L, 4 L, 0 L, 0 L, −1 L, 0 L, 0 L);     /* a graph with rook moves */
    gg = board(3 L, 4 L, 0 L, 0 L, −2 L, 0 L, 0 L);     /* a graph with bishop moves */
    ggg = gunion(g, gg, 0 L, 0 L);     /* a graph with queen moves */
    save_graph(ggg, "queen.gb");       /* generate an ASCII file for ggg */
    ⟨Print the vertices and edges of ggg 2⟩;
    return 0;      /* normal exit */
  }
```

2. ⟨Print the vertices and edges of *ggg* 2⟩ ≡

```
  if (ggg ≡ Λ) printf("Something went wrong (panic code %ld)!\n", panic_code);
  else {
    register Vertex *v;      /* current vertex being visited */
    printf("Queen Moves on a 3x4 Board\n\n");
    printf("  The graph whose official name is\n%s\n", ggg→id);
    printf("  has %ld vertices and %ld arcs:\n\n", ggg→n, ggg→m);
    for (v = ggg→vertices; v < ggg→vertices + ggg→n; v++) {
      register Arc *a;      /* current arc from v */
      printf("%s\n", v→name);
      for (a = v→arcs; a; a = a→next) printf("  -> %s, length %ld\n", a→tip→name, a→len);
    }
  }
```

This code is used in section 1.

QUEEN.W _____

% This file is part of the Stanford GraphBase (c) Stanford University 1993
@i boilerplate.w %<< legal stuff: PLEASE READ IT BEFORE MAKING ANY CHANGES!
@i gb_types.w

\def\title{QUEEN}

@* Queen moves.
This is a short demonstration of how to generate and traverse graphs
with the Stanford GraphBase. It creates a graph with 12 vertices,
representing the cells of a 3×4 rectangular board; two
cells are considered adjacent if you can get from one to another
by a queen move. Then it prints a description of the vertices and
their neighbors, on the standard output file.

An ASCII file called \.{queen.gb} is also produced. Other programs
can obtain a copy of the queen graph by calling |restore_graph("queen.gb")|.
You might find it interesting to compare the output of {\sc QUEEN} with
the contents of \.{queen.gb}; the former is intended to be readable
by human beings, the latter by computers.

@p
#include "gb_graph.h" /* we use the {\sc GB_\,GRAPH} data structures */
#include "gb_basic.h" /* we test the basic graph operations */
#include "gb_save.h" /* and we save our results in ASCII format */
@#
main()
{@+Graph *g,*gg,*ggg;
  g=board(3L,4L,0L,0L,-1L,0L,0L); /* a graph with rook moves */
  gg=board(3L,4L,0L,0L,-2L,0L,0L); /* a graph with bishop moves */
  ggg=gunion(g,gg,0L,0L); /* a graph with queen moves */
  save_graph(ggg,"queen.gb"); /* generate an ASCII file for |ggg| */
  @<Print the vertices and edges of |ggg|@>;
  return 0; /* normal exit */
}

@ @<Print the vertices and edges of |ggg|@>=
if (ggg==NULL) printf("Something went wrong (panic code %ld)!\n",panic_code);
else {
  register Vertex *v; /* current vertex being visited */
  printf("Queen Moves on a 3x4 Board\n\n");
  printf(" The graph whose official name is\n%s\n", ggg->id);
  printf(" has %ld vertices and %ld arcs:\n\n", ggg->n, ggg->m);
  for (v=ggg->vertices; v<ggg->vertices+ggg->n; v++) {
    register Arc *a; /* current arc from |v| */
    printf("%s\n", v->name);

```
    for (a=v->arcs; a; a=a->next)
      printf("  -> %s, length %ld\n", a->tip->name, a->len);
  }
}
```

@* Index.

QUEEN_WRAP. CH

% This file is part of the Stanford GraphBase (c) Stanford University 1993
It's a demonstration "change file", which converts the demonstration program
called "queen" into a similar demonstration program called "queen_wrap".

Change files make it easy to modify CWEB source programs without
touching the master files, thereby remaining totally compatible with
all other users. Anybody can make whatever modifications they like
in change files, but everybody is supposed to leave the master files
intact. Please also leave the present file intact, so that it remains
as a useful demonstration of the change-file idea.

The format of change files is simple: First comes a line that begins with @x,
then comes a line that is a verbatim copy of some line from the master file,
followed by zero or more additional lines that should match the subsequent
lines of the master file. Then you say @y, and then you give replacement
lines for everything between @x and @y in the master file. Then you say @z.
All changes must occur in the order of replaced text in the master file,
and must be uniquely identifiable by the first line that follows @x.

Optional comments may follow @x, @y, or @z on a line, and may occur outside
of @x-@y-@z groups. In fact, you are now reading such an optional comment.

@x replace the copyright notice by a change notice
@i boilerplate.w %<< legal stuff: PLEASE READ IT BEFORE MAKING ANY CHANGES!
@y
\let\maybe=\iffalse % tell CWEB to print only sections that change
\def\prerequisite#1{} \def\prerequisites#1#2{} % disable boilerplate macros
\def\botofcontents{\vskip 0pt plus 1filll \parskip=0pt
  This program was obtained by modifying {\sc QUEEN} in the Stanford
  GraphBase.\par   Only sections that have changed are listed here.\par}
@z
@x change the program title
\def\title{QUEEN}
@y
\def\title{QUEEN_WRAP}
@z

@x now we modify the introductory remarks of section 1
An ASCII file called \.{queen.gb} is also produced. Other programs
can obtain a copy of the queen graph by calling |restore_graph("queen.gb")|.
You may find it interesting to compare the output of {\sc QUEEN} with
the contents of \.{queen.gb}; the former is intended to be readable
by human beings, the latter by computers.
@y
Unlike an ordinary chessboard, the board considered here ``wraps around''
at the left and right edges, so that it is essentially a cylinder.

It does not, however, wrap around at the top and bottom; double wrapping
would actually allow a lowly bishop to move from any given cell to any other,
in two different ways.

An ASCII file called \.{queen_wrap.gb} is also produced. Other programs
can obtain a copy of the graph by calling |restore_graph("queen_wrap.gb")|.
You may find it interesting to compare the output of {\sc QUEEN_WRAP} with
the contents of \.{queen_wrap.gb}; the former is intended to be readable
by human beings, the latter by computers.
@z

@x changes to the code of section 1
  g=board(3L,4L,0L,0L,-1L,0L,0L); /* a graph with rook moves */
  gg=board(3L,4L,0L,0L,-2L,0L,0L); /* a graph with bishop moves */
  ggg=gunion(g,gg,0L,0L); /* a graph with queen moves */
  save_graph(ggg,"queen.gb"); /* generate an ASCII file for |ggg| */
@y we add wraparound
  g=board(3L,4L,0L,0L,-1L,2L,0L); /* a graph with rook moves and wrapping */
    /* we set |wrap=2| because only the second coordinate wraps */
  gg=board(3L,4L,0L,0L,-2L,2L,0L); /* a graph with bishop moves and wrapping */
  ggg=gunion(g,gg,0L,0L); /* a graph with queen moves and wrapping */
  save_graph(ggg,"queen_wrap.gb"); /* generate an ASCII file for |ggg| */
@z

@x change to the code of section 2
  printf("Queen Moves on a 3x4 Board\n\n");
@y
  printf("Queen Moves on a Cylindrical 3x4 Board\n\n");
@z

A change file is usually much shorter than the master file, but the
present one is an exception because the master file itself is short.
You can use many different change files with the same master file.

To run the queen_wrap program on a UNIX system, you can say
  ctangle queen.w queen_wrap.ch queen_wrap.c
and then compile and go. (The .w is optional in the first argument to ctangle;
the .ch is optional in the second; the .c is optional in the third.)

The C compiler and debugger will refer to appropriate lines of the original
source file queen.w and/or the change file queen_wrap.ch when you are
troubleshooting. You need never look at the file queen_wrap.c that was
output by ctangle, although the compiler and debugger will want to see it.

To obtain a TeXed documentation, you can say
  cweave queen queen_wrap
  tex queen
  rm queen.tex
after which you print the file queen.dvi output by TeX.

ROGET_COMPONENTS

Important: Before reading ROGET_COMPONENTS, please read or at least skim the program for GB_ROGET.

1. Strong components. This demonstration program computes the strong components of GraphBase graphs derived from Roget's Thesaurus, using a variant of Tarjan's algorithm [R. E. Tarjan, "Depth-first search and linear graph algorithms," *SIAM Journal on Computing* **1** (1972), 146–160]. We also determine the relationships between strong components.

Two vertices belong to the same strong component if and only if they are reachable from each other via directed paths.

We will print the strong components in "reverse topological order"; that is, if v is reachable from u but u is not reachable from v, the strong component containing v will be listed before the strong component containing u.

Vertices from the *roget* graph are identified both by name and by category number.

#define *specs*(*v*) (*filename* ? *v* − *g*⃗*vertices* + 1 L : *v*⃗*cat_no*), *v*⃗*name*
    /∗ category number and category name ∗/

2. We permit command-line options in UNIX style so that a variety of graphs can be studied: The user can say '-n⟨number⟩', '-d⟨number⟩', '-p⟨number⟩', and/or '-s⟨number⟩' to change the default values of the parameters in the graph *roget*(*n*, *d*, *p*, *s*). Or '-g⟨filename⟩' to change the graph itself.

```
#include "gb_graph.h"      /* the GraphBase data structures */
#include "gb_roget.h"      /* the roget routine */
#include "gb_save.h"       /* restore_graph */
⟨ Preprocessor definitions ⟩
⟨ Global variables 5 ⟩;
main(argc, argv)
      int argc;        /* the number of command-line arguments */
      char *argv[];      /* an array of strings containing those arguments */
{ Graph *g;      /* the graph we will work on */
  register Vertex *v;      /* the current vertex of interest */
  unsigned long n = 0;      /* the desired number of vertices (0 means infinity) */
  unsigned long d = 0;      /* the minimum distance between categories in arcs */
  unsigned long p = 0;      /* 65536 times the probability of rejecting an arc */
  long s = 0;      /* the random number seed */
  char *filename = Λ;      /* external graph substituted for roget */
  ⟨ Scan the command-line options 3 ⟩;
  g = (filename ? restore_graph(filename) : roget(n, d, p, s));
  if (g ≡ Λ) {
      fprintf(stderr, "Sorry,␣can't␣create␣the␣graph!␣(error␣code␣%ld)\n", panic_code);
      return −1;
  }
  printf("Reachability␣analysis␣of␣%s\n\n", g⃗id);
  ⟨ Perform Tarjan's algorithm on g 10 ⟩;
  return 0;      /* normal exit */
}
```

3. ⟨ Scan the command-line options 3 ⟩ ≡
  while ($-\!-argc$) {
    if ($sscanf(argv[argc], \texttt{"-n\%lu"}, \&n) \equiv 1$) ;
    else if ($sscanf(argv[argc], \texttt{"-d\%lu"}, \&d) \equiv 1$) ;
    else if ($sscanf(argv[argc], \texttt{"-p\%lu"}, \&p) \equiv 1$) ;
    else if ($sscanf(argv[argc], \texttt{"-s\%ld"}, \&s) \equiv 1$) ;
    else if ($strncmp(argv[argc], \texttt{"-g"}, 2) \equiv 0$) *filename* $= argv[argc] + 2$;
    else {
      $fprintf(stderr, \texttt{"Usage:\_\%s\_[-nN]\ [-dN]\ [-pN]\ [-sN]\ [-gfoo]\textbackslash n"}, argv[0])$; **return** -2;
    }
  }

This code is used in section 2.

4. Tarjan's algorithm is inherently recursive. We will implement the recursion explicitly via linked lists, instead of using C's runtime stack, because some computer systems bog down in the presence of deeply nested recursion.

Each vertex goes through three stages during the algorithm: First it is "unseen"; then it is "active"; finally it becomes "settled," when it has been assigned to a strong component.

The data structures that represent the current state of the algorithm are implemented by using five of the utility fields in each vertex: *rank*, *parent*, *untagged*, *link*, and *min*. We will consider each of these in turn.

5. First is the integer *rank* field, which is zero when a vertex is unseen. As soon as the vertex is first examined, it becomes active and its *rank* becomes and remains nonzero. Indeed, the kth vertex to become active will receive rank k. When a vertex finally becomes settled, its rank is reset to infinity.

It's convenient to think of Tarjan's algorithm as a simple adventure game in which we want to explore all the rooms of a cave. Passageways between the rooms allow one-way travel only. When we come into a room for the first time, we assign a new number to that room; this is its rank. Later on we might happen to enter the same room again, and we will notice that it has nonzero rank. Then we'll be able to make a quick exit, saying "we've already been here." (The extra complexities of computer games, like dragons that might need to be vanquished, do not arise.)

#define *rank* $z.I$ /∗ the *rank* of a vertex is stored in utility field z ∗/

⟨ Global variables 5 ⟩ ≡
  long nn;   /∗ the number of vertices that have been seen ∗/

See also sections 8 and 11.

This code is used in section 2.

---

6. The active vertices will always form an oriented tree, whose arcs are a subset of the arcs in the original graph. A tree arc from u to v will be represented by $v \text{-} parent \equiv u$. Every active vertex has a parent, which is usually another active vertex; the only exception is the root of the tree, whose *parent* is Λ.

In the cave analogy, the "parent" of room v is the room we were in immediately before entering v the first time. By following parent pointers, we will be able to leave the cave whenever we want.

As soon as a vertex becomes settled, its *parent* field changes significance. Then $v \text{-} parent$ is set equal to the unique representative of the strong component containing vertex v. Thus two settled vertices will belong to the same strong component if and only if they have the same *parent*.

#define *parent* $y.V$ /* the *parent* of a vertex is stored in utility field y */

7. All arcs in the original directed graph are explored systematically during a depth-first search. Whenever we look at an arc, we tag it so that we won't need to explore it again. In a cave, for example, we might mark each passageway between rooms once we've tried to go through it.

The algorithm doesn't actually place a tag on its **Arc** records; instead, each vertex v has a pointer $v \text{-} untagged$ that leads to all hitherto-unexplored arcs from v. The arcs of the list that appear between $v \text{-} arcs$ and $v \text{-} untagged$ are the ones already examined.

#define *untagged* $x.A$ /* the *untagged* field points to an **Arc** record, or Λ */

8. The algorithm maintains two special stacks: *active_stack* contains all the currently active vertices, and *settled_stack* contains all the currently settled vertices. Each vertex has a *link* field that points to the vertex that is next lower on its stack, or to Λ if the vertex is at the bottom. The vertices on *active_stack* always appear in increasing order of rank from bottom to top.

#define *link* $w.V$ /* the *link* field of a vertex occupies utility field w */

⟨ Global variables 5 ⟩ +≡
    Vertex *∗active_stack*; /* the top of the stack of active vertices */
    Vertex *∗settled_stack*; /* the top of the stack of settled vertices */

9. Finally there's a *min* field, which is the tricky part that makes everything work. If vertex v is unseen or settled, its *min* field is irrelevant. Otherwise $v \text{-} min$ points to the active vertex u of smallest rank having the following property: Either $u \equiv v$ or there is a directed path from v to u consisting of zero or more mature tree arcs followed by a single non-tree arc.

What is a tree arc, you ask. And what is a mature arc? Good questions. At the moment when arcs of the graph are tagged, we classify them either as tree arcs (if they correspond to a new *parent* link in the tree of active nodes) or non-tree arcs (otherwise). A tree arc becomes mature when it is no longer on the path from the root to the current vertex being explored. We also say that a vertex becomes mature when it is no longer on that path. All arcs from a mature vertex have been tagged.

We said before that every vertex is initially unseen, then active, and finally settled. With our new definitions, we see further that every arc starts out untagged, then it becomes either a non-tree arc or a tree arc. In the latter case, the arc begins as an immature tree arc and eventually matures.

Just believe these definitions, for now. All will become clear soon.

#define *min* $v.V$ /* the *min* field of a vertex occupies utility field v */

10.   Depth-first search explores a graph by systematically visiting all vertices and seeing what they can lead to. In Tarjan's algorithm, as we have said, the active vertices form an oriented tree. One of these vertices is called the current vertex.

If the current vertex still has an arc that hasn't been tagged, we tag one such arc and there are two cases: Either the arc leads to an unseen vertex, or it doesn't. If it does, the arc becomes a tree arc; the previously unseen vertex becomes active, and it becomes the new current vertex. On the other hand if the arc leads to a vertex that has already been seen, the arc becomes a non-tree arc and the current vertex doesn't change.

Finally there will come a time when the current vertex v has no untagged arcs. At this point, the algorithm might decide that v and all its descendants form a strong component. Indeed, this condition turns out to be true if and only if $v\text{-}min \equiv v$; a proof appears below. If so, v and all its descendants become settled, and they leave the tree. If not, the tree arc from v's parent u to v becomes mature, so the value of $v\text{-}min$ is used to update the value of $u\text{-}min$. In both cases, v becomes mature and the new current vertex will be the parent of v. Notice that only the value of $u\text{-}min$ needs to be updated, when the arc from u to v matures; all other values $w\text{-}min$ stay the same, because a newly mature arc has no mature predecessors.

The cave analogy helps to clarify the situation: If there's no way out of the subcave starting at v unless we come back through v itself, and if we can get back to v from all its descendants, then room v and its descendants will become a strong component. Once such a strong component is identified, we close it off and don't explore that subcave any further.

If v is the root of the tree, it always has $v\text{-}min \equiv v$, so it will always define a new strong component at the moment it matures. Then the depth-first search will terminate, since v has no parent. But Tarjan's algorithm will press on, trying to find a vertex u that is still unseen. If such a vertex exists, a new depth-first search will begin with u as the root. This process keeps on going until at last all vertices are happily settled.

The beauty of this algorithm is that it all works very efficiently when we organize it as follows:

⟨ Perform Tarjan's algorithm on g 10 ⟩ ≡
  ⟨ Make all vertices unseen and all arcs untagged 12 ⟩;
  for ($vv = g\text{-}vertices$; $vv < g\text{-}vertices + g\text{-}n$; $vv{+}{+}$)
    if ($vv\text{-}rank \equiv 0$)   /∗ vv is still unseen ∗/
      ⟨ Perform a depth-first search with vv as the root, finding the strong components of all unseen
         vertices reachable from vv 13 ⟩;
  ⟨ Print out one representative of each arc that runs between strong components 17 ⟩;
This code is used in section 2.

11.   ⟨ Global variables 5 ⟩ +≡
  Vertex ∗vv;   /∗ sweeps over all vertices, making sure none is left unseen ∗/

A: **Arc** ∗, GB_GRAPH §8.
Arc = **struct**, GB_GRAPH §10.
arcs: **Arc** ∗, GB_GRAPH §9.
g: **Graph** ∗, §2.
n: **long**, GB_GRAPH §20.

$rank = z.I$, §5.
V: **Vertex** ∗, GB_GRAPH §8.
v: **util**, GB_GRAPH §9.
Vertex = **struct**, GB_GRAPH §9.

vertices: **Vertex** ∗, GB_GRAPH §20.
w: **util**, GB_GRAPH §9.
x: **util**, GB_GRAPH §9.
y: **util**, GB_GRAPH §9.

12. It's easy to get the data structures started, according to the conventions stipulated above.

$\langle$ Make all vertices unseen and all arcs untagged $12 \rangle \equiv$
    for $(v = g\text{→}vertices + g\text{→}n - 1;\ v \geq g\text{→}vertices;\ v\text{--})$ {
        $v\text{→}rank = 0;$
        $v\text{→}untagged = v\text{→}arcs;$
    }
    $nn = 0;$
    $active\_stack = settled\_stack = \Lambda;$

This code is used in section 10.

13. The task of starting a depth-first search isn't too bad either. Throughout this part of the algorithm, variable v will point to the current vertex.

$\langle$ Perform a depth-first search with vv as the root, finding the strong components of all unseen vertices reachable from vv $13 \rangle \equiv$
    {
      $v = vv;$
      $v\text{→}parent = \Lambda;$
      $\langle$ Make vertex v active $14 \rangle;$
      do $\langle$ Explore one step from the current vertex v, possibly moving to another current vertex and calling it v $15 \rangle$ **while** $(v \neq \Lambda);$
    }

This code is used in section 10.

14. $\langle$ Make vertex v active $14 \rangle \equiv$
    $v\text{→}rank = {++}nn;\ v\text{→}link = active\_stack;$
    $active\_stack = v;$
    $v\text{→}min = v;$

This code is used in sections 13 and 15.

15. Now things get interesting. But we're just doing what any well-organized spelunker would do when calmly exploring a cave. There are three main cases, depending on whether the current vertex stays where it is, moves to a new child, or backtracks to a parent.

$\langle$ Explore one step from the current vertex v, possibly moving to another current vertex and calling it v $15 \rangle \equiv$
    { **register Vertex** $*u;$ /* a vertex adjacent to v */
      register Arc $*a = v\text{→}untagged;$ /* v's first remaining untagged arc, if any */
      if (a) {
        $u = a\text{→}tip;$
        $v\text{→}untagged = a\text{→}next;$ /* tag the arc from v to u */
        if $(u\text{→}rank)$ {   /* we've seen u already */
          if $(u\text{→}rank < v\text{→}min\text{→}rank)\ v\text{→}min = u;$ /* non-tree arc, just update $v\text{→}min$ */
        } **else** {   /* u is presently unseen */
          $u\text{→}parent = v;$ /* the arc from v to u is a new tree arc */
          $v = u;$ /* u will now be the current vertex */
          $\langle$ Make vertex v active $14 \rangle;$
        }
      } **else** {   /* all arcs from v are tagged, so v matures */

$u = v\text{-}parent;$ /∗ prepare to backtrack in the tree ∗/
if $(v\text{-}min \equiv v)$ ⟨Remove v and all its successors on the active stack from the tree, and mark
      them as a strong component of the graph 16⟩
else   /∗ the arc from u to v has just matured, making $v\text{-}min$ visible from u ∗/
  if $(v\text{-}min\text{-}rank < u\text{-}min\text{-}rank)$ $u\text{-}min = v\text{-}min;$
  $v = u;$ /∗ the former parent of v is the new current vertex v ∗/
  }
}

This code is used in section 13.

active_stack: **Vertex** ∗, §8.	*n*: **long**, GB_GRAPH §20.	*tip*: **Vertex** ∗, GB_GRAPH §10.
Arc = **struct**, GB_GRAPH §10.	*next*: **Arc** ∗, GB_GRAPH §10.	*untagged* = *x.A*, §7.
arcs: **Arc** ∗, GB_GRAPH §9.	*nn*: **long**, §5.	*v*: **register Vertex** ∗, §2.
g: **Graph** ∗, §2.	*parent* = *y.V*, §6.	**Vertex** = **struct**, GB_GRAPH §9.
link = *w.V*, §8.	*rank* = *z.I*, §5.	*vertices*: **Vertex** ∗, GB_GRAPH §20.
min = *v.V*, §9.	*settled_stack*: **Vertex** ∗, §8.	*vv*: **Vertex** ∗, §11.

16. The elements of the active stack are always in order by rank, and all children of a vertex v in the tree have rank higher than v. Tarjan's algorithm relies on a converse property: *All active nodes whose rank exceeds that of the current vertex v are descendants of v.* (This property holds because the algorithm has constructed the tree by assigning ranks in preorder, "the order of succession to the throne." First come v's firstborn and descendants, then the nextborn, and so on.) Therefore the descendants of the current vertex always appear consecutively at the top of the stack.

Another fundamental property of Tarjan's algorithm is more subtle: *There is always a way to get from any active vertex to the current vertex.* This follows from the fact that all mature active vertices u have $u\text{-}min\text{-}rank < u\text{-}rank$. If some active vertex does not lead to the current vertex v, let u be the counterexample with smallest rank. Then u isn't an ancestor of v, hence u must be mature; hence it leads to the active vertex $u\text{-}min$, from which there *is* a path to v, contradicting our assumption.

Therefore v and its active descendants are all reachable from each other, and they must belong to the same strong component. Moreover, if $v\text{-}min = v$, this component can't be made any larger. For there is no arc from any of these vertices to an unseen vertex; all arcs from v and its descendants have already been tagged. And there is no arc from any of these vertices to an active vertex that is below v on the stack; otherwise $v\text{-}min$ would have smaller rank than v. Hence all arcs, if any, that lead from these vertices to some other vertex must lead to settled vertices. And we know from previous steps of the computation that the settled vertices all belong to other strong components.

Therefore we are justified in settling v and its active descendants now. Removing them from the tree of active vertices does not remove any vertex from which there is a path to a vertex of rank less than $v\text{-}rank$. Hence their removal does not affect the validity of the $u\text{-}min$ value for any vertex u that remains active.

We print out enough information for a reader to verify the strength of the claimed component easily.

#define *infinity* $g\text{-}n$ /* infinite rank (or close enough) */

⟨ Remove v and all its successors on the active stack from the tree, and mark them as a strong
      component of the graph 16 ⟩ ≡
  { **register Vertex** *t; /* runs through the vertices of the new strong component */

    $t = active\_stack$;
    $active\_stack = v\text{-}link$;
    $v\text{-}link = settled\_stack$;
    $settled\_stack = t$; /* we've moved the top of one stack to the other */
    printf ("Strong␣component␣'%ld␣%s'", *specs*(v));
    if $(t \equiv v)$ *putchar* ('\n'); /* single vertex */
    else {
      printf ("␣also␣includes:\n");
      while $(t \neq v)$ {
        printf ("␣%ld␣%s␣(from␣%ld␣%s;␣..to␣%ld␣%s)\n", *specs*(t), *specs*($t\text{-}parent$), *specs*($t\text{-}min$));
        $t\text{-}rank = infinity$; /* now t is settled */
        $t\text{-}parent = v$; /* and v represents the new strong component */
        $t = t\text{-}link$;
      }

```
    }
  v⃗rank = infinity;     /* v too is settled */
  v⃗parent = v;        /* and represents its own strong component */
}
```
This code is used in section 15.

17. After all the strong components have been found, we can also compute the relations between them, without mentioning any cross-connection more than once. In fact, we built the *settled_stack* precisely so that this task could be done easily without sorting or searching. If only the strong components themselves were of interest, this part of the algorithm wouldn't be necessary.

For this step we use the name *arc_from* for the field we previously called *untagged*. The trick here relies on the fact that all vertices of the same strong component appear together in *settled_stack*.

#define *arc_from* *x.V*   /* utility field *x* will now point to a vertex */
⟨ Print out one representative of each arc that runs between strong components 17 ⟩ ≡
```
  printf("\nLinks␣between␣components:\n");
  for (v = settled_stack; v; v = v⃗link) { register Vertex *u = v⃗parent;
    register Arc *a;

    u⃗arc_from = u;
    for (a = v⃗arcs; a; a = a⃗next) { register Vertex *w = a⃗tip⃗parent;
      if (w⃗arc_from ≠ u) {
        w⃗arc_from = u;
        printf("%ld␣%s␣->␣%ld␣%s␣(e.g.,␣%ld␣%s␣->␣%ld␣%s)\n", specs(u), specs(w), specs(v),
            specs(a⃗tip));
      }
    }
  }
```
This code is used in section 10.

TAKE_RISC _____

Important: Before reading TAKE_RISC, please read or at least skim the program for GB_GATES.

1. Introduction. This demonstration program uses graphs constructed by the *risc* procedure in the GB_GATES module to produce an interactive program called `take_risc`, which multiplies and divides small numbers the slow way—by simulating the behavior of a logical circuit, one gate at a time.

The program assumes that UNIX conventions are being used. Some code in sections listed under 'UNIX dependencies' in the index might need to change if this program is ported to other operating systems.

To run the program under UNIX, say '`take_risc` ⟨trace⟩', where ⟨trace⟩ is nonempty if and only if you want the machine computations to be printed out.

The program will prompt you for two numbers, and it will use the simulated RISC machine to compute their product and quotient. Then it will ask for two more numbers, and so on.

2. Here is the general layout of this program, as seen by the C compiler:

```
#include "gb_graph.h"      /* the standard GraphBase data structures */
#include "gb_gates.h"      /* routines for gate graphs */
⟨ Preprocessor definitions ⟩
⟨ Global variables 3 ⟩
main(argc, argv)
    int argc;        /* the number of command-line arguments */
    char *argv[];       /* an array of strings containing those arguments */
{
  trace = (argc > 1 ? 8 : 0);      /* we'll show registers 0–7 if tracing */
  if ((g = risc(8ᴸ)) ≡ Λ) {
    printf("Sorry,␣I␣couldn't␣generate␣the␣graph␣(trouble␣code␣%ld)!\n", panic_code);
    return (−1);
  }
  printf("Welcome␣to␣the␣world␣of␣microRISC.\n");
  while (1) {
    ⟨ Prompt for two numbers; break if unsuccessful 4 ⟩;
    ⟨ Use the RISC machine to compute the product, p 7 ⟩;
    printf("The␣product␣of␣%ld␣and␣%ld␣is␣%ld%s.\n", m, n, p, o ? "␣(overflow␣occurred)" : "");
    ⟨ Use the RISC machine to compute the quotient and remainder, q and r 8 ⟩;
    printf("The␣quotient␣is␣%ld,␣and␣the␣remainder␣is␣%ld.\n", q, r);
  }
  return 0;      /* normal exit */
}
```

3. ⟨ Global variables 3 ⟩ ≡
```
  Graph *g;      /* graph that defines a simple RISC machine */
  long o, p, q, r;      /* overflow, product, quotient, remainder */
  long trace;      /* number of registers to trace */
  long m, n;      /* numbers to be multiplied and divided */
```

char *buffer*[100];   /∗ input buffer ∗/

See also section 6.

This code is used in section 2.

4.   **#define** *prompt*(*s*)
         { *printf*(*s*); *fflush*(*stdout*);   /∗ make sure the user sees the prompt ∗/
              if (*fgets*(*buffer*, 99, *stdin*) ≡ Λ) **break**; }

⟨ Prompt for two numbers; **break** if unsuccessful 4 ⟩ ≡
    prompt("\nGimme␣a␣number:␣");
step0:
    if (*sscanf*(*buffer*, "%ld", &*m*) ≠ 1) **break**;
step1:
    if (*m* ≤ 0) {
        prompt("Excuse␣me,␣I␣meant␣a␣positive␣number:␣");
        if (*sscanf*(*buffer*, "%ld", &*m*) ≠ 1) **break**;
        if (*m* ≤ 0) **break**;
    }
    while (*m* > #7fff) {
        prompt("That␣number's␣too␣big;␣please␣try␣again:␣");
        if (*sscanf*(*buffer*, "%ld", &*m*) ≠ 1) **goto** *step0*;   /∗ *step0* will **break** out ∗/
        if (*m* ≤ 0) **goto** *step1*;
    }
    ⟨ Now do the same thing for *n* instead of *m* 5 ⟩;

This code is used in section 2.

5.   ⟨ Now do the same thing for *n* instead of *m* 5 ⟩ ≡
    prompt("OK,␣now␣gimme␣another:␣");
    if (*sscanf*(*buffer*, "%ld", &*n*) ≠ 1) **break**;
step2:
    if (*n* ≤ 0) {
        prompt("Excuse␣me,␣I␣meant␣a␣positive␣number:␣");
        if (*sscanf*(*buffer*, "%ld", &*n*) ≠ 1) **break**;
        if (*n* ≤ 0) **break**;
    }
    while (*n* > #7fff) {
        prompt("That␣number's␣too␣big;␣please␣try␣again:␣");
        if (*sscanf*(*buffer*, "%ld", &*n*) ≠ 1) **goto** *step0*;   /∗ *step0* will **break** out ∗/
        if (*n* ≤ 0) **goto** *step2*;
    }

This code is used in section 4.

---

fflush: **int** ( ), <stdio.h>.   *panic_code*: **long**, GB_GRAPH §5.   *sscanf*: **int** ( ), <stdio.h>.
fgets: **char** ∗( ), <stdio.h>.   *printf*: **int** ( ), <stdio.h>.   *stdin*: **FILE** ∗, <stdio.h>.
Graph = **struct**, GB_GRAPH §20.   *risc*: **Graph** ∗( ), GB_GATES §8.   *stdout*: **FILE** ∗, <stdio.h>.

6. A RISC program.

Here is the little program we will run on the little computer. It consists mainly of a subroutine called *tri*, which computes the value of the ternary operation $x\lfloor y/z \rfloor$, assuming that $y \geq 0$ and $z > 0$; the inputs x, y, z appear in registers $1, 2, 3$, respectively, and the exit address is assumed to be in register 7. As special cases we can compute the product xy (letting $z = 1$) or the quotient $\lfloor y/z \rfloor$ (letting $x = 1$). When the subroutine returns, it leaves the result in register 4; it also leaves the value $(y \bmod z) - z$ in register 2. Overflow will be set if and only if the true result was not between -2^{15} and $2^{15} - 1$, inclusive.

It would not be difficult to modify the code to make it work with unsigned 16-bit numbers, or to make it deliver results with 32 or 48 or perhaps even 64 bits of precision.

#define *div* 7 /* location '*div*' in the program below */
#define *mult* 10 /* location '*mult*' in the program below */
#define *memry_size* 34 /* the number of instructions in the program below */
⟨ Global variables 3 ⟩ +≡
  unsigned long *memry*[*memry_size*] = { /* a "read-only memory" used by *run_risc* */
  #2ff0, /* *start*: $r2 = m$ (contents of next word) */
  #1111, /* (we will put the value of m here, in *memry*[1]) */
  #1a30, /* $r1 = n$ (contents of next word) */
  #3333, /* (we will put the value of n here, in *memry*[3]) */
  #7f70, /* **jumpto** (contents of next word), $r7 =$ return address */
  #5555, /* (we will put either *mult* or *div* here, in *memry*[5]) */
  #0f8f, /* halt without changing any status bits */
  #3a21, /* *div*: $r3 = r1$ */
  #1a01, /* $r1 = 1$ */
  #0a12, /* **goto** *tri* (literally, $r0 \mathrel{+}= 2$) */
  #3a01, /* *mult*: $r3 = 1$ */
  #4000, /* *tri*: $r4 = 0$ */
  #5000, /* $r5 = 0$ */
  #6000, /* $r6 = 0$ */
  #2a63, /* $r2 \mathrel{-}= r3$ */
  #0f95, /* **goto** *l2* */
  #3063, /* *l1*: $r3 \mathrel{\ll}= 1$ */
  #1061, /* $r1 \mathrel{\ll}= 1$ */
  #6ac1, /* **if** (overflow) $r6 = 1$ */
  #5fd1, /* $r5\mathord{+}\mathord{+}$ */
  #2a63, /* *l2*: $r2 \mathrel{-}= r3$ */
  #039b, /* **if** (≥ 0) **goto** *l1* */
  #0843, /* **goto** *l4* */
  #3463, /* *l3*: $r3 \mathrel{\gg}= 1$ */
  #1561, /* $r1 \mathrel{\gg}= 1$ */
  #2863, /* *l4*: $r2 \mathrel{+}= r3$ */
  #0c94, /* **if** (< 0) **goto** *l5* */
  #4861, /* $r4 \mathrel{+}= r1$ */
  #6ac1, /* **if** (overflow) $r6 = 1$ */
  #2a63, /* $r2 \mathrel{-}= r3$ */
  #5a41, /* *l5*: $r5\mathord{-}\mathord{-}$ */
  #0398, /* **if** (≥ 0) **goto** *l3* */
  #6666, /* **if** ($r6$) force overflow (literally $r6 \mathrel{\gg}= 4$) */

$^{\#}$0fa7};   /*   **return** (literally, $r0 = r7$, preserving overflow) */

7.   ⟨ Use the RISC machine to compute the product, p 7 ⟩ ≡
  $memry[1] = m;$
  $memry[3] = n;$
  $memry[5] = mult;$
  $run\_risc(g, memry, memry\_size, trace);$
  $p = (\textbf{long})\ risc\_state[4];$
  $o = (\textbf{long})\ risc\_state[16]\ \&\ 1;$ /* the overflow bit */
This code is used in section 2.

8.   ⟨ Use the RISC machine to compute the quotient and remainder, q and r 8 ⟩ ≡
  $memry[5] = div;$
  $run\_risc(g, memry, memry\_size, trace);$
  $q = (\textbf{long})\ risc\_state[4];$
  $r = ((\textbf{long})\ (risc\_state[2] + n))\ \&\ ^{\#}7\texttt{fff};$
This code is used in section 2.

---

g: **Graph** *, §3.
m: **long**, §3.
n: **long**, §3.
o: **long**, §3.

p: **long**, §3.
q: **long**, §3.
r: **long**, §3.
$risc\_state$: **unsigned long** [],

GB_GATES §48.
run_risc: **long** ( ), GB_GATES §43.
$trace$: **long**, §3.

TEST_SAMPLE _____

1. Introduction. This GraphBase program is intended to be used only when the Stanford GraphBase is being installed. It invokes the most critical subroutines and creates a file that can be checked against the correct output. The testing is not exhaustive by any means, but it is designed to detect errors of portability—cases where different results might occur on different systems. Thus, if nothing goes wrong, one can assume that the GraphBase routines are probably installed satisfactorily.

The basic idea of TEST_SAMPLE is quite simple: We generate a graph, then print out a few of its salient characteristics. Then we recycle the graph and generate another, etc. The test is passed if the output file matches a "correct" output file generated at Stanford by the author.

Actually there are two output files. The main one, containing samples of graph characteristics, is the standard output. The other, called `test.gb`, is a graph that has been saved in ASCII format with *save_graph*.

```
#include "gb_graph.h"      /* we use the GB_GRAPH data structures */
#include "gb_io.h"        /* and the GraphBase input/output routines */
⟨ Include headers for all of the GraphBase generation modules 2 ⟩

⟨ Private variables 7 ⟩
⟨ Procedures 13 ⟩
int main( )
{ Graph *g, *gg; long i; Vertex *v;      /* temporary registers */
  printf("GraphBase␣samples␣generated␣by␣test_sample:\n");
  ⟨ Save a graph to be restored later 6 ⟩;
  ⟨ Print samples of generated graphs 3 ⟩;
  return 0;     /* normal exit */
}
```

2. ⟨ Include headers for all of the GraphBase generation modules 2 ⟩ ≡
```
#include "gb_basic.h"     /* we test the basic graph operations */
#include "gb_books.h"     /* and the graphs based on literature */
#include "gb_econ.h"      /* and the graphs based on economic data */
#include "gb_games.h"     /* and the graphs based on football scores */
#include "gb_gates.h"     /* and the graphs based on logic circuits */
#include "gb_lisa.h"      /* and the graphs based on Mona Lisa */
#include "gb_miles.h"     /* and the graphs based on mileage data */
#include "gb_plane.h"     /* and the planar graphs */
#include "gb_raman.h"     /* and the Ramanujan graphs */
#include "gb_rand.h"      /* and the random graphs */
#include "gb_roget.h"     /* and the graphs based on Roget's Thesaurus */
#include "gb_save.h"      /* and we save results in ASCII format */
#include "gb_words.h"     /* and we also test five-letter-word graphs */
```
This code is used in section 1.

3. The subroutine *print_sample*(g, n) will be specified later. It prints global characteristics of g and local characteristics of the nth vertex.

We begin the test cautiously by generating a graph that requires no input data and no pseudo-random numbers. If this test fails, the fault must lie either in GB_GRAPH or GB_RAMAN.

⟨ Print samples of generated graphs 3 ⟩ ≡
  $print\_sample(raman(31_L, 3_L, 0_L, 4_L), 4)$;
See also sections 4, 5, 8, 9, 10, and 11.
This code is used in section 1.

4. Next we test part of GB_BASIC that relies on a particular interpretation of the operation '$w \gg= 1$'. If this part of the test fails, please look up 'system dependencies' in the index to GB_BASIC, and correct the problem on your system by making a change file `gb_basic.ch`. (See `queen_wrap.ch` for an example of a change file.)

On the other hand, if TEST_SAMPLE fails only in this particular test while passing all those that follow, chances are excellent that you have a pretty good implementation of the GraphBase anyway, because the bug detected here will rarely show up in practice. Ask yourself: Can I live comfortably with such a bug?

⟨ Print samples of generated graphs 3 ⟩ +≡
  $print\_sample(board(1_L, 1_L, 2_L, -33_L, 1_L, -{}^\#40000000_L - {}^\#40000000_L, 1_L), 2000)$;
    /∗ coordinates 32 and 33 (only) should wrap around ∗/

5. Another system-dependent part of GB_BASIC is tested here, this time involving character codes.

⟨ Print samples of generated graphs 3 ⟩ +≡
  $print\_sample(subsets(32_L, 18_L, 16_L, 0_L, 999_L, -999_L, {}^\#80000000_L, 1_L), 1)$;

6. If `test.gb` fails to match `test.correct`, the most likely culprit is *vert_offset*, a "pointer hack" in GB_BASIC. That macro absolutely must be made to work properly, because it is used heavily. In particular, it is used in the *complement* routine tested here, and in the *gunion* routine tested below.

⟨ Save a graph to be restored later 6 ⟩ ≡
  $g = random\_graph(3_L, 10_L, 1_L, 1_L, 0_L, \Lambda, dst, 1_L, 2_L, 1_L)$;
    /∗ a random multigraph with 3 vertices, 10 edges ∗/
  $gg = complement(g, 1_L, 1_L, 0_L)$;   /∗ a copy of g, without multiple edges ∗/
  $v = gb\_typed\_alloc(1, \textbf{Vertex}, gg{\rightarrow}data)$;   /∗ we create a stray vertex too ∗/
  $v{\rightarrow}name = gb\_save\_string(\texttt{"Testing"})$;
  $gg{\rightarrow}util\_types[10] = \texttt{'V'}$;
  $gg{\rightarrow}ww.V = v$;   /∗ the stray vertex is now part of gg ∗/
  $save\_graph(gg, \texttt{"test.gb"})$;   /∗ so it will appear in `test.gb` (we hope) ∗/
  $gb\_recycle(g)$; $gb\_recycle(gg)$;
This code is used in section 1.

---

board: **Graph** ∗( ), GB_BASIC §8.
complement: **Graph** ∗( ),
  GB_BASIC §74.
data: **Area**, GB_GRAPH §20.
dst: **static long** [], §7.
gb_recycle: **void** ( ), GB_GRAPH §40.
gb_save_string: **char** ∗( ),
  GB_GRAPH §35.
gb_typed_alloc = macro ( ),

  GB_GRAPH §11.
Graph = **struct**, GB_GRAPH §20.
gunion: **Graph** ∗( ), GB_BASIC §78.
name: **char** ∗, GB_GRAPH §9.
print_sample: **static void** ( ), §13.
printf: **int** ( ), <stdio.h>.
raman: **Graph** ∗( ), GB_RAMAN §4.
random_graph: **Graph** ∗( ),
  GB_RAND §5.

save_graph: **long** ( ), GB_SAVE §20.
subsets: **Graph** ∗( ), GB_BASIC §37.
util_types: **char** [], GB_GRAPH §20.
V: **Vertex** ∗, GB_GRAPH §8.
vert_offset = macro ( ),
  GB_BASIC §75.
Vertex = **struct**, GB_GRAPH §9.
ww: **util**, GB_GRAPH §20.

7. ⟨Private variables 7⟩ ≡

   static long $dst[\,] = \{^{\#}20000000, ^{\#}10000000, ^{\#}10000000\}$;

   /* a probability distribution with frequencies 50%, 25%, 25% */

See also section 12.

This code is used in section 1.

8. Now we try to reconstruct the graph we saved before, and we also randomize its lengths.

⟨Print samples of generated graphs 3⟩ +≡

   $g = restore\_graph(\texttt{"test.gb"})$;

   if $(i = random\_lengths(g, 0_L, 10_L, 12_L, dst, 2_L))$

      $printf(\texttt{"\\nFailure}_\sqcup\texttt{code}_\sqcup\texttt{\%ld}_\sqcup\texttt{returned}_\sqcup\texttt{by}_\sqcup\texttt{random\_lengths!\\n"}, i)$;

   else {

      $gg = random\_graph(3_L, 10_L, 1_L, 1_L, 0_L, \Lambda, dst, 1_L, 2_L, 1_L)$;   /* same as before */

      $print\_sample(gunion(g, gg, 1_L, 0_L), 2)$;

      $gb\_recycle(g)$; $gb\_recycle(gg)$;

   }

9. Partial evaluation of a RISC circuit involves fairly intricate pointer manipulation, so this step should help to test the portability of the author's favorite programming tricks.

⟨Print samples of generated graphs 3⟩ +≡

   $print\_sample(partial\_gates(risc(0_L), 1_L, 43210_L, 98765_L, \Lambda), 79)$;

10. Now we're ready to test the mechanics of reading data files, sorting with GB_SORT, and heavy randomization. Lots of computation takes place in this section.

⟨Print samples of generated graphs 3⟩ +≡

   $print\_sample(book(\texttt{"homer"}, 500_L, 400_L, 2_L, 12_L, 10000_L, -123456_L, 789_L), 81)$;

   $print\_sample(econ(40_L, 0_L, 400_L, -111_L), 11)$;

   $print\_sample(games(60_L, 70_L, 80_L, -90_L, -101_L, 60_L, 0_L, 999999999_L), 14)$;

   $print\_sample(miles(50_L, -500_L, 100_L, 1_L, 500_L, 5_L, 314159_L), 20)$;

   $print\_sample(plane\_lisa(100_L, 100_L, 50_L, 1_L, 300_L, 1_L, 200_L, 50_L * 299_L * 199_L, 200_L * 299_L * 199_L), 1294)$;

   $print\_sample(plane\_miles(50_L, 500_L, -100_L, 1_L, 1_L, 40000_L, 271818_L), 14)$;

   $print\_sample(random\_bigraph(300_L, 3_L, 1000_L, -1_L, 0_L, dst, -500_L, 500_L, 666_L), 3)$;

   $print\_sample(roget(1000_L, 3_L, 1009_L, 1009_L), 40)$;

11. Finally, here's a picky, picky test that is supposed to fail the first time, succeed the second. (The weight vector just barely exceeds the maximum weight threshold allowed by GB_WORDS. That test is ultraconservative, but eminently reasonable nevertheless.)

⟨Print samples of generated graphs 3⟩ +≡

   $print\_sample(words(100_L, wt\_vector, 70000000_L, 69_L), 5)$;

   $wt\_vector[1]\mathrel{++}$;

   $print\_sample(words(100_L, wt\_vector, 70000000_L, 69_L), 5)$;

   $print\_sample(words(0_L, \Lambda, 0_L, 69_L), 5555)$;

12. ⟨Private variables 7⟩ +≡

   static long $wt\_vector[\,] = \{100, -80589, 50000, 18935, -18935, 18935, 18935, 18935, 18935\}$;

13. Printing the sample data.   Given a graph g in GraphBase format and an integer n, the subroutine *print_sample*(g, n) will output global characteristics of g, such as its name and size, together with detailed information about its nth vertex. Then g will be shredded and recycled; the calling routine should not refer to it again.

⟨ Procedures 13 ⟩ ≡
   static void *pr_vert*( );   /* a subroutine for printing a vertex is declared below */
   static void *pr_arc*( );   /* likewise for arcs */
   static void *pr_util*( );   /* and for utility fields in general */
   static void *print_sample*(g, n)
      Graph *$*g$;   /* graph to be sampled and destroyed */
      int n;   /* index to the sampled vertex */
   {
     printf("\n");
     if $(g \equiv \Lambda)$ {
     printf("Ooops,␣we␣just␣ran␣into␣panic␣code␣%ld!\n", *panic_code*);
     if (*io_errors*) *printf*("(The␣I/O␣error␣code␣is␣0x%lx)\n", (**unsigned long**) *io_errors*);
     } **else** {
     ⟨ Print global characteristics of g 18 ⟩;
     ⟨ Print information about the nth vertex 17 ⟩;
     gb_recycle(g);
     }
   }
See also sections 14, 15, and 16.

This code is used in section 1.

---

14. The graph's *util_types* are used to determine how much information should be printed. A level parameter also helps control the verbosity of printout. In the most verbose mode, each utility field that points to a vertex or arc, or contains integer or string data, will be printed.

⟨ Procedures 13 ⟩ +≡

```
    static void pr_vert(v, l, s)
        Vertex *v;      /* vertex to be printed */
        int l;      /* ≤ 0 if the output should be terse */
        char *s;      /* format for graph utility fields */
    {
        if (v ≡ Λ)  printf("NULL");
        else if (is_boolean(v))  printf("ONE");      /* see GB_GATES */
        else {
            printf("\"%s\"", v⃗name);
            pr_util(v⃗u, s[0], l − 1, s);
            pr_util(v⃗v, s[1], l − 1, s);
            pr_util(v⃗w, s[2], l − 1, s);
            pr_util(v⃗x, s[3], l − 1, s);
            pr_util(v⃗y, s[4], l − 1, s);
            pr_util(v⃗z, s[5], l − 1, s);
            if (l > 0) { register Arc *a;
                for (a = v⃗arcs; a; a = a⃗next) {
                    printf("\n␣␣␣"); pr_arc(a, 1, s);
                }
            }
        }
    }
```

15. ⟨ Procedures 13 ⟩ +≡

```
    static void pr_arc(a, l, s)
        Arc *a;      /* non-null arc to be printed */
        int l;      /* ≤ 0 if the output should be terse */
        char *s;      /* format for graph utility fields */
    {
        printf("->");
        pr_vert(a⃗tip, 0, s);
        if (l > 0) {
            printf(",␣%ld", a⃗len);
            pr_util(a⃗a, s[6], l − 1, s);
            pr_util(a⃗b, s[7], l − 1, s);
        }
    }
```

16. ⟨ Procedures 13 ⟩ +≡

```
    static void pr_util(u, c, l, s)
        util u;      /* a utility field to be printed */
        char c;      /* its type code */
        int l;      /* 0 if output should be terse, −1 if pointers omitted */
        char *s;      /* utility types for overall graph */
```

```
    {
      switch (c) {
      case 'I': printf("[%ld]", u.I); break;
      case 'S': printf("[\"%s\"]", u.S ? u.S : "(null)"); break;
      case 'A':
        if (l < 0) break;
        printf("[");
        if (u.A ≡ Λ) printf("NULL");
        else pr_arc(u.A, l, s);
        printf("]");
        break;
      case 'V':
        if (l < 0) break;       /* avoid infinite recursion */
        printf("["); pr_vert(u.V, l, s); printf("]");
      default: break;       /* case 'Z' does nothing, other cases won't occur */
      }
    }
```

17. ⟨ Print information about the nth vertex 17 ⟩ ≡
$printf$ (`"V%d:␣"`, n);
if ($n ≥ g\text{-}n ∨ n < 0$) $printf$ (`"index␣is␣out␣of␣range!\n"`);
else {
  pr_vert ($g\text{-}vertices + n, 1, g\text{-}util\_types$);
  $printf$ (`"\n"`);
}
This code is used in section 13.

18. ⟨ Print global characteristics of g 18 ⟩ ≡
$printf$ (`"\"%s\"\n%ld␣vertices,␣%ld␣arcs,␣util_types␣%s"`, $g\text{-}id, g\text{-}n, g\text{-}m, g\text{-}util\_types$);
pr_util ($g\text{-}uu, g\text{-}util\_types[8], 0, g\text{-}util\_types$);
pr_util ($g\text{-}vv, g\text{-}util\_types[9], 0, g\text{-}util\_types$);
pr_util ($g\text{-}ww, g\text{-}util\_types[10], 0, g\text{-}util\_types$);
pr_util ($g\text{-}xx, g\text{-}util\_types[11], 0, g\text{-}util\_types$);
pr_util ($g\text{-}yy, g\text{-}util\_types[12], 0, g\text{-}util\_types$);
pr_util ($g\text{-}zz, g\text{-}util\_types[13], 0, g\text{-}util\_types$);
$printf$ (`"\n"`);
This code is used in section 13.

---

a: **util**, GB_GRAPH §10.
A: **Arc** *, GB_GRAPH §8.
Arc = **struct**, GB_GRAPH §10.
$arcs$: **Arc** *, GB_GRAPH §9.
b: **util**, GB_GRAPH §10.
g: **Graph** *, §13.
I: **long**, GB_GRAPH §8.
id: **char** [], GB_GRAPH §20.
$is\_boolean$ = macro ( ), GB_GATES §2.
len: **long**, GB_GRAPH §10.
m: **long**, GB_GRAPH §20.
n: **int**, §13.

n: **long**, GB_GRAPH §20.
$name$: **char** *, GB_GRAPH §9.
$next$: **Arc** *, GB_GRAPH §10.
$printf$: **int** ( ), <stdio.h>.
S: **char** *, GB_GRAPH §8.
tip: **Vertex** *, GB_GRAPH §10.
u: **util**, GB_GRAPH §9.
util = **union**, GB_GRAPH §8.
$util\_types$: **char** [], GB_GRAPH §20.
uu: **util**, GB_GRAPH §20.
v: **util**, GB_GRAPH §9.
V: **Vertex** *, GB_GRAPH §8.

Vertex = **struct**, GB_GRAPH §9.
$vertices$: **Vertex** *, GB_GRAPH §20.
vv: **util**, GB_GRAPH §20.
w: **util**, GB_GRAPH §9.
ww: **util**, GB_GRAPH §20.
x: **util**, GB_GRAPH §9.
xx: **util**, GB_GRAPH §20.
y: **util**, GB_GRAPH §9.
yy: **util**, GB_GRAPH §20.
z: **util**, GB_GRAPH §9.
zz: **util**, GB_GRAPH §20.

WORD_COMPONENTS ⸻

Important: Before reading WORD_COMPONENTS, please read or at least skim the program for GB_WORDS.

1. Components. This simple demonstration program computes the connected components of the GraphBase graph of five-letter words. It prints the words in order of decreasing weight, showing the number of edges, components, and isolated vertices present in the graph defined by the first n words for all n.

```
#include "gb_graph.h"      /* the GraphBase data structures */
#include "gb_words.h"      /* the words routine */
  ⟨ Preprocessor definitions ⟩
  main( )
  { Graph *g = words(0 L, 0 L, 0 L, 0 L);     /* the graph we love */
    Vertex *v;     /* the current vertex being added to the component structure */
    Arc *a;     /* the current arc of interest */
    long n = 0;     /* the number of vertices in the component structure */
    long isol = 0;     /* the number of isolated vertices in the component structure */
    long comp = 0;     /* the current number of components */
    long m = 0;     /* the current number of edges */
    printf("Component␣analysis␣of␣%s\n", g→id);
    for (v = g→vertices; v < g→vertices + g→n; v++) {
      n++, printf("%4ld:␣%5ld␣%s", n, v→weight, v→name);
      ⟨ Add vertex v to the component structure, printing out any components it joins 2 ⟩;
      printf(";␣c=%ld,i=%ld,m=%ld\n", comp, isol, m);
    }
    ⟨ Display all unusual components 5 ⟩;
    return 0;     /* normal exit */
  }
```

2. The arcs from v to previous vertices all appear on the list $v\rightarrow arcs$ after the arcs from v to future vertices. In this program, we aren't interested in the future, only the past; so we skip the initial arcs.

```
⟨ Add vertex v to the component structure, printing out any components it joins 2 ⟩ ≡
  ⟨ Make v a component all by itself 3 ⟩;
  a = v→arcs;
  while (a ∧ a→tip > v) a = a→next;
  if (¬a) printf("[1]");     /* indicate that this word is isolated */
  else { long c = 0;     /* the number of merge steps performed because of v */
    for ( ; a; a = a→next) { register Vertex *u = a→tip;
      m++;
      ⟨ Merge the components of u and v, if they differ 4 ⟩;
    }
    printf("␣in␣%s[%ld]", v→master→name, v→master→size);     /* show final component */
  }
```

This code is used in section 1.

3.   We keep track of connected components by using circular lists, a procedure that is known to take average time $O(n)$ on truly random graphs [Knuth and Schönhage, *Theoretical Computer Science* **6** (1978), 281–315].

Namely, if v is a vertex, all the vertices in its component will be in the list

$$v, \quad v\text{-}link, \quad v\text{-}link\text{-}link, \quad \ldots,$$

eventually returning to v again. There is also a master vertex in each component, $v\text{-}master$; if v is the master vertex, $v\text{-}size$ will be the number of vertices in its component.

#define *link* $z.V$ /* link to next vertex in component (occupies utility field z) */
#define *master* $y.V$ /* pointer to master vertex in component */
#define *size* $x.I$ /* size of component, kept up to date for master vertices only */
⟨ Make v a component all by itself 3 ⟩ ≡
  $v\text{-}link = v$;
  $v\text{-}master = v$;
  $v\text{-}size = 1$;
  $isol\text{++}$;
  $comp\text{++}$;

This code is used in section 2.

Arc = **struct**, GB_GRAPH §10.
arcs: **Arc** *, GB_GRAPH §9.
Graph = **struct**, GB_GRAPH §20.
I: **long**, GB_GRAPH §8.
id: **char** [ ], GB_GRAPH §20.
n: **long**, GB_GRAPH §20.
name: **char** *, GB_GRAPH §9.
next: **Arc** *, GB_GRAPH §10.
printf: **int** ( ), <stdio.h>.
tip: **Vertex** *, GB_GRAPH §10.
V: **Vertex** *, GB_GRAPH §8.
Vertex = **struct**, GB_GRAPH §9.
vertices: **Vertex** *, GB_GRAPH §20.
weight = $u.I$, GB_WORDS §26.
words: **Graph** *( ), GB_WORDS §7.
x: **util**, GB_GRAPH §9.
y: **util**, GB_GRAPH §9.
z: **util**, GB_GRAPH §9.

4. When two components merge together, we change the identity of the master vertex in the smaller component. The master vertex representing v itself will change if v is adjacent to any prior vertex.

⟨ Merge the components of u and v, if they differ 4 ⟩ ≡

```
u = u→master;
if (u ≠ v→master) { register Vertex *w = v→master, *t;
    if (u→size < w→size) {
        if (c++ > 0) printf("%s␣%s[%ld]", (c ≡ 2 ? "␣with" : ","), u→name, u→size);
        w→size += u→size;
        if (u→size ≡ 1) isol −−;
        for (t = u→link; t ≠ u; t = t→link) t→master = w;
        u→master = w;
    } else {
        if (c++ > 0) printf("%s␣%s[%ld]", (c ≡ 2 ? "␣with" : ","), w→name, w→size);
        if (u→size ≡ 1) isol −−;
        u→size += w→size;
        if (w→size ≡ 1) isol −−;
        for (t = w→link; t ≠ w; t = t→link) t→master = u;
        w→master = u;
    }
    t = u→link;
    u→link = w→link;
    w→link = t;
    comp −−;
}
```

This code is used in section 2.

5. The *words* graph has one giant component and lots of isolated vertices. We consider all other components unusual, so we print them out when the other computation is done.

⟨ Display all unusual components 5 ⟩ ≡

```
printf("\nThe␣following␣non-isolated␣words␣didn't␣join␣the␣giant␣component:\n");
for (v = g→vertices; v < g→vertices + g→n; v++)
    if (v→master ≡ v ∧ v→size > 1 ∧ v→size + v→size < g→n) { register Vertex *u;
        long c = 1;        /* count of number printed on current line */
        printf("%s", v→name);
        for (u = v→link; u ≠ v; u = u→link) {
            if (c++ ≡ 12) putchar('\n'), c = 1;
            printf("␣%s", u→name);
        }
        putchar('\n');
    }
```

This code is used in section 1.

INDEX

The following list, a compilation of the indexes produced by **CWEAVE** from the GraphBase programs, shows the section numbers where the identifiers of each program appear. Underlined numbers indicate a place of definition. Single-letter identifiers are indexed only when defined. Types are indexed only in the program where they are defined.

Other characteristics of the program segments, such as 'system dependencies', can also be found here. But the main index at the close of this book should be consulted for references to concepts such as 'hashing' or 'depth-first search' or 'strong components'.

APPENDIX A: *ERROR CODES*

Normal operation of the Stanford GraphBase should be error-free, unless parameters are pushed to unreasonable limits. Therefore the programs do not include elaborate error-recovery mechanisms. They do, however, check their calculations carefully and record errors whenever an anomaly is detected. Each error is given an identifying number, which can be decoded symbolically by the user program (as in GIRTH §5) or simply reported as a numeric error code (as in QUEEN §2).

In general, a GraphBase generator will return Λ (NULL) when it is unable to generate the requested graph, and the reason for failure will be the value of external variable *panic_code*. Panic codes are defined in GB_GRAPH §7, as follows:

-1	*alloc_fault*	(out of memory while building graph)
$1..9$	*no_room* $+ (0..8)$	(out of memory before building graph)
10	*early_data_fault*	(problem opening the .dat or .gb file)
11	*late_data_fault*	(problem closing the .dat or .gb file)
$12..29$	*syntax_error* $+ (-8..9)$	(problem parsing the .dat or .gb file)
$30..39$	*bad_specs* $+ (0..9)$	(parameter has illegal value)
$40..49$	*very_bad_specs* $+ (0..9)$	(parameter has incredibly illegal value)
$50..59$	*missing_operand* $+ (0..9)$	(graph parameter is Λ)
$60..69$	*invalid_operand* $+ (0..9)$	(graph or dist parameter not up to snuff)
$90..99$	*impossible* $+ (0..9)$	(internal consistency failure)

The index to programs will help pinpoint the exact nature of errors that are exceptionally mysterious.

An *early_data_fault* or *late_data_fault* is further qualified by the value of *io_errors*, which is a binary sum of up to twelve bits indicating various things that might go wrong in the realm of input/output. These error bits are explained in GB_IO §3. The most common case is an *early_data_fault* with *io_errors* $= 1$ (*cant_open_file*); this happens when the program can't find a requested .dat file. The file might not have been installed properly, or the DATADIR specified in Makefile might not have suitable permissions (see Chapter 3).

The *save_graph* subroutine has its own conventions. It returns a positive sum of binary codes if it was obliged to correct certain problems in the graph being saved. The bit codes are defined in GB_SAVE §19.

APPENDIX B: *SUMMARY OF FUNCTION CALLS*

This appendix lists all the external functions defined in the GraphBase library. It has two parts. First come the generators, like *board* and *words*; these are the routines that create GraphBase graphs. Then come all the others; these are routines for allocating storage, for generating pseudo-random numbers, for transforming graphs, for reading files, etc. In each part, the listing is alphabetical by function name and a cross reference is given to the program section or sections where the function is explained.

B.1 GENERATORS

The value returned by a GraphBase generator is always of type **Graph** * (pointer to a graph record). The parameters are almost always of type **long** or **unsigned long**, and they are **unsigned long** only in cases where a negative value would be meaningless. Therefore the parameter type is shown here only when it is neither **long** nor **unsigned long**.

Note: If you are using these routines on a machine for which **int** is different from **long**, you should use long constants (e.g., '3L') or cast the parameters to be type **long**.

#include "gb_basic.h" (see GB_BASIC §54)
all_parts(n, *directed*)   /* all partitions of n */

#include "gb_basic.h" (see GB_BASIC §41)
all_perms(n, *directed*)   /* all permutations of $\{0, 1, \ldots, n-1\}$ */

#include "gb_basic.h" (see GB_BASIC §63)
all_trees(n, *directed*)   /* all binary trees with $n+1$ leaves */

#include "gb_books.h" (see GB_BOOKS §4)
bi_book(**char** *title*, n, x, *first_chapter*, *last_chapter*, *in_weight*, *out_weight*, *seed*);
                    /* bipartite relation between chapters and characters of a novel */

#include "gb_basic.h" (see GB_BASIC §101)
bi_complete(*n1*, *n2*);   /* complete bipartite graph with *n1* · *n2* edges */

#include "gb_lisa.h" (see GB_LISA §33)
bi_lisa(m, n, *m0*, *m1*, *n0*, *n1*, *thresh*, *c*);
                            /* bipartite brightness relation between rows and columns */

#include "gb_basic.h" (see GB_BASIC §63)
binary(n, *max_height*, *directed*);   /* binary trees related by rotation */

#include "gb_basic.h" (see GB_BASIC §6)
board(*n1*, *n2*, *n3*, *n4*, *piece*, *wrap_mask*, *directed*);   /* grid or game board */

#include "gb_books.h" (see GB_BOOKS §2)
book(**char** *title*, n, x, *first_chapter*, *last_chapter*, *in_weight*, *out_weight*, *seed*);
                            /* encounters between characters in a novel */

#include "gb_basic.h" (see GB_BASIC §7)
circuit(n)   /* undirected graph with x adjacent to $x \pm 1$ (mod n) */

#include "gb_basic.h" (see GB_BASIC §73)
complement(**Graph** *g, copy, self, directed*);   /* complement or copy of *g* */

#include "gb_basic.h" (see GB_BASIC §7)
complete(*n*)   /* the complete graph K_n */

#include "gb_basic.h" (see GB_BASIC §7)
cycle(*n*)   /* directed graph arcs from *x* to *x* + 1 (mod *n*) */

#include "gb_basic.h" (see GB_BASIC §36)
disjoint_subsets(*n, k*)   /* *k*-subsets related by disjointness */

#include "gb_econ.h"   (see GB_ECON §2, §3)
econ(*n, omit, threshold, seed*);   /* directed flow between economic sectors */

#include "gb_basic.h" (see GB_BASIC §7)
empty(*n*)   /* graph with no edges */

#include "gb_games.h" (see GB_GAMES §2)
games(*n, ap0_weight, upi0_weight, ap1_weight, upi1_weight, first_day, last_day, seed*);
                                                             /* football scores */

#include "gb_basic.h" (see GB_BASIC §77)
gunion(**Graph** *g,* **Graph** *gg, multi, directed*);
                        /* vertices of *g* with edges that are present in either *g* or *gg* */

#include "gb_basic.h" (see GB_BASIC §100, §102)
induced(**Graph** *g,* **char** *description, self, multi, directed*);
                            /* graph *g* with vertices deleted, identified, split, etc. */

#include "gb_basic.h" (see GB_BASIC §77)
intersection(**Graph** *g,* **Graph** *gg, multi, directed*);
                        /* vertices of *g* with edges that are common to both *g* and *gg* */

#include "gb_basic.h" (see GB_BASIC §87)
lines(**Graph** *g, directed*);   /* the line graph of *g* (its adjacent edges or arcs) */

#include "gb_miles.h" (see GB_MILES §2, §3)
miles(*n, north_weight, west_weight, pop_weight, max_distance, max_degree, seed*)
                                                  /* undirected mileage between cities */

#include "gb_gates.h" (see GB_GATES §84)
partial_gates(**Graph** *g, r, prob, seed,* **char** *buf*);
                        /* simplification of gate graph *g* when some inputs are constant */

#include "gb_basic.h" (see GB_BASIC §54)
parts(*n, max_parts, max_size, directed*);
                                /* representations of *n* as a sum of positive integers */

#include "gb_basic.h" (see GB_BASIC §41)
perms(*n0, n1, n2, n3, n4, max_inv, directed*);   /* permutations of a multiset */

#**include** "gb_basic.h" (see GB_BASIC §36)
petersen( )   /∗ the famous Petersen graph on 10 vertices ∗/

#**include** "gb_plane.h" (see GB_PLANE §2)
plane(*n*, *x_range*, *y_range*, *extend*, *prob*, *seed*);
                                    /∗ Delaunay edges between random points in a rectangle ∗/

#**include** "gb_lisa.h" (see GB_LISA §23)
plane_lisa(*m*, *n*, *d*, *m0*, *m1*, *n0*, *n1*, *d0*, *d1*);
                                    /∗ adjacent regions of constant color in a digitized image ∗/

#**include** "gb_plane.h" (see GB_PLANE §41)
plane_miles(*n*, *north_weight*, *west_weight*, *pop_weight*, *extend*, *prob*, *seed*);
                                                    /∗ Delaunay edges between cities ∗/

#**include** "gb_gates.h" (see GB_GATES §66)
prod(*m*, *n*);   /∗ gate graph for parallel binary multiplication ∗/

#**include** "gb_basic.h" (see GB_BASIC §94)
product(**Graph** ∗*g*, **Graph** ∗*gg*, *type*, *directed*);
                                        /∗ cartesian, direct, or strong product of *g* and *gg* ∗/

#**include** "gb_raman.h" (see GB_RAMAN §2)
raman(*p*, *q*, *type*, *reduce*);   /∗ Ramanujan graph based on integer quaternions ∗/

#**include** "gb_rand.h" (see GB_RAND §22)
random_bigraph(*n1*, *n2*, *m*, *multi*, **long** ∗*dist1*, **long** ∗*dist2*, *min_len*, *max_len*, *seed*);
                                                            /∗ random bipartite graph ∗/

#**include** "gb_rand.h" (see GB_RAND §3, §4)
random_graph(*n*, *m*, *multi*, *self*, *directed*, **long** ∗*dist_from*, **long** ∗*dist_to*, *min_len*, *max_len*, *seed*);   /∗ random graph ∗/

#**include** "gb_save.h" (see GB_SAVE §4)
restore_graph(**char** ∗*file_name*);   /∗ graph represented by text file ∗/

#**include** "gb_gates.h" (see GB_GATES §8, §10)
risc(*regs*);   /∗ gate graph for simple computer chip ∗/

#**include** "gb_roget.h" (see GB_ROGET §2)
roget(*n*, *min_distance*, *prob*, *seed*);   /∗ cross references in Roget's *Thesaurus* ∗/

#**include** "gb_basic.h" (see GB_BASIC §24, §25)
simplex(*n*, *n0*, *n1*, *n2*, *n3*, *n4*, *directed*);   /∗ generalized triangular game board ∗/

#**include** "gb_basic.h" (see GB_BASIC §36)
subsets(*n*, *n0*, *n1*, *n2*, *n3*, *n4*, *size_bits*, *directed*);
                                        /∗ submultisets related by size of intersection ∗/

#**include** "gb_basic.h" (see GB_BASIC §7)
transitive(*n*)   /∗ transitive tournament on *n* vertices ∗/

#include "gb_basic.h" (see GB_BASIC §103)
wheel(*n*, *n1*, *directed*);   /* wheel with *n* points on rim and *n1* in center */

#include "gb_words.h" (see GB_WORDS §2, §5)
words(*n*, **long** **wt_vector*, *wt_threshold*, *seed*);
                                        /* five-letter words that agree in four places */

B.2 NONGENERATORS

For the other routines, we show the type of the result and the type of each parameter.

#include "gb_dijk.h"   (see GB_DIJK §6, §20)
Vertex **del_128*( );   /* delete smallest from priority queue with key range < 128 */

#include "gb_dijk.h"   (see GB_DIJK §6, §15)
Vertex **del_first*( );   /* delete smallest from simple priority queue */

#include "gb_plane.h" (see GB_PLANE §9)
void *delaunay*(**Graph** **g*, **void** (**f*)(**Vertex** **u*, **Vertex** **v*));
                                        /* call *f* with all Delaunay edges of *g* */

#include "gb_dijk.h"   (see GB_DIJK §1)
long *dijkstra*(**Vertex** **uu*, **Vertex** **vv*, **Graph** **gg*, **long** (**hh*)(**Vertex** **v*));
                                /* find shortest path from *uu* to *vv* in *gg* with heuristic *hh* */

#include "gb_dijk.h"   (see GB_DIJK §6, §15)
void *enlist*(**Vertex** **v*, **long** *d*);   /* insert *v* with key *d* in simple priority queue */

#include "gb_dijk.h"   (see GB_DIJK §6, §20)
void *enq_128*(**Vertex** **v*, **long** *d*);
                                /* insert *v* with key *d* in priority queue with key range < 128 */

#include "gb_words.h" (see GB_WORDS §30)
Vertex **find_word* (**char** **q*, **void** (**f* (**Vertex** ***))) ;
                                /* find *q* or neighbors of *q* in most recent *words* graph */

#include "gb_gates.h" (see GB_GATES §3)
long *gate_eval*(**Graph** **g*, **char** **in_vec*, **char** **out_vec*);
                                /* evaluate the gate graph *g* with given inputs */

#include "gb_graph.h" (see GB_GRAPH §11)
char **gb_alloc*(**long** *n*, **Area** *s*);   /* allocate *n* bytes in storage area *s* */

#include "gb_io.h"   (see GB_IO §21)
void *gb_backup*( );   /* backspace in current line of input */

#include "gb_io.h"   (see GB_IO §21)
char *gb_char*( );   /* return next character in current line of input */

#include "gb_io.h"   (see GB_IO §38)
long *gb_close*( );   /* close the data file just read */

```
#include "gb_io.h"     (see GB_IO §23)
long gb_digit(char d);                          /* return next character as a radix-d digit */
```

```
#include "gb_io.h"     (see GB_IO §19)
long gb_eof( );                     /* have all lines of the current data file been read? */
```

```
#include "gb_graph.h"  (see GB_GRAPH §11)
void gb_free(Area s);                               /* deallocate all storage in s */
```

```
#include "gb_flip.h"   (see GB_FLIP §1)
void gb_init_rand(long seed);               /* initialize the random number generator */
```

```
#include "gb_sort.h"   (see GB_SORT §2)
void gb_linksort(node *l);                       /* sort a linked list of nodes */
```

```
#include "gb_graph.h"  (see GB_GRAPH §30)
void gb_new_arc(Vertex *u, Vertex *v, long len);
                                 /* append an arc from u to v of length len */
```

```
#include "gb_graph.h"  (see GB_GRAPH §31)
void gb_new_edge(Vertex *u, Vertex *v, long len);
                                 /* append an edge between u and v of length len */
```

```
#include "gb_graph.h"  (see GB_GRAPH §23)
Graph *gb_new_graph(long n);          /* create a new graph with n nameless vertices */
```

```
#include "gb_io.h"     (see GB_IO §16)
void gb_newline( );                                 /* advance to next line of input */
```

```
#include "gb_flip.h"   (see GB_FLIP §1)
long gb_next_rand( );                               /* the next pseudo-random number */
```

```
#include "gb_io.h"     (see GB_IO §28)
long gb_open(char *file_name);
                            /* open a GraphBase data file and check its preamble */
```

```
#include "gb_io.h"     (see GB_IO §40)
long gb_raw_close( );          /* close the current input file and return its checksum */
```

```
#include "gb_io.h"     (see GB_IO §28)
void gb_raw_open(char *file_name);               /* open a file for GraphBase input */
```

```
#include "gb_graph.h"  (see GB_GRAPH §40)
void gb_recycle(Graph *g);          /* remove g and its data areas from memory */
```

```
#include "gb_graph.h"  (see GB_GRAPH §35)
char *gb_save_string(char *s);        /* make a copy of s for use in the current graph */
```

```
#include "gb_io.h"     (see GB_IO §25)
char *gb_string(char *p, char c);        /* input a string delimited by c to location p */
```

#include "gb_graph.h" (see GB_GRAPH §11)
whatever **gb_typed_alloc*(**long** *n*, **type whatever**, **Area** *s*)
                                          /∗ allocate *n* records of type **whatever**† ∗/

#include "gb_graph.h" (see GB_GRAPH §29)
Arc **gb_virgin_arc*( );   /∗ a new **Arc** record for the current graph ∗/

#include "gb_graph.h" (see GB_GRAPH §42)
void *hash_in*(**Vertex** **v*);   /∗ insert name of *v* in current graph's hash table ∗/

#include "gb_graph.h" (see GB_GRAPH §42)
Vertex **hash_lookup*(**char** **s*, **Graph** **g*);   /∗ find vertex named *s* in graph *g* ∗/

#include "gb_graph.h" (see GB_GRAPH §42)
Vertex **hash_out*(**char** **s*);   /∗ find vertex named *s* in current graph ∗/

#include "gb_graph.h" (see GB_GRAPH §42)
void *hash_setup*(**Graph** **g*);   /∗ create hash table for all vertices of *g* ∗/

#include "gb_io.h"   (see GB_IO §11, §12)
char *imap_chr*(**long** *d*);   /∗ the *d*th character of GraphBase's internal *icode* ∗/

#include "gb_io.h"   (see GB_IO §11, §12)
long *imap_ord*(**char** *c*);   /∗ the internal *icode* for system character *c* ∗/

#include "gb_dijk.h" (see GB_DIJK §6, §20)
void *init_128*(**long** *d*);   /∗ initialize priority queue for keys in $[d, d + 128)$ ∗/

#include "gb_graph.h" (see GB_GRAPH §11)
void *init_area*(**Area** *s*);   /∗ clear a non-static **Area** variable ∗/

#include "gb_dijk.h" (see GB_DIJK §6, §15)
void *init_dlist*(**long** *d*);   /∗ initialize simple priority queue for keys $\geq d$ ∗/

#include "gb_lisa.h" (see GB_LISA §2, §3)
long **lisa*(**unsigned long** *m*, **unsigned long** *n*, **unsigned long** *d*, **unsigned long** *m0*,
      unsigned long *m1*, **unsigned long** *n0*, **unsigned long** *n1*, **unsigned long** *d0*,
      unsigned long *d1*, **Area** *area*);   /∗ create an $m \times n$ matrix of pixel values ∗/

#include "gb_io.h"   (see GB_IO §23)
unsigned long *gb_number*(**char** *d*);   /∗ scan a radix-*d* number ∗/

#include "gb_graph.h" (see GB_GRAPH §26)
void *make_compound_id*(**Graph** **g*, **char** **s1*, **Graph** **gg*, **char** **s2*);
                                          /∗ *sprintf*(*g*→*id*, "%s%s%s", *s1*, *gg*→*id*, *s2*) ∗/

#include "gb_graph.h" (see GB_GRAPH §26)
void *make_double_compound_id*(**Graph** **g*, **char** **s1*, **Graph** **gg*, **char** **s2*, **Graph** **ggg*,
      char **s3*);   /∗ *sprintf*(*g*→*id*, "%s%s%s%s%s", *s1*, *gg*→*id*, *s2*, *ggg*→*id*, *s3*) ∗/

---

† The second parameter of this macro is used as an argument to the **sizeof** operator, and it
also determines the result type.

`#include "gb_graph.h"` (see GB_GRAPH §22)
void *mark_bipartite*(**Graph** *g, **long** *n1*)
/* specify that graph g is bipartite with *n1* vertices in first part */

`#include "gb_miles.h"` (see GB_MILES §20)
long *miles_distance*(**Vertex** *u, **Vertex** *v);
/* distance between u and v in most recent *miles* graph */

`#include "gb_io.h"` (see GB_IO §10, §17)
long *new_checksum*(**char** *s, **long** *old_checksum*);
/* update checksum with respect to additional input s */

`#include "gb_dijk.h"` (see GB_DIJK §1)
void *print_dijkstra_result*(**Vertex** *vv);   /* print path to vv just found by *dijkstra* */

`#include "gb_gates.h"` (see GB_GATES §49)
void *print_gates*(**Graph** *g);   /* print a symbolic representation of gate graph g */

`#include "gb_rand.h"` (see GB_RAND §24)
long *random_lengths*(**Graph** *g, **long** *directed*, **long** *min_len*, **long** *max_len*,
   long *dist, **long** *seed*);   /* assign random lengths to all arcs or edges of g */

`#include "gb_dijk.h"` (see GB_DIJK §6, §15)
void *reenlist*(**Vertex** *v, **long** *d*);   /* decrease key of v to d in simple priority queue */

`#include "gb_dijk.h"` (see GB_DIJK §6, §20)
void *req_128*(**Vertex** *v, **long** *d*);
/* decrease key of v to d in priority queue with key range < 128 */

`#include "gb_gates.h"` (see GB_GATES §43)
long *run_risc*(**Graph** *g, **unsigned** **long** *rom, **unsigned** **long** *size*,
   unsigned **long** *trace_regs*);
/* emulate the *risc* graph g using instructions in *rom* */

`#include "gb_save.h"` (see GB_SAVE §1, §19)
long *save_graph*(**Graph** *g, **char** *file_name);   /* convert g to text format */

`#include "gb_graph.h"` (see GB_GRAPH §39)
void *switch_to_graph*(**Graph** *g);   /* make g the current graph */

APPENDIX C: *EXAMPLE GRAPH PARAMETERS*

This appendix lists a few of the GraphBase graphs g in order of size—according to the number of vertices, $g\text{-}n$, and the number of arcs, $g\text{-}m$, that they contain. In an undirected graph, the number of edges is half the number of arcs.

Several examples involve nonuniform random numbers, where the *dist* array is defined by

$$dist[k] = \begin{cases} 2^{17}(k+1), & \text{for } 0 \le k \le 126, \\ 2^{16}(k+1), & \text{for } k = 127. \end{cases}$$

Thus the kth vertex occurs with probability proportional to k, except that the last vertex occurs only half that often. (It is not difficult to verify that $\sum_{k=0}^{127} dist[k] = 2^{30}$, as required by the program.)

C.1 UNDIRECTED GRAPHS

A dagger (†) following the number of vertices means that the graph is known to be bipartite and that its first $g\text{-}n1$ vertices are the vertices of its first part. A double dagger (‡) indicates that the graph is known to be planar.

$g\text{-}n$	$g\text{-}m$	$g =$	
10 ‡	0	*empty*(10)	/* $\overline{K}_{10}$ */
10 ‡	20	*circuit*(10)	/* C_{10} */
10	30	*petersen*( )	/* Figure 13 in Section 1.12 */
10	90	*complete*(10)	/* K_{10} */
11 ‡	40	*wheel*(10, 1, 0)	/* wheel with 10 spokes */
19 ‡	66	*plane_lisa*(10, 10, 5, 0, 250, 0, 250, 0, 0)	/* 10 × 10 Mona Lisa with 6 shades of gray */
20 †	200	*bi_complete*(10, 10, 0)	/* $K_{10,10}$ */
30	124	*games*(30, 0, 0, 1, 2, 0, 0, 0)	/* football games between top 30 teams */
37	126	*book*("huck", 37, 0, 1, 43, 0, 0, 0)	/* half the characters of Huck Finn */
41 ‡	160	*plane_lisa*(10, 10, 11, 0, 250, 0, 250, 0, 0)	/* 10 × 10 Mona Lisa with 12 shades of gray */
42	228	*all_parts*(10)	/* partitions of 10 */
48 †	406	*bi_lisa*(16, 32, *smile*, 18000, 1)	/* her inscrutable smile */
49	250	*book*("anna", 50, 1, 10, 120, 1, 1, 0)	/* top 50 people in *Anna Karenina*, except Levin */
50	218	*book*("david", 50, 0, 0, 0, 0, 0, 0)	/* random 50 people in *David Copperfield* */
50	290	*book*("anna", 50, 0, 10, 120, 1, 1, 0)	/* top 50 people in *Anna Karenina* */
53	378	*games*(53, 1, 1, 1, 1, 0, 0, 0)	/* football games between nationally ranked teams */
64 ‡	224	*board*(8, 8, 0, 0, 1, 0, 0)	/* 8 × 8 grid */
64	256	*board*(8, 8, 0, 0, 1, −1, 0)	/* 8 × 8 torus */
64	336	*board*(8, 8, 0, 0, 5, 0, 0)	/* knight moves on chessboard */

64	1456	$gunion(board(8,8,0,0,-1,0,0), board(8,8,0,0,-2,0,0),0,0)$
		/* queen moves */
64	3696	$complement(board(8,8,0,0,5,0,0),0,0,0)$
		/* non-knight moves on chessboard */
67	448	$games(67,-1,-1,-1,-1,0,0,0)$
		/* football games between nationally unranked teams */
73	566	$book("huck",0,1,0,0,0,0,0)$
		/* encounters in *Huckleberry Finn*, excluding Huck */
75	134	$book("jean",0,5,0,178,1,0,0)$
		/* encounters between all but the top five characters of *Les Misérables*, in first half */
75	228	$book("jean",0,5,179,0,0,1,0)$ /* same, in second half */
80	508	$book("jean",0,0,1,356,0,0,0)$ /* all encounters in *Les Misérables* */
86	800	$book("david",0,1,1,64,0,0,0)$
		/* all non-Davidic encounters in *David Copperfield* */
97	184	$parts(50,8,8,0)$ /* partitions of 50 into ≤ 8 parts ≤ 8 */
100	1812	$random\_graph(100,1000,-1,0,0,0,0,0,0,0)$
		/* multiple edges collapsed; no loops */
100	1814	$random\_graph(100,1000,-1,1,0,0,0,0,0,0)$
		/* multiple edges collapsed; loops OK */
100	2000	$random\_graph(100,1000,0,0,0,0,0,0,0,0)$
		/* no multiple edges or self-loops */
100	9900	$miles(100,0,0,1,0,99,0)$ /* all mileages between 100 largest cities */
101 ‡	594	$plane\_miles(100,0,0,1,1,0,0)$
		/* Delaunay edges between 100 largest cities and ∞ */
120	480	$all\_perms(5)$
		/* permutations of $\{0,1,2,3,4\}$ related by adjacent transposition */
120	634	$games(0,0,0,0,0,50,0,0)$ /* all football games in the last half season */
120	642	$games(0,0,0,0,0,0,50,0)$ /* all football games in the first half season */
120	1276	$games(0,0,0,0,0,0,0,0)$ /* all football games in games.dat */
121	420	$subsets(20,10,20,10,0,0,{}^{\#}7f,0)$
		/* 20-element submultisets of $\{10 \cdot 0, 20 \cdot 1, 10 \cdot 2\}$ with ≤ 7 common elements */
121 ‡	640	$simplex(20,10,20,10,0,0,0)$ /* board for the game of Hex */
121	7200	$subsets(20,10,20,10,0,0,{}^{\#}55555,0)$
		/* 20-element submultisets of $\{10 \cdot 0, 20 \cdot 1, 10 \cdot 2\}$ with even intersections */
128 ‡	736	$plane\_miles(0,0,0,0,0,0,0)$ /* Delaunay edges for all cities of miles.dat */
128	1016	$miles(0,0,0,0,0,10,0)$ /* mileages with respect to ≤ 10 nearest neighbors */
128	1648	$miles(0,0,0,0,400,0,0)$ /* intercity mileages that are ≤ 400 */
128	1784	$random\_graph(128,1000,-1,0,0,dist,dist,0,0,0)$
		/* nonuniform distribution, multiple edges collapsed */
128	1788	$random\_graph(128,1000,-1,1,0,dist,dist,0,0,0)$
		/* same, with self-loops permitted */

128	1876	$random\_graph(128, 1000, -1, 0, 0, 0, 0, 0, 0, 0)$
		/* multiple edges collapsed; no loops */
128	1878	$random\_graph(128, 1000, -1, 1, 0, 0, 0, 0, 0, 0)$
		/* multiple edges collapsed; loops OK */
128	2000	$random\_graph(128, 1000, 0, 0, 0, 0, 0, 0, 0, 0)$
		/* no multiple edges or self-loops */
128	2054	$miles(0, 0, 0, 0, 0, 20, 0)$ /* mileages with respect to ≤ 20 nearest neighbors */
128	6432	$miles(128, 0, 0, 0, 1000, 0, 0)$ /* intercity mileages that are ≤ 1000 */
138	986	$book(\texttt{"anna"}, 0, 0, 0, 0, 0, 0, 0)$ /* all encounters in *Anna Karenina* */
161 †	712	$bi\_book(\texttt{"anna"}, 50, 0, 10, 120, 1, 1, 0)$
		/* appearances of the top 50 characters in chapters 10–120 */
168	1672	$lines(board(8, 8, 0, 0, 5, 0, 0), 0)$ /* adjacencies between knight moves */
256	1882	$random\_bigraph(128, 128, 1000, -1, dist, dist, 0, 0, 0)$
		/* multiple edges collapsed; nonuniform */
256	1924	$random\_bigraph(128, 128, 1000, -1, 0, 0, 0, 0, 0)$
		/* multiple edges collapsed; uniform */
256	2000	$random\_bigraph(128, 128, 1000, 0, dist, dist, 0, 0, 0)$
		/* no multiple edges; nonuniform */
280	122	$book(\texttt{"homer"}, 280, 0, 0, 0, -1, -1, 0)$
		/* encounters between the least important 280 characters */
280	774	$book(\texttt{"homer"}, 280, 0, 0, 0, 0, 0, 0)$
		/* encounters between 280 random characters */
280	776	$book(\texttt{"homer"}, 280, 0, 0, 0, 0, 0, 1)$
		/* encounters between another such selection */
280	2148	$book(\texttt{"homer"}, 280, 0, 1, 0, 0, 1, 1, 0)$
		/* encounters between the most important 280 characters */
331 ‡	1860	$simplex(30, 20, 20, 20, 0, 0, 0)$ /* hexagonal board with 60 boundary cells */
377 †	1924	$bi\_book(\texttt{"anna"}, 0, 0, 0, 0, 0, 0, 0)$ /* all appearances in *Anna Karenina* */
400	5720	$product(book(\texttt{"david"}, 21, 1, 0, 0, 1, 1, 0), book(\texttt{"huck"}, 21, 1, 0, 0, 1, 1, 0), 0, 0)$
400	19920	$product(book(\texttt{"david"}, 21, 1, 0, 0, 1, 1, 0), book(\texttt{"huck"}, 21, 1, 0, 0, 1, 1, 0), 1, 0)$
435	5040	$raman(11, 29, 2, 1)$
		/* expander graph of degree 12, multiple edges and loops reduced */
435	5400	$raman(11, 29, 2, 0)$ /* same, before reduction */
462	2772	$disjoint\_subsets(11, 5)$ /* the $\binom{11}{5}$ subsets of $\{0, 1, \ldots, 10\}$ */
462	13860	$simplex(5, 1, -10, 0, 0, 0, 0)$ /* another relation between $\binom{11}{5}$ objects */
500 ‡	2958	$plane(500, 16384, 16384, 0, 0, 0)$
		/* Delaunay edges between 500 random points in a square */
501 ‡	1512	$plane(500, 16384, 16384, 1, 32768, 0)$
		/* half of the Delaunay edges, including ∞ */
501 ‡	2994	$plane(500, 16384, 16384, 1, 0, 0)$ /* Delaunay edges, including ∞ */

561	294	*intersection*(*book*("homer", 0, 0, 1, 12, 0, 0, 0), *book*("homer", 0, 0, 13, 24, 0, 0, 0), 0, 0)	
			/* encounters found in both halves */
561	3258	*book*("homer", 0, 0, 0, 0, 0, 0, 0)	/* all encounters in the *Iliad* */
585 †	2456	*bi_book*("homer", 0, 0, 0, 0, 0, 0, 0)	/* all appearances in the *Iliad* */
610 †	14876	*bi_lisa*(0, 0, 0, 0, 0, 0, 32768, 0)	/* the brightest pixels of Mona Lisa */
610 †	70930	*bi_lisa*(0, 0, 0, 0, 0, 0, 5000, 1)	/* the darkest pixels of Mona Lisa */
610 †	165124	*bi_lisa*(0, 0, 0, 0, 0, 0, 32768, 1)	/* the not-too-bright pixels of Mona Lisa */
627	7090	*all_parts*(20, 0)	/* partitions of 20 */
763	3804	*perms*(1, 2, 3, 4, −8, 4, 0)	/* multiset permutations with ≤ 4 inversions */
891	636	*words*(0, {−1, −1, −1, −1, −1, −1, −1, −1, −1}, 0, 0)	
			/* five-letter words of zero weight */
1010	11988	*raman*(11, 1009, 1, 1)	
			/* expander graph of degree 12, multiple edges and loops reduced */
1010	12120	*raman*(11, 1009, 1, 0)	/* same, before reduction */
1024	10240	*board*(2, −10, 0, 0, 1, 0, 0)	/* 10-dimensional hypercube */
1024	122880	*board*(2, −10, 0, 0, 3, 0, 0)	/* 10-bit vectors differing in 3 places */
1050	3148	*raman*(2, 1049, 1, 1)	
			/* expander graph of degree 3, multiple edges and loops reduced */
1050	3152	*raman*(2, 1049, 1, 0)	/* same, before reduction */
1057	7122	*parts*(50, 8, 10, 0)	/* partitions of 50 into ≤ 8 parts ≤ 10 */
1057	8056	*parts*(50, 10, 8, 0)	/* partitions of 50 into ≤ 10 parts ≤ 8 */
1092	4368	*raman*(3, 13, 0, 0)	/* expander graph of degree 4 */
1430	10010	*all_trees*(8, 0)	/* binary trees with 9 leaves */
1446 ‡	6324	*plane_lisa*(100, 100, 22, 0, 0, 0, 0, 0, 0)	
			/* 100 × 100 Mona Lisa with 23 shades of gray */
1619 ‡	7034	*plane_lisa*(100, 100, 22, 0, 250, 0, 250, 0, 0)	
			/* same, from square top of picture */
1800	657920	*board*(2, 3, 5, −7, 7, 5, 0)	
			/* weird hyperboard with two wraparound coordinates */
1860	11180	*binary*(10, 5, 0)	/* binary trees with 11 leaves and height ≤ 5 */
2000	2792	*words*(2000, {−100, −10, −4, −2, −2, −1, −1, −1, −1}, −2147483647, 0)	
			/* the least common 2000 words */
2000	3314	*words*(2000, {0, 0, 0, 0, 0, 0, 0, 0, 0}, 0, 0)	/* random sample of 2000 words */
2000	3370	*words*(2000, {0, 0, 0, 0, 0, 0, 0, 0, 0}, 0, 1)	
			/* another random sample of 2000 words */
2000	5146	*words*(2000, 0, 0, 1)	/* the most common 2000 words */
2000	5150	*words*(2000, 0, 0, 0)	/* same, with another way to break ties */
2016	4000	*binary*(125, 7, 0)	/* binary trees with 126 leaves and height 7 */
2184	13104	*raman*(5, 13, 0, 0)	/* expander graph of degree 6 */
2184	17472	*raman*(7, 13, 0, 0)	/* expander graph of degree 8 */

2184	26208	$raman(11,13,0,0)$	/* expander graph of degree 12 */
3300	13192	$words(0,\{1,0,0,0,0,0,0,0,0,0\},1,0)$	/* the 'common' words in `words.dat` */
3317	13234	$words(0,0,100,0)$	/* the words with default weight ≥ 100 */
4493	27238	$restore\_graph(\texttt{"word\_giant.gb"})$	/* largest component of $words(0,0,0,0)$ */
4494	19624	$words(0,\{1,1,0,0,0,0,0,0,0,0\},1,0)$	/* the 'common' or 'advanced' words */
4866	21680	$words(0,0,1,0)$	/* the words with nonzero weight */
5040	30240	$all\_perms(7)$	/* permutations of $\{0,\ldots,6\}$ related by adjacent transposition */
5448	72210	$parts(50,10,10,0)$	/* partitions of 50 into ≤ 10 parts ≤ 10 */
5757	28270	$words(0,0,0,0)$	/* all "real" five-letter words of English */
6671 ‡	28132	$plane\_lisa(0,0,26,0,0,0,0,0,0)$	/* 360×250 Mona Lisa with 27 shades of gray */
16796	151164	$all\_trees(10,0)$	/* the $\binom{20}{10}/11$ binary trees with 11 leaves */

C.2 DIRECTED GRAPHS

Here an arrow (↑) following the number of vertices means that the graph is known to be acyclic.

$g\text{-}n$	$g\text{-}m$	$g =$	
10	10	$cycle(10)$	/* oriented 10-cycle */
10↑	15	$subsets(2,1,-4,0,0,0,1,1)$	/* oriented Petersen graph */
10↑	45	$transitive(10)$	/* $u \to v$ when $u < v$ */
10	100	$complement(empty(10),0,1,1)$	/* $u \to v$ for all u and v */
42↑	114	$all\_parts(10,1)$	/* partitions of 10 under refinement */
45↑	120	$lines(transitive(10),1)$	/* arc graph of transitive tournament */
50	1702	$econ(50,2,10,2)$	/* random 50-sector flows */
50	1751	$econ(50,2,10,1)$	/* same, for 50 different sectors */
50	1776	$econ(50,2,10,0)$	/* same, for 50 balanced sectors */
64↑	112	$board(8,8,0,0,1,0,1)$	/* directed 8×8 grid */
64	128	$board(8,8,0,0,1,-1,1)$	/* directed 8×8 torus */
79	2206	$econ(79,2,100,0)$	/* strong flows between sectors */
79	3559	$econ(79,2,10,0)$	/* mildly strong flows between sectors */
79	4473	$econ(79,2,1,0)$	/* nontrivial flows between sectors */
79	4602	$econ(79,2,0,0)$	/* nonzero flows between sectors */
81	4902	$econ(81,0,0,0)$	/* all flows between sectors and pseudo-sectors */
97↑	92	$parts(50,8,8,1)$	/* partitions of 50 into ≤ 8 parts ≤ 8 */
100	20	$roget(100,3,0,1)$	/* cross-references between 100 random categories */
100	26	$roget(100,3,0,0)$	/* same, another sample; distance must be ≥ 3 */
100	957	$random\_graph(100,1000,-1,1,1,0,0,0,0,0)$	/* multiple edges collapsed */
100	1000	$random\_graph(100,1000,1,1,1,0,0,0,0,0)$	/* multiple edges allowed */

120↑ 240 *all_perms*(5, 1)
/∗ permutations of {0, 1, 2, 3, 4} under adjacent transposition ∗/

121↑ 210 *subsets*(20, 10, 20, 10, 0, 0, #7f, 1)
/∗ 20-element submultisets of {10 · 0, 20 · 1, 10 · 2} with ≤ 7 common elements ∗/

121↑ 320 *simplex*(20, 10, 20, 10, 0, 0, 1) /∗ oriented Hex game board ∗/

128 944 *random_graph*(128, 1000, −1, 1, 1, *dist*, *dist*, 0, 0, 0)
/∗ nonuniform distribution, multiple edges collapsed ∗/

128 962 *random_graph*(128, 1000, −1, 0, 1, 0, 0, 0, 0, 0)
/∗ same, but self-loops not allowed ∗/

128 963 *random_graph*(128, 1000, −1, 1, 1, 0, 0, 0, 0, 0)
/∗ uniform distribution, multiple edges collapsed ∗/

128 1000 *random_graph*(128, 1000, 1, 1, 1, 0, 0, 0, 0, 0) /∗ same, without collapsing ∗/

151↑ 238 *partial_gates*(*risc*(16), 1, 32768, 0, Λ)
/∗ freeze about half the inputs of *risc*(16) ∗/

331↑ 930 *simplex*(30, 20, 20, 20, 0, 0, 1)
/∗ directed hexagonal grid with 60 boundary cells ∗/

462↑ 1386 *subsets*(5, 1, −10, 0, 0, 0, 1, 1) /∗ disjoint 5-element subsets of {0, 1, . . . , 10} ∗/

462↑ 6930 *simplex*(5, 1, −10, 0, 0, 0, 1) /∗ oriented 10-dimensional simplex ∗/

627↑ 3545 *all_parts*(20, 1) /∗ partitions of 20 under refinement ∗/

763↑ 1902 *perms*(1, 2, 3, 4, −8, 4, 1) /∗ multiset permutations with ≤ 4 inversions ∗/

1000 2871 *product*(*roget*(100, 0, 0, 0), *econ*(10, 0, 0, 0), 1, 1) /∗ direct product ∗/

1000 10190 *product*(*roget*(100, 0, 0, 0), *econ*(10, 0, 0, 0), 0, 1) /∗ cartesian product ∗/

1022 2563 *roget*(0, 0, 32768, 0) /∗ half the cross-references in **roget.dat** ∗/

1022 3823 *roget*(0, 3, 0, 0) /∗ the cross-references of distance ≥ 3 ∗/

1022 5075 *roget*(0, 0, 0, 0) /∗ all cross-references in Roget's *Thesaurus* ∗/

1024↑ 5120 *board*(2, −10, 0, 0, 1, 0, 1) /∗ oriented 10-dimensional hypercube ∗/

1430↑ 5005 *all_trees*(8, 1) /∗ binary trees with 9 leaves under right rotation ∗/

1630↑ 3972 *risc*(2) /∗ picocomputer with minimum number of built in registers ∗/

1800 328960 *board*(2, 3, 5, −7, 7, 5, 1)
/∗ weird hyperboard with two wraparound coordinates ∗/

1860↑ 5590 *binary*(10, 5, 1) /∗ binary trees with 11 leaves and height ≤ 5 ∗/

2016↑ 2000 *binary*(125, 7, 1) /∗ binary trees with 126 leaves and height 7 ∗/

2896↑ 6251 *partial_gates*(*risc*(16), 1, 50000, 0, Λ)
/∗ freeze about 24% of the inputs of *risc*(16) ∗/

3024↑ 6773 *partial_gates*(*risc*(16), 1, 60000, 0, Λ)
/∗ freeze about 8% of the inputs of *risc*(16) ∗/

3240↑ 7878 *risc*(16) /∗ picocomputer with 16 built in registers ∗/

4996↑ 9952 *partial_gates*(*prod*(20, 80), 20, 0, 1, Λ)
/∗ multiply 20 bits by a random 80-bit constant ∗/

5040↑ 15120 *all_perms*(7, 1)
/∗ permutations of {0, 1, . . . , 6} under adjacent transposition ∗/

 5448 ↑   36105   *parts*(50, 10, 10, 1)   /* partitions of 50 into $\leq$ 10 parts $\leq$ 10 */

 6316 ↑   12592   *partial_gates*(*prod*(20, 80), 20, 0, 0, Λ)
                                                /* multiply 20 bits by a different random 80-bit constant */

 6739 ↑   13318   *partial_gates*(*prod*(80, 20), 80, 0, 1, Λ)
                                                /* multiply 80 bits by a random 20-bit constant */

 7197 ↑   14234   *partial_gates*(*prod*(80, 20), 80, 0, 0, Λ)   /* same, with a different constant */

10038 ↑   19976   *partial_gates*(*prod*(50, 50), 50, 0, 0, Λ)
                                                /* multiply 50 bits by a random 50-bit constant */

10815 ↑   21430   *prod*(20, 80)   /* multiply 20 bits by 80 bits */

11821 ↑   23442   *prod*(80, 20)   /* multiply 80 bits by 20 bits */

16789 ↑   33378   *prod*(50, 50)   /* multiply 50 bits by 50 bits */

16796 ↑   75582   *all_trees*(10, 1)   /* all trees with 11 leaves, under right rotation */

APPENDIX D: *3000 FIVE-LETTER WORDS*

When the words of `words.dat` are ranked by the default weights of GB_WORDS, the first 3000 are those shown here. A complete list of all 5757 words in rank order can be obtained by running the demo program WORD_COMPONENTS.

2476 aback	1354 ample	1688 barks	2921 bluer	1130 buggy	2561 chasm	1012 clung	1673 crook
2877 abate	1512 amuse	1195 barns	1229 blues	1171 bugle	2937 chats	635 coach	2659 croon
2410 abbey	754 angel	1820 baron	1359 bluff	145 build	850 cheap	1126 coals	251 crops
2592 abbot	493 anger	266 based	1344 blunt	94 built	1735 cheat	224 coast	246 cross
2451 abhor	265 angle	2743 baser	2933 blurs	1245 bulbs	127 check	709 coats	2998 croup
1594 abide	292 angry	605 bases	1756 blush	1657 bulge	783 cheek	1787 cobra	293 crowd
2995 abler	1177 ankle	208 basic	194 board	1395 bulky	821 cheer	2735 cocci	676 crown
2839 abort	1814 annex	971 basin	2756 boars	1283 bulls	2310 cheep	1980 cocks	1213 crows
4 about	1843 annoy	366 basis	1312 boast	1649 bully	2984 chefs	2545 cocky	918 crude
49 above	1214 anode	2379 baste	315 boats	1542 bumps	1585 chess	1060 cocoa	755 cruel
1453 abuse	2713 antic	1752 batch	2723 bogey	736 bunch	2099 chews	2959 cocos	2212 crumb
2025 abyss	1369 anvil	1360 bathe	1495 boils	1662 bunks	2764 chewy	2238 coded	1523 crush
1409 ached	261 apart	1441 baths	1256 bolts	2422 bunny	1614 chick	1153 codes	628 crust
2028 aches	2306 aorta	1799 baton	1045 bombs	2217 buoys	244 chief	1381 coils	2619 cubed
1053 acids	2550 aphid	2898 bayed	914 bonds	1861 burly	140 child	599 coins	954 cubes
2036 acorn	408 apple	2650 bayou	260 bones	800 burns	1808 chili	1785 colds	798 cubic
556 acres	429 apply	377 beach	2944 bongs	703 burnt	1003 chill	1500 colon	1416 cuffs
2617 acrid	802 apron	661 beads	1362 bonus	1120 burro	2491 chime	116 color	2303 curbs
771 acted	2136 aptly	1438 beaks	2121 booby	2157 burrs	2413 chimp	1363 colts	1654 curds
1031 actor	2228 arbor	956 beams	2523 booed	506 burst	1289 china	1870 combs	1353 cured
1109 acute	2455 ardor	537 beans	139 books	982 buses	2345 chins	88 comes	1924 cures
2454 adage	163 areas	786 beard	2098 booms	1536 bushy	1218 chips	1554 comet	1205 curls
1415 adapt	1720 argon	510 bears	1605 boost	2678 busts	1949 chirp	1146 comic	1145 curly
115 added	927 argue	662 beast	1322 booth	2530 butte	1329 choir	826 comma	1786 curry
2668 adder	773 arise	521 beats	548 boots	2292 butts	1690 choke	2660 conch	1200 curse
2247 adept	670 armed	1609 beech	1899 booty	1333 buyer	1556 chops	1048 cones	475 curve
2847 adios	994 armor	2300 beeps	2707 booze	314 cabin	402 chord	2501 conic	613 cycle
749 admit	1875 aroma	2830 beers	1051 bored	941 cable	1754 chore	2514 cooed	2848 cynic
1268 adobe	698 arose	1212 beets	2421 bores	1504 cacao	518 chose	1240 cooks	2865 cysts
1299 adopt	959 array	2903 befit	1572 borne	2352 cache	1478 chuck	1858 cooky	2992 czars
2255 adore	530 arrow	44 began	1378 bosom	2147 cadet	2519 chugs	948 cools	1780 daddy
2019 adorn	2584 arson	129 begin	2140 bossy	1889 cafes	2589 chums	2869 coops	387 daily
582 adult	2606 ashen	445 begun	1620 bough	2143 caged	1627 chunk	2727 coped	653 dairy
1772 affix	790 ashes	1760 beige	448 bound	1208 cages	1645 churn	2339 copra	2087 daisy
1881 afire	446 aside	2758 belch	2002 bouts	1959 caked	1731 chute	2993 copse	2852 dales
1805 afoot	35 asked	397 bells	859 bowed	664 cakes	1642 cider	882 coral	2654 damps
2221 afore	2608 askew	1025 belly	2895 bowel	416 calls	1314 cigar	1006 cords	201 dance
13 after	2065 aspen	862 belts	1033 bowls	2234 calms	2229 cinch	1791 cores	1705 dandy
22 again	2917 aspic	607 bench	1926 boxed	961 camel	1514 cited	2716 corks	915 dared
2627 agate	2753 assay	938 bends	1510 boxer	1055 camps	2092 cites	2802 corns	2022 dares
756 agent	1597 asses	2403 beret	329 boxes	623 canal	2793 civet	2368 corny	2630 darks
1895 agile	2667 aster	1679 berry	1293 brace	450 candy	1461 civic	1469 corps	2934 darns
1823 aging	2213 astir	1812 berth	2553 brags	2201 canes	645 civil	415 costs	1327 darts
1280 agony	1262 atlas	1894 beset	1630 braid	2633 canny	481 claim	1152 couch	1485 dated
372 agree	2354 atoll	1463 bevel	373 brain	616 canoe	1227 clamp	1325 cough	686 dates
186 ahead	363 atoms	2461 bible	952 brake	1577 canon	1149 clams	9 could	2809 datum
1345 aided	2842 atone	1697 bikes	977 brand	2490 caper	2056 clang	283 count	2569 dawns
2155 aides	1080 attic	723 bills	516 brass	2190 capes	2326 clank	2601 coupe	2246 dazed
2994 ailed	1828 auger	2091 binds	2803 brats	2409 carat	2620 clans	318 court	1036 deals
996 aimed	1465 aunts	2843 bingo	407 brave	515 cards	2030 claps	195 cover	1043 dealt
2731 aired	2052 autos	1368 birch	2273 bravo	932 cared	1617 clash	2585 coves	1183 debts
1656 aisle	2070 avert	106 birds	2549 brawl	1221 cares	1943 clasp	2172 covet	1739 debut
712 alarm	388 avoid	491 birth	2771 brawn	769 cargo	98 class	2873 covey	897 decay
1856 album	1460 await	1717 bison	2925 brays	1575 carol	660 claws	2982 cower	1449 decks
2177 alder	639 awake	2235 bitch	221 bread	135 carry	2250 clays	2546 coyly	2398 decoy
879 alert	1062 award	1181 bites	203 break	1243 carts	227 clean	1265 crabs	832 deeds
997 algae	453 aware	81 black	1018 breed	1396 carve	137 clear	632 crack	2445 defer
2971 alias	1637 awful	648 blade	2587 brews	249 cases	2887 cleat	739 craft	2355 deity
2158 alibi	1021 awoke	763 blame	2575 briar	2436 casks	2638 clefs	2142 crags	955 delay
1230 alien	2119 axles	1981 bland	1920 bribe	1669 caste	1811 cleft	1964 cramp	2954 dells
279 alike	2282 babes	417 blank	659 brick	1440 casts	715 clerk	1250 crane	1516 delta
298 alive	778 backs	2430 blare	909 bride	192 catch	1341 click	1307 crank	2901 delve
2556 allay	767 bacon	864 blast	494 brief	2394 cater	805 cliff	738 crash	1425 demon
1163 alley	1831 badge	1261 blaze	2915 brims	204 cause	409 climb	1587 crate	2010 denim
2444 allot	547 badly	1343 bleak	1864 brine	931 caves	1274 cling	2675 crave	851 dense
391 allow	1850 baggy	2132 bleat	99 bring	1093 cease	2269 clink	967 crawl	2526 dents
1244 alloy	809 baked	1846 bleed	1712 brink	1204 cedar	1419 clips	2503 craze	1595 depot
1452 aloft	2024 bakes	580 blend	1423 brisk	2905 ceded	1147 cloak	770 crazy	449 depth
131 alone	2672 baled	1165 bless	361 broad	1476 cello	367 clock	1866 creak	2438 derby
39 along	1506 bales	478 blind	2059 broil	174 cells	2245 clods	392 cream	1275 desks
1822 aloof	586 balls	1704 blink	281 broke	277 cents	2225 clogs	1916 creed	2211 deter
459 aloud	2547 balmy	1042 bliss	1450 brood	2981 chafe	89 close	788 creek	840 devil
680 altar	2329 balsa	2840 blitz	894 brook	2336 chaff	301 cloth	1216 creep	1982 dials
1176 alter	620 bands	2328 bloat	933 broom	467 chain	357 cloud	830 crept	1023 diary
2558 altos	1960 bangs	2614 blobs	1606 broth	275 chair	2417 clout	1166 crest	2973 diced
2653 amass	1392 banjo	343 block	226 brown	768 chalk	2571 clove	1233 crews	1151 didst
2590 amaze	473 banks	1387 blond	1750 brows	1947 champ	1046 clown	2596 cribs	1691 diets
1723 amber	1758 barbs	147 blood	2642 brunt	1052 chant	820 clubs	126 cried	546 digit
2805 amble	1825 bared	985 bloom	432 brush	1351 chaos	700 clues	789 cries	1320 dikes
2284 amend	2855 bares	2615 blots	1773 brute	1830 chaps	1303 clump	835 crime	713 dimes
2338 amigo	1228 barge	818 blown	1938 bucks	950 charm		1008 crisp	1287 dimly
1795 amino		688 blows	1829 buddy	337 chart		2067 croak	1730 dined
2340 amiss			1592 budge	810 chase		2685 crock	
71 among			2720 buffs			2857 crone	

2145 diner	76 early	271 farms	891 flung	853 geese	200 guess	1058 hooks	886 jumps
2706 dines	1535 earns	2652 fasts	2928 flunk	1310 genes	718 guest	2529 hooky	2215 jumpy
1863 dingy	43 earth	1074 fatal	1407 flush	1788 genie	341 guide	1927 hoops	2350 junks
2889 dirge	1323 eased	2645 fates	920 flute	1408 genus	2180 guild	2634 hoots	2525 juror
597 dirty	1871 easel	642 fault	2296 flyer	650 germs	2150 guile	443 hoped	2827 karat
1638 discs	2559 eases	1672 fauna	883 focus	646 ghost	1291 guilt	604 hopes	2782 kayak
1420 disks	420 eaten	574 favor	1300 foggy	382 giant	2348 guise	585 horns	2985 keels
1000 ditch	1558 eater	2281 fawns	2745 foils	2202 giddy	2302 gulfs	2831 horny	389 keeps
2646 ditty	1991 eaves	890 fears	1112 folds	655 gifts	1020 gulls	101 horse	2769 kelps
2611 divan	2242 ebony	647 feast	618 folks	1094 gills	1834 gully	2005 hoses	2252 ketch
1010 dived	1455 edged	1716 feats	1401 folly	113 girls	2237 gulps	875 hosts	2333 keyed
876 diver	394 edges	1123 feeds	1297 fonts	2153 girth	2527 gummy	541 hotel	2205 khaki
1745 dives	2375 edict	492 feels	250 foods	59 given	2012 guppy	1827 hotly	1527 kicks
1815 dizzy	2890 edits	2907 feign	1404 fools	158 gives	1733 gusts	1174 hound	1168 kills
1196 docks	1531 eerie	2607 felts	2425 foray	2286 glade	2387 gusty	118 hours	2875 kilns
1769 dodge	2927 egret	355 fence	166 force	1365 gland	1158 gypsy	32 house	2845 kilts
2531 dogie	2649 eider	1301 ferns	2625 fords	1242 glare	624 habit	2673 hovel	86 kinds
2327 dogma	182 eight	1238 ferry	1590 forge	141 glass	2695 hacks	2007 hover	512 kings
121 doing	2755 eject	957 fetch	2747 forgo	2020 glaze	1318 haiku	1954 howls	2380 kinks
2724 doled	1073 elbow	2957 feted	1568 forks	1361 gleam	2728 hails	2910 huffs	2800 kinky
2911 doles	1167 elder	2508 fetus	123 forms	2513 glens	926 hairs	2913 hulks	1696 kites
929 dolls	1129 elect	2464 feuds	232 forth	1442 glide	1332 hairy	2224 hulls	1579 kitty
2952 dolts	2669 elegy	627 fever	1328 forts	2073 glint	1125 halls	142 human	1956 knack
2301 domed	2851 elfin	573 fewer	513 forty	2505 gloat	2593 halos	1285 humid	2679 knave
2054 domes	2729 elude	991 fiber	2174 forum	1209 gloom	2069 halts	842 humor	2468 knead
2063 donor	1847 elves	111 field	2637 fouls	496 glory	2579 halve	1873 humps	1865 kneel
426 doors	2902 embed	1958 fiend	26 found	2008 gloss	79 hands	1505 humus	464 knees
2772 doped	2371 ember	1081 fiery	1162 fours	1185 glove	1032 handy	2042 hunch	2871 knell
2763 dosed	1836 emery	2534 fifes	2018 fowls	1932 glows	1041 hangs	1486 hunts	990 knelt
1778 doses	2626 emits	485 fifth	2976 foxed	1391 glued	156 happy	379 hurry	354 knife
2689 doted	278 empty	321 fifty	1078 foxes	2477 glues	1431 hardy	1482 hurts	2318 knits
340 doubt	2166 enact	213 fight	2208 foyer	2759 gnash	1915 hares	1885 husks	1564 knobs
1054 dough	401 ended	1251 filed	1324 frail	2001 gnats	2600 harps	1484 husky	813 knock
2932 douse	2407 endow	1134 files	396 frame	2623 gnaws	2194 harry	2481 hutch	1037 knots
1497 doves	335 enemy	2870 filet	2384 franc	2812 gnome	960 harsh	2552 hydra	73 known
2923 dowdy	284 enjoy	1068 fills	1471 frank	901 goals	1034 haste	2101 hyena	219 knows
2233 dowel	2861 enrol	1837 filly	1898 fraud	772 goats	1816 hasty	1375 hymns	2567 koala
1901 downy	2841 ensue	904 films	2818 frays	1770 godly	742 hatch	2815 icily	532 label
2395 dowry	360 enter	2524 filmy	2314 freak	36 going	699 hated	1685 icing	419 labor
1694 dozed	489 entry	2015 filth	993 freed	2997 golds	1434 hates	785 ideal	1763 laced
406 dozen	2583 envoy	212 final	2264 freer	1235 gonna	2183 hauls	125 ideas	1872 laces
2768 dozes	2486 epics	2334 finch	2492 frees	274 goods	2021 haunt	1807 idiom	1346 lacks
949 draft	1675 epoch	527 finds	2648 frets	2231 goody	2071 havoc	1841 idiot	1517 laden
1682 drags	197 equal	1969 fined	2656 friar	2586 gooey	1121 hawks	2977 idled	2077 ladle
1067 drain	2390 equip	1286 finer	1029 fried	2164 goofy	1887 hazel	2838 idler	2834 lairs
757 drama	1639 erase	2080 fines	2806 fries	571 goose	289 heads	1239 idols	447 lakes
687 drank	1170 erect	2258 fiord	2705 frill	1845 gouge	2401 heals	2230 igloo	946 lambs
2786 drape	2497 erode	619 fired	2896 frisk	1936 gowns	1458 heaps	457 image	2636 lamed
2137 drawl	2449 erred	609 fires	2424 frock	1853 grabs	56 heard	2565 impel	940 lamps
305 drawn	561 error	907 firms	666 frogs	746 grace	873 hears	1424 imply	1665 lance
1030 draws	1961 erupt	11 first	2945 frond	545 grade	95 heart	2938 inane	207 lands
1194 dread	1039 essay	2322 firth	84 front	2330 graft	2304 heath	583 index	1422 lanes
398 dream	2958 ether	2460 fishy	872 frost	344 grain	1337 heats	2386 inept	1604 lanky
2573 dregs	2988 etude	1219 fists	2431 froth	1302 grams	1607 heave	1612 inert	2884 lapel
296 dress	2312 evade	2876 fitly	1231 frown	669 grand	124 heavy	1860 infer	2049 lapse
460 dried	24 every	1410 fives	1467 froze	902 grant	1309 hedge	2946 ingot	2909 larch
1503 drier	2947 evict	414 fixed	252 fruit	1680 grape	2879 heeds	2097 inked	37 large
1489 dries	1543 evils	2093 fixes	1599 fudge	339 graph	707 heels	1747 inlet	2307 larks
838 drift	2196 evoke	2779 fjord	1248 fuels	863 grasp	2288 hefty	528 inner	1518 larva
589 drill	431 exact	989 flags	466 fully	172 grass	1668 heirs	795 input	1913 laser
259 drink	1728 exalt	2243 flail	2414 fumed	2127 grate	600 hello	2725 irate	1738 lasso
2710 drips	2165 exams	2179 flair	1414 fumes	747 grave	240 helps	2916 irked	1247 lasts
215 drive	2563 excel	2134 flake	744 funds	1608 gravy	487 hence	1687 irons	1817 latch
1839 drone	1488 exert	2076 flaky	866 fungi	2528 grays	1338 herbs	1686 irony	65 later
2953 drool	1269 exile	587 flame	376 funny	1193 graze	697 herds	2244 isles	1994 latex
1937 droop	564 exist	1681 flank	1427 furry	21 great	2088 heron	490 issue	1295 lathe
470 drops	1804 exits	1411 flaps	1993 fused	1753 greed	2450 hewed	440 items	2966 laths
313 drove	2471 expel	1789 flare	1935 fuses	104 green	825 hides	1225 ivory	393 laugh
1377 drown	2216 extol	705 flash	1842 fussy	1157 greet	2200 highs	2504 jacks	2674 lavas
692 drugs	351 extra	1027 flask	1631 fuzzy	2472 greys	1973 hiked	2742 jaded	1246 lawns
683 drums	2265 exult	1139 flats	2274 gable	2148 grids	2323 hikes	2295 jails	434 layer
1001 drunk	2341 eying	2133 flaws	1296 gaily	984 grief	303 hills	2465 jambs	602 leads
2045 dryer	1308 fable	1854 fleas	1044 gains	1569 grill	1114 hilly	2885 jaunt	1402 leafy
2006 dryly	565 faced	2719 fleck	2188 gales	2437 grime	2859 hinds	1550 jeans	1767 leaks
652 ducks	312 faces	2621 flees	2000 galls	2576 grimy	1615 hinge	2075 jeeps	2013 leaky
1709 ducts	2388 facet	796 fleet	347 games	1271 grind	1376 hints	2467 jeers	1661 leans
2825 dudes	230 facts	499 flesh	2027 gangs	1551 grins	2956 hippy	836 jelly	1277 leaps
2880 duels	916 faded	1748 flick	1971 gaped	2930 gripe	856 hired	2094 jerks	1878 leapt
2572 duets	2034 fades	2950 flied	425 gases	1970 grips	2462 hires	1765 jerky	72 learn
2631 dukes	1154 fails	2118 flier	1928 gasps	2790 grist	1727 hitch	2781 jests	1355 lease
2849 dulls	644 faint	1678 fling	741 gates	2349 grits	2499 hoard	2389 jetty	1729 leash
1986 dully	1833 fairs	1340 flint	1909 gaudy	1371 groan	2363 hoary	1524 jewel	110 least
1670 dummy	936 fairy	1923 flips	1017 gauge	2378 groin	913 hobby	2961 jibed	114 leave
2112 dumps	437 faith	2540 flirt	1473 gaunt	1603 groom	2072 hogan	2744 jibes	1009 ledge
2319 dumpy	2551 faked	719 float	911 gauss	2272 grope	2033 hoist	2227 jiffy	2494 leech
1119 dunes	2692 fakes	637 flock	2883 gauze	1022 gross	413 holds	1115 joins	2967 leers
2864 duped	356 falls	2888 flood	1840 gauzy	70 group	405 holes	577 joint	674 legal
2014 dusky	403 false	128 floor	2440 gavel	1276 grove	1566 holly	2163 joked	999 lemon
2739 dusts	2493 famed	2603 flops	2502 gawky	1429 growl	2066 homer	2718 joker	2817 lemur
839 dusty	733 fancy	469 flour	2536 gayer	248 grown	222 homes	870 jokes	1710 lends
1621 dwarf	1544 fangs	1072 flown	2405 gayly	369 grows	2796 homey	1206 jolly	2391 leper
777 dwell	2090 farce	498 flows	953 gazed	1813 grubs	672 honey	2828 jolts	2712 levee
986 dwelt	2210 fared	1975 fluff	2232 gazes	2432 gruel	423 honor	439 judge	184 level
685 dying	1768 fares	975 fluid	1613 gears	2038 gruff	1929 hoods	568 juice	689 lever
578 eager				2062 grunt	981 hoofs	1241 juicy	2385 liars
787 eagle				497 guard		2613 jumbo	2100 licks
2824 earls							

1190 lifts
48 light
209 liked
2816 liken
461 likes
1917 lilac
951 limbs
2197 limes
649 limit
1888 lined
852 linen
1553 liner
68 lines
1066 links
665 lions
656 lists
2162 liter
2203 lithe
91 lived
1016 liver
161 lives
2095 livid
1737 llama
869 loads
1132 loans
2207 loath
1549 lobby
2241 lobes
217 local
1210 locks
2975 lodes
1156 lodge
1279 lofty
1267 logic
1658 loins
2562 longs
159 looks
1702 looms
2807 loons
1179 loops
428 loose
2459 loped
2970 lopes
1118 lords
2046 loser
874 loses
2478 lotus
2448 louse
1855 lousy
310 loved
1057 lover
775 loves
175 lower
1593 lowly
1098 loyal
2219 lucid
552 lucky
2698 lulls
1192 lumps
1070 lunar
336 lunch
2199 lunge
508 lungs
2343 lurch
2168 lured
2313 lures
2479 lurid
2754 lurks
2103 lusty
2761 lutes
365 lying
2159 lymph
2835 lynch
2475 lyres
1447 lyric
889 macro
1342 madam
1628 madly
399 magic
1636 magma
1372 maids
1940 mails
1997 mains
1974 maize
183 major
1191 maker
87 makes
1059 males
2980 malts
2912 mamas
2439 mamma
2254 manes
2257 mango
2320 mangy
1967 manly
1655 manor
831 maple

641 march
1883 mares
268 marks
567 marry
1101 marsh
1347 masks
2240 mason
1197 masts
331 match
2640 mated
1490 mates
2582 mauve
2223 maxim
155 maybe
793 mayor
2962 mazes
610 meals
2919 mealy
60 means
214 meant
1077 meats
2578 meaty
1172 medal
1480 media
1373 meets
1734 melon
987 melts
2850 memos
2629 mends
1953 menus
716 mercy
1777 merge
1358 merit
781 merry
2191 mesas
1790 messy
170 metal
2560 meted
525 meter
2610 mewed
2991 micas
2331 middy
631 midst
30 might
2664 mikes
62 miles
2427 milks
575 mills
1908 milky
2366 mimic
2116 mince
596 minds
1079 mined
1580 miner
566 mines
2261 minks
436 minor
2605 mints
903 minus
1819 mirth
2544 miser
1487 mists
1562 misty
2299 miter
2419 mites
2866 mitts
346 mixed
1743 mixes
2443 moans
2892 moats
2358 mocks
319 model
1182 modes
673 moist
2602 molar
1014 molds
2383 moldy
2048 moles
2109 molts
67 money
2141 monks
237 month
1284 moods
2176 moody
2704 mooed
1038 moons
2305 moors
1143 moose
2797 moped
608 moral
2152 mores
2655 morns
2776 moron
1903 mossy
1224 motel
1076 moths

462 motor
1632 motto
1104 mound
1082 mount
1451 mourn
522 mouse
150 mouth
119 moved
263 moves
614 movie
1798 mowed
1851 mower
1635 mucus
834 muddy
410 mulch
1106 mules
2651 mulls
1826 mummy
1835 mumps
2609 munch
1859 murky
1996 mused
2506 mushy
61 music
2810 musky
2039 musty
1922 muted
2686 mutes
1998 myrrh
1207 myths
584 nails
1890 naive
884 naked
138 named
122 names
1583 nasal
1508 nasty
964 naval
2433 navel
2171 nears
1226 necks
180 needs
1477 needy
2496 neigh
668 nerve
760 nests
28 never
1223 newer
740 newly
1774 nicer
1930 niche
2639 nicks
1475 niece
47 night
2289 nines
1064 ninth
849 noble
299 noise
814 noisy
2298 nomad
2701 nooks
2151 noose
152 north
2325 nosed
1065 noses
1403 notch
504 noted
234 notes
257 nouns
2906 novas
701 novel
1721 nudge
1350 nurse
1069 nylon
1591 nymph
2285 oaken
1596 oases
1494 oasis
1741 oaths
2023 obeys
2016 oboes
400 occur
167 ocean
2711 odder
1249 oddly
1263 odors
384 offer
42 often
1963 olden
270 older
937 olive
2170 omens
2192 omits
1148 onion
1757 onset
2248 oozed
2635 oozes

2543 opals
794 opens
517 opera
1879 opium
2676 optic
472 orbit
69 order
804 organ
7 other
1405 otter
385 ought
1155 ounce
2209 outdo
410 outer
2400 ovals
1456 ovary
2814 ovule
1316 owing
500 owned
519 owner
1135 oxide
2489 ozone
1629 paced
1499 paces
1466 packs
2736 pacts
1762 paddy
1715 pagan
178 pages
1571 pails
1136 pains
288 paint
282 pairs
2267 paled
2434 paler
2832 pales
1047 palms
2657 palsy
2117 panda
935 panel
1666 panes
1939 pangs
1005 panic
1882 pansy
919 pants
50 paper
2518 parka
731 parks
51 parts
144 party
2968 pasta
807 paste
759 patch
743 paths
2122 patio
2787 patty
696 pause
1160 paved
2271 pawed
236 peace
942 peach
784 peaks
2854 peals
868 pearl
1406 pears
2291 pecan
2186 pecks
1211 pedal
2752 peeks
2853 peels
1800 peers
1684 pelts
2293 pence
2924 penis
675 penny
2829 peons
2931 peony
1304 perch
1546 peril
2703 perky
2083 pesos
1567 pests
2124 petal
1784 petty
877 phase
2929 phlox
782 phone
2131 phony
1137 photo
308 piano
963 picks
103 piece
1893 piers
1948 piety
2382 piggy
2811 pigmy

2488 pikes
819 piled
995 piles
1600 pills
549 pilot
1331 pinch
2702 pined
1290 pines
2510 pinks
1749 pinto
1144 pints
1802 pious
1507 piped
721 pipes
411 pitch
2715 pithy
1547 pivot
1824 pixel
1711 pizza
19 place
258 plain
2979 plait
132 plane
1288 plank
322 plans
120 plant
352 plate
359 plays
1957 plaza
1821 plead
2402 pleas
1950 pleat
2120 plied
2632 plies
2821 plods
1443 plots
1386 plows
1116 pluck
1806 plugs
1714 plumb
1983 plume
1201 plump
1533 plums
2266 plush
531 poems
801 poets
54 point
1674 poise
1188 poked
1962 poker
2085 pokes
865 polar
2532 poled
427 poles
1097 polio
1907 polka
1581 polls
2125 polyp
972 ponds
976 pools
2344 popes
1766 poppy
562 porch
2457 pored
1671 pores
898 ports
1640 posed
1966 poses
1951 posse
855 posts
1330 pouch
463 pound
1198 pours
90 power
2181 prays
2666 preen
364 press
2749 preys
326 price
543 pride
2914 pried
2804 pries
320 prime
633 print
1095 prior
939 prism
553 prize
1266 probe
2894 prods
1742 prone
2709 prong
691 proof
2081 props
1278 prose
317 proud
324 prove

1978 prowl
2978 prows
2570 proxy
2082 prune
1904 psalm
1448 puffs
2435 puffy
681 pulls
2670 pulpy
1071 pulse
2628 pumas
1131 pumps
1061 punch
2874 punks
1706 pupae
2926 pupas
752 pupil
885 puppy
2426 purer
1972 purge
965 purse
2037 pussy
2011 putty
2408 pygmy
1457 quack
1611 quail
1955 quake
888 quart
2568 quays
484 queen
776 queer
2751 quell
2428 query
1380 quest
2206 queue
295 quick
228 quiet
1623 quill
1259 quilt
97 quite
2665 quits
1740 quota
1294 quote
2123 rabbi
2996 rabid
569 raced
2987 racer
704 races
1999 racks
923 radar
2084 radii
205 radio
1744 rafts
1428 raged
2311 rages
1557 raids
1024 rails
728 rains
730 rainy
330 raise
944 rally
2822 ramps
456 ranch
267 range
958 ranks
2278 rarer
2899 rasps
1530 rated
588 rater
595 ratio
2813 ravel
2362 raven
2788 raves
1186 rayon
2680 razor
187 reach
1090 react
651 reads
96 ready
1184 realm
2533 reams
1270 rebel
2377 recur
980 reeds
1736 reefs
2364 reels
465 refer
2872 refit
2279 regal
1400 reign
1063 reins
1089 relax
1178 relay

2130 relic
1619 renew
1519 rents
1472 repay
1483 repel
625 reply
1445 resin
1002 rests
2361 revel
495 rhyme
734 rider
833 rides
837 ridge
603 rifle
2969 rifts
15 right
966 rigid
2473 rigor
2801 rills
2774 rinds
581 rings
2941 rinks
1509 rinse
1699 riots
1793 ripen
988 risen
1349 risks
1676 risky
1809 rites
1138 rival
108 river
1601 rivet
1979 roach
333 roads
2863 roams
1700 roars
1049 roast
2498 robed
1202 robes
1092 robin
979 robot
181 rocks
640 rocky
1141 rodeo
2541 rogue
1173 roles
992 roofs
455 rooms
2260 roomy
1781 roost
287 roots
1848 roped
779 ropes
847 roses
1439 rotor
2370 rouge
386 rough
100 round
1725 rouse
507 route
2406 rover
2844 roves
1217 rowed
511 royal
2074 ruddy
2882 ruffs
2470 rugby
943 ruins
626 ruled
412 ruler
233 rules
1874 rumor
2507 rumps
2480 rungs
2798 runts
808 rural
2515 rusts
1086 rusty
2297 saber
2858 sable
1292 sacks
732 sadly
1096 safer
2867 safes
751 sails
1462 saint
1667 sakes
858 salad
550 sales
1726 salts
1100 salty
2500 salve

729 sandy
2785 saner
2178 sassy
1646 satin
970 sauce
2160 saucy
380 saved
1236 saves
2566 savor
1857 sawed
2737 scabs
2846 scald
173 scale
1298 scalp
2113 scaly
2270 scamp
2359 scans
1852 scant
1087 scare
1124 scarf
1555 scars
1537 scary
338 scene
934 scent
945 schwa
2102 scoff
1703 scold
1719 scoop
2683 scoot
1035 scope
285 score
1584 scorn
2594 scour
1108 scout
2236 scowl
2671 scows
2721 scram
1150 scrap
765 screw
1264 scrub
1905 scuba
2365 scuff
1105 seals
881 seams
593 seats
2110 sects
2249 sedan
239 seeds
1203 seeks
146 seems
2009 seeps
2699 seers
1113 seize
947 sells
843 sends
164 sense
1810 serfs
2799 serge
1437 serum
291 serve
149 seven
2276 sever
1352 sewed
1933 sewer
1724 sexes
1234 shack
480 shade
1399 shady
906 shaft
611 shake
1582 shaky
1651 shale
40 shall
2511 shalt
702 shame
1844 shank
160 shape
297 share
861 shark
241 sharp
1624 shave
1253 shawl
2047 sheaf
1364 shear
1598 sheds
2055 sheen
256 sheep
983 sheer
327 sheet
2780 sheik
629 shelf
454 shell
750 shift
606 shine
2681 shins
711 shiny
179 ships

2860 shirk	1944 sling	2908 spews	2734 strew	1573 tenor	1616 trash	2691 veers	358 wheat
468 shirt	2392 slink	1493 spice	444 strip	536 tense	2029 trawl	1995 veils	294 wheel
2351 shoal	1254 slips	2032 spicy	1502 strum	824 tenth	1570 trays	871 veins	2955 whelk
677 shock	1529 slits	1382 spied	1976 strut	845 tents	1652 tread	2175 venom	2886 whelp
273 shoes	1977 sloop	1633 spies	2218 stubs	2193 tepee	671 treat	2411 vents	14 where
720 shone	615 slope	1708 spike	529 stuck	2537 tepid	77 trees	238 verbs	1 which
290 shook	2951 slops	2542 spiky	2061 studs	191 terms	1026 trend	2204 verge	2222 whiff
474 shoot	2548 slosh	1522 spill	46 study	2741 terse	2823 tress	638 verse	31 while
612 shops	1650 sloth	2404 spilt	594 stuff	438 tests	514 trial	2897 vests	2762 whims
286 shore	1647 slots	1273 spine	1237 stump	1430 texts	476 tribe	2051 vexed	1794 whine
2360 shorn	1622 slows	1133 spins	1432 stung	370 thank	501 trick	2939 vials	1692 whips
78 short	2064 slugs	1797 spiny	1397 stunt	3 their	109 tried	2035 vicar	1454 whirl
780 shots	1446 slums	2214 spire	280 style	483 theme	667 tries	2612 vices	2972 whirs
622 shout	1515 slung	471 spite	2989 sucks	2 there	2487 trill	2557 video	1941 whisk
1764 shove	2935 slunk	2321 spits	2581 suede	6 these	2687 trims	576 views	58 white
82 shown	2622 slurs	538 split	229 sugar	225 thick	2963 trios	2108 vigil	57 whole
102 shows	2775 slush	887 spoil	2595 suing	917 thief	2708 tripe	1335 vigor	1689 whoop
1779 showy	2060 slyly	202 spoke	1257 suite	1317 thigh	601 trips	2089 villa	112 whose
2517 shred	25 small	2538 spook	725 suits	1602 thine	2757 trite	962 vines	1713 widen
2522 shrew	726 smart	1521 spool	2789 sulfa	63 thing	2128 troll	2416 vinyl	745 wider
1511 shrub	1532 smash	828 spoon	2990 sulks	16 think	1255 troop	1552 viola	822 widow
1910 shrug	2893 smear	579 spots	2376 sulky	2647 thins	2324 trots	2447 viper	505 width
2690 shuns	328 smell	1412 spout	717 sunny	117 third	1122 trout	1107 virus	2446 wield
2773 shush	2026 smelt	896 spray	1952 sunup	2220 thong	2043 truce	3000 visas	1912 wilds
2004 shuts	375 smile	2357 spree	1426 super	1389 thorn	307 truck	242 visit	2337 wiles
2949 shyer	2420 smirk	2696 sprig	2381 surer	27 those	2369 truer	2287 visor	1987 wilts
1588 shyly	2495 smite	2616 spunk	1481 surge	17 three	535 truly	2658 vista	2862 wilts
2396 sided	2105 smith	1496 spurs	2423 surly	334 threw	560 trunk	761 vital	2509 wince
148 sides	2346 smock	1869 spurt	1011 swamp	2290 throb	1914 truss	893 vivid	342 winds
2714 sidle	345 smoke	2965 squab	1803 swans	390 throw	572 trust	2792 vixen	1110 windy
1501 siege	1868 smoky	1538 squad	1470 swarm	2466 thuds	332 truth	766 vocal	1220 wines
1707 sieve	1413 smote	1663 squat	2746 sways	621 thumb	2808 tsars	2315 vogue	247 wings
2820 sifts	1385 snack	1771 squaw	1084 swear	1040 thump	2643 tuber	85 voice	2520 winks
1618 sighs	2135 snags	1683 squid	724 sweat	2482 thyme	555 tubes	2777 voile	817 wiped
190 sight	1175 snail	2456 stabs	827 sweep	2795 tiara	2107 tucks	1759 volts	2374 wipes
302 signs	458 snake	1117 stack	349 sweet	1698 ticks	1911 tufts	2442 vomit	1474 wired
1540 silks	2999 snaky	418 staff	1099 swell	1050 tidal	1589 tulip	930 voted	524 wires
1366 silky	1525 snaps	255 stage	551 swept	748 tides	1305 tumor	1965 voter	1435 wiser
2294 sills	1281 snare	2733 stags	643 swift	2722 tiers	1007 tuned	921 votes	2114 wisps
598 silly	1776 snarl	1526 stain	1187 swims	753 tiger	905 tunes	2794 vouch	774 witch
2738 silos	1498 sneak	1659 stair	1921 swine	477 tight	2316 tunic	1801 vowed	1902 witty
2819 silts	2256 sneer	1028 stake	533 swing	2086 tiled	262 turns	92 vowel	799 wives
45 since	1421 sniff	1746 stale	2918 swipe	1164 tiles	1282 tusks	2732 vying	2778 woken
1925 sinew	2765 snipe	978 stalk	1783 swirl	2185 tilts	1539 tutor	1348 waded	130 woman
2948 singe	1886 snips	791 stall	1319 swish	1610 timed	2268 twain	2418 wades	169 women
694 sings	2700 snobs	806 stamp	1877 swiss	55 times	2144 twang	2904 wafer	243 woods
1357 sinks	2079 snore	105 stand	2837 swoon	1436 timid	2044 tweed	1862 waged	1576 woody
2856 sinus	1732 snort	188 stars	1761 swoop	2347 tines	276 twice	2053 wager	2694 wooed
1918 siren	1545 snout	107 start	762 sword	2644 tinge	998 twigs	690 wages	2942 woofs
2964 sirup	1383 snows	80 state	1056 swore	2126 tints	1695 twine	348 wagon	8 words
2003 sisal	1085 snowy	2184 stave	1491 sworn	269 tired	973 twins	2580 wails	2599 wordy
1075 sites	2041 snuff	816 stays	544 swung	860 tires	2342 twirl	846 waist	193 works
1626 sixes	1792 soaks	1384 stead	1140 syrup	2068 tithe	823 twist	1390 waits	33 world
714 sixth	1884 soaps	1356 steak	75 table	353 title	1528 tying	2826 waive	722 worms
679 sixty	2139 soapy	844 steal	2399 taboo	878 toads	1433 typed	2096 waked	2891 wormy
1892 sized	2393 soars	362 steam	1513 tacks	968 toast	304 types	2412 waken	441 worry
503 sizes	1311 sober	1897 steed	2662 tacos	64 today	2661 udder	1578 wakes	452 worse
1367 skate	899 socks	231 steel	2564 taffy	2356 toils	2138 ulcer	630 walks	654 worst
2688 skein	2415 sodas	558 steep	695 tails	708 token	2577 ultra	245 walls	306 worth
2791 skied	2588 sofas	895 steer	2539 taint	2129 tolls	2766 umbra	1336 waltz	5 would
1990 skier	1718 soggy	542 stems	93 taken	1326 tombs	509 uncle	2697 wands	557 wound
908 skies	969 soils	162 steps	143 takes	2484 tommy	29 under	2277 waned	854 woven
1722 skiff	534 solar	829 stern	2878 talcs	922 tonal	2031 undid	2760 wanes	2943 wowed
404 skill	1561 soles	2516 stews	634 tales	422 tones	1775 undue	2881 wanly	1891 wraps
2920 skimp	300 solid	253 stick	684 talks	1896 tongs	1849 unfit	254 wants	857 wrath
2641 skims	1989 solos	2986 sties	1693 tally	1468 tonic	2283 unify	2280 wards	2833 wreak
591 skins	151 solve	636 stiff	1388 tamed	325 tools	486 union	1574 wares	910 wreck
1560 skips	1464 sonar	2332 stile	2900 tames	797 tooth	1315 unite	1398 warms	2182 wrens
663 skirt	196 songs	23 still	2308 tangy	2512 toots	235 units	1677 warns	2452 wrest
2663 skits	2104 sonic	1272 sting	693 tanks	2618 topaz	735 unity	2251 warts	2195 wring
925 skull	2115 sores	2275 stink	2146 taped	350 topic	1751 untie	1818 wasps	974 wrist
2253 skunk	381 sorry	1945 stint	1643 taper	1334 torch	38 until	442 waste	10 write
1541 slabs	815 sorts	1755 stirs	1660 tapes	2373 torso	2770 upped	134 watch	177 wrong
1634 slack	1128 souls	378 stock	2353 tapir	185 total	309 upper	12 water	136 wrote
2974 slags	20 sound	1013 stole	2574 tardy	2784 toted	727 upset	2263 watts	2309 wryly
1004 slain	1985 soups	2187 stomp	1919 tarry	1992 totem	900 urban	570 waved	1548 yacht
2922 slake	171 south	198 stone	2040 tarts	2682 totes	758 urged	2239 waver	2555 yanks
2317 slams	2058 sowed	1479 stony	880 tasks	264 touch	1946 urges	199 waves	323 yards
1459 slang	74 space	83 stood	368 taste	523 tough	1563 urine	592 waxed	1015 yarns
812 slant	2057 spade	1019 stool	1417 tasty	1906 tours	1083 usage	2453 waxen	2397 yearn
1984 slaps	2535 spank	1418 stoop	2078 taunt	1664 towed	2106 usher	218 waxes	18 years
1215 slash	2017 spans	657 stops	2050 tawny	1127 towel	66 using	928 wears	1169 yeast
1559 slate	710 spare	189 store	1648 taxed	526 tower	311 usual	706 weary	1520 yells
2226 slats	892 spark	2167 stork	678 taxes	316 towns	1159 utter	1103 weave	2717 yelps
563 slave	2111 spars	371 storm	1586 taxis	2169 toxic	1102 vague	1321 wedge	792 yield
2748 slays	2156 spasm	53 story	395 teach	2983 toyed	2726 vales	764 weeds	2740 yodel
1370 sleds	2868 spats	1111 stout	617 teams	590 trace	2483 valet	2367 weedy	2693 yoked
1258 sleek	2149 spawn	554 stove	430 tears	374 track	1091 valid	206 weeks	2485 yokes
176 sleep	165 speak	1232 strap	1393 tease	1313 tract	1900 valor	2604 weeps	52 young
1565 sleet	803 spear	488 straw	2441 teddy	210 trade	153 value	540 weigh	502 yours
479 slept	1374 speck	1189 stray	2767 teems	383 trail	1222 valve	1252 weird	421 youth
2783 slews	168 speed		1306 teens	220 train	1867 vanes	2624 welds	1931 yucca
811 slice	133 spell		223 teeth	1444 trait	520 vapor	924 wells	1534 zebra
1379 slick	272 spend		157 tells	1260 tramp	1782 vases	2677 welts	1180 zeros
559 slide	216 spent		841 tempo	1088 traps	1701 vault	1934 whack	1161 zones
1880 slime	2335 sperm		2189 tempt			435 whale	2591 zooms
2154 slimy			867 tends			1199 wharf	

BIBLIOGRAPHY

[1] Michael Beeler, "Information theory and the game of Jotto," *A.I. Memo 218* (Cambridge, MA: M.I.T. Artificial Intelligence Laboratory, August 1971), 3 pp.

[2] Michael Beeler, Word ladders. Unpublished memorandum (Cambridge, MA: Bolt, Beranek, and Newman, 1975). 15 pp.

[3] John B. Carroll, Peter Davies, and Barry Richman, *The American Heritage Word Frequency Book* (Boston, MA: Houghton Mifflin, 1971).

[4] C. L. Dodgson, letter to William Warner dated March 12, 1892, in Morton N. Cohen (ed.), *The Letters of Lewis Carroll* **2** (London: Macmillan, 1979), 896.

[5] Jack J. Dongarra and Eric Grosse, "Distribution of mathematical software via electronic mail," *Communications of the ACM* **30**,5 (May 1987), 403–407. [The current status of netlib is described regularly in the *ACM SIGNUM Newsletter*; see, for example, [9].]

[6] Martin Gardner, "The games and puzzles of Lewis Carroll," in *New Mathematical Diversions from Scientific American* (New York: Simon and Schuster, 1966), Chapter 4.

[7] Solomon W. Golomb and Leonard D. Baumert, "Backtrack programming," *Journal of the ACM* **12** (1965), 516–524.

[8] Martin Grötschel, Michael Jünger, and Gerhard Reinelt, "A cutting plane algorithm for the linear ordering problem," *Operations Research* **32** (1984), 1195–1220.

[9] Eric Grosse, "Netlib news: Searching for files," *ACM SIGNUM Newsletter* **27**,3 (July 1992), 2–4. [Mentions `lp/generators`, containing transportation networks, assignment and generalized network flow problems, contributed by Darwin Klingman, Fred Glover, and M. Ramamurti.]

[10] John D. Hobby, "A user's manual for MetaPost," Computer Science Technical Report 162, AT&T Bell Laboratories, Murray Hill, NJ 07974-2070, 87 pp. [Can be retrieved electronically in PostScript form by mailing the one-line message '`send 162 from research/cstr`' to `netlib@research.att.com`.]

[11] Richard M. Karp, "Reducibility among combinatorial problems," in Raymond E. Miller and James W. Thatcher (eds.), *Complexity of Computer Computations* (New York: Plenum, 1972), 85–103.

[12] Kenneth C. Knowlton, "Representation of designs," *U.S. Patent 4,398,890* (August 16, 1983), 13 pp.

[13] Ken Knowlton and Leon Harmon, "Computer-produced grey scales," *Computer Graphics and Image Processing* **1** (1972), 1–20.

[14] Donald E. Knuth, *Sorting and Searching*, Volume 3 of *The Art of Computer Programming* (Reading, MA: Addison-Wesley, 1973), xii + 725 pp.

[15] Donald E. Knuth, "Are toy problems useful?" *Popular Computing* **5**,1 (January 1977), 1, 3–10; **5**,2 (February 1977), 3–7.

[16] Donald E. Knuth, "Lewis Carroll's WORD - WARD - WARE - DARE - DAME - GAME,"
     GAMES **2**,4 (July 1978), 22–23.

[17] Donald E. Knuth, "Fonts for digital halftones," *TUGboat* **8** (1987), 135–160.

[18] Donald E. Knuth, "Digital halftones by dot diffusion," *ACM Transactions on Graphics*
     6 (1987), 245–273.

[19] Donald E. Knuth, *Axioms and Hulls* (Heidelberg: Springer-Verlag, 1992), ix + 109 pp.
     (Lecture Notes in Computer Science, no. 606.)

[20] Donald E. Knuth, *Literate Programming* (Stanford, California: Center for the Study
     of Language and Information, 1992), xvi + 368 pp. (CSLI Lecture Notes, no. 27.)
     Distributed by the University of Chicago Press.

[21] Donald E. Knuth and Arvind Raghunathan, "The problem of compatible representa-
     tives," *SIAM Journal on Discrete Mathematics* **5** (1992), 422–427.

[22] Henry Kučera and W. Nelson Francis, *Computational Analysis of Present-Day Amer-
     ican English* (Providence, RI: Brown University Press, 1967).

[23] Wassily W. Leontief, "The structure of the U.S. economy," *Scientific American* **212**,4
     (April 1965), 25–35.

[24] Silvio Levy and Donald E. Knuth, "The CWEB system of structured documentation,"
     Report STAN-CS-90-1336 (Stanford, CA: Department of Computer Science, Oc-
     tober 1990), 200 pp. A revised version is available online as part of the CWEB
     distribution, in file ~ftp/pub/cweb/cwebman.tex at labrea.stanford.edu. The
     revision includes full support for ANSI C and for C++.

[25] M. D. McIlroy, "7x7 computer-generated word squares," *Word Ways* **8** (1975), 195–
     197. [See also the sequel about unsymmetrical 6 × 6 squares, *Word Ways* **9** (1976),
     80–84.]

[26] Roy McMullen, *Mona Lisa: The Picture and the Myth* (Boston: Houghton Mifflin,
     1975).

[27] William Morris (ed.), *The Xerox Intermediate Dictionary* (Middletown, CT: Xerox
     Family Education Series, 1973).

[28] Gerhard Reinelt, "TSPLIB — A traveling salesman problem library," *ORSA Journal
     on Computing* **3** (1991), 376–384.

[29] Kenneth A. Ross and Donald E. Knuth, "A programming and problem solving semi-
     nar," *Report STAN-CS-89-1269* (Stanford, CA: Department of Computer Science,
     July 1989), 87 pp.

[30] U.S. Bureau of Economic Analysis, Interindustry Economics Division, "The Input-
     Output structure of the U.S. economy, 1977," *Survey of Current Business* **64**,5
     (May 1984), 42–84.

[31] U.S. Bureau of Economic Analysis, Interindustry Economics Division, "Annual input-
     output accounts of the U.S. economy, 1985," *Survey of Current Business* **70**,1
     (January 1990), 41–56.

[32] U.S. Department of the Interior, Geological Survey, *The National Atlas of the United
     States of America* (Washington, DC: 1970).

GENERAL INDEX

About this book

This book was typeset with the Computer Modern family of typefaces developed by the author, using the author's TEX typesetting system. Illustrations were prepared with a preliminary version of the METAPOST system developed by John Hobby at AT&T Bell Laboratories. The computer programs and mini-indexes were typeset by a system called CTWILL developed by the author. Camera-ready copy was produced from PostScript® files on a Linotronic phototypesetter with a resolution of 50 dots per mm.

About the cover

The cover illustration shows a solution to the Steiner tree problem of connecting together the four words {tools, games, bench, marks} using the smallest possible number of auxiliary words from the list in Appendix D. Jill Knuth designed a graphic representation of this solution, which was then incorporated into the overall cover design by Joe Svadlenka, using Adobe Illustrator® and Quark XPress®.

About the author

Donald E. Knuth is a prominent computer scientist and author of the best-selling series *The Art of Computer Programming*. As creator of the TEX typesetting system, he is also author of the five-volume set *Computers & Typesetting*. In 1971 he was the first recipient of ACM's Grace Murray Hopper award; he received ACM's Alan M. Turing award in 1974, and the ACM SIGCSE and Software Systems awards in 1986. He is a member of the National Academy of Sciences and the National Academy of Engineering, an associate of the scientific academies of Norway and France, and has received honorary doctorates from nineteen colleges and universities in six countries. In 1979 he received the nation's highest scientific honor, the National Medal of Science, from President Carter. Knuth is presently Professor Emeritus of The Art of Computer Programming at Stanford University, spending full time at home as he completes the remaining volumes of his pioneering series of books.

This book is published as part of ACM Press Books—a collaboration between the Association for Computing Machinery and Addison-Wesley Publishing Company. ACM is the oldest and largest educational and scientific society in the information technology field. Through its high-quality publications and services, ACM is a major force in advancing the skills and knowledge of IT professionals throughout the world. For further information about ACM, contact:

ACM Member Services
1515 Broadway, 17th Floor
New York, NY 10036-5701
Phone: 1-212-626-0500
Fax: 1-212-944-1318
E-mail: ACMHELP@ACM.org

ACM European Service Center
Avenue Marcel Thiry 204
1200 Brussels, Belgium
Phone: 32-2-774-9602
Fax: 32-2-774-9690
E-mail: ACM_Europe@ACM.org